Teacher's Edition for
Beginning Algebra

Books by Streeter

Streeter and Alexander: Basic Mathematical Skills
Streeter, Hutchison, and Alexander: Beginning Algebra
Streeter and Hutchison: Intermediate Algebra

Also Available From McGraw-Hill

Schaum's Outline Series in Mathematics & Statistics

Each outline includes basic theory, definitions, and hundreds of solved problems and supplementary problems with answers.

Current List Includes:
Analytic Geometry
College Algebra
Elementary Algebra
Review of Elementary Mathematics
First Year College Mathematics
Mathematical Handbook
Modern Elementary Algebra
Technical Mathematics
Trigonometry

AVAILABLE AT YOUR COLLEGE BOOKSTORE

Second Edition

Teacher's Edition for

Beginning Algebra

James Streeter
Clackamas Community College

Donald Hutchison
Clackamas Community College

Gerald Alexander
Clackamas Community College

McGraw-Hill Book Company
New York St. Louis San Francisco Auckland Bogotá Caracas
Colorado Springs Hamburg Lisbon London Madrid Mexico
Milan Montreal New Delhi Oklahoma City Panama Paris
San Juan São Paulo Singapore Sydney Tokyo Toronto

Teacher's Edition for Beginning Algebra

1 2 3 4 5 6 7 8 9 0 VNH VNH 8 9 4 3 2 1 0 9 8

ISBN 0-07-062491-7

This book was set in Aster by York Graphic Services, Inc.
The editors were Robert A. Weinstein, David Damstra, and Jack Maisel;
the production supervisor was Friederich W. Schulte.
Design was done by Caliber Design Planning, Inc.
Von Hoffmann Press, Inc. was printer and binder.

Library of Congress Cataloging-in-Publication Data

Streeter, James (James A.)
Teacher's edition for beginning algebra.

Includes index.
1. Algebra. I. Alexander, Gerald. II. Hutchison, Donald (date). III. Title.
QA152.2.S773 1989 512 88-13268
ISBN 0-07-062491-7

Contents

Chapter 7 Graphing Linear Equations and Inequalities 351

Chapter 8 Systems of Linear Equations 427

Chapter 9 Exponents and Radicals 483

Chapter 10 Quadratic Equations 531

Preface

Beginning Algebra is the second in a series of three developmental level college texts which include *Basic Mathematical Skills* and *Intermediate Algebra*. The design of these texts was based on our experiences with students at Clackamas Community College. This second edition of *Beginning Algebra* retains the format and style of the first edition.

It is clear that students come to an introductory program at the college level with a considerable variety of needs and preparation. Some have had no previous exposure to the study of algebra; some have had previous algebra experience, but with limited success; others have had previous positive experience at this level, but in returning to college classes find that they need a review of these basic skills before they continue on to further mathematics classes or courses in business, computer science, the social sciences, or the traditional sciences.

A primary objective for both editions was to accommodate this wide range of student needs. Two years of class-testing the original manuscript and our students' use of the first edition have brought forth many helpful comments from students of all ages and from a wide variety of backgrounds. These comments, along with suggestions from our teaching colleagues, have guided us in the process of revision for this second edition.

To encourage the involvement of all levels of students, we have directed our attention to readability. Each topic is developed in a straightforward fashion, with numerous examples to clarify the ideas being presented. All important rules and algorithms are boxed for easy identification and reference. Examples are generally followed by parallel "Check Yourself" exercises to provide the student with immediate feedback as to his or her success in mastering each new concept. In addition, we have an open format with clarifying marginal notes to help students follow the steps of an example.

We have given special emphasis to a structured approach in solving word problems. This technique is introduced through the translation of simple word phrases to algebraic expressions. As complete examples of solving word problems are developed, each solution is carefully broken down into a five-step process to encourage the student to work toward a thoughtful and organized approach. Also, we have integrated these applications throughout the text, using further applications whenever new equation-solving methods are shown. It is our firm belief that this cyclical approach is the most effective in building these most necessary skills.

New to the second edition are sections on pattern recognition in which we emphasize strategies in equation solving and graphing. We feel that this feature will also enhance the students' ability to choose the appropriate problem solving techniques.

This text covers all the topics traditionally included in a college-level course in beginning algebra. Changes for the second edition include earlier coverage of the properties of the real numbers in Chapter 1; inclusion of the material dealing with the slope of a line and the forms for a linear equation in Chapter 7; a new section in Chapter 9 expanding the coverage of the laws of exponents; and a more extensive presentation on operations with expressions involving radicals in that same chapter.

The text material is organized into 10 chapters, each of which incorporates the following features. Examples are generally followed by parallel *Check Yourself* exercises to provide a student with a chance for immediate practice over the concepts just presented.

Section exercises within each chapter have been carefully constructed on a graded level of increasing difficulty. Generally odd and even exercises parallel each other. We have provided answers for the odd-numbered problems with the problem set. Answers for all the even-numbered exercises appear in the accompanying Instructor's Manual.

Two other features new to this second edition are *Skillscan* and *Going Beyond* exercises that appear in many section exercise sets. *Skillscan* are problems drawn from previous sections of the text and are designed to aid the student in the process of reviewing concepts that will be applied in the following section. *Going Beyond* exercises have also been incorporated into many section exercises. As the title suggests, these are true extension exercises that will provide further challenge to the student. In all cases the necessary concepts have been covered in the text section.

A careful and concise summary of all the important terms, definitions, and algorithms appears at the end of each chapter. Further examples illustrate all important techniques to facilitate the student's review.

Summary exercises keyed to the appropriate chapter sections, follow each chapter summary. Answers to these problems are provided in the accompanying Instructor's Manual. This allows an instructor the option of using these problems for homework assignments.

A *Self-Test,* with all answers keyed to chapter sections, concludes each chapter and will give the student guidance in preparing for a parallel in-class test. Also new to the second edition are *Cumulative Tests* after Chapters 3, 6, 8, and 10. These are designed to give the student further opportunity for building skills in the process of a cumulative review. These should be especially useful aids to students preparing for midterm and final examinations.

Supplements

Available for this text is an Instructor's Manual containing the answers to all of the even-numbered exercises and to the chapter summary exercises. Additional sample examinations for each chapter are also included, as are sample midterm and final examinations.

This Teacher's Edition includes answers to all exercises. These answers are printed in a second color for easier use by the teacher. Other supplements aimed especially at the instructor include both print and computerized testing. The computerized testing provides the instructor with over 1800 test questions from throughtout the text. Several types of test questions are used, including multiple-choice, open-ended, matching, true-false, and vocabulary. The testing system enables the teacher to find the questions by section, topic, question type, difficulty level, and other criteria. In addition, instructors may add their own criteria and edit their own questions. As noted above, sample tests are provided in the *Instructor's Manual.* The Print Test Bank is a hard-copy listing of the questions found in the computerized version.

Several supplements are available for the student. First, a *Student Solutions Manual,* which includes answers and, when appropriate, solutions to all odd-numbered exercises. Students may purchase this manual through their local bookstores.

Videotapes are also available to adopters for student use in the learning lab. For the entire series, twenty-four hours of video lessons have been developed, broken into smaller sections that follow the three texts. These videos provide the student with additional instructional and visual support of the lessons.

Finally, a computerized study guide is, in addition, available to adopters. This tutorial provides additional coverage and support for all sections of each text in the series. Students can work additional problems of many different types and receive constructive feedback based on their answers. Virtually no computer training is needed for the student to work with this supplement.

We trust that this range of supplements will support both teachers and students in a variety of instructional settings. At the same time, both the authors and McGraw-Hill welcome any recommendations from students and teachers for the continual development of the package as the teaching environment and technology change.

Acknowledgments

Many of the features discussed have been developed and refined in response to the comments of our colleagues at Clackamas Community College, and we would like to especially acknowledge their helpful support. In particular we thank Miriam Coulter for her keen eye in checking all of the problems for this second edition.

We are grateful to the following reviewers for their many valuable suggestions which have been incorporated into this second edition: Ann Bartholomay, Southwest Virginia Community College; Susan L. Friedman, Bernard M. Baruch College; Jack W. Kotman, Lansing Community College; Virginia Lee, Brookdale Community College; Shirley Markus, University of Louisville; Peggy Rejto, Normandale Community College; Sylvester Roebuck, City College of Chicago, Olive-Harvey College; Jack Rotman, Lansing Community College; Amber Steinmetz, Mathematics Consultant-Campbell, MO; Alexa Stiegemeier, Elgin Community College; Eleanor Strauss, Community College of Philadelphia and Tommy Thompson, Brookhaven College.

Our work in this revision has been greatly eased by the helpful staff at McGraw-Hill. In particular, our special appreciation goes to Robert Weinstein, David Damstra, and Jack Maisel.

Also to our wives, Sharon, Ann, and Claudia, our thanks for their patience, and to Micol Hutchison our appreciation for her help in preparing the manuscript. Finally to our students, our acknowledgment of their many useful ideas and comments and of their support during the processes of writing and revision.

James Streeter
Donald Hutchison
Gerald Alexander

To the Student

You are about to begin a course in algebra. We have made every attempt to provide a text that will help you understand what algebra is about and how to effectively use it. We have made no assumptions about your previous experience with algebra. Your rate of progress through the course will depend both upon the amount of time and effort that you give to the course and to your previous background in mathematics. There are some specific features in this textbook that will aid you in your studies. Here are some suggestions about how to use those features.

Keep in mind that a review of *all* of the chapter material will further enhance your ability to grasp later topics and to move more effectively through the following chapters.

1. If you are in a lecture class, make sure that you take the time to read the appropriate text section *before* your instructor's lecture on the subject. Then take careful notes on the examples that your instructor presents during class.
2. After class, work through similar examples in the text, making sure that you understand each of the steps shown. Examples are followed by *Check Yourself* exercises. Algebra is best learned by being involved in the process and that is the purpose of these exercises. Always have a pencil and paper at hand and work out the problems that are presented and check your result immediately. If you have difficulty, go back and carefully review the previous exercises. Make sure that you understand what you are doing and why. The best test of whether you do understand a concept lies in your ability to explain that concept to one of your fellow students. Try working together.
3. At the end of each chapter section you will find a set of exercises. Work these carefully in order to check your progress on the section you have just finished. You will find the answers for the odd-numbered exercises following the problem set. If you have had difficulties with any of the exercises, review the appropriate parts of the chapter section. If your questions are not completely cleared up, by all means do not become discouraged. Ask your instructor or an available tutor for further assistance. A word of caution: Work the exercises on a regular (preferably daily) basis. Again, learning algebra requires becoming involved. As is the case with learning any skill, the main ingredient is practice.
4. When you have completed a chapter, review by using the *Chapter*

Summary. You will find all the important terms and definitions in this section, along with examples illustrating all the techniques that have been developed in the chapter. Following the summary are *Summary Exercises* for further practice. The exercises are keyed to chapter sections, so you will know where to turn if you are still having problems.

5. When done with the *Summary Exercises*, try the *Self-Test* that appears at the end of each chapter. This will give you an actual practice test to work as you review for in-class testing. Again, answers with section references are provided.

6. Finally, an important element of success in studying algebra is the process of regular review. We have provided a series of *Cumulative Tests* throughout the textbook (they are located after Chapters 3, 6, 8, and 10). These will help you review not only the concepts of the chapter that you have just completed, but those of previous chapters. Use these tests in preparation for any midterm or final examinations. If it appears that you have forgotten some concepts that are being tested, don't worry. Go back and review the sections where the idea was initially explained or the appropriate chapter summary. That is the purpose of these cumulative tests.

We hope that you will find our suggestions helpful as you work through this material, and we wish you the best of luck in the course.

J. S.
D. H.
G. A.

Teacher's Edition for
Beginning Algebra

Chapter One

The Language of Algebra

1.1 From Arithmetic to Algebra

OBJECTIVE
To represent the operations of addition, subtraction, multiplication, and division by using the notation of algebra.

In arithmetic you learned how to do calculations with numbers by using the basic operations of addition, subtraction, multiplication, and division.

In algebra you will still be using numbers and the same four operations. However, you will also be using letters to represent numbers. Letters such as x, y, L, or W can be given various numerical values and are called *variables* for this reason.

Example 1

We can represent the length and width of a rectangle by the letters L and W.

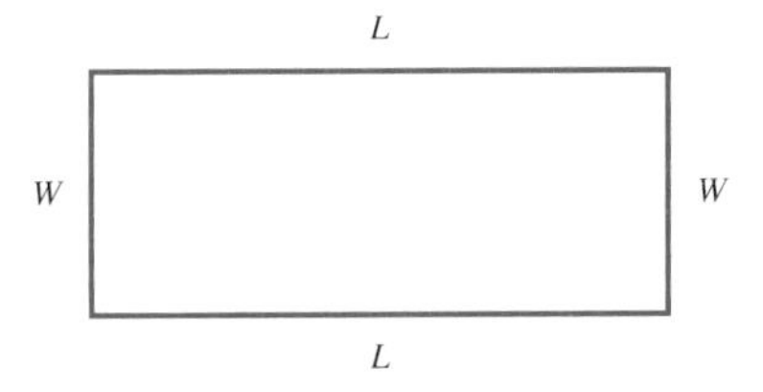

In arithmetic: + denotes addition, − denotes subtraction, × denotes multiplication, ÷ denotes division.

You are familiar with the symbols used to indicate the four fundamental operations of arithmetic.

Let's look at how these operations are indicated in algebra.

ADDITION

$x + y$ means the *sum* of x and y or x *plus* y.

Example 2

Some other words that tell you to add are "more than" and "increased by."

(*a*) The *sum* of a and 3 is written as $a + 3$.
(*b*) L *plus* W is written as $L + W$.
(*c*) 5 *more than* m is written as $m + 5$.
(*d*) x *increased by* 7 is written as $x + 7$.

CHECK YOURSELF 1*

Write, using symbols.

1. The sum of y and 4
2. a plus b
3. 3 more than x
4. n increased by 6

Let's look at how subtraction is indicated in algebra.

SUBTRACTION

$x - y$ means the *difference* of x and y or x *minus* y.

Example 3

Some other words that mean to subtract are "decreased by" and "less than."

(*a*) r *minus* s is written $r - s$.
(*b*) The *difference* of m and 5 is written $m - 5$.
(*c*) x *decreased by* 8 is written $x - 8$.
(*d*) 4 *less than* a is written $a - 4$.

CHECK YOURSELF 2

Write, using symbols.

1. w minus z
2. The difference of a and 7
3. y decreased by 3
4. 5 less than b

You have seen that the operations of addition and subtraction are written exactly the same way in algebra as they were in arithmetic. Multiplication presents a slight difficulty because the sign × looks like the letter x. So in algebra we use other symbols to show multiplication to avoid any confusion. Here are some ways to write multiplication.

Note: x and y are called the *factors* of the product xy.

MULTIPLICATION

A raised dot	$x \cdot y$	These all indicate the *product* of x and y.
Parentheses	$(x)(y)$	
Writing the letters next to each other	xy	

*Check Yourself Answers appear at the end of each section throughout the book.

Example 4

(*a*) The product of 5 and a is written as $5 \cdot a$, $(5)(a)$, or $5a$. The last expression, $5a$, is the shortest and the most common way of writing the product.

Note: You can place two letters next to each other or a number and a letter next to each other to show multiplication. But you *cannot* place two numbers side by side to show multiplication: 37 means the number "thirty-seven," not 3 times 7.

(*b*) 3 times 7 can be written as $3 \cdot 7$ or $(3)(7)$.
(*c*) Twice z is written as $2z$.
(*d*) The product of 2, s, and t is written as $2st$.
(*e*) 4 more than the product of 6 and x is written as $6x + 4$.

CHECK YOURSELF 3

Write, using symbols.

1. m times n
2. The product of h and b
3. The product of 8 and 9
4. The product of 5, w, and y
5. 3 more than the product of 8 and a

To write more complicated products in algebra, we need some "punctuation marks." Parentheses ( ) mean that an expression is to be thought of as a single quantity. Brackets [ ] and braces { } are used in exactly the same way as parentheses in algebra. Look at the following example showing the use of these signs of grouping.

Example 5

(*a*) 3 times the sum of a and b is written as

This can be read as "3 times the quantity *a* plus *b*."

$3(a + b)$

The sum of a and b is a single quantity, so it is enclosed in parentheses.

(*b*) The sum of 3 times a and b is written as

No parentheses are needed here since the 3 multiplies *only* the *a*.

$3a + b$

(*c*) 2 times the difference of m and n is written as

$2(m - n)$

(*d*) The product of s plus t and s minus t is written as

$(s + t)(s - t)$

(*e*) The product of b and 3 less than b is written as

$b(b - 3)$

CHECK YOURSELF 4

Write, using symbols.

1. Twice the sum of p and q
2. The sum of twice p and q
3. The product of a and the quantity $b - c$
4. The product of x plus 2 and x minus 2
5. The product of x and 4 more than x

Now let's look at the operation of division. Think of the variety of ways you showed division in arithmetic. Besides the division sign ÷, you used the long division symbol $\overline{)\quad}$ and the fraction notation. For example, to indicate the quotient when 9 is divided by 3, you could write

$$9 \div 3 \qquad \text{or} \qquad 3\overline{)9} \qquad \text{or} \qquad \frac{9}{3}$$

In algebra the fraction form is usually used.

DIVISION

$\frac{x}{y}$ means x *divided by* y or the *quotient* when x is divided by y.

Example 6

(*a*) m divided by 3 is written as $\frac{m}{3}$.

(*b*) The quotient of a plus b divided by 5 is written as

$$\frac{a + b}{5}$$

(*c*) The sum p plus q divided by the difference p minus q is written as

$$\frac{p + q}{p - q}$$

CHECK YOURSELF 5

Write, using symbols.

1. r divided by s
2. The quotient when x minus y is divided by 7
3. The difference a minus 2 divided by the sum a plus 2

CHECK YOURSELF ANSWERS

1. (1) $y + 4$; (2) $a + b$; (3) $x + 3$; (4) $n + 6$.
2. (1) $w - z$; (2) $a - 7$; (3) $y - 3$; (4) $b - 5$.
3. (1) mn; (2) hb; (3) $8 \cdot 9$ or $(8)(9)$; (4) $5wy$; (5) $8a + 3$.
4. (1) $2(p + q)$; (2) $2p + q$; (3) $a(b - c)$; (4) $(x + 2)(x - 2)$; (5) $x(x + 4)$.
5. (1) $\frac{r}{s}$; (2) $\frac{x - y}{7}$; (3) $\frac{a - 2}{a + 2}$.

1.1 Exercises

Write each of the following phrases, using symbols.

1. The sum of c and d $c + d$
2. a plus 7 $a + 7$
3. w plus z $w + z$
4. The sum of m and n $m + n$
5. x increased by 2 $x + 2$
6. 3 more than b $b + 3$
7. 10 more than y $y + 10$
8. m increased by 4 $m + 4$
9. a minus b $a - b$
10. 5 less than s $s - 5$
11. b decreased by 7 $b - 7$
12. r minus 3 $r - 3$
13. 6 less than r $r - 6$
14. x decreased by 3 $x - 3$
15. w times z wz
16. The product of 3 and c $3c$
17. The product of 5 and t $5t$
18. 8 times a $8a$
19. The product of 8, m, and n $8mn$
20. The product of 7, r, and s $7rs$
21. The product of 3 and the quantity p plus q $3(p + q)$
22. The product of 5 and the sum of a and b $5(a + b)$
23. Twice the sum of x and y $2(x + y)$
24. 3 times the sum of m and n $3(m + n)$

25. The sum of twice x and y $2x + y$

26. The sum of 3 times m and n $3m + n$

27. Twice the difference of x and y $2(x - y)$

28. 3 times the difference of c and d $3(c - d)$

29. The quantity a plus b times the quantity a minus b $(a + b)(a - b)$

30. The product of x plus y and x minus y $(x + y)(x - y)$

31. The product of m and 3 less than m $m(m - 3)$

32. The product of a and 7 more than a $a(a + 7)$

33. x divided by 5 $\frac{x}{5}$

34. The quotient when b is divided by 8 $\frac{b}{8}$

35. The quotient of a plus b, divided by 7 $\frac{a + b}{7}$

36. The difference x minus y, divided by 9 $\frac{x - y}{9}$

37. The difference of p and q, divided by 4 $\frac{p - q}{4}$

38. The sum of a and 5, divided by 9 $\frac{a + 5}{9}$

39. The sum of a and 3, divided by the difference of a and 3 $\frac{a + 3}{a - 3}$

40. The difference of m and n, divided by the sum of m and n $\frac{m - n}{m + n}$

Write each of the following phrases, using symbols. Use the variable x to represent the number in each case.

41. 5 more than a number
$x + 5$

42. A number increased by 8
$x + 8$

43. 7 less than a number
$x - 7$

44. A number decreased by 10
$x - 10$

45. 9 times a number
$9x$

46. Twice a number
$2x$

47. 6 more than 3 times a number
$3x + 6$

48. 5 times a number decreased by 10
$5x - 10$

49. Twice the sum of a number and 5
$2(x + 5)$

50. 3 times the difference of a number and 4
$3(x - 4)$

51. The product of 2 more than a number and 2 less than that same number
$(x + 2)(x - 2)$

52. The product of 5 less than a number and 5 more than that same number
$(x - 5)(x + 5)$

53. The quotient of a number and 7
$\frac{x}{7}$

54. A number divided by 3
$\frac{x}{3}$

55. The sum of a number and 5, divided by 8
$\frac{x+5}{8}$

56. The quotient when 7 less than a number is divided by 3
$\frac{x-7}{3}$

57. 6 more than a number divided by 6 less than that same number
$\frac{x+6}{x-6}$

58. The quotient when 3 less than a number is divided by 3 more than that same number
$\frac{x-3}{x+3}$

Skillscan (Appendix 1)

The following set of exercises has been chosen to help you review the skills that you will need in the next section. Where possible, a reference to similar examples and problems is provided (here, Appendix 1). Go back to the indicated section for further review if you would like.

Perform the indicated operations.

a. $\frac{5}{2}+\frac{3}{2}$ 4 **b.** $\frac{5}{6}+\frac{2}{3}$ $\frac{3}{2}$ **c.** $\frac{4}{3}+\frac{3}{5}$ $\frac{29}{15}$

d. $\frac{3}{8}+\frac{7}{12}$ $\frac{23}{24}$ **e.** $\left(\frac{5}{3}\right)\left(\frac{9}{2}\right)$ $\frac{15}{2}$ **f.** $\left(\frac{8}{5}\right)\left(\frac{5}{12}\right)$ $\frac{2}{3}$

g. $\left(\frac{12}{25}\right)\left(\frac{15}{20}\right)$ $\frac{9}{25}$ **h.** $\left(\frac{9}{16}\right)\left(\frac{24}{27}\right)$ $\frac{1}{2}$ **i.** $(3)\left(\frac{1}{3}\right)$ 1

Answers

We will provide the solutions for the odd-numbered exercises at the end of each exercise set. The solutions for the even-numbered exercises are provided at the back of the book.

1. $c+d$ **3.** $w+z$ **5.** $x+2$ **7.** $y+10$ **9.** $a-b$ **11.** $b-7$ **13.** $r-6$ **15.** wz **17.** $5t$ **19.** $8mn$ **21.** $3(p+q)$ **23.** $2(x+y)$ **25.** $2x+y$ **27.** $2(x-y)$ **29.** $(a+b)(a-b)$ **31.** $m(m-3)$ **33.** $\frac{x}{5}$ **35.** $\frac{a+b}{7}$ **37.** $\frac{p-q}{4}$ **39.** $\frac{a+3}{a-3}$ **41.** $x+5$ **43.** $x-7$ **45.** $9x$ **47.** $3x+6$ **49.** $2(x+5)$ **51.** $(x+2)(x-2)$ **53.** $\frac{x}{7}$ **55.** $\frac{x+5}{8}$ **57.** $\frac{x+6}{x-6}$ **a.** 4 **b.** $\frac{3}{2}$ **c.** $\frac{29}{15}$ **d.** $\frac{23}{24}$ **e.** $\frac{15}{2}$ **f.** $\frac{2}{3}$ **g.** $\frac{9}{25}$ **h.** $\frac{1}{2}$ **i.** 1

1.2 The Properties of Addition and Multiplication

OBJECTIVE
To recognize applications of the commutative, associative, and distributive laws.

Everything that we do in algebra will be based on certain rules for the operations introduced in Section 1.1. We call these rules *properties of the real numbers*. In this section we will consider those properties that we will be using in the remainder of this chapter.

The first two properties tell us that we can add or multiply in any order.

These rules should come as no surprise. In arithmetic you naturally add or multiply in whichever order is most convenient.

THE COMMUTATIVE PROPERTIES

If a and b are any numbers,

1. $a + b = b + a$
2. $a \cdot b = b \cdot a$

Example 1

(*a*) $5 + 9 = 9 + 5$ and $x + 7 = 7 + x$

are applications of the commutative property of addition.

(*b*) $5 \cdot 9 = 9 \cdot 5$

is an application of the commutative property of multiplication.

We will also want to be able to change the grouping in simplifying expressions. This is possible because of the associative properties. Numbers or variables can be grouped in any manner to find a sum or a product.

THE ASSOCIATIVE PROPERTIES

If a, b, and c are any numbers,

1. $a + (b + c) = (a + b) + c$
2. $a \cdot (b \cdot c) = (a \cdot b) \cdot c$

Example 2

(*a*) Find

Always do the operation in the parentheses first. More about this in Section 1.6.

$2 + \underbrace{(3 + 8)}_{\text{Add first.}}$ and $\underbrace{(2 + 3)}_{\text{Add first.}} + 8$

$= 2 + 11$ $\qquad$ $= 5 + 8$

$= 13$ $\qquad$ $= 13$

So

$$2 + (3 + 8) = (2 + 3) + 8$$

(*b*) Find

$$\underbrace{2 \cdot (3 \cdot 8)}_{\text{Multiply first.}} \quad \text{and} \quad \underbrace{(2 \cdot 3)}_{\text{Multiply first.}} \cdot 8$$

$= 2 \cdot 24$ $\qquad$ $= 6 \cdot 8$

$= 48$ $\qquad$ $= 48$

So

$$2 \cdot (3 \cdot 8) = (2 \cdot 3) \cdot 8$$

(*c*) Find

$$\frac{1}{3} \cdot \underbrace{(6 \cdot 5)}_{\text{Multiply first.}} \quad \text{and} \quad \underbrace{\left(\frac{1}{3} \cdot 6\right)}_{\text{Multiply first.}} \cdot 5$$

$= \frac{1}{3} \cdot (30)$ $\qquad$ $= (2) \cdot 5$

$= 10$ $\qquad$ $= 10$

So

$$\left(\frac{1}{3} \cdot 6\right) \cdot 5 = \frac{1}{3}(6 \cdot 5)$$

CHECK YOURSELF 1

Show that the following statements are true.

1. $3 + (4 + 7) = (3 + 4) + 7$
2. $3 \cdot (4 \cdot 7) = (3 \cdot 4) \cdot 7$
3. $\left(\frac{1}{5} \cdot 10\right) \cdot 4 = \frac{1}{5} \cdot (10 \cdot 4)$

Another important property involves both addition and multiplication. We can illustrate this property with an application. Suppose that we want to find the area of the two rooms shown.

Remember: The area of a rectangle is the product of its length and width:

$A = L \cdot W$

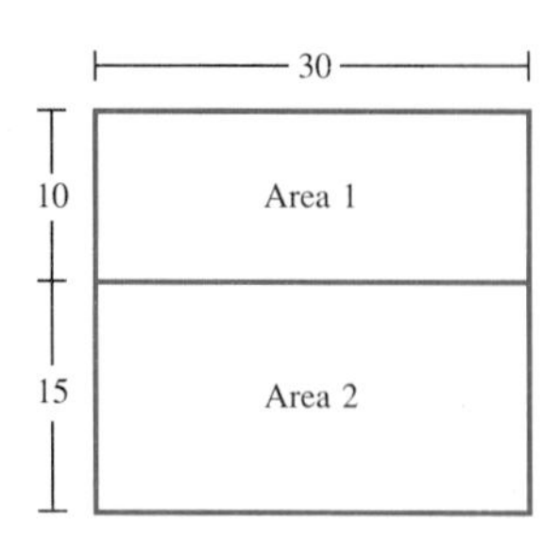

We can find the total area by adding first to find the overall width. Then we multiply by the length.

[or]

We can find the total area as a sum of the two areas.

$$\overbrace{30}^{\text{Length}} \cdot \overbrace{(10 + 15)}^{\text{Overall width}}$$
$$= 30 \cdot 25$$
$$= 750$$

$$\overbrace{30 \cdot 10}^{\substack{\text{(Area 1)} \\ \text{Length} \cdot \text{Width}}} + \overbrace{30 \cdot 15}^{\substack{\text{(Area 2)} \\ \text{Length} \cdot \text{Width}}}$$
$$= 300 + 450$$
$$= 750$$

So

$$30 \cdot (10 + 15) = 30 \cdot 10 + 30 \cdot 15$$

This leads us to the following property.

Note the pattern.

$a(b + c) = a \cdot b + a \cdot c$

We "distributed" the multiplication "over" the addition.

THE DISTRIBUTIVE PROPERTY

If a, b, and c are any numbers,

$a(b + c) = a \cdot b + a \cdot c$

Example 3

(*a*) $5(3 + 4) = 5 \cdot 3 + 5 \cdot 4$

Note: $5(3 + 4) = 5 \cdot 7 = 35$

or

$5 \cdot 3 + 5 \cdot 4 = 15 + 20 = 35$

$8x + 8y$ cannot be simplified further.

(*b*) $8(x + y) = 8x + 8y$

(*c*) $2(3x + 5) = 2 \cdot 3x + 2 \cdot 5$
$= 6x + 10$

(*d*) $\frac{1}{3}(9 + 12) = \frac{1}{3} \cdot 9 + \frac{1}{3} \cdot 12$
$= 3 + 4 = 7$

Note: It is also true that

$$\frac{1}{3}(9 + 12) = \frac{1}{3}(21) = 7$$

CHECK YOURSELF 2

Use the distributive property to simplify (remove the parentheses).

1. $4(6 + 7)$

2. $9(m + n)$

3. $3(5a + 7)$

4. $\frac{1}{5}(10 + 15)$

Note: The distributive property can also be used in a "right-hand" manner:

We will use the distributive property in this fashion in Section 1.4.

$$(2 + 3)x = (2 + 3)x = 2x + 3x = 5x$$

CHECK YOURSELF ANSWERS

1. (1) $3 + (4 + 7) = 3 + 11 = 14$
$(3 + 4) + 7 = 7 + 7 = 14$
(2) $3 \cdot (4 \cdot 7) = 3 \cdot 28 = 84$
$(3 \cdot 4) \cdot 7 = 12 \cdot 7 = 84$
(3) $\left(\frac{1}{5} \cdot 10\right) \cdot 4 = 2 \cdot 4 = 8$
$\frac{1}{5} \cdot (10 \cdot 4) = \frac{1}{5} \cdot 40 = 8$

2. (1) $4 \cdot 6 + 4 \cdot 7$, or 52; (2) $9m + 9n$; (3) $15a + 21$; (4) 5.

1.2 Exercises

Identify the property that is illustrated by each of the following statements.

1. $5 + 9 = 9 + 5$
Commutative property of addition

2. $6 + 3 = 3 + 6$
Commutative property of addition

3. $2 \cdot (3 \cdot 5) = (2 \cdot 3) \cdot 5$
Associative property of multiplication

4. $3 \cdot (5 \cdot 6) = (3 \cdot 5) \cdot 6$
Associative property of multiplication

5. $10 \cdot 5 = 5 \cdot 10$
Commutative property of multiplication

6. $8 \cdot 4 = 4 \cdot 8$
Commutative property of multiplication

7. $8 + 12 = 12 + 8$
Commutative property of addition

8. $6 + 2 = 2 + 6$
Commutative property of addition

9. $(5 \cdot 7) \cdot 2 = 5 \cdot (7 \cdot 2)$
Associative property of multiplication

10. $(8 \cdot 9) \cdot 2 = 8 \cdot (9 \cdot 2)$
Associative property of multiplication

11. $9 \cdot 8 = 8 \cdot 9$
Commutative property of multiplication

12. $6 \cdot 4 = 4 \cdot 6$
Commutative property of multiplication

13. $2(3 + 5) = 2 \cdot 3 + 2 \cdot 5$
Distributive property

14. $5 \cdot (4 + 6) = 5 \cdot 4 + 5 \cdot 6$
Distributive property

15. $5 + (7 + 8) = (5 + 7) + 8$
Associative property of addition

16. $8 + (2 + 9) = (8 + 2) + 9$
Associative property of addition

17. $(10 + 5) + 9 = 10 + (5 + 9)$
Associative property of addition

18. $(5 + 5) + 3 = 5 + (5 + 3)$
Associative property of addition

19. $7 \cdot (3 + 8) = 7 \cdot 3 + 7 \cdot 8$
Distributive property

20. $5 \cdot (6 + 8) = 5 \cdot 6 + 5 \cdot 8$
Distributive property

Verify that each of the following statements is true by evaluating each side of the equation separately and comparing the results.

21. $7 \cdot (3 + 4) = 7 \cdot 3 + 7 \cdot 4$ $49 = 49$

22. $4 \cdot (5 + 1) = 4 \cdot 5 + 4 \cdot 1$ $24 + 24$

23. $2 + (9 + 8) = (2 + 9) + 8$ $19 = 19$

24. $6 + (15 + 3) = (6 + 15) + 3$ $24 = 24$

25. $5 \cdot (6 \cdot 3) = (5 \cdot 6) \cdot 3$ $90 = 90$

26. $2 \cdot (9 \cdot 10) = (2 \cdot 9) \cdot 10$ $180 = 180$

27. $5 \cdot (2 + 8) = 5 \cdot 2 + 5 \cdot 8$ $50 = 50$

28. $3 \cdot (10 + 2) = 3 \cdot 10 + 3 \cdot 2$ $36 = 36$

29. $(3 + 12) + 8 = 3 + (12 + 8)$ $23 = 23$

30. $(8 + 12) + 7 = 8 + (12 + 7)$ $27 = 27$

31. $(4 \cdot 7) \cdot 2 = 4 \cdot (7 \cdot 2)$ $56 = 56$

32. $(6 \cdot 5) \cdot 3 = 6 \cdot (5 \cdot 3)$ $90 = 90$

33. $\frac{1}{2} \cdot (2 + 6) = \frac{1}{2} \cdot 2 + \frac{1}{2} \cdot 6$ $4 = 4$

34. $\frac{1}{3} \cdot (6 + 9) = \frac{1}{3} \cdot 6 + \frac{1}{3} \cdot 9$ $5 = 5$

35. $\left(\frac{2}{3} + \frac{1}{6}\right) + \frac{1}{3} = \frac{2}{3} + \left(\frac{1}{6} + \frac{1}{3}\right)$ $\frac{7}{6} = \frac{7}{6}$

36. $\frac{3}{4} + \left(\frac{5}{8} + \frac{1}{2}\right) = \left(\frac{3}{4} + \frac{5}{8}\right) + \frac{1}{2}$ $\frac{15}{8} = \frac{15}{8}$

37. $2.5 + (4.6 + 3.2) = (2.5 + 4.6) + 3.2$
$10.3 = 10.3$

38. $8.1 + (1.9 + 3.4) = (8.1 + 1.9) + 3.4$
$13.4 = 13.4$

39. $\frac{1}{2} \cdot (2 \cdot 8) = \left(\frac{1}{2} \cdot 2\right) \cdot 8$ $8 = 8$

40. $\frac{1}{5} \cdot (5 \cdot 3) = \left(\frac{1}{5} \cdot 5\right) \cdot 3$ $3 = 3$

41. $\left(\frac{3}{5}\cdot\frac{5}{6}\right)\cdot\frac{4}{3}=\frac{3}{5}\cdot\left(\frac{5}{6}\cdot\frac{4}{3}\right)$ $\quad \frac{2}{3}=\frac{2}{3}$

42. $\frac{4}{7}\cdot\left(\frac{21}{16}\cdot\frac{8}{3}\right)=\left(\frac{4}{7}\cdot\frac{21}{16}\right)\cdot\frac{8}{3}$ $\quad 2=2$

43. $2.5\cdot(4\cdot 5)=(2.5\cdot 4)\cdot 5$ $\quad 50=50$

44. $4.2\cdot(5\cdot 2)=(4.2\cdot 5)\cdot 2$ $\quad 42=42$

Use the distributive property to remove the parentheses in each of the following expressions. Then simplify your result where possible.

45. $2(3+5)$ $\quad 16$

46. $5(4+6)$ $\quad 50$

47. $3(x+5)$ $\quad 3x+15$

48. $5(y+8)$ $\quad 5y+40$

49. $4(w+v)$ $\quad 4w+4v$

50. $7(c+d)$ $\quad 7c+7d$

51. $2(3x+5)$ $\quad 6x+10$

52. $3(7a+4)$ $\quad 21a+12$

53. $\frac{1}{3}\cdot(15+9)$ $\quad 8$

54. $\frac{1}{6}\cdot(36+24)$ $\quad 10$

Use the properties of addition and multiplication to complete each of the following statements.

55. $5+7=$ 7 $+5$

56. $(5+3)+4=5+($ 3 $+4)$

57. $(8)(3)=(3)($ 8 $)$

58. $8(3+4)=8\cdot 3+$ 8 $\cdot 4$

59. $7(2+5)=7\cdot$ 2 $+7\cdot 5$

60. $4\cdot(2\cdot 4)=($ 4 $\cdot 2)\cdot 4$

Use the indicated property to write an expression that is equivalent to each of the following expressions.

61. $3+7$ (commutative property) $\quad 7+3$

62. $2(3+4)$ (distributive property) $\quad 2\cdot 3+2\cdot 4$

63. $5\cdot(3\cdot 2)$ (associative property) $\quad (5\cdot 3)\cdot 2$

64. $(3+5)+2$ (associative property) $\quad 3+(5+2)$

65. $2 \cdot 4 + 2 \cdot 5$ (distributive property) $2 \cdot (4 + 5)$

66. $7 \cdot 9$ (commutative property) $9 \cdot 7$

Evaluate each of the following pairs of expressions. Then answer the given question.

67. $8 - 5$ and $5 - 8$
Is subtraction commutative? No

68. $12 \div 3$ and $3 \div 12$
Is division commutative? No

69. $(12 - 8) - 4$ and $12 - (8 - 4)$
Is subtraction associative? No

70. $(48 \div 16) \div 4$ and $48 \div (16 \div 4)$
Is division associative? No

Skillscan

Perform each of the indicated operations.

a. $3 + 3 + 3$ 9

b. $3 \cdot 3 \cdot 3$ 27

c. $5 + 5 + 5 + 5$ 20

d. $5 \cdot 5 \cdot 5 \cdot 5$ 625

e. $4 + 4 + 4 + 4 + 4$ 20

f. $4 \cdot 4 \cdot 4 \cdot 4 \cdot 4$ 1024

Answers

1. Commutative property of addition **3.** Associative property of multiplication **5.** Commutative property of multiplication **7.** Commutative property of addition **9.** Associative property of multiplication **11.** Commutative property of multiplication **13.** Distributive property **15.** Associative property of addition **17.** Associative property of addition **19.** Distributive property **21.** $49 = 49$ **23.** $19 = 19$ **25.** $90 = 90$ **27.** $50 = 50$ **29.** $23 = 23$ **31.** $56 = 56$ **33.** $4 = 4$ **35.** $\frac{7}{6} = \frac{7}{6}$ **37.** $10.3 = 10.3$ **39.** $8 = 8$ **41.** $\frac{2}{3} = \frac{2}{3}$ **43.** $50 = 50$ **45.** 16 **47.** $3x + 15$ **49.** $4w + 4v$ **51.** $6x + 10$ **53.** 8 **55.** $7 + 5$ **57.** $(3)(8)$ **59.** $7 \cdot 2 + 7 \cdot 5$ **61.** $7 + 3$ **63.** $(5 \cdot 3) \cdot 2$ **65.** $2 \cdot (4 + 5)$ **67.** No **69.** No **a.** 9 **b.** 27 **c.** 20 **d.** 625 **e.** 20 **f.** 1024

1.3 Exponents

OBJECTIVE
To write a product of factors in exponential form.

In Section 1.1, we showed how symbols are used to denote the four basic operations. Often in mathematics we define other symbols that will allow us to write statements in a more compact or "shorthand" form. This is an idea that you have encountered before. For example, given a statement with repeated addition, such as

$5 + 5 + 5$

how could you rewrite the expression? You might respond by noting that the statement

$5 + 5 + 5 = 15$
and
$3 \cdot 5 = 15$

$3 \cdot 5$

has the same meaning. You have learned that multiplication is shorthand for repeated addition.

In algebra we frequently have a *factor* that is repeated in an expression several times. For instance, we might have

$5 \cdot 5 \cdot 5$

To abbreviate this product, we write

$5 \cdot 5 \cdot 5 = 5^3$

This is called *exponential notation* or *form*. The exponent or power, here 3, indicates the number of times that the factor or base, here 5, appears in a product.

Exponent or power

$5 \cdot 5 \cdot 5 = 5^3$

Base

Be careful: 5^3 is *not* the same as $5 \cdot 3$ since $5^3 = 5 \cdot 5 \cdot 5 = 125$ and $5 \cdot 3 = 15$.

Example 1

(*a*) Write $3 \cdot 3 \cdot 3 \cdot 3$, using exponential form. The number 3 appears 4 times in the product, so

Four factors of 3

$3 \cdot 3 \cdot 3 \cdot 3 = 3^4$

This is read "3 to the fourth power."

(*b*) Write $x \cdot x \cdot x$, using exponential form. The same idea works for letters or variables, so since x appears three times in the product

$x \cdot x \cdot x = x^3$

This is read as "x to the third power" or "x cubed."

CHECK YOURSELF 1

Write in exponential form.

1. $4 \cdot 4 \cdot 4 \cdot 4 \cdot 4 \cdot 4$ **2.** $y \cdot y \cdot y \cdot y$

If a product involves a combination of numbers and letters or different letters, the exponential form can also be used.

Example 2

(*a*) Write $5 \cdot m \cdot m$, using exponents.

Note that the exponent applies *only* to *m*, *not* to 5.

$$5 \cdot m \cdot m = 5m^2$$

Two factors of *m*

This is read "5*m* to the second power" or "5*m* squared."

(*b*) Write $a \cdot a \cdot b \cdot b \cdot b \cdot b$, using exponents.

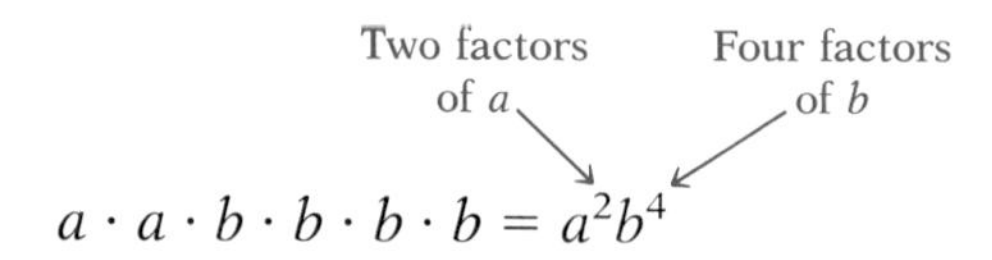

If more than one letter appears, we usually write the product in alphabetical order.

$$a \cdot a \cdot b \cdot b \cdot b \cdot b = a^2b^4$$

(*c*) Write $3 \cdot x \cdot x \cdot y \cdot y \cdot y$, using exponents.

$$3 \cdot x \cdot x \cdot y \cdot y \cdot y$$
$$= 3x^2y^3$$

One factor of 3 | Two factors of *x* | Three factors of *y*

CHECK YOURSELF 2

Write in exponential form.

1. $6 \cdot b \cdot b \cdot b$ **2.** $p \cdot p \cdot p \cdot q \cdot q \cdot q \cdot q \cdot q$ **3.** $5 \cdot m \cdot m \cdot m \cdot n \cdot n$

There may also be situations in which it will be useful to change an expression from exponential form to an expanded form (written as a product of factors).

Example 3

Write each expression in expanded form.

The exponent 3 applies *only* to *x*.

(*a*) $4x^3 = 4 \cdot x \cdot x \cdot x$

Now the exponent applies to $4x$ because of the parentheses. So $4x$ is repeated as a factor 3 times.

(*b*) $(4x)^3 = (4x) \cdot (4x) \cdot (4x)$
(*c*) $5x^2y^3 = 5 \cdot x \cdot x \cdot y \cdot y \cdot y$

CHECK YOURSELF 3

Write each expression in expanded form.

1. $5x^4$ **2.** $(5x)^4$ **3.** $3a^3b^3$

We mentioned earlier that an expression will have different values depending on the number assigned to the variables in the expression. Finding the value of an expression is called *evaluating an expression*. This involves replacing each variable with a given value, as illustrated in our final example.

Example 4

Evaluate each expression.

(*a*) $5x^2$ when $x = 2$

We replace x with the value 2.

$$5x^2 = 5(2)^2 = 5 \cdot 2 \cdot 2 = 20$$

2 appears as a factor 2 times

(*b*) $2xy^3$ when $x = 2, y = 3$

Now let x be 2 and y be 3. Then

$$2xy^3 = 2(2)(3)^3 = 2 \cdot 2 \cdot 3 \cdot 3 \cdot 3 = 108$$

CHECK YOURSELF 4

Evaluate each expression.

1. $3a^3$ when $a = 2$
2. $5m^2n$ when $m = 3$ and $n = 4$

CHECK YOURSELF ANSWERS

1. (1) 4^6; (2) y^4.
2. (1) $6b^3$; (2) p^3q^5; (3) $5m^3n^2$.
3. (1) $5 \cdot x \cdot x \cdot x \cdot x$; (2) $(5x) \cdot (5x) \cdot (5x) \cdot (5x)$; (3) $3 \cdot a \cdot a \cdot a \cdot b \cdot b \cdot b$.
4. (1) 24; (2) 180.

1.3 Exercises

Write each expression, using exponential form.

1. $4 \cdot 4 \cdot 4$
4^3

2. $3 \cdot 3 \cdot 3 \cdot 3 \cdot 3$
3^5

3. $5 \cdot 5 \cdot 5 \cdot 5 \cdot 5 \cdot 5$
5^6

4. $a \cdot a \cdot a \cdot a$
a^4

5. $m \cdot m \cdot m \cdot m \cdot m$
m^5

6. $2 \cdot 2 \cdot 2 \cdot 2 \cdot 2 \cdot 2 \cdot 2$
2^7

7. $2 \cdot x \cdot x \cdot x$
$2x^3$

8. $3 \cdot y \cdot y \cdot y \cdot y$
$3y^4$

9. $a \cdot a \cdot a \cdot a \cdot b \cdot b$
a^4b^2

10. $m \cdot m \cdot n \cdot n \cdot n$
m^2n^3

11. $9 \cdot r \cdot r \cdot r \cdot s \cdot s \cdot s$
$9r^3s^3$

12. $6 \cdot a \cdot b \cdot b \cdot b \cdot b$
$6ab^4$

13. $5 \cdot w \cdot z \cdot z \cdot z \cdot z \cdot z$
$5wz^5$

14. $7 \cdot x \cdot x \cdot x \cdot x \cdot y \cdot y \cdot y$
$7x^4y^3$

Write each expression in expanded form.

15. $5x^3$
$5 \cdot x \cdot x \cdot x$

16. $2y^5$
$2 \cdot y \cdot y \cdot y \cdot y \cdot y$

17. $2x^2y$
$2 \cdot x \cdot x \cdot y$

18. $3ab^3$
$3 \cdot a \cdot b \cdot b \cdot b$

19. $12a^3b^4$
$12 \cdot a \cdot a \cdot a \cdot b \cdot b \cdot b \cdot b$

20. $9x^5y^2$
$9 \cdot x \cdot x \cdot x \cdot x \cdot x \cdot y \cdot y$

21. $(2x)^3$
$(2x) \cdot (2x) \cdot (2x)$

22. $(3y)^4$
$(3y) \cdot (3y) \cdot (3y) \cdot (3y)$

23. $(5p)^5$
$(5p) \cdot (5p) \cdot (5p) \cdot (5p) \cdot (5p)$

24. $(8s)^3$
$(8s) \cdot (8s) \cdot (8s)$

Evaluate each expression if $x = 3$ and $y = 4$.

25. $3x^2$ 27

26. $2y^2$ 32

27. $(3x)^2$ 81

28. $(2y)^2$ 64

29. $3xy^2$ 144

30. $2x^3y$ 216

Evaluate each expression if $a = 2$ and $b = 5$.

31. a^2b^2 100

32. a^3b^3 1000

33. $(ab)^2$ 100

34. $(ab)^3$ 1000

Skillscan (Section 1.2)

Use the distributive property to remove the parentheses. Then simplify each expression.

a. $3(2 + 5)$ 21 **b.** $5(7 + 3)$ 50 **c.** $(2 + 5)3$ 21

d. $(7 + 3)5$ 50 **e.** $(2 + 5)x$ $7x$ **f.** $(7 + 3)y$ $10y$

g. $(8 + 3)a$ $11a$ **h.** $(9 + 4)m$ $13m$ **i.** $(10 + 2)s$ $12s$

j. $(12 + 3)w$ $15w$

Answers

1. 4^3 **3.** 5^6 **5.** m^5 **7.** $2x^3$ **9.** a^4b^2 **11.** $9r^3s^3$ **13.** $5wz^5$ **15.** $5 \cdot x \cdot x \cdot x$ **17.** $2 \cdot x \cdot x \cdot y$ **19.** $12 \cdot a \cdot a \cdot a \cdot b \cdot b \cdot b \cdot b$ **21.** $(2x) \cdot (2x) \cdot (2x)$ **23.** $(5p) \cdot (5p) \cdot (5p) \cdot (5p) \cdot (5p)$ **25.** 27 **27.** 81 **29.** 144 **31.** 100 **33.** 100 **a.** $3 \cdot 2 + 3 \cdot 5$, or 21 **b.** 50 **c.** 21 **d.** 50 **e.** $2x + 5x$, or $7x$ **f.** $10y$ **g.** $11a$ **h.** $13m$ **i.** $12s$

1.4 Adding and Subtracting Expressions

OBJECTIVES

1. To add algebraic expressions.
2. To subtract algebraic expressions.

To find the perimeter of (or the distance around) a rectangle, we add 2 times the length and 2 times the width. In the language of algebra, this can be written as

Perimeter $= 2L + 2W$

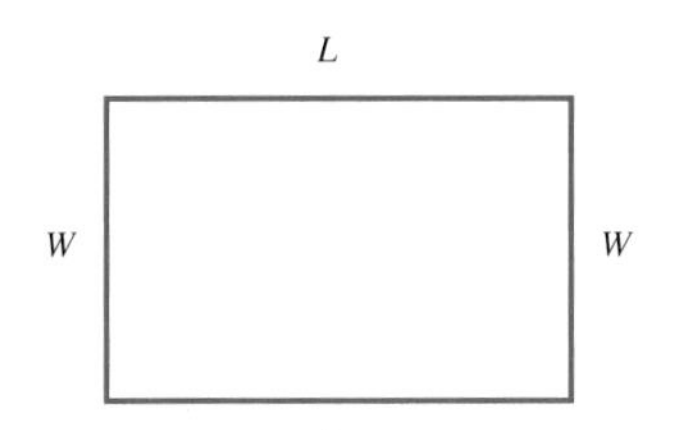

We call $2L + 2W$ an *algebraic expression,* or more simply an *expression*. An expression allows us to write a mathematical idea in symbols. It can be thought of as any meaningful collection of letters and numbers connected by the operations signs, as is discussed in Section 1.1.

Example 1

Some expressions are

(*a*) $5x^2$
(*b*) $3a + 2b$
(*c*) $4x^3 - 2y + 1$

In the expressions of Example 1, the addition and subtraction signs break the expressions into smaller parts called *terms*.

A *term* is a number, or the product of a number and one or more variables, raised to a power.

In an expression, each sign (+ or −) is a part of the term that follows the sign.

Example 2

(*a*) $5x^2$ has one term.

(*b*) $3a + 2b$ has two terms, $3a$ and $2b$.

Term Term

Note that each term "owns" the sign that precedes it.

(*c*) $4x^3 - 2y + 1$ has three terms: $4x^3$, $-2y$, and 1.

Term Term Term

CHECK YOURSELF 1

List the terms of each expression.

1. $2b^4$ **2.** $5m + 3n$ **3.** $2s^2 - 3t - 6$

Note that a term in an expression may have any number of factors. For instance, $5xy$ is a term. It has factors of 5, x, and y. The number factor of a term is called the *numerical coefficient*. So for the term $5xy$, the numerical coefficient is 5.

Example 3

(*a*) $4a$ has the numerical coefficient 4.

(*b*) $6a^3b^4c^2$ has the numerical coefficient 6.

(*c*) $-7m^2n^3$ has the numerical coefficient -7.

(*d*) Since $1 \cdot x = x$, the numerical coefficient of x is understood to be 1.

CHECK YOURSELF 2

Give the numerical coefficient for the following terms.

1. $8a^2b$ **2.** $-5m^3n^4$ **3.** y

If terms contain exactly the *same letters* (or variables) raised to the *same power,* they are called *like terms.*

Example 4

(*a*) The following are like terms.

$6a$ and $7a$
$5b^2$ and b^2
$10x^2y^3z$ and $-6x^2y^3z$
$5m^3$, $-3m^3$, and m^3

These terms have the same letters raised to the same power—the numerical coefficients can be any number.

(*b*) The following are *not* like terms.

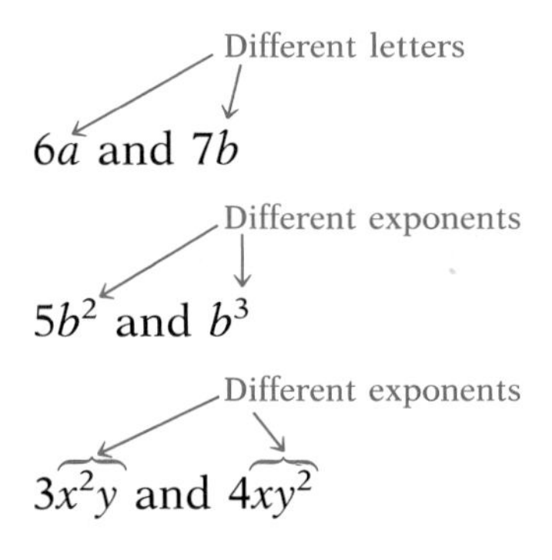

CHECK YOURSELF 3

Circle the like terms.

$5a^2b$ $\quad ab^2$ $\quad a^2b$ $\quad -3a^2$ $\quad 4ab$ $\quad 3b^2$ $\quad -7a^2b$

Like terms of an expression can always be combined into a single term. Look at the following:

$$\overbrace{x + x}^{2x} + \overbrace{x + x + x + x + x}^{5x} = \overbrace{x + x + x + x + x + x + x}^{7x}$$

Rather than having to write out all those x's, try

Here we use the distributive property from Section 1.2.

$$2x + 5x = (2 + 5)x = 7x$$

In the same way,

You don't have to write all this out—just do it mentally!

$$9b + 6b = (9 + 6)b = 15b$$
and $$10a - 4a = (10 - 4)a = 6a$$

This leads us to the following rule.

COMBINING LIKE TERMS

To combine like terms use the following steps.

STEP 1 Add or subtract the numerical coefficients.

STEP 2 Attach the common variables.

Example 5

(*a*) $8m + 5m = (8 + 5)m = 13m$

(*b*) $10a^2 - 3a^2 = 7a^2$

(*c*) $5pq^3 - 4pq^3 = 1pq^3 = pq^3$

(*d*) $7a^3b^2 - 7a^3b^2 = 0a^3b^2 = 0$

CHECK YOURSELF 4

Combine the like terms.

1. $6b + 8b$ **2.** $12x^2 - 3x^2$ **3.** $8xy^3 - 7xy^3$
4. $9a^2b^4 - 9a^2b^4$

Let's look at some expressions involving more than two terms. The idea is just the same.

The distributive property can be used over any number of terms.

Example 6

(*a*) $5ab - 2ab + 3ab = (5 - 2 + 3)ab = 6ab$

Only the like terms can be combined.

(*b*) $\overbrace{8x - 2x} + 5y$
$= 6x + 5y$

Like terms Like terms

(*c*) $5m + 8n + 4m - 3n$
$= (5m + 4m) + (8n - 3n)$
$= 9m + 5n$

Here we have used the associative and commutative properties.

With practice you won't be writing out these steps, but doing them mentally.

(*d*) $4x^2 + 2x - 3x^2 + x$
$= (4x^2 - 3x^2) + (2x + x)$
$= x^2 + 3x$

As these examples illustrate, combining like terms often means changing the grouping and the order in which terms are written. Again all this is possible because of the properties of addition that we introduced in Section 1.2.

CHECK YOURSELF 5

Combine like terms.

1. $4m^2 - 3m^2 + 8m^2$ **2.** $9ab + 3a - 5ab$
3. $4p + 7q + 5p - 3q$

CHECK YOURSELF ANSWERS

1. (1) $2b^4$; (2) $5m$, $3n$; (3) $2s^2$, $-3t$, -6.
2. (1) 8; (2) -5; (3) 1.
3. The like terms are $5a^2b$, a^2b, and $-7a^2b$.
4. (1) $14b$; (2) $9x^2$; (3) xy^3; (4) 0.
5. (1) $9m^2$; (2) $4ab + 3a$; (3) $9p + 4q$.

1.4 Exercises

List the terms of the following expressions.

1. $5a + 2$ $5a, 2$

2. $7a - 4b$ $7a, -4b$

3. $4x^3$ $4x^3$

4. $3x^2$ $3x^2$

5. $3x^2 + 3x - 7$ $3x^2, 3x, -7$

6. $2a^3 - a^2 + a$ $2a^3, -a^2, a$

Circle the like terms in the following groups of terms.

7. $5ab$, $3b$, $3a$, $4ab$ $5ab, 4ab$

8. $9m^2$, $8mn$, $5m^2$, $7m$ $9m^2, 5m^2$

9. $4xy^2$, $2x^2y$, $5x^2$, $-3x^2y$, $5y$, $6x^2y$
$2x^2y, -3x^2y, 6x^2y$

10. $8a^2b$, $4a^2$, $3ab^2$, $-5a^2b$, $3ab$, $5a^2b$
$8a^2b, -5a^2b, 5a^2b$

Combine the like terms.

11. $4m + 6m$
$10m$

12. $5a^2 + 9a^2$
$14a^2$

13. $5b^3 + 12b^3$
$17b^3$

14. $8rs + 12rs$
$20rs$

15. $21xyz + 7xyz$
$28xyz$

16. $4mn^2 + 15mn^2$
$19mn^2$

17. $9z^2 - 3z^2$
$6z^2$

18. $7m - 6m$
m

19. $5a^3 - 5a^3$
0

20. $13xy - 9xy$
$4xy$

21. $19n^2 - 18n^2$
n^2

22. $7cd - 7cd$
0

23. $18p^2q - 3p^2q$
$15p^2q$

24. $15r^3s^2 - 6r^3s^2$
$9r^3s^2$

25. $7x^2 - 5x^2 + 4x^2$
$6x^2$

26. $9uv + 5uv - 8uv$
$6uv$

27. $5a - 3a + 4b$
$2a + 4b$

28. $7m^2 - 3m + 4m^2$
$11m^2 - 3m$

29. $5x + 3y - 2x - 2y$
$3x + y$

30. $8a^2 + 9a + 5a^2 - 7a$
$13a^2 + 2a$

31. $4a + 7b + 3 - 2a + 3b - 2$
$2a + 10b + 1$

32. $5p^2 + 2p + 8 + 4p^2 + 5p - 6$
$9p^2 + 7p + 2$

33. $\frac{2}{3}m + 3 + \frac{4}{3}m$
$2m + 3$

34. $\frac{1}{5}a - 2 + \frac{4}{5}a$
$a - 2$

35. $\frac{13}{5}x - 2 - \frac{3}{5}x + 5$
$2x + 3$

36. $\frac{17}{12}y + 7 + \frac{7}{12}y - 3$
$2y + 4$

37. $2.3a - 7 + 4.7a + 3$
$7a - 4$

38. $5.8m + 4 - 2.8m + 11$
$3m + 15$

Perform the indicated operations.

39. The sum of $5a^4$ and $8a^4$ is $13a^4$

40. The sum of $9p^2$ and $12p^2$ is $21p^2$

41. Subtract $12a^3$ from $15a^3$. $3a^3$

42. Subtract $5m^3$ from $18m^3$. $13m^3$

43. Subtract $4x$ from the sum of $8x$ and $3x$. $7x$

44. Subtract $8ab$ from the sum of $7ab$ and $5ab$. $4ab$

45. Subtract $3mn^2$ from the sum of $9mn^2$ and $5mn^2$. $11mn^2$

46. Subtract $4x^2y$ from the sum of $6x^2y$ and $12x^2y$. $14x^2y$

Use the distributive property to remove the parentheses in each expression. Then simplify by combining like terms.

47. $2(3x + 2) + 4$
$6x + 8$

48. $3(4z + 5) - 9$
$12z + 6$

49. $5(6a - 2) + 12a$
$42a - 10$

50. $7(4w - 3) - 25w$
$3w - 21$

51. $4s + 2(s + 4) + 4$
$6s + 12$

52. $5p + 4(p + 3) - 8$
$9p + 4$

Evaluate each of the following expressions if $a = 2$, $b = 3$, and $c = 5$. Be sure to combine like terms where possible as the first step.

53. $5a^2 + 2a$
24

54. $7b^2 - 7b$
42

55. $6c^2 + 2c^2$
200

56. $6b^3 - 2b^3$
108

57. $5b + 3a - 2b$
15

58. $7c - 2b + 3c$
44

59. $5ac^2 - 2ac^2$
150

60. $5a^3b - 2a^3b$
72

Skillscan (Section 1.3)

Write each expression in expanded form.

a. 2^2
$2 \cdot 2$

b. 2^3
$2 \cdot 2 \cdot 2$

c. 2^4
$2 \cdot 2 \cdot 2 \cdot 2$

d. 2^5
$2 \cdot 2 \cdot 2 \cdot 2 \cdot 2$

e. $2^3 \cdot 2^2$
$2 \cdot 2 \cdot 2 \cdot 2 \cdot 2$

f. $2^2 \cdot 2^5$
$2 \cdot 2 \cdot 2 \cdot 2 \cdot 2 \cdot 2 \cdot 2$

g. $\dfrac{2^5}{2^2}$
$2 \cdot 2 \cdot 2$

h. $\dfrac{2^4}{2^3}$
2

Answers

1. $5a$, 2 **3.** $4x^3$ **5.** $3x^2, 3x, -7$ **7.** $5ab, 4ab$ **9.** $2x^2y, -3x^2y, 6x^2y$ **11.** $10m$ **13.** $17b^3$ **15.** $28xyz$ **17.** $6z^2$ **19.** 0 **21.** n^2 **23.** $15p^2q$ **25.** $6x^2$ **27.** $2a + 4b$ **29.** $3x + y$ **31.** $2a + 10b + 1$ **33.** $2m + 3$ **35.** $2x + 3$ **37.** $7a - 4$ **39.** $13a^4$ **41.** $3a^3$ **43.** $7x$ **45.** $11mn^2$ **47.** $6x + 8$ **49.** $42a - 10$ **51.** $6s + 12$ **53.** 24 **55.** 200 **57.** 15 **59.** 150 **a.** $2 \cdot 2$ **b.** $2 \cdot 2 \cdot 2$ **c.** $2 \cdot 2 \cdot 2 \cdot 2$ **d.** $2 \cdot 2 \cdot 2 \cdot 2 \cdot 2$ **e.** $2 \cdot 2 \cdot 2 \cdot 2 \cdot 2$ **f.** $2 \cdot 2 \cdot 2 \cdot 2 \cdot 2 \cdot 2 \cdot 2$ **g.** $\dfrac{2 \cdot 2 \cdot 2 \cdot 2 \cdot 2}{2 \cdot 2}$ or $2 \cdot 2 \cdot 2$ **h.** 2

1.5 Multiplying and Dividing Expressions

OBJECTIVES

1. To find the product of certain algebraic expressions.
2. To find the quotient of certain algebraic expressions.

In Section 1.2 we introduced the exponential notation. Remember that the exponent tells us how many times the base is to be used as a factor.

Exponent

$$2^5 = 2 \cdot 2 \cdot 2 \cdot 2 \cdot 2 = 32$$

Base — The fifth power of two

In general,

$$x^m = \underbrace{x \cdot x \cdot \cdots \cdot x}_{m \text{ factors}}$$

where m is a natural number. Natural numbers are the numbers we use for counting: 1, 2, 3, and so on.

The notation can also be used when you are working with letters or variables.

$$x^4 = \underbrace{x \cdot x \cdot x \cdot x}_{4 \text{ factors}}$$

Now look at the product $x^2 \cdot x^3$.

$$x^2 \cdot x^3 = \underbrace{(x \cdot x)}\underbrace{(x \cdot x \cdot x)} = \underbrace{x \cdot x \cdot x \cdot x \cdot x} = x^5$$

2 factors + 3 factors = 5 factors

So

$$x^2 \cdot x^3 = x^{2+3} = x^5$$

Note that the exponent of x^5 is the *sum* of the exponents in x^2 and x^3.

This leads us to the following rule for exponents.

THE FIRST PROPERTY OF EXPONENTS

For any natural numbers m and n and any number x,

$$x^m \cdot x^n = x^{m+n}$$

In words, to multiply expressions with the same base, keep the base and add the exponents.

Example 1

If no exponent is written, it is understood to be 1. In symbols,

$$x = x^1$$

(*a*) $a^5 \cdot a^7 = a^{5+7} = a^{12}$

(*b*) $x \cdot x^8 = x^1 \cdot x^8 = x^{1+8} = x^9$

(*c*) $3^2 \cdot 3^4 = 3^{2+4} = 3^6$

Be careful! The product is *not* 9^6. The base does not change.

(*d*) $y^2 \cdot y^3 \cdot y^5 = y^{2+3+5} = y^{10}$

You can extend the first law to find the product of any number of factors.

(*e*) $x^3 \cdot y^4$ *cannot* be simplified. The bases are not the same.

CHECK YOURSELF 1

Multiply.

1. $b^6 \cdot b^8$ **2.** $y^7 \cdot y$ **3.** $2^3 \cdot 2^4$ **4.** $a^2 \cdot a^4 \cdot a^3$

Suppose that numerical coefficients (other than 1) are involved in a product. For instance, look at the product

$$2x^3 \cdot 3x^5$$

Note that although we have several factors, this is still a single *term*.

To find the product, multiply the numbers and then use the first property of exponents to combine the variables.

Multiply.

$$2x^3 \cdot 3x^5 = (2 \cdot 3)(x^3 \cdot x^5)$$

Add the exponents.

$$= 6x^{3+5}$$
$$= 6x^8$$

You may have noticed that we have again changed the order and grouping. This method uses the commutative and associative properties of Section 1.2.

Example 2

Again we have written out all the steps. You can do the multiplication mentally with practice.

(*a*) $5a^4 \cdot 7a^6 = (5 \cdot 7)(a^4 \cdot a^6)$
$= 35a^{10}$

(*b*) $y^2 \cdot 3y^3 \cdot 6y^4 = (1 \cdot 3 \cdot 6)(y^2 \cdot y^3 \cdot y^4)$
$= 18y^9$

(*c*) $2x^2y^3 \cdot 3x^5y^2 = (2 \cdot 3)(x^2 \cdot x^5)(y^3 \cdot y^2)$
$= 6x^7y^5$

CHECK YOURSELF 2

Multiply.

1. $4x^3 \cdot 7x^5$ **2.** $3a^2 \cdot 2a^4 \cdot 2a^5$ **3.** $3m^2n^4 \cdot 5m^3n$

What about dividing expressions when exponents are involved? For instance, what if we want to divide x^5 by x^2? We can use the following approach to the division:

$$\frac{x^5}{x^2} = \frac{\overbrace{x \cdot x \cdot x \cdot x \cdot x}^{5 \text{ factors}}}{\underbrace{x \cdot x}_{2 \text{ factors}}} = \frac{x \cdot x \cdot x \cdot x \cdot x}{x \cdot x}$$

We can divide by two factors of x.

$$= \overbrace{x \cdot x \cdot x}^{3 \text{ factors}} = x^3$$

So

Note that the exponent of x^3 is the *difference* of the exponents in x^5 and x^2.

$$\frac{x^5}{x^2} = x^{5-2} = x^3$$

This leads us to a second rule for exponents.

THE SECOND PROPERTY OF EXPONENTS

For any natural numbers m and n where m is greater than n and any number x not equal to zero,

$$\frac{x^m}{x^n} = x^{m-n}$$

In words, to divide expressions with the *same base,* keep the base and subtract the exponents.

Example 3

(*a*) $\dfrac{y^7}{y^3} = y^{7-3} = y^4$

(*b*) $\dfrac{m^6}{m} = \dfrac{m^6}{m^1} = m^{6-1} = m^5$

(*c*) $\dfrac{x^4y^5}{y^3} = x^4y^{5-3} = x^4y^2$

(*d*) $\dfrac{a^3b^5}{a^2b^2} = a^{3-2} \cdot b^{5-2} = ab^3$

Apply the second law to each variable separately.

CHECK YOURSELF 3

Divide.

1. $\dfrac{m^9}{m^6}$ **2.** $\dfrac{a^8}{a}$ **3.** $\dfrac{a^3b^5}{a^2}$ **4.** $\dfrac{r^5s^6}{r^3s^2}$

If numerical coefficients are involved, just divide the numbers and then use the second law of exponents to divide the variables. Look at the following example.

Example 4

Subtract the exponents.

(*a*) $\dfrac{6x^5}{3x^2} = 2x^{5-2} = 2x^3$

6 divided by 3

20 divided by 5

(b) $\dfrac{20a^7b^5}{5a^3b^4} = 4a^{7-3} \cdot b^{5-4}$

Again apply the second law to each variable separately.

$= 4a^4b$

CHECK YOURSELF 4

Divide.

1. $\dfrac{4x^3}{2x}$ 2. $\dfrac{20a^6}{5a^2}$ 3. $\dfrac{24x^5y^3}{4x^2y^2}$

Be careful! Later in this text you will encounter expressions such as

$$\frac{x^3 + 2}{x}$$

Students are sometimes tempted to apply the second property of exponents (*incorrectly*) to the first *term* of the numerator and the denominator to write

$$\frac{x^3 + 2}{x} = x^2 + 2$$

Again this is *not correct*. The given expression cannot be further simplified with this approach.

CHECK YOURSELF ANSWERS

1. (1) b^{14}; (2) y^8; (3) 2^7; (4) a^9.
2. (1) $28x^8$; (2) $12a^{11}$; (3) $15m^5n^5$.
3. (1) m^3; (2) a^7; (3) ab^5; (4) r^2s^4.
4. (1) $2x^2$; (2) $4a^4$; (3) $6x^3y$.

1.5 Exercises

Multiply.

1. $x^4 \cdot x^8$
   x^{12}
2. $b \cdot b^5$
   b^6
3. $5^3 \cdot 5^3$
   5^6
4. $y^7 \cdot y^3$
   y^{10}
5. $a^9 \cdot a$
   a^{10}
6. $3^4 \cdot 3^5$
   3^9

7. $z^{10} \cdot z^3$
z^{13}

8. $x^7 \cdot x$
x^8

9. $p^5 \cdot p^7$
p^{12}

10. $s^6 \cdot s^9$
s^{15}

11. $x^3y \cdot x^2y^4$
x^5y^5

12. $m^2n^3 \cdot mn^4$
m^3n^7

13. $w^5 \cdot w^2 \cdot w$
w^8

14. $x^5 \cdot x^4 \cdot x^6$
x^{15}

15. $m^3 \cdot m^2 \cdot m^4$
m^9

16. $r^3 \cdot r \cdot r^5$
r^9

17. $a^3b \cdot a^2b^2 \cdot ab^3$
a^6b^6

18. $w^2z^3 \cdot wz \cdot w^3z^4$
w^6z^8

19. $p^2q \cdot p^3q^5 \cdot pq^4$
p^6q^{10}

20. $c^3d \cdot c^4d^2 \cdot cd^5$
c^8d^8

21. $3a^5 \cdot 2a^4$
$6a^9$

22. $5s^7 \cdot s^3$
$5s^{10}$

23. $x^4 \cdot 3x^3$
$3x^7$

24. $4m^5 \cdot 3m^6$
$12m^{11}$

25. $5m^3n^2 \cdot 4mn^3$
$20m^4n^5$

26. $7x^2y^5 \cdot 6xy^4$
$42x^3y^9$

27. $6x^3y \cdot 9xy^5$
$54x^4y^6$

28. $5a^3b \cdot 10ab^4$
$50a^4b^5$

29. $2a^2 \cdot a^3 \cdot 3a^7$
$6a^{12}$

30. $4x^5 \cdot 2x^3 \cdot 3x^2$
$24x^{10}$

31. $3c^2d \cdot 4cd^3 \cdot 2c^5d$
$24c^8d^5$

32. $5p^2q \cdot p^3q^2 \cdot 3pq^3$
$15p^6q^6$

33. $5m^2 \cdot m^3 \cdot 2m \cdot 3m^4$
$30m^{10}$

34. $3a^3 \cdot 2a \cdot a^4 \cdot 2a^5$
$12a^{13}$

35. $2r^3s \cdot rs^2 \cdot 3r^2s \cdot 5rs$
$30r^7s^5$

36. $6a^2b \cdot ab \cdot 3ab^3 \cdot 2a^2b$
$36a^6b^6$

Divide.

37. $\frac{a^8}{a^5}$ a^3

38. $\frac{m^7}{m}$ m^6

39. $\frac{y^7}{y}$ y^6

40. $\frac{b^8}{b^3}$ b^5

41. $\frac{p^{15}}{p^{10}}$ p^5

42. $\frac{s^{18}}{s^{12}}$ s^6

43. $\frac{x^5y^3}{x^2y^2}$ x^3y

44. $\frac{s^5t^4}{s^3t^2}$ s^2t^2

45. $\frac{6m^3}{3m}$ $2m^2$

46. $\frac{8x^5}{4x}$ $2x^4$

47. $\frac{12a^5}{3a^2}$ $4a^3$

48. $\frac{20x^7}{4x^6}$ $5x$

49. $\frac{18m^5n}{9m^3}$ $2m^2n$

50. $\frac{50a^4b^3}{10b^2}$ $5a^4b$

51. $\frac{28w^3z^5}{7wz}$ $4w^2z^4$

52. $\frac{48p^6q^7}{8p^4q}$ $6p^2q^6$

53. $\dfrac{18x^3y^4z^5}{9xy^2z^2}$ $2x^2y^2z^3$ **54.** $\dfrac{25a^5b^4c^3}{5a^4bc^2}$ $5ab^3c$

Simplify each of the following expressions where possible.

55. $2a^3b \cdot 3a^2b$ $6a^5b^2$

56. $2xy^3 \cdot 3xy^2$ $6x^2y^5$

57. $2a^3b + 3a^2b$ Cannot be simplified

58. $2xy^3 + 3xy^2$ Cannot be simplified

59. $2x^2y^3 \cdot 3x^2y^3$ $6x^4y^6$

60. $5a^3b^2 \cdot 10a^3b^2$ $50a^6b^4$

61. $2x^2y^3 + 3x^2y^3$ $5x^2y^3$

62. $5a^3b^2 + 10a^3b^2$ $15a^3b^2$

63. $\dfrac{8a^2b \cdot 6a^2b}{2ab}$ $24a^3b$

64. $\dfrac{6x^2y^3 \cdot 9x^2y^3}{3x^2y^2}$ $18x^2y^4$

65. $\dfrac{8a^2b + 6a^2b}{2ab}$ $7a$

66. $\dfrac{6x^2y^3 + 9x^2y^3}{3x^2y^2}$ $5y$

Skillscan (Section 1.3)

Evaluate each of the following expressions for the indicated value of the variable.

a. a^2 $(a = 4)$ 16 **b.** x^3 $(x = 4)$ 64 **c.** $2w^3$ $(w = 3)$ 54

d. $3z^3$ $(z = 5)$ 375 **e.** $5p^2$ $(p = 2)$ 20 **f.** $4s^3$ $(s = 2)$ 32

g. $(5p)^2$ $(p = 2)$ 100 **h.** $(4s)^3$ $(s = 2)$ 512

Answers

1. x^{12} **3.** 5^6 **5.** a^{10} **7.** z^{13} **9.** p^{12} **11.** x^5y^5 **13.** w^8 **15.** m^9 **17.** a^6b^6 **19.** p^6q^{10} **21.** $6a^9$ **23.** $3x^7$ **25.** $20m^4n^5$ **27.** $54x^4y^6$ **29.** $6a^{12}$ **31.** $24c^8d^5$ **33.** $30m^{10}$ **35.** $30r^7s^5$ **37.** a^3 **39.** y^6 **41.** p^5 **43.** x^3y **45.** $2m^2$ **47.** $4a^3$ **49.** $2m^2n$ **51.** $4w^2z^4$ **53.** $2x^2y^2z^3$ **55.** $6a^5b^2$ **57.** Cannot be simplified **59.** $6x^4y^6$ **61.** $5x^2y^3$ **63.** $24a^3b$ **65.** $7a$ **a.** 16 **b.** 64 **c.** 54 **d.** 375 **e.** 20 **f.** 32 **g.** 100 **h.** 512

1.6 Evaluating Algebraic Expressions

OBJECTIVES

1. To determine the order in which operations should be done
2. To evaluate algebraic expressions

In applying algebra to problem solving, you will often want to find the value of algebraic expressions when you know certain values for the letters (or variables) in the expressions. As we pointed out earlier, finding the value of an expression is called *evaluating the expression* and uses the following steps.

TO EVALUATE AN EXPRESSION

STEP 1 Replace each variable by the given number value.

STEP 2 Do the necessary arithmetic operations.

Example 1

Suppose that $a = 5$ and $b = 7$.

(*a*) To evaluate $a + b$,

Replace a with 5. Replace b with 7.

$$a + b = 5 + 7 = 12$$

(*b*) To evaluate $3ab$,

Replace a with 5. Replace b with 7.

$$3ab = 3 \cdot 5 \cdot 7 = 105$$

CHECK YOURSELF 1

If $x = 6$ and $y = 7$, evaluate.

1. $y - x$ **2.** $5xy$

It is step 2 *(do the necessary arithmetic operations)* that needs some clarification if you want to evaluate more complicated expressions. For instance, to simplify $5 + 2 \cdot 3$, we could have

Only one of the statements can be correct.

(1)	or	(2)
$\underbrace{5 + 2} \cdot 3$		$5 + \underbrace{2 \cdot 3}$
Add first.		Multiply first.
$= 7 \cdot 3$		$= 5 + 6$
$= 21$		$= 11$

Since we get different answers depending on how we do the problem, the language of algebra would not be clear if there were no agreement on which method is correct. The following rules tell us the order in which operations should be done.

THE ORDER OF OPERATIONS

STEP 1 Do any operations in parentheses first.

STEP 2 Evaluate all expressions involving exponents.

STEP 3 Do any multiplication or division in order, working from left to right.

STEP 4 Do any addition or subtraction in order, working from left to right.

Example 2

Evaluate $5 + 2 \cdot 3$.

There are no parentheses or exponents, so start with step 3. You should multiply and then add.

$5 + 2 \cdot 3$ ← Multiply first.

$= 5 + 6$ ← Then add.

$= 11$

Note: Method (2) shown above is the correct one.

CHECK YOURSELF 2

Evaluate the following expressions.

1. $20 - 3 \cdot 4$ **2.** $9 + 6 \div 3$

Example 3

Evaluate $5 \cdot 3^2$.

$$5 \cdot 3^2$$
$$= 5 \cdot 9$$ Evaluate the power first.
$$= 45$$ Multiply.

CHECK YOURSELF 3

Evaluate $4 \cdot 2^4$.

Example 4

Evaluate $7 \cdot 6 - 5 \cdot 4$.

Do the multiplication in order from left to right; then subtract.

$$7 \cdot 6 - 5 \cdot 4$$
$$= 42 - 20$$
$$= 22$$

CHECK YOURSELF 4

Evaluate $8 \div 4 + 3 \cdot 2$.

Note: We could also choose to apply the distributive property so that

$(5 + 2) \cdot 3$
$= 5 \cdot 3 + 2 \cdot 3$
$= 15 + 6$
$= 21$

The result will be the same.

Example 5

Evaluate $(5 + 2) \cdot 3$.

Do the operation inside the parentheses as the first step.

$$(5 + 2) \cdot 3$$ Add.
$$= 7 \cdot 3$$
$$= 21$$

CHECK YOURSELF 5

Evaluate $4(9 - 3)$.

The principle is the same when more than two "levels" of operations are involved.

Example 6

Evaluate $5 \cdot 2^4 + 3$.

$5 \cdot 2^4 + 3$

Evaluate the power.

$= 5 \cdot 16 + 3$

Multiply.

$= 80 + 3$

Add.

$= 83$

CHECK YOURSELF 6

Evaluate $4 \cdot 3^3 - 8 \cdot 11$.

Example 7

Evaluate:

(*a*) $4(2 + 3)^3$

Add inside the parentheses first.

$= 4(5)^3$

Evaluate the power.

$= 4 \cdot 125$

Multiply.

$= 500$

(*b*) $5(7 - 3)^2 - 10$

Subtract.

$= 5(4)^2 - 10$

Evaluate the power.

$= 5 \cdot 16 - 10$

Multiply.

$= 80 - 10$

Subtract.

$= 70$

CHECK YOURSELF 7

Evaluate $12 + 4(2 + 3)^2$.

We are now ready to return to the evaluation of algebraic expressions. Again, replace the letters with their number values. Then do the arithmetic, following the rules for the order of operations.

Example 8

Evaluate the following expressions if $a = 2$, $b = 3$, $c = 4$, and $d = 5$.

Replace a with 2 and b with 3.

(a) $5a + 7b = 5 \cdot 2 + 7 \cdot 3$

Multiply.

$= 10 + 21$

Add.

$= 31$

Be Careful! This is different from $(3c)^2 = (3 \cdot 4)^2$
$= 12^2 = 144$

(b) $3c^2 = 3 \cdot 4^2$

Evaluate the power.

$= 3 \cdot 16$

Multiply.

$= 48$

(c) $7(c + d) = 7(4 + 5)$

Add inside the parentheses.

$= 7 \cdot 9$
$= 63$

Replace a with 2 and d with 5.

(d) $5a^4 - 2d^2 = 5 \cdot 2^4 - 2 \cdot 5^2$

Evaluate the powers.

$= 5 \cdot 16 - 2 \cdot 25$

Multiply.

$= 80 - 50$

Subtract.

$= 30$

CHECK YOURSELF 8

If $x = 3$, $y = 2$, $z = 4$, and $w = 5$, evaluate the following expressions.

1. $4x^2 + 2$ **2.** $5(z + w)$ **3.** $7(z^2 - y^2)$

To evaluate algebraic expressions when a fraction bar is used, you can do the following: Start by doing all the work in the numerator, then do the work in the denominator. Divide as the last step.

Example 9

If $p = 2$, $q = 3$, and $r = 4$, evaluate

(*a*) $\dfrac{8p}{r}$

Replace p with 2 and r with 4 as before.

The fraction bar is another example of a grouping symbol, like parentheses. Work first in the numerator and then in the denominator.

$$\frac{8p}{r} = \frac{8 \cdot 2}{4} = \frac{16}{4} = 4$$

Divide as the last step.

(*b*) $$\frac{4p + r}{q} = \frac{4 \cdot 2 + 4}{3} = \frac{8 + 4}{3} = \frac{12}{3} = 4$$

(*c*) $$\frac{7q + r}{p + q} = \frac{7 \cdot 3 + 4}{2 + 3} = \frac{21 + 4}{2 + 3} = \frac{25}{5} = 5$$

Now evaluate the top and bottom separately.

CHECK YOURSELF 9

Evaluate the following if $c = 5$, $d = 8$, and $e = 3$.

1. $\dfrac{6c}{e}$ **2.** $\dfrac{4d + e}{c}$ **3.** $\dfrac{10d - e}{d + e}$

CHECK YOURSELF ANSWERS

1. (1) 1; (2) 210.
2. (1) 8; (2) 11.
3. 64.
4. 8.
5. 24.
6. 20.
7. 112.
8. (1) 38; (2) 45; (3) 84.
9. (1) 10; (2) 7; (3) 7.

1.6 Exercises

Evaluate the following expressions.

1. $4 + 3 \cdot 5$ 19

2. $8 - 2 \cdot 3$ 2

3. $(4 + 3) \cdot 5$ 35

4. $(8 - 2) \cdot 3$ 18

5. $12 - 8 \div 4$ 10

6. $10 + 20 \div 5$ 14

7. $(12 - 8) \div 4$ 1

8. $(10 + 20) \div 5$ 6

9. $8 \cdot 7 + 2 \cdot 2$ 60

10. $48 \div 8 - 4 \div 2$ 4

11. $8 \cdot (7 + 2) \cdot 2$ 144

12. $48 \div (8 - 4) \div 2$ 6

13. $3 \cdot 5^2$ 75

14. $5 \cdot 2^3$ 40

15. $(3 \cdot 5)^2$ 225

16. $(5 \cdot 2)^3$ 1000

17. $4 \cdot 3^2 - 2$ 34

18. $3 \cdot 2^4 - 8$ 40

19. $4(3^2 - 2)$ 28

20. $3(2^4 - 8)$ 24

21. $2 \cdot 4^2 - 8 \cdot 3$ 8

22. $3 \cdot 2^3 - 7 \cdot 3$ 3

23. $(2 \cdot 4)^2 - 8 \cdot 3$ 40

24. $(3 \cdot 2)^3 - 7 \cdot 3$ 195

25. $4(2 + 6)^2$ 256

26. $3(8 - 4)^2$ 48

27. $(4 \cdot 2 + 6)^2$ 196

28. $(3 \cdot 8 - 4)^2$ 400

29. $3(4 + 3)^2$ 147

30. $5(4 - 2)^3$ 40

31. $3 \cdot 4 + 3^2$ 21

32. $5 \cdot 4 - 2^3$ 12

33. $4(2 + 3)^2 - 25$ 75

34. $8 + 2(3 + 3)^2$ 80

35. $(4 \cdot 2 + 3)^2 - 25$ 96

36. $8 + (2 \cdot 3 + 3)^2$ 89

Evaluate the following expressions if $a = 2$, $b = 3$, $c = 5$, and $d = 6$.

37. $6bd$ 108

38. $3ac$ 30

39. $a + d$ 8

40. $c - a$ 3

41. $5a + 2b$ 16

42. $3d - 7a$ 4

43. $2c^3$ 250

44. $3d^2$ 108

45. $2b^2 + 3c^2$ 93

46. $d^3 - 3b^2$ 189

47. $3(a + c)$ 21

48. $5(d - a)$ 20

49. $3a + c$ 11

50. $5d - a$ 28

51. $6(a + 2d)$ 84

52. $3(3b + c)$ 42

53. $6a + 2d$ 24

54. $3(3b) + c$ 32

55. $5(4d - 3c)$ 45

56. $7(2a + 3b)$ 91

57. $c(4a + 3d)$ 130

58. $d(4c - 3b)$ 66

59. $2a^2 + b^2$ 17

60. $5a^2 + c^2$ 45

61. $2(a^2 + b^2)$ 26

62. $5(a^2 + c^2)$ 145

63. $(2a^2 + b)^2$ 121

64. $(5a^2 + c)^2$ 625

65. $\dfrac{10a}{c}$ 4

66. $\dfrac{9d}{b}$ 18

67. $\dfrac{5cd}{b}$ 50

68. $\dfrac{8ab}{d}$ 8

69. $\dfrac{7b - d}{c}$ 3

70. $\dfrac{3c + d}{b}$ 7

71. $\dfrac{a + 2d}{2c - b}$ 2

72. $\dfrac{2b + d}{c - b}$ 6

73. $\dfrac{3a^3 - 2d}{c - a}$ 4

74. $\dfrac{2b^2 + 3c}{2b + c}$ 3

Insert parentheses where necessary so that each of the following statements will be true.

75. $2 \cdot 3 + 4 = 14$
$2 \cdot (3 + 4) = 14$

76. $4 \cdot 5 - 4 = 16$
true

77. $5 \cdot 4 - 2 = 18$
true

78. $4 \cdot 7 - 3 = 16$
$4 \cdot (7 - 3) = 16$

79. $5 \cdot 6 - 2 \cdot 3 = 24$
true

80. $4 \cdot 5 + 2 \cdot 3 = 84$
$4 \cdot (5 + 2) \cdot 3 = 84$

81. $3 \cdot 6 + 2 \cdot 2 = 48$
$3 \cdot (6 + 2) \cdot 2 = 48$

82. $5 \cdot 3 - 2 \cdot 7 = 1$
true

83. $2 + 3^2 = 25$
$(2 + 3)^2 = 25$

84. $5 - 2^2 = 9$
$(5 - 2)^2 = 9$

85. $3 + 2 \cdot 4^2 = 80$
$(3 + 2) \cdot 4^2 = 80$

86. $5 + 3 \cdot 2^3 = 64$
$(5 + 3) \cdot 2^3 = 64$

Solve each of the following problems.

87. The perimeter of a rectangle is given by the formula $P = 2L + 2W$. Find the perimeter of a rectangle with $L = 8$ centimeters (cm) and $W = 5$ cm. 26 cm

88. The area of a triangle is given by the formula $A = \frac{1}{2} \cdot b \cdot h$. Find the area of a triangle with $b = 4$ feet (ft) and $h = 6$ ft. 12 ft^2

89. The volume of a square prism is given by $V = LW^2$. Find the volume of a prism with $L = 5$ cm and $W = 3$ cm. 45 cm^3

90. The area of a trapezoid is given by $A = \frac{1}{2}h(B + b)$. Find the area of a trapezoid with $h = 6$ inches (in), $B = 8$ in, and $b = 5$ in. 39 in^2

91. The surface area of a square prism is given by $S = 2W^2 + 4LW$. Find the surface area of a prism with $L = 8$ in and $W = 3$ in. 114 in^2

92. The kinetic energy (KE) of a moving body is given by

$$\text{KE} = \frac{wv^2}{2g}$$

Find the kinetic energy of a moving body where $w = 20$ pounds (lb), $v = 60$ feet per second (ft/s), and $g = 32$ ft/s^2. 1125 ft · lb

Answers

1. 19 **3.** 35 **5.** 10 **7.** 1 **9.** 60 **11.** 144 **13.** 75 **15.** 225 **17.** 34 **19.** 28 **21.** 8 **23.** 40 **25.** 256 **27.** 196 **29.** 147 **31.** 21 **33.** 75 **35.** 96 **37.** 108 **39.** 8 **41.** 16 **43.** 250 **45.** 93 **47.** 21 **49.** 11 **51.** 84 **53.** 24 **55.** 45 **57.** 130 **59.** 17 **61.** 26 **63.** 121 **65.** 4 **67.** 50 **69.** 3 **71.** 2 **73.** 4 **75.** $2 \cdot (3 + 4) = 14$ **77.** True **79.** True **81.** $3 \cdot (6 + 2) \cdot 2 = 48$ **83.** $(2 + 3)^2 = 25$ **85.** $(3 + 2) \cdot 4^2 = 80$ **87.** 26 cm **89.** 45 cm^3 **91.** 114 in^2

Summary

From Arithmetic to Algebra [1.1]

Addition $x + y$ means the *sum* of x and y or x plus y. Some other words indicating addition are "more than" and "increased by."

The sum of x and 5 is $x + 5$.
7 more than a is $a + 7$.
b increased by 3 is $b + 3$.

Subtraction $x - y$ means the *difference* of x and y or x *minus* y. Some other words indicating subtraction are "less than" and "decreased by."

The difference of x and 3 is $x - 3$.
5 less than p is $p - 5$.
a decreased by 4 is $a - 4$.

Multiplication $x \cdot y$, $(x)(y)$, xy — These all mean the *product* of x and y or x *times* y.

The product of m and n is mn.
The product of 2 and the sum of a and b is $2(a + b)$.

Division $\frac{x}{y}$ means x *divided by* y or the *quotient* when x is divided by y.

n divided by 5 is $\frac{n}{5}$.
The sum of a and b, divided by 3, is $\frac{a + b}{3}$.

The Properties of Addition and Multiplication [1.2]

The Commutative Properties If a and b are any numbers,

1. $a + b = b + a$
2. $a \cdot b = b \cdot a$

The Associative Properties If a, b, and c are any numbers,

1. $a + (b + c) = (a + b) + c$
2. $a \cdot (b \cdot c) = (a \cdot b) \cdot c$

The Distributive Properties If a, b, and c are any numbers,

$a(b + c) = a \cdot b + a \cdot c$

Exponents [1.3]

The Notation

$$\overset{\text{Exponent}}{\underset{\text{Base}}{a}}{}^{4} = \underbrace{a \cdot a \cdot a \cdot a}_{\text{4 factors}}$$

$5^3 = 5 \cdot 5 \cdot 5$
$= 125$
$a^2b^3 = a \cdot a \cdot b \cdot b \cdot b$
$6m^2 = 6 \cdot m \cdot m$

The number or letter used as a factor, here a, is called the *base*. The *exponent,* which is written above and to the right of the base, tells us how many times the base is used as a factor.

Adding and Subtracting Expressions [1.4]

Terms A term is a number, or the product of a number and one or more variables, raised to a power.

Like Terms Terms that contain exactly the same letters raised to the same powers.

$4a^2$ and $3a^2$ are like terms.
$5x^2y$ and $2xy^2$ are not like terms.

Combining Like Terms

1. Add or subtract the numerical coefficients.
2. Attach the common variables.

$5a + 3a = 8a$
$7xy - 3xy = 4xy$

Multiplying and Dividing Expressions [1.5]

The First Property of Exponents

$$x^m \cdot x^n = x^{m+n}$$

$x^3 \cdot x^5 = x^{3+5}$
$= x^8$

Multiplying Expressions Multiply the numerical coefficients and use the first property of exponents to combine the variables.

$3a^2b^3 \cdot 4a^3b^4 = 12a^5b^7$

The Second Property of Exponents

$$\frac{x^m}{x^n} = x^{m-n}$$

where x is not 0 and m is greater than n.

$\frac{m^8}{m^4} = m^{8-4}$
$= m^4$

Dividing Expressions Divide the numerical coefficients and use the second property of exponents to combine the variables.

$\frac{28x^5y^3}{7xy^2} = 4x^4y$

Evaluating Algebraic Expressions [1.6]

The Order of Operations

1. Do any operations in parentheses first.
2. Evaluate all expressions involving exponents.
3. Do any multiplication or division in order, working from left to right.
4. Do any addition or subtraction in order, working from left to right.

$5 + 3 \cdot 2^2$ (Evaluate the power.)
$= 5 + 3 \cdot 4$ (Multiply.)
$= 5 + 12$ (Add.)
$= 17$

To Evaluate an Expression

1. Replace each variable by the given number values.
2. Do the necessary arithmetic operations following the rules for the order of operations.

If $a = 2$ and $b = 4$,
$5a^2 - 3b$
$= 5 \cdot 2^2 - 3 \cdot 4$
$= 5 \cdot 4 - 3 \cdot 4$
$= 20 - 12$
$= 8$

Summary Exercises Chapter 1

This exercise set is provided to give you practice with each of the objectives of the chapter. Each exercise is keyed to the appropriate chapter section. The answers are provided in the instructor's manual. Your instructor will give you guidelines on how to best use these exercises in your instructional setting.

[1.1] Write, using symbols.

1. 5 more than y $y + 5$

2. c decreased by 10 $c - 10$

3. The product of 8 and a $8a$

4. The quotient when y is divided by 3 $\frac{y}{3}$

5. 5 times the product of m and n $5mn$

6. The product of a and 5 less than a $a(a - 5)$

7. 3 more than the product of 17 and x $7x + 3$

8. The quotient when a plus 2 is divided by a minus 2 $\frac{a + 2}{a + 2}$

[1.2] Identify the property that is illustrated by each of the following statements.

9. $5 + (7 + 12) = (5 + 7) + 12$ Associative property of addition

10. $2(8 + 3) = 2 \cdot 8 + 2 \cdot 3$ Distributive property

11. $4 \cdot (5 \cdot 3) = (4 \cdot 5) \cdot 3$ Associative property of multiplication

12. $4 \cdot 7 = 7 \cdot 4$ Commutative property of multiplication

[1.2] Verify that each of the following statements is true by evaluating each side of the equation separately and comparing the results.

13. $8(5 + 4) = 8 \cdot 5 + 8 \cdot 4$ $72 = 72$

14. $2(3 + 7) = 2 \cdot 3 + 2 \cdot 7$ $20 = 20$

15. $(7 + 9) + 4 = 7 + (9 + 4)$ $20 = 20$

16. $(2 + 3) + 6 = 2 + (3 + 6)$ $11 = 11$

17. $(8 \cdot 2) \cdot 5 = 8(2 \cdot 5)$ $80 = 80$

18. $(3 \cdot 7) \cdot 2 = 3 \cdot (7 \cdot 2)$ $42 = 42$

[1.2] Use the distributive law to remove parentheses.

19. $3(7 + 4)$ $3 \cdot 7 + 3 \cdot 4$ **20.** $4(2 + 6)$ $4 \cdot 2 + 4 \cdot 6$ **21.** $4(w + v)$ $4w + 4v$

22. $6(x + y)$ $6x + 6y$ **23.** $3(5a + 2)$ $15a + 6$ **24.** $2(4x^2 + 3x)$ $8x^2 + 6x$

Write, using exponents.

25. $5 \cdot 5 \cdot 5 \cdot 5$ 5^4 **26.** $6 \cdot 6 \cdot 6 \cdot 6 \cdot 6$ 6^5 **27.** $y \cdot y \cdot y \cdot y \cdot y$ y^5

28. $b \cdot b \cdot b \cdot b$ b^4 **29.** $8 \cdot a \cdot a \cdot a$ $8a^3$ **30.** $c \cdot c \cdot d \cdot d \cdot d \cdot d$ c^2d^4

[1.3] Write in expanded form.

31. x^3 $x \cdot x \cdot x$ **32.** $y^3 \cdot y^2$ $y \cdot y \cdot y \cdot y \cdot y$

33. $2x^3y^4$ $2 \cdot x \cdot x \cdot x \cdot y \cdot y \cdot y \cdot y$ **34.** $(2x)^3y^4$ $(2x) \cdot (2x) \cdot (2x) \cdot y \cdot y \cdot y \cdot y$

[1.4] List the terms of the expressions.

35. $4a^3 - 3a^2$ $4a^3, -3a^2$ **36.** $5x^2 - 7x + 3$ $5x^2, -7x, 3$

[1.4] Circle the like terms.

37. $5m^2, -3m, -4m^2, 5m^3, m^2$ $5m^2, -4m^2, m^2$

38. $4ab^2, 3b^2, -5a, ab^2, 7a^2, -3ab^2, 4a^2b$ $4ab^2, ab^2, -3ab^2$

[1.4] Combine the like terms.

39. $5c + 7c$ $12c$ **40.** $2x + 5x$ $7x$

41. $4a - 2a$ $2a$ **42.** $6c - 3c$ $3c$

43. $9xy - 6xy$ $3xy$

44. $5ab^2 + 2ab^2$ $7ab^2$

45. $7a + 3b + 12a - 2b$ $19a + b$

46. $6x - 2x + 5y - 3x$ $x + 5y$

47. $5x^3 + 17x^2 - 2x^3 - 8x^2$ $3x^3 + 9x^2$

48. $3a^3 + 5a^2 + 4a - 2a^3 - 3a^2 - a$ $a^3 + 2a^2 + 3a$

49. Subtract $4a^3$ from the sum of $2a^3$ and $12a^3$. $10a^3$

50. Subtract the sum of $3x^2$ and $5x^2$ from $15x^2$. $7x^2$

[1.4] Multiply.

51. $a^7 \cdot a^{10}$ a^{17}

52. $x^6 \cdot x^4$ x^{10}

53. $2m^2 \cdot 5m^7$ $10m^9$

54. $6x^5 \cdot 3x^2$ $18x^7$

55. $3y^2 \cdot y^3 \cdot 8y^4$ $24y^9$

56. $2a^2 \cdot 3a \cdot 5a^3$ $30a^6$

57. $3p^2q^3 \cdot 7pq^5$ $21p^3q^8$

58. $6a^2b^3 \cdot 5a^3b^2$ $30a^5b^5$

[1.5] Divide.

59. $\dfrac{x^{10}}{x^3}$ x^7

60. $\dfrac{a^5}{a^4}$ a

61. $\dfrac{x^2 \cdot x^3}{x^4}$ x

62. $\dfrac{m^2 \cdot m^3 \cdot m^4}{m^5}$ m^4

63. $\dfrac{18p^7}{9p^5}$ $2p^2$

64. $\dfrac{24x^{17}}{8x^{13}}$ $3x^4$

65. $\dfrac{30m^7n^5}{6m^2n^3}$ $5m^5n^2$

66. $\dfrac{108x^9y^4}{9xy^4}$ $12x^8$

67. $\dfrac{48p^5q^3}{6p^3q}$ $8p^2q^2$

68. $\dfrac{52a^5b^3c^5}{13a^4c}$ $4ab^3c^4$

[1.6] Evaluate the following expressions.

69. $18 - 3 \cdot 5$ 3

70. $(18 - 3) \cdot 5$ 75

71. $5 \cdot 4^2$ 80

72. $(5 \cdot 4)^2$ 400

73. $5 \cdot 3^2 - 4$ 41

74. $5(3^2 - 4)$ 25

75. $5(4 - 2)^2$ 20

76. $5 \cdot 4 - 2^2$ 16

77. $(5 \cdot 4 - 2)^2$ 324

78. $3(5 - 2)^2$ 27

79. $3 \cdot 5 - 2^2$ 11

80. $(3 \cdot 5 - 2)^2$ 169

[1.6] Evaluate the following expressions if $x = 2$, $y = 3$, $z = 4$, and $w = 5$.

81. $x + y$ 5

82. $x + w + z$ 11

83. $8xw$ 80

84. $3xyz$ 72

85. $2xw$ 20

86. $5xyzw$ 600

87. $4w - 3z$ 8

88. $3z - 2x$ 8

89. $5x - 2y$ 4

90. $3y - 2z$ 1

91. $3z^2 - 4x^3$ 16

92. $5x^2 - 2y^2$ 2

93. $5(2z - w)$ 15

94. $4(6x + y)$ 60

95. $2(x + y)^2$ 50

96. $(2x + y)^2$ 49

97. $\dfrac{5yz}{2w}$ 6

98. $\dfrac{wx^3y}{10z}$ 3

99. $\dfrac{2w + 3z}{3x + w}$ 2

100. $\dfrac{5x + 2y + z}{wz}$ 1

Self-Test for Chapter One

The purpose of this self-test is to help you check your progress and to review for a chapter test in class. Allow yourself about an hour to take the test. When you are done, check your answers in the back of the book. If you missed any problems, be sure to go back and review the appropriate sections in the chapter and the exercises that are provided.

Write, using symbols.

1. 5 less than a $\quad a - 5$

2. The product of 6 and m $\quad 6m$

3. 4 times the sum of m and n $\quad 4(m + n)$

4. The quotient when the sum of a and b is divided by 3 $\quad \frac{a + b}{3}$

Identify the property that is illustrated by each of the following statements.

5. $6 \cdot 7 = 7 \cdot 6$ $\quad$ Commutative property of multiplication

6. $2(6 + 7) = 2 \cdot 6 + 2 \cdot 7$ $\quad$ Distributive property

7. $4 + (3 + 7) = (4 + 3) + 7$ $\quad$ Associative property of addition

Use the distributive law to remove parentheses. Then simplify your result.

8. $3(5 + 2)$ $\quad 21$

9. $4(5x + 3)$ $\quad 20x + 12$

Write, using exponents.

10. $4 \cdot 4 \cdot 4 \cdot 4$ $\quad 4^4$

11. $7 \cdot b \cdot b \cdot b$ $\quad 7b^3$

Combine the like terms.

12. $8a + 7a$ $\quad 15a$

13. $8x^2y - 5x^2y$ $\quad 3x^2y$

14. $10x + 8y + 9x - 3y$ $\quad 19x + 5y$

15. Subtract $9a^2$ from the sum of $12a^2$ and $5a^2$. $\quad 8a^2$

Multiply.

16. $a^5 \cdot a^9$ a^{14}

17. $3x^2y^3 \cdot 5xy^4$ $15x^3y^7$

Divide.

18. $\dfrac{4x^5}{2x^2}$ $2x^3$

19. $\dfrac{20a^3b^5}{5a^2b^2}$ $4ab^3$

Evaluate the following expressions.

20. $23 - 4 \cdot 5$ 3

21. $4 \cdot 5^2 - 35$ 65

22. $4(2 + 4)^2$ 144

Evaluate the following expressions if $a = 3$, $b = 4$, and $c = 5$.

23. $2a^3 - 3b^2$ 6

24. $5(4a - 2c)$ 10

25. $\dfrac{4a + 3c}{b + c}$ 3

Chapter Two

Signed Numbers

2.1 Signed Numbers—An Introduction

OBJECTIVES

1. To understand the meaning of a negative number.
2. To recognize the set of integers.
3. To evaluate an expression involving the absolute value notation.

You probably remember that the numbers you use to count things, 1, 2, 3, 4, 5, and so on, are called the *natural* (or *counting*) *numbers.* The *whole numbers* consist of the natural numbers and zero—0, 1, 2, 3, 4, 5, and so on. They can be represented on a number line.

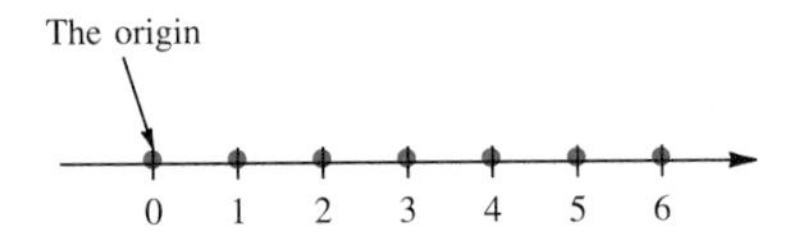

When numbers are used to represent physical quantities (altitudes, temperatures, and amounts of money are examples), it may be necessary to distinguish between *positive* and *negative* numbers. It is convenient to represent these numbers with positive (+) or negative (−) signs. For instance,

The altitude of Mount Whitney is 14,495 ft *above* sea level (+14,495).

The altitude of Death Valley is 282 ft *below* sea level (−282).

The temperature in Chicago is 10° *below* zero (−10°).

The temperature in Miami is 60° *above* zero (+60°).

A *gain* of \$100 (+100).

A *loss* of \$100 (−100).

To represent all these numbers, we extend our number line to the *left* of the origin.

Numbers used to name points to the right of the origin are positive numbers. They are written with a positive (+) sign or with no sign at all.

+6 and 9 are positive numbers

Numbers used to name points to the left of the origin are negative numbers. They are always written with a negative (−) sign.

-3 and -20 are negative numbers

↑ Read "negative 3."

Positive and negative numbers considered together are *signed numbers.*

Here is the number line extended to include negative numbers.

The arrow always points in the *positive direction.*

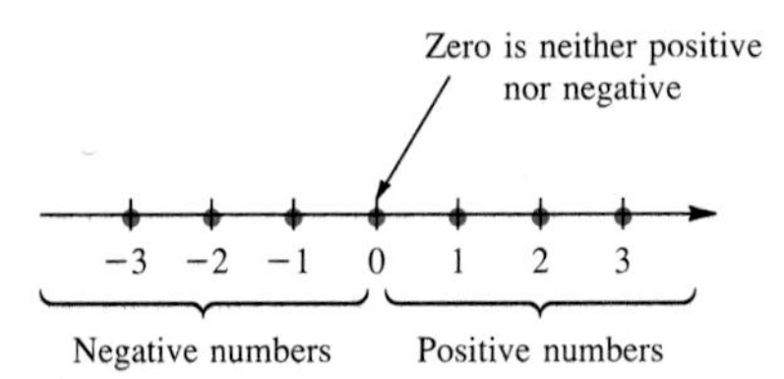

The numbers used to name the points shown on the number line above are called the *integers.* The integers consist of the natural numbers, their negatives, and the number 0. We can represent the set of integers by

$$\{\ldots, -3, -2, -1, 0, 1, 2, 3, \ldots\}$$

Example 1

(*a*) 145 is an integer.
(*b*) -28 is an integer.
(*c*) 0.35 is not an integer.
(*d*) $-\dfrac{2}{3}$ is not an integer.

CHECK YOURSELF 1

Which of the following numbers are integers?

-23 1054 -0.23 0 -500

An important idea for our work in this chapter is the *absolute value* of a number. This represents the distance of the point named by the number from the origin on the number line. For instance, 5 and -5 represent points the same distance from the origin but in opposite directions.

Five units Five units

−5 0 5

The absolute value of 5 is 5. The absolute value of -5 is also 5.

In symbols we write

$|5| = 5$ and $|-5| = 5$

Read "the absolute value of 5." Read "the absolute value of negative 5."

The absolute value of a number does *not* depend on whether the number is to the right or to the left of the origin, but on its *distance* from the origin.

Example 2

(*a*) $|7| = 7$
(*b*) $|-7| = 7$
(*c*) $-|-7| = -7$

This is the *negative* of the absolute value of negative 7.

(*d*) $|-10| + |10| = 10 + 10 = 20$
(*e*) $|8 - 3| = |5| = 5$

Absolute value bars serve as another set of grouping symbols, so do the operation *inside* first.

(*f*) $|8| - |3| = 8 - 3 = 5$

Here, evaluate the absolute values, then subtract.

CHECK YOURSELF 2

Evaluate.

1. $\lvert 8\rvert$	**2.** $\lvert -8\rvert$	**3.** $-\lvert -8\rvert$
4. $\lvert -9\rvert + \lvert 4\rvert$	**5.** $\lvert 9 - 4\rvert$	**6.** $\lvert 9\rvert - \lvert 4\rvert$

CHECK YOURSELF ANSWERS

1. −23, 1054, 0, and −500.
2. (1) 8; (2) 8; (3) −8; (4) 13; (5) 5; (6) 5.

2.1 Exercises

Represent each quantity with a signed number.

1. An altitude of 300 ft above sea level 300

2. An altitude of 50 ft below sea level −50

3. A loss of $200 −200

4. A profit of $200 200

5. A temperature of 30° above zero 30

6. A temperature of $5°$ below zero -5

Represent the integers on the number lines shown.

7. $5, -15, 18, -8, 3$

−15 −8 3 5 18
−20 −10 0 10 20

8. $-18, 4, -5, 13, 9$

−18 −5 4 9 13
−20 −10 0 10 20

Which numbers in the following sets are integers?

9. $\left\{5, -\frac{2}{9}, 175, -234, -0.64\right\}$
5, 175, −234

10. $\left\{-45, 0.35, \frac{3}{5}, 700, -26\right\}$
−45, 700, −26

Evaluate.

11. $|12|$ 12

12. $|-12|$ 12

13. $|-14|$ 14

14. $-|-3|$ −3

15. $-|-12|$ −12

16. $|-8| + |8|$ 16

17. $|-7| + |5|$ 12

18. $|7 - 5|$ 2

19. $|9| + |-4|$ 13

20. $|9 - 4|$ 5

21. $|8| - |-8|$ 0

22. $|-8 + 8|$ 0

23. $|13 - 5|$ 8

24. $|13| - |5|$ 8

25. $|-9 + 7|$ 2

26. $|9| + |-7|$ 16

27. $|9| - |7|$ 2

28. $|9| - |-7|$ 2

29. $|-8| - |-7|$ 1

30. $|-15| - |-10|$ 5

Skillscan (Section 1.2)

Find each sum.

a. $3 + (8 + 9)$ 20

b. $(6 + 12) + 3$ 21

c. $(3 + 8) + 9$ 20

d. $6 + (12 + 3)$ 21

e. $\frac{2}{3} + \left(3 + \frac{1}{3}\right)$ 4

f. $\left(\frac{3}{4} + 1\right) + \frac{5}{4}$ 3

g. $\left(\frac{2}{3}+\frac{1}{3}\right)+3$ 4 **h.** $\left(\frac{3}{4}+\frac{5}{4}\right)+1$ 3 **i.** $\left(\frac{1}{2}+\frac{1}{3}\right)+\frac{1}{6}$ 1

j. $\left(\frac{1}{4}+\frac{3}{8}\right)+\frac{1}{2}$ $\frac{9}{8}$

Answers

We will provide the solutions for the odd-numbered exercises at the end of each exercise set. The solutions for the even-numbered exercises are at the back of the book.

1. 300 or (+300) **3.** −200 **5.** 30

7.

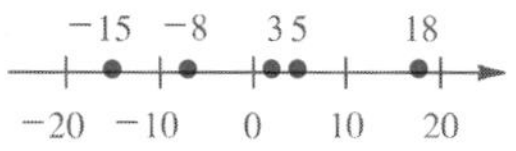

9. 5, 175, −234 **11.** 12 **13.** 14 **15.** −12 **17.** 12 **19.** 13 **21.** 0 **23.** 8 **25.** 2 **27.** 2 **29.** 1 **a.** 20 **b.** 21 **c.** 20 **d.** 21 **e.** 4 **f.** 3 **g.** 4 **h.** 3

2.2 Adding Signed Numbers

OBJECTIVE
To find the sum of signed numbers.

In the previous section, we introduced the idea of signed numbers. Now we will examine the four arithmetic operations (addition, subtraction, multiplication, and division) and see how those operations are performed when signed numbers are involved. We will start by considering addition.

An application may help. As before, let's represent a gain of money as a positive number and a loss as a negative number.

If you gain $3 and then gain $4, the result is a gain of $7:

$$3 + 4 = 7$$

If you lose $3 and then lose $4, the result is a loss of $7:

$$-3 + (-4) = -7$$

If you gain $3 and then lose $4, the result is a loss of $1:

$$3 + (-4) = -1$$

If you lose $3 and then gain $4, the result is a gain of $1:

$$-3 + 4 = 1$$

The number line can be used to illustrate the addition of integers. Starting at the origin, we move to the *right* for positive numbers but to the *left* for negative numbers.

Example 1

To add $3 + 4$:

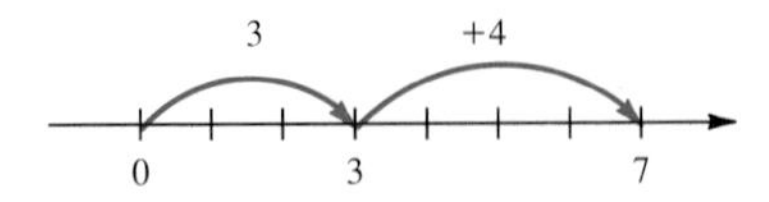

Start at the origin and move 3 units to the right. Then move 4 more units to the right to find the sum. Here we have

$3 + 4 = 7.$

The number line will also help you visualize the sum of two negative numbers. The following example illustrates.

Example 2

(*a*) To add $(-3) + (-4)$:

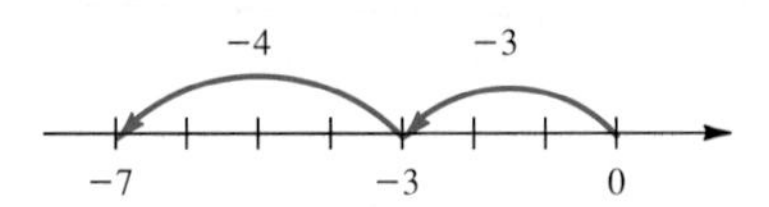

Start at the origin and move 3 units to the left. Then move 4 more units to the left to find the sum. From the graph we see that the sum is

$(-3) + (-4) = -7$

(*b*) To add $(-6) + (-5)$:

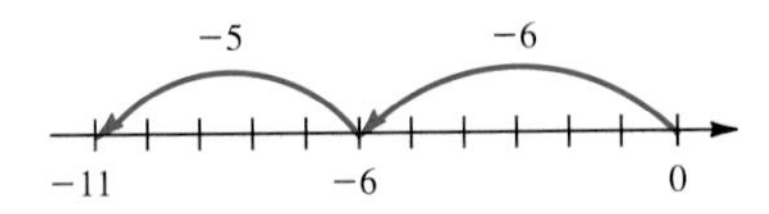

Again we start at the origin, this time moving 6 units to the left. We then move 5 more units to the left to find the sum. Here

$(-6) + (-5) = -11$

(*c*) To add $\left(-\frac{3}{2}\right) + \left(-\frac{1}{2}\right)$:

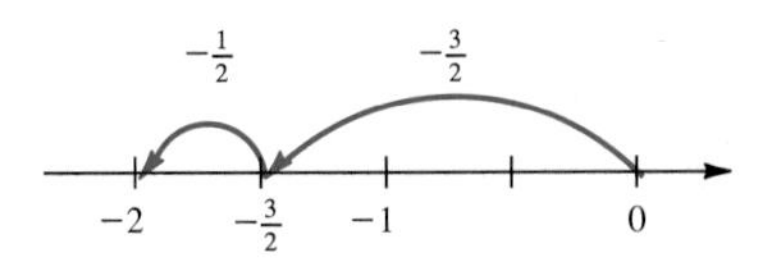

As before, we start at the origin. From that point move $\frac{3}{2}$ units left.

Then move another $\frac{1}{2}$ unit left to find the sum. In this case

$$\left(-\frac{3}{2}\right) + \left(-\frac{1}{2}\right) = -2$$

CHECK YOURSELF 1

Add.

1. $5 + 7$

2. $(-3) + (-7)$

3. $(-5) + (-15)$

4. $\left(-\frac{5}{2}\right) + \left(-\frac{3}{2}\right)$

You have probably noticed some helpful patterns in the previous examples. These patterns will allow you to do the work mentally without having to use the number line. Look at the following rule.

This means that the sum of two positive numbers is positive and the sum of two negative numbers is negative.

ADDING SIGNED NUMBERS 1

If two numbers have the same sign, add their absolute values. Give the sum the sign of the original numbers.

Example 3

(*a*) $8 + 5 = 13$

The sum of two positive numbers is positive.

(*b*) $(-8) + (-5) = -13$

Add the absolute values $(8 + 5 = 13)$ and give the sum the sign $(-)$ of the original numbers.

(*c*) $[(-3) + (-4)] + (-6)$ Add inside the brackets as your first step

$= (-7) + (-6)$

$= -13$

CHECK YOURSELF 2

Add mentally.

1. $7 + 9$

2. $(-7) + (-9)$

3. $(-5.8) + (-3.2)$

4. $[(-5) + (-2)] + (-3)$

Let's again use the number line to illustrate the addition of two numbers. This time the numbers will have *different* signs.

Example 4

To add $5 + (-2)$:

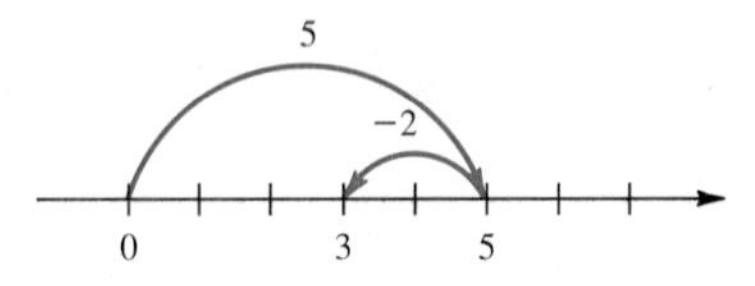

First move 5 units to the right of the origin. Then move 2 units to the left.

$5 + (-2) = 3$

Example 5

To add $3 + (-6)$:

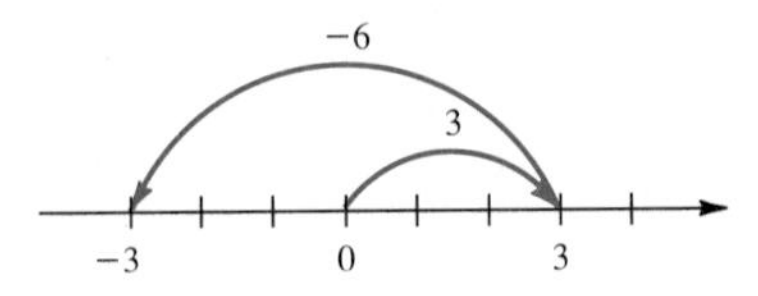

First move 3 units to the right of the origin. Then move 6 units to the left.

$3 + (-6) = -3$

Example 6

To add $-4 + 7$:

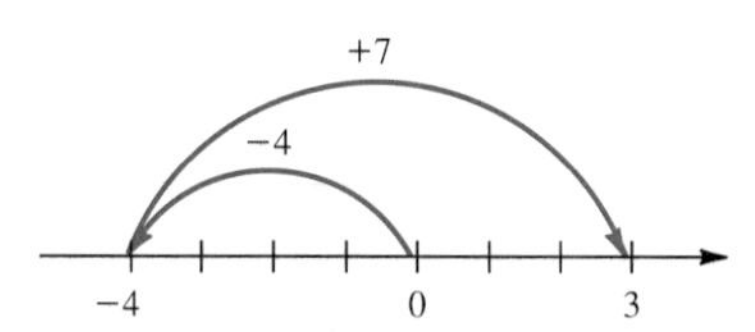

This time move 4 units to the left of the origin as the first step. Then move 7 units to the right.

$-4 + 7 = 3$

CHECK YOURSELF 3

Add.

1. $7 + (-5)$ **2.** $4 + (-8)$ **3.** $-4 + 9$ **4.** $-7 + 3$

You have no doubt also noticed that in adding a positive number and a negative number, sometimes the sum is positive and sometimes it is negative. This depends on which of the numbers has the larger absolute value. This leads us to the second part of our addition rule.

ADDING SIGNED NUMBERS 2

If two numbers have different signs, subtract their absolute values, the smaller from the larger. Give the sum the sign of the number with the larger absolute value.

Example 7

(*a*) $12 + (-9) = 3$

Since the numbers have different signs, subtract their absolute values ($12 - 9 = 3$). The sum has the sign (+) of the number with the larger absolute value, 12.

(*b*) $7 + (-19) = -12$

Since the two numbers have different signs, again subtract the absolute values ($19 - 7 = 12$). The sum has the sign (−) of the number with the larger absolute value, −19.

(*c*) $-13 + 7 = -6$

Subtract the absolute values ($13 - 7 = 6$). The sum has the sign (−) of the number with the larger absolute value, −13.

CHECK YOURSELF 4

Add mentally.

1. $5 + (-14)$ **2.** $-7 + (-8)$ **3.** $-8 + 15$

4. $7 + (-8)$ **5.** $-\frac{2}{3} + \left(-\frac{7}{3}\right)$ **6.** $5.3 + (-2.3)$

There are two other properties of addition that we should mention before concluding this section. First, the sum of any number and 0 is always that number. In symbols,

ADDITIVE IDENTITY PROPERTY

For any number *a*,

$a + 0 = 0 + a = a$

No number loses its identity after addition with 0.

Example 8

(*a*) $9 + 0 = 9$
(*b*) $0 + (-8) = -8$
(*c*) $(-25) + 0 = -25$

The opposite of a number is also called the *additive inverse* of that number.

We'll need one further definition to state our second property. Every number has an *opposite*. It corresponds to a point the same distance from the origin as the given integer, but in the opposite direction.

The opposite of 9 is -9.

The opposite of -15 is 15.

Our second property states that the sum of any number and its opposite is 0. In symbols,

Here $-a$ represents the opposite of the number a. The sum of any number and its opposite, or additive inverse, is 0.

ADDITIVE INVERSE PROPERTY

For any number a, there exists a number $-a$ such that

$$a + (-a) = (-a) + a = 0$$

Example 9

(*a*) $9 + (-9) = 0$
(*b*) $-15 + 15 = 0$

CHECK YOURSELF 5

Add.

1. $(-17) + 0$ **2.** $(-17) + 17$
3. $0 + (-10)$ **4.** $12 + (-12)$

We can now use the associative and commutative properties of addition, introduced in Section 1.2, to find the sum when more than two signed numbers are involved. Our final example illustrates.

Example 10

We use the associative property to "group" 4 and 5. We then add 4 and 5 and proceed as before. Of course, the grouping step can be done mentally.

(*a*) $4 + 5 + (-8)$
$= (4 + 5) + (-8)$
$= 9 + (-8)$
$= 1$

We use the commutative property to reverse the order of addition for −3 and 5. We then group −5 and 5. Do you see why?

(*b*) $(-5) + (-3) + 5$
$= (-5) + 5 + (-3)$
$= [(-5) + 5] + (-3)$
$= 0 + (-3) = -3$

CHECK YOURSELF 6

Add.

1. $(-4) + 5 + (-3)$ **2.** $(-8) + 4 + 8$

CHECK YOURSELF ANSWERS

1. (1) 12; (2) −10; (3) −20; (4) −4. **2.** (1) 16; (2) −16; (3) −9; (4) −10. **3.** (1) 2; (2) −4; (3) 5; (4) −4 **4.** (1) −9; (2) −15; (3) 7; (4) −1; (5) −3; (6) 3. **5.** (1) −17; (2) 0; (3) −10; (4) 0. **6.** (1) −2; (2) 4.

2.2 Exercises

Add.

1. $-6 + (-5)$ −11
2. $8 + 7$ 15
3. $12 + 9$ 21
4. $-3 + (-17)$ −20
5. $-8 + (-10)$ −18
6. $-12 + (-8)$ −20
7. $9 + (-3)$ 6
8. $6 + (-12)$ −6
9. $5 + (-9)$ −4
10. $10 + (-4)$ 6
11. $15 + (-10)$ 5
12. $17 + (-8)$ 9
13. $-9 + 5$ −4
14. $-8 + 7$ −1
15. $-6 + 8$ 2
16. $-10 + 12$ 2
17. $-10 + 10$ 0
18. $8 + (-8)$ 0
19. $-9 + 0$ −9
20. $-17 + 17$ 0
21. $-18 + (-12)$ −30
22. $-15 + 0$ −15
23. $24 + (-24)$ 0
24. $-28 + (-22)$ −50
25. $-10 + (-15)$ −25
26. $7 + (-20)$ −13
27. $-\frac{3}{4} + \left(-\frac{5}{4}\right)$ −2

28. $\frac{2}{3} + \left(-\frac{11}{3}\right)$ −3

29. $\frac{5}{8} + \left(-\frac{13}{8}\right)$ −1

30. $-\frac{5}{6} + \left(-\frac{7}{6}\right)$ −2

31. $-3.7 + (-5.3)$ −9

32. $4.3 + (-9.3)$ −5

33. $-6.5 + 8.5$ 2

34. $-9.8 + (-3.2)$ −13

Perform each of the indicated operations.

35. $(-3 + 1) + (-4)$ −6

36. $(-8 + 7) + (-5)$ −6

37. $-2 + (-3) + (-8)$ −13

38. $7 + (-9) + (-10)$ −12

39. $-9 + (-17) + 9$ −17

40. $15 + (-3) + (-15)$ −3

41. $5 + 2 + (-9) + 1$ −1

42. $6 + (-8) + (-4) + 5$ −1

43. $(-3) + 0 + (-6) + 8$ −1

44. $5 + (-6) + 7 + (-8)$ −2

45. $1 + (-2) + 3 + (-4)$ −2

46. $(-9) + 0 + (-2) + 12$ 1

47. $\frac{5}{3} + \left(-\frac{4}{3}\right) + \frac{5}{3}$ 2

48. $-\frac{6}{5} + \left(-\frac{13}{5}\right) + \frac{4}{5}$ −3

49. $-\frac{3}{2} + \left(-\frac{7}{4}\right) + \frac{1}{4}$ −3

50. $\frac{1}{3} + \left(-\frac{5}{6}\right) + \left(-\frac{1}{2}\right)$ −1

51. $2.3 + (-5.4) + (-2.9)$ −6

52. $-(5.4) + (-2.1) + (-3.5)$ −11

Evaluate each of the following expressions.

53. $|2 + (-3)|$ 1

54. $|(-4) + 2|$ 2

55. $|-15 + 6|$ 9

56. $|-19 + 16|$ 3

57. $|-3 + 2 + (-5)|$ 6

58. $|-2 + 8 + (-6)|$ 0

59. $|2 + (-3)| + |(-3) + 2|$ 2

60. $|8 + (-10)| + |-12 + 14|$ 4

Skillscan (Appendix 1)

Subtract as indicated.

a. $\frac{5}{2} - \frac{3}{2}$ 1 **b.** $\frac{7}{2} - 2$ $\frac{3}{2}$ **c.** $\frac{4}{3} - \frac{1}{6}$ $\frac{7}{6}$ **d.** $\frac{9}{4} - \frac{3}{2}$ $\frac{3}{4}$

e. $5 - \frac{2}{3}$ $\frac{13}{3}$ **f.** $\frac{9}{10} - \frac{1}{4}$ $\frac{13}{20}$ **g.** $\frac{1}{5} - \frac{1}{7}$ $\frac{2}{35}$ **h.** $\frac{12}{5} - \frac{2}{3}$ $\frac{26}{15}$

Answers

1. −11 **3.** 21 **5.** −18 **7.** 6 **9.** −4 **11.** 5 **13.** −4 **15.** 2 **17.** 0 **19.** −9 **21.** −30 **23.** 0 **25.** −25 **27.** −2 **29.** −1 **31.** −9 **33.** 2 **35.** −6 **37.** −13 **39.** −17 **41.** −1 **43.** −1 **45.** −2 **47.** 2 **49.** −3 **51.** −6 **53.** 1 **55.** 9 **57.** 6 **59.** 2 **a.** 1 **b.** $\frac{3}{2}$ **c.** $\frac{7}{6}$ **d.** $\frac{3}{4}$ **e.** $\frac{13}{3}$ **f.** $\frac{13}{20}$ **g.** $\frac{2}{35}$ **h.** $\frac{26}{15}$

2.3 Subtracting Signed Numbers

OBJECTIVE
To find the difference of two integers.

To begin our discussion of subtraction when signed numbers are involved, we can look back at a problem using natural numbers. Of course, we know that

$$8 - 5 = 3 \tag{1}$$

From our work in adding signed numbers in the last section, we know that it is also true that

$$8 + (-5) = 3 \tag{2}$$

Comparing Equations (1) and (2), we see that the results are the same. This leads us to an important pattern. Any subtraction problem can be written as a problem in addition. Subtracting 5 is the same as adding the opposite of 5, or −5. Let's try it again.

Example 1

$$9 - 7 = 2$$
$$9 + (-7) = 2$$

Again we see that subtracting a number, here 7, is the same as adding its opposite, −7.

This leads us to the following rule for subtracting signed numbers.

SUBTRACTING SIGNED NUMBERS

STEP 1 Rewrite the subtraction problem as an addition problem by

(*a*) Changing the subtraction symbol to an addition symbol
(*b*) Replacing the number being subtracted with its opposite

STEP 2 Add the resulting signed numbers as before.

In symbols,

This is the *definition* of subtraction.

$$a - b = a + (-b)$$

The following example illustrates the use of this definition in performing subtraction.

Example 2

Subtraction *Addition*

Change the subtraction symbol (−) to an addition symbol (+).

(*a*) $15 - 7 = 15 + (-7)$

Replace 7 with its opposite, −7.

$= 8$

(*b*) $9 - 12 = 9 + (-12)$
$= -3$

(*c*) $-6 - 7 = -6 + (-7)$
$= -13$

(*d*) $-\frac{3}{5} - \frac{7}{5} = -\frac{3}{5} + \left(-\frac{7}{5}\right)$
$= \frac{-10}{5} = -2$

(*e*) Subtract 5 from −2. We write the statement as

$-2 - 5$

and proceed as before:

$$-2 - 5 = -2 + (-5)$$
$$= -7$$

CHECK YOURSELF 1

Subtract.

1. $18 - 7$ **2.** $5 - 13$ **3.** $-7 - 9$

4. $-\frac{5}{6} - \frac{7}{6}$ **5.** $-2 - 7$

The subtraction rule is used in the same way when the number being subtracted is negative. Again change the subtraction to addition. Replace the negative number being subtracted with its opposite—that will now be positive. The following example illustrates.

Example 3

Subtraction *Addition*

Change the subtraction to addition.

(*a*) $5 - (-2) = 5 + (+2)$

Replace -2 with its opposite, $+2$ or 2.

$= 5 + 2$
$= 7$

(*b*) $7 - (-8) = 7 + (+8)$
$= 7 + 8$
$= 15$

(*c*) $-9 - (-5) = -9 + 5$
$= -4$

(*d*) $-12.7 - (-3.7) = -12.7 + 3.7 = -9$

(*e*) Subtract -4 from -5. We write

$-5 - (-4) = -5 + 4 = -1$

CHECK YOURSELF 2

Subtract.

1. $8 - (-2)$ **2.** $3 - (-10)$ **3.** $-7 - (-2)$

4. $-9.8 - (-5.8)$ **5.** $7 - (-7)$

As before, when parentheses are involved in an expression, you should do any operations inside parentheses as the first step. Our final example illustrates.

Example 4

Evaluate.

$5 - (-5 - 4)$

Evaluate as the first step.

$= 5 - (-9)$
$= 5 + 9$
$= 14$

CHECK YOURSELF 3

Evaluate.

$7 - (-9 + 6)$

CHECK YOURSELF ANSWERS

1. (1) 11; (2) −8; (3) −16; (4) −2; (5) −9.
2. (1) 10; (2) 13; (3) −5; (4) − 4; (5) 14.
3. 10.

2.3 Exercises

Subtract.

1. $17 - 9$ 8
2. $5 - 7$ −2
3. $7 - 12$ −5
4. $-5 - 3$ −8
5. $-6 - 8$ −14
6. $19 - 5$ 14
7. $9 - 18$ −9
8. $9 - 13$ −4
9. $-9 - 14$ −23
10. $-8 - 12$ −20
11. $5 - (-2)$ 7
12. $7 - (-9)$ 16
13. $3 - (-9)$ 12
14. $8 - (-2)$ 10
15. $7 - (-12)$ 19
16. $3 - (-10)$ 13
17. $-6 - (-2)$ −4
18. $-5 - (-8)$ 3

19. $-9-(-12)$ 3
20. $-12-(-3)$ -9
21. $-7-(-7)$ 0

22. $0-(-5)$ 5
23. $0-(-9)$ 9
24. $20-(-18)$ 38

25. $15-(-10)$ 25
26. $-15-(-15)$ 0
27. $-17-(-19)$ 2

28. $15-(-25)$ 40
29. $6-(-31)$ 37
30. $-16-(-24)$ 8

31. $-\frac{2}{3}-\left(-\frac{8}{3}\right)$ 2
32. $\frac{5}{6}-\left(-\frac{7}{6}\right)$ 2
33. $\frac{3}{4}-\left(-\frac{3}{2}\right)$ $\frac{9}{4}$

34. $-\frac{5}{8}-\left(-\frac{3}{4}\right)$ $\frac{1}{8}$
35. $-2.7-(-5.7)$ 3
36. $8.3-(-5.7)$ 14

37. $6.9-(-10.1)$ 17
38. $-5.6-(-2.6)$ -3

Do the indicated operations.

39. $2-5-7$ -10
40. $3-8-4$ -9
41. $5+9-12$ 2

42. $-9+6-12$ -15
43. $-7-(-5)-2$ -4
44. $-8-(-2)-3$ -9

45. $-8-(-2)+7$ 1
46. $-7-(-5)+9$ 7
47. $-3-8+4$ -7

48. $-12-6+14$ -4
49. $-3-(8+4)$ -15
50. $-12-(6+14)$ -32

51. $-4-6-9$ -19
52. $10-8-7$ -5
53. $-4-(6-9)$ -1

54. $10-(8-7)$ 9
55. $7-(3-5)+2$ 11
56. $8-(2-6)+10$ 22

57. Subtract -8 from 5. 13
58. Subtract -7 from -2. 5

59. Subtract 10 from -3. -13
60. Subtract -9 from 3. 12

61. Subtract -12 from -8. 4
62. Subtract 12 from -8. -20

63. Subtract 12 from the sum of -6 and 3.
-15

64. Subtract -9 from the sum of -3 and 8.
14

65. Subtract the sum of 8 and -14 from -6.
0

66. Subtract the sum of -9 and -3 from 2.
14

Evaluate each expression.

67. $|6 - (-9)|$ 15

68. $|4 - (-3)|$ 7

69. $|-22 - (-2)|$ 20

70. $|23 - (-6)|$ 29

71. $|18| - |-2|$ 16

72. $|23| - |-6|$ 17

73. $|-28 - (-34)|$ 6

74. $|-64 - (-81)|$ 17

75. $|-28| - |-34|$ -6

76. $|-64| - |-81|$ -17

Skillscan (Section 2.2)

Add.

a. $(-1) + (-1) + (-1) + (-1)$ -4

b. $3 + 3 + 3 + 3 + 3$ 15

c. $9 + 9 + 9$ 27

d. $(-10) + (-10) + (-10)$ -30

e. $(-5) + (-5) + (-5) + (-5) + (-5)$ -25

f. $(-8) + (-8) + (-8) + (-8)$ -32

Answers

1. 8 **3.** -5 **5.** -14 **7.** -9 **9.** -23 **11.** 7 **13.** 12 **15.** 19 **17.** -4 **19.** 3 **21.** 0
23. 9 **25.** 25 **27.** 2 **29.** 37 **31.** 2 **33.** $\frac{9}{4}$ **35.** 3 **37.** 17 **39.** -10 **41.** 2 **43.** -4
45. 1 **47.** -7 **49.** -15 **51.** -19 **53.** -1 **55.** 11 **57.** 13 **59.** -13 **61.** 4 **63.** -15
65. 0 **67.** 15 **69.** 20 **71.** 16 **73.** 6 **75.** -6 **a.** -4 **b.** 15 **c.** 27 **d.** -30 **e.** -25
f. -32

2.4 Multiplying Signed Numbers

OBJECTIVE
To find the product of two or more signed numbers.

When you first considered multiplication in arithmetic, it was thought of as repeated addition. Let's see what our work with the addition of signed numbers can tell us about multiplication when signed numbers are involved.

Example 1

$3 \cdot 4 = \underbrace{4 + 4 + 4} = 12$

We interpret multiplication as repeated addition to find the product, 12.

Example 2

(*a*) $(3)(-4) = (-4) + (-4) + (-4) = -12$

(*b*) $(4)(-5) = (-5) + (-5) + (-5) + (-5)$
$= -20$

Looking at the products found by repeated addition in Example 2 should suggest the first portion of our rule for multiplying signed numbers. In each case the product of a positive number and a negative number was negative.

MULTIPLYING SIGNED NUMBERS 1

The product of two numbers with different signs is negative.

To use this rule in multiplying two numbers with different signs, multiply their absolute values and attach a negative sign.

Example 3

(*a*) $(5)(-6) = -30$

The product is negative.

(*b*) $(-10)(10) = -100$

(*c*) $(8)(-12) = -96$

CHECK YOURSELF 1

Multiply.

1. $(-7)(5)$ **2.** $(-12)(9)$ **3.** $(-15)(8)$

The product of two negative numbers is harder to visualize.

The following pattern may help you see how we can determine the sign of the product.

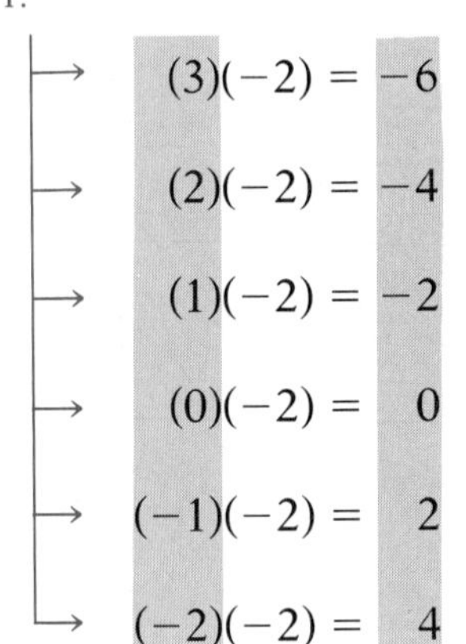
This number is decreasing by 1.
(3)(−2) = −6
(2)(−2) = −4
(1)(−2) = −2
(0)(−2) = 0
(−1)(−2) = 2
(−2)(−2) = 4

Do you see that the product is *increasing* by 2 each time?

What should the product $(-3)(-2)$ be? By the pattern shown,

$$(-3)(-2) = 6$$

This suggests that the product of two negative numbers is positive, and that is in fact the case. We can extend our multiplication rule.

If you would like a more detailed explanation, see the discussion at the end of this section.

MULTIPLYING SIGNED NUMBERS 2

The product of two integers with the same sign is positive.

Example 4

(*a*) $9 \cdot 7 = 63$ The product of two positive numbers (same sign, +) is positive.

(*b*) $(-8)(-5) = 40$ The product of two negative numbers (same sign, −) is positive.

CHECK YOURSELF 2

Multiply.

1. $10 \cdot 12$ **2.** $(-8)(-9)$

Two numbers, 0 and 1, have special properties in multiplication, as will be discussed in the following:

The number 1 is called the *multiplicative identity* for this reason.

MULTIPLICATIVE IDENTITY PROPERTY

The product of 1 and any number is that number. In symbols,

$a \cdot 1 = 1 \cdot a = a$

MULTIPLICATIVE PROPERTY OF ZERO

The product of 0 and any number is 0. In symbols,

$a \cdot 0 = 0 \cdot a = 0$

Example 5

(*a*) $(1)(-7) = -7$
(*b*) $(15)(1) = 15$
(*c*) $(-7)(0) = 0$
(*d*) $0 \cdot 12 = 0$

CHECK YOURSELF 3

Multiply.

1. $(-10)(1)$ **2.** $(0)(-17)$

To complete our discussion of the properties of multiplication, we state the following.

$\frac{1}{a}$ is called the *multiplicative inverse,* or the *reciprocal,* of *a*. The product of any nonzero number and its reciprocal is 1.

MULTIPLICATIVE-INVERSE PROPERTY

For any number *a*, where $a \neq 0$, there is a number $\frac{1}{a}$ such that

$$a \cdot \frac{1}{a} = 1$$

The following example illustrates.

Example 6

(*a*) $3 \cdot \frac{1}{3} = 1$ The reciprocal of 3 is $\frac{1}{3}$.

(*b*) $-5\left(-\frac{1}{5}\right) = 1$ The reciprocal of -5 is $\frac{1}{-5}$ or $-\frac{1}{5}$.

(*c*) $\frac{2}{3} \cdot \frac{3}{2} = 1$ The reciprocal of $\frac{2}{3}$ is $\frac{1}{\frac{2}{3}}$, or $\frac{3}{2}$.

CHECK YOURSELF 4

Find the multiplicative inverse (or the reciprocal) of each of the following numbers.

1. 6 **2.** -4 **3.** $\frac{1}{4}$ **4.** $-\frac{3}{5}$

In addition to the properties just mentioned, we can extend the commutative and associative properties for multiplication to signed numbers. Our final example illustrates an application of the associative property of multiplication.

Example 7

Find the following product:

$(-3)(2)(-7)$

Applying the associative property, we can group the first two factors to write

Once again, this "grouping" can be done mentally.

$[(-3)(2)](-7)$

Evaluate first.

$= (-6)(-7)$
$= 42$

CHECK YOURSELF 5

Find the product

$(-5)(-8)(-2)$

When symbols of grouping, or operations of more than one level, are involved in an expression, we must again follow our rules for the order of operations. Consider the following example.

Example 8

Evaluate each expression.

(*a*) $7(-9 + 12)$

$= 7(3) = 21$ Evaluate inside the parentheses first.

(*b*) $(-8)(-7) - 40$ Multiply first, then subtract.
$= 56 - 40$
$= 16$

(*c*) $(-5)^2 - 3$ — Evaluate the power first.

$= (-5)(-5) - 3$ — Note that $(-5)^2 = (-5)(-5) = 25$

$= 25 - 3$

$= 22$

(*d*) $-5^2 - 3$ — Note that $-5^2 = -25$

$= -25 - 3$

$= -28$ — The power applies *only* to the 5.

CHECK YOURSELF 6

Evaluate each expression.

1. $8(-9 + 7)$

2. $(-3)(-5) + 7$

3. $(-4)^2 - (-4)$

4. $-4^2 - (-4)$

Earlier we mentioned that we would give a more detailed explanation of why the product of two negative numbers is positive. That argument appears to the right.

THE PRODUCT OF TWO NEGATIVE NUMBERS

The following argument shows why the product of two negative numbers is positive.

From our earlier work, we know that the sum of a number and its opposite is 0:

$5 + (-5) = 0$

Multiply both sides of the equation by -3:

$(-3)[5 + (-5)] = (-3)(0)$

Since the product of 0 and any number is 0, on the right we have 0.

$(-3)[5 + (-5)] = 0$

We use the distributive law on the left.

$(-3)(5) + (-3)(-5) = 0$

We know that $(-3)(5) = -15$, so the equation becomes

$-15 + (-3)(-5) = 0$

We now have a statement of the form

$-15 + \square = 0$

where $\square$ is the value of $(-3)(-5)$. We also know that $\square$ is the number that must be added to -15 to get 0, so $\square$ is the opposite of -15, or 15. This means that

$(-3)(-5) = 15$ — The product is positive!

It doesn't matter what numbers we use in this argument. The resulting product of two negative numbers will always be positive.

CHECK YOURSELF ANSWERS

1. (1) -35; (2) -108; (3) -120. **2.** (1) 120; (2) 72.
3. (1) -10; (2) 0. **4.** (1) $\frac{1}{6}$; (2) $-\frac{1}{4}$; (3) 4; (4) $-\frac{5}{3}$.
5. -80. **6.** (1) -16; (2) 22; (3) 20; (4) -12.

2.4 Exercises

Multiply.

1. $8 \cdot 5$ 40

2. $(6)(-10)$ -60

3. $(5)(-4)$ -20

4. $20 \cdot 5$ 100

5. $(-8)(9)$ -72

6. $(-12)(3)$ -36

7. $(-8)(-7)$ 56

8. $(-9)(-8)$ 72

9. $(-12)(4)$ -48

10. $(-10)(1)$ -10

11. $(-5)(0)$ 0

12. $(-10)(-6)$ 60

13. $(-12)(-12)$ 144

14. $(8)(-15)$ -120

15. $(-15)(-7)$ 105

16. $(0)(-18)$ 0

17. $(30)(-3)$ -90

18. $(-7)(12)$ -84

19. $(1)(-12)$ -12

20. $(-25)(-8)$ 200

21. $(-6)\left(-\frac{1}{3}\right)$ 2

22. $(8)\left(-\frac{1}{4}\right)$ -2

23. $(4)\left(-\frac{3}{2}\right)$ -6

24. $(-9)\left(-\frac{2}{3}\right)$ 6

25. $\left(-\frac{3}{2}\right)\left(-\frac{2}{3}\right)$ 1

26. $\left(-\frac{4}{5}\right)\left(-\frac{5}{4}\right)$ 1

27. $(-6)\left(-\frac{3}{2}\right)$ 9

28. $(-5)\left(-\frac{3}{10}\right)$ $\frac{3}{2}$

29. $(8)\left(-\frac{3}{4}\right)$ -6

30. $(3)\left(-\frac{5}{6}\right)$ $-\frac{5}{2}$

31. $(-2.5)(-4)$ 10

32. $(3.25)(-4)$ -13

33. $(5.4)(-5)$ -27

34. $(-7.5)(-6)$ 45

35. $(-5)(3)(-2)$ 30

36. $(4)(-2)(3)$ -24

37. $(-2)(-4)(-2)$ −16

38. $(-7)(5)(-2)$ 70

39. $(2)(-3)(-8)$ 48

40. $(4)(-1)(-6)$ 24

41. $(-9)(-12)(0)$ 0

42. $(-13)(0)(-7)$ 0

43. $\left(-\frac{1}{3}\right)\left(\frac{6}{5}\right)(-10)$ 4

44. $\left(-\frac{1}{2}\right)\left(-\frac{4}{3}\right)(-6)$ −4

Do the indicated operations. Remember the rules for the order of operations.

45. $5(7 - 2)$ 25

46. $3(8 - 10)$ −6

47. $-2(8 - 5)$ −6

48. $-3(-2 - 5)$ 21

49. $(-2)(3) - 5$ −11

50. $(-4)(-7) - 5$ 23

51. $(-5)(-2) - 12$ −2

52. $(3)(-7) + 20$ −1

53. $-3 + (-2)(-9)$ 15

54. $-4 + (-3)(6)$ −22

55. $5 - (-2)(-3)$ −1

56. $-8 - (-2)(-5)$ −18

57. $-12 - (-6)(5)$ 18

58. $10 - (-2)(7)$ 24

59. $(-2)(-7) + (2)(-3)$ 8

60. $(-7)(3) - (-2)(-8)$ −37

61. $(-6)^2 - 4$ 32

62. $(-5)^2 + 3$ 28

63. $-6^2 - 4$ −40

64. $-5^2 + 3$ −22

65. $(-4)^2 - (-2)(-5)$ 6

66. $(-3)^3 - (-8)(2)$ −11

67. $(-8)^2 - 5^2$ 39

68. $(-6)^2 - (-3)^2$ 27

69. $-8^2 - 5^2$ −89

70. $-6^2 - 3^2$ −45

71. $-8^2 - (-5)^2$ −89

72. $-6^2 - (-3)^2$ −45

Skillscan (Appendix A)

Reduce each of the following fractions to lowest terms.

a. $\frac{25}{5}$ 5

b. $\frac{49}{7}$ 7

c. $\frac{66}{11}$ 6

d. $\frac{84}{12}$ 7

e. $\frac{90}{15}$ 6 **f.** $\frac{144}{24}$ 6 **g.** $\frac{81}{18}$ $\frac{9}{2}$ **h.** $\frac{80}{15}$ $\frac{16}{3}$

Answers

1. 40 **3.** −20 **5.** −72 **7.** 56 **9.** −48 **11.** 0 **13.** 144 **15.** 105 **17.** −90 **19.** −12 **21.** 2 **23.** −6 **25.** 1 **27.** 9 **29.** −6 **31.** 10 **33.** −27 **35.** 30 **37.** −16 **39.** 48 **41.** 0 **43.** 4 **45.** 25 **47.** −6 **49.** −11 **51.** −2 **53.** 15 **55.** −1 **57.** 18 **59.** 8 **61.** 32 **63.** −40 **65.** 6 **67.** 39 **69.** −89 **71.** −89 **a.** 5 **b.** 7 **c.** 6 **d.** 7 **e.** 6 **f.** 6 **g.** $\frac{9}{2}$ **h.** $\frac{16}{3}$

2.5 Dividing Signed Numbers

OBJECTIVE
To find the quotient of two integers.

You know from your work in arithmetic that multiplication and division are related operations. We can use that fact, and our work of the last section, to determine rules for the division of signed numbers. Every division problem can be stated as an equivalent multiplication problem. For instance,

$$\frac{15}{5} = 3 \quad \text{since} \quad 15 = 5 \cdot 3$$

$$\frac{-24}{6} = -4 \quad \text{since} \quad -24 = (6)(-4)$$

$$\frac{-30}{-5} = 6 \quad \text{since} \quad -30 = (-5)(6)$$

The examples above illustrate that because the two operations are related, the rule of signs that we stated in the last section for multiplication is also true for division.

DIVIDING SIGNED NUMBERS

1. The quotient of two numbers with different signs is negative.
2. The quotient of two numbers with the same sign is positive.

Again, the rule is easy to use. To divide two signed numbers, divide their absolute values. Then attach the proper sign according to the rule above.

(*a*) Positive → 28 / Positive → 7: $\frac{28}{7} = 4$ ← Positive

(*b*) Negative → −36 / Negative → −4: $\frac{-36}{-4} = 9$ ← Positive

(*c*) Negative → −42 / Positive → 7: $\frac{-42}{7} = -6$ ← Negative

(*d*) Positive → 75 / Negative → −3: $\frac{75}{-3} = -25$ ← Negative

(*e*) Positive → 15.2 / Negative → −3.8: $\frac{15.2}{-3.8} = -4$ ← Negative

CHECK YOURSELF 1

Divide.

1. $\frac{-55}{11}$ **2.** $\frac{80}{20}$ **3.** $\frac{-48}{-8}$ **4.** $\frac{144}{-12}$ **5.** $\frac{-13.5}{-2.7}$

You should be very careful when 0 is involved in a division problem. Remember that 0 divided by any nonzero number is just 0.

Example 2

$\frac{0}{-7} = 0$ because $0 = (-7)(0)$

However, if zero is the *divisor,* we have a special problem.

Example 3

$\frac{9}{0} = ?$

This means that $9 = 0 \cdot ?$.

Can 0 times a number ever be 9? No, so there is no solution.

To solve the problem, we agree that *division by 0 is not allowed.* We say that

Division by 0 is undefined.

Example 4

(*a*) $\frac{7}{0}$ is undefined.

(*b*) $\frac{-9}{0}$ is undefined.

Note: The expression $\frac{0}{0}$ is called an *indeterminate form.* You will learn more about this in later mathematics classes.

CHECK YOURSELF 2

Divide if possible.

1. $\frac{0}{3}$ **2.** $\frac{5}{0}$ **3.** $\frac{-7}{0}$ **4.** $\frac{0}{-9}$

Recall that we can incorporate division into our rules for the order of operations by noting that the fraction bar serves as a grouping symbol. This simply means that all operations in the numerator and denominator should be performed separately. In this case, the division is done as the last step. Our final example illustrates.

Example 5

Evaluate each expression.

Multiply in the numerator, then divide.

(*a*) $\frac{(-6)(-7)}{-3} = \frac{42}{-3} = -14$

Add in the numerator, then divide.

(*b*) $\frac{3 + (-12)}{3} = \frac{-9}{3} = -3$

Multiply in the numerator. Then add in the numerator and subtract in the denominator.

(*c*) $\frac{-4 + (2)(-6)}{-6 - 2} = \frac{-4 + (-12)}{-6 - 2}$

Divide as the last step.

$= \frac{-16}{-8} = 2$

CHECK YOURSELF 3

Evaluate each expression.

1. $\frac{-4 + (-8)}{6}$ **2.** $\frac{3 - (2)(-6)}{-5}$ **3.** $\frac{(-2)(-4) - (-6)(-5)}{(-4)(11)}$

CHECK YOURSELF ANSWERS

1. (1) −5; (2) 4; (3) 6; (4) −12; (5) 5.
2. (1) 0; (2) undefined; (3) undefined; (4) 0.
3. (1) −2; (2) −3; (3) $\frac{1}{2}$.

2.5 Exercises

Divide.

1. $\frac{-15}{-3}$ 5
2. $\frac{60}{12}$ 5
3. $\frac{72}{9}$ 8
4. $\frac{-27}{9}$ −3
5. $\frac{50}{-5}$ −10
6. $\frac{-32}{-8}$ 4
7. $\frac{-52}{4}$ −13
8. $\frac{56}{-7}$ −8
9. $\frac{-75}{-3}$ 25
10. $\frac{-60}{15}$ −4
11. $\frac{0}{-8}$ 0
12. $\frac{-125}{-25}$ 5
13. $\frac{-9}{-1}$ 9
14. $\frac{-10}{0}$ Undefined
15. $\frac{-84}{-7}$ 12
16. $\frac{-10}{1}$ −10
17. $\frac{15}{0}$ Undefined
18. $\frac{0}{-15}$ 0
19. $\frac{-17}{1}$ −17
20. $\frac{-27}{-1}$ 27
21. $\frac{-144}{-16}$ 9
22. $\frac{-150}{6}$ −25
23. $\frac{-22.2}{3.7}$ −6
24. $\frac{-16.8}{-2.4}$ 7
25. $\frac{-5}{20}$ $-\frac{1}{4}$
26. $\frac{-7}{-35}$ $\frac{1}{5}$
27. $\frac{24}{-16}$ $-\frac{3}{2}$
28. $\frac{-25}{10}$ $-\frac{5}{2}$
29. $\frac{-28}{-42}$ $\frac{2}{3}$
30. $\frac{-125}{-75}$ $\frac{5}{3}$

Perform indicated operations.

31. $\frac{(-6)(-3)}{2}$ 9
32. $\frac{(-9)(5)}{-3}$ 15
33. $\frac{(-8)(2)}{-4}$ 4

34. $\dfrac{(7)(-8)}{-14}$ 4

35. $\dfrac{18}{-3-6}$ −2

36. $\dfrac{27}{-10+7}$ −9

37. $\dfrac{-8-8}{-2}$ 8

38. $\dfrac{15-24}{-3}$ 3

39. $\dfrac{43-15}{-10-4}$ −2

40. $\dfrac{-7-2}{-12+9}$ 3

41. $\dfrac{7-5}{2-2}$ Undefined

42. $\dfrac{10-6}{4-4}$ Undefined

43. $\dfrac{-15-(-3)}{3-(-1)}$ −3

44. $\dfrac{21-(-4)}{-3-2}$ −5

45. $\dfrac{(-5)(-8)-4}{15-(-3)}$ 2

46. $\dfrac{6-2(-10)}{-18-(-5)}$ −2

47. $\dfrac{(-3)(-6)-(-4)(8)}{6-(-4)}$ 5

48. $\dfrac{(5)(-2)-(-4)(-5)}{-4-2}$ 5

49. $\dfrac{2(-5)+4(6-8)}{3(-4+2)}$ 3

50. $\dfrac{(-3)(-5)-3(5-8)}{4(-8+6)}$ −3

51. $\dfrac{(-5)^2-(-4)(5)}{3(5-8)}$ −5

52. $\dfrac{(-3)^2-(-9)(-5)}{4-(-2)}$ −6

Skillscan (Section 1.6)

Evaluate each expression if $x = 2$, $y = 3$, and $z = 5$.

a. $5x + 3z$ 25

b. $y^2 - 4x$ 1

c. $3x^2 - 2z$ 2

d. $x^3 + 2z^2$ 58

e. $\dfrac{2x+4z}{2y}$ 4

f. $\dfrac{x(z-y)}{y+z}$ $\frac{1}{2}$

g. $\dfrac{2(x+y+2z)}{y^2-2x}$ 6

h. $\dfrac{3x-2y+5z}{3y-2x}$ 5

Answers

1. 5 **3.** 8 **5.** −10 **7.** −13 **9.** 25 **11.** 0 **13.** 9 **15.** 12 **17.** Undefined **19.** −17 **21.** 9 **23.** −6 **25.** $-\frac{1}{4}$ **27.** $-\frac{3}{2}$ **29.** $\frac{2}{3}$ **31.** 9 **33.** 4 **35.** −2 **37.** 8 **39.** −2 **41.** Undefined **43.** −3 **45.** 2 **47.** 5 **49.** 3 **51.** −5 **a.** 25 **b.** 1 **c.** 2 **d.** 58 **e.** 4 **f.** $\frac{1}{2}$ **g.** 6 **h.** 5

2.6 More on Evaluating Algebraic Expressions

OBJECTIVE

To evaluate algebraic expressions given any signed-number values for the variables.

In Section 1.6 you learned how to evaluate algebraic expressions. That is, given certain values for the letters or variables in an expression, you were able to find the value of that expression.

With our work in this chapter, we can now let the variables take on negative values. Let's review the method of Section 1.6.

TO EVALUATE AN EXPRESSION

STEP 1 Replace each variable by the given number value.

STEP 2 Do the necessary arithmetic operations.

Example 1

Evaluate $5a + 4b$ if $a = -2$ and $b = 3$.

Replace a with -2 and b with 3.

$$5a + 4b = 5(-2) + 4(3)$$
$$= -10 + 12$$
$$= 2$$

Remember the rules for the order of operations. Multiply first, then add.

CHECK YOURSELF 1

Evaluate $3x + 5y$ if $x = -2$ and $y = -5$.

Example 2

Evaluate the following expressions if $a = -4$, $b = 2$, $c = -5$, and $d = 6$.

This becomes $-(-20)$, or $+20$.

(*a*) $7a - 4c = 7(-4) - 4(-5)$
$$= -28 + 20$$
$$= -8$$

Evaluate the power first, then multiply by 7.

(*b*) $7c^2 = 7(-5)^2 = 7 \cdot 25$
$$= 175$$

Note:

$$(-5)^2 = (-5)(-5) = 25$$

The exponent applies to -5!

$-5^2 = -(5 \cdot 5) = -25$

The exponent applies only to 5!

(c) $b^2 - 4ac = 2^2 - 4(-4)(-5)$
$= 4 - 4(-4)(-5)$
$= 4 - 80$
$= -76$

Add inside the parentheses first.

(d) $b(a + d) = 2(-4 + 6)$
$= 2(2)$
$= 4$

CHECK YOURSELF 2

Evaluate if $p = -4$, $q = 3$, and $r = -2$.

1. $5p - 3r$ **2.** $2p^2 + q$ **3.** $p(q + r)$
4. $-q^2$ **5.** $(-q)^2$

If an expression involves a fraction, remember that the fraction bar is a grouping symbol. This means that you should do the required operations first in the numerator and then the denominator. Divide as the last step.

Example 3

Evaluate the following expressions if $x = 4$, $y = -5$, $z = 2$, and $w = -3$.

(a) $\dfrac{z - 2y}{x} = \dfrac{2 - 2(-5)}{4}$
$= \dfrac{2 + 10}{4}$
$= \dfrac{12}{4}$
$= 3$

(b) $\dfrac{3x - w}{2x + w} = \dfrac{3(4) - (-3)}{2(4) + (-3)}$
$= \dfrac{12 + 3}{8 + (-3)}$
$= \dfrac{15}{5}$
$= 3$

CHECK YOURSELF 3

Evaluate if $m = -6$, $n = 4$, and $p = -3$.

1. $\dfrac{m + 3n}{p}$ **2.** $\dfrac{4m + n}{m + 4n}$

When an expression is evaluated by a computer, the same order-of-operation algorithm that we introduced in Section 1.6 is followed.

Let's look at some common symbols used in computer languages to represent those operations.

	ALGEBRAIC NOTATION	COMPUTER NOTATION
Addition	$6 + 2$	6 + 2
Subtraction	$4 - 8$	4 − 8
Multiplication	$(3)(-5)$	3 * (−5)
Division	$\frac{8}{6}$	8/6
Exponential	3^4	3^4

The use of this notation is illustrated in our final example.

Example 4

Evaluate each of the following expressions if $A = 2$, $B = 8$, and $C = 4$.

(*a*) $A + B * (-C)$

Letting A, B, and C take on the given values, we have

$$2 + 8 * (-4) = 2 + (-32)$$
$$= -30$$

(*b*) $-B + (-A) * C\text{^}2$

Substituting the given values, we have

$$-8 + (-2) * 4\text{^}2$$
$$= -8 + (-2) * 16$$
$$= -8 + (-32) = -40$$

(*c*) $(-B + C)/(-A * 4)$

Substituting, we have

$$= (-8 + 4)/(-2 * 4)$$

$$= -4/(-8) = \frac{1}{2}$$

CHECK YOURSELF 4

Evaluate each of the following expressions where $A = -2$, $B = 3$, and $C = 5$.

1. $A + B * (-C)$ **2.** $C + B * A\^3$ **3.** $4 * (B - C)/(2 * A)$

CHECK YOURSELF ANSWERS

1. −31. **2.** (1) −14; (2) 35; (3) −4; (4) −9; (5) 9.
3. (1) −2; (2) −2. **4.** (1) −17; (2) −19; (3) 2.

2.6 Exercises

Evaluate the expressions if $a = -2$, $b = 5$, $c = -4$, and $d = 6$.

1. $3a + 4c$ −22 **2.** $3b + 5c$ −5 **3.** $4b - 3c$ 32

4. $5c - 7a$ −6 **5.** $6a + 2c$ −20 **6.** $4a - 3d$ −26

7. $3a^2$ 12 **8.** $6c^2$ 96 **9.** $c^2 - 2d$ 4

10. $3a^2 + 4c$ −4 **11.** $2a^2 + 3b^2$ 83 **12.** $4b^2 - 2c^2$ 68

13. $2(a + b)$ 6 **14.** $5(b - c)$ 45 **15.** $4(2a - d)$ −40

16. $6(3c - d)$ −108 **17.** $a(b + 3c)$ 14 **18.** $c(3a - d)$ 48

19. $\frac{6d}{c}$ −9 **20.** $\frac{8b}{5c}$ −2 **21.** $\frac{2b - a}{d}$ 2

22. $\frac{3d + 2b}{c}$ −7 **23.** $\frac{2b - 3a}{c + 2d}$ 2 **24.** $\frac{3d - 2b}{5a + d}$ −2

25. $d^2 - b^2$ 11 **26.** $c^2 - a^2$ 12 **27.** $(d - b)^2$ 1

28. $(c - a)^2$ 4 **29.** $(d - b)(d + b)$ 11 **30.** $(c - a)(c + a)$ 12

31. $d^3 - b^3$ 91 **32.** $c^3 + a^3$ −72 **33.** $(d - b)^3$ 1

34. $(c + a)^3$ −216 **35.** $(d - b)(d^2 + db + b^2)$ 91 **36.** $(c + a)(c^2 - ac + a^2)$ −72

37. $a^2 + d^2$ 40 **38.** $b^2 - c^2$ 9 **39.** $(a + d)^2$ 16

40. $(b - c)^2$ 81 **41.** $a^2 + 2ad + d^2$ 16 **42.** $b^2 - 2bc + c^2$ 81

Evaluate each expression if $X = -2$, $Y = -5$, and $Z = 3$.

43. $X + Y * Z$ −17 **44.** $Y - 2 * Z$ −11 **45.** $X\text{^}2 - Z\text{^}2$ −5

46. $X\text{^}2 + Y\text{^}2$ 29 **47.** $(X * Y)/(Z - X)$ 2 **48.** $Y\text{^}2/(Z * Y)$ $-\frac{5}{3}$

49. $(2 * X + Y)/(2 * X + Z)$ 9 **50.** $(X\text{^}2 * Y\text{^}2)/(X * Y)$ 10

Answers

1. −22 **3.** 32 **5.** −20 **7.** 12 **9.** 4 **11.** 83 **13.** 6 **15.** −40 **17.** 14 **19.** −9 **21.** 2 **23.** 2 **25.** 11 **27.** 1 **29.** 11 **31.** 91 **33.** 1 **35.** 91 **37.** 40 **39.** 16 **41.** 16 **43.** −17 **45.** −5 **47.** 2 **49.** 9

Summary

Signed Numbers—The Terms [2.1]

Positive numbers Numbers used to name points to the right of the origin on the number line.

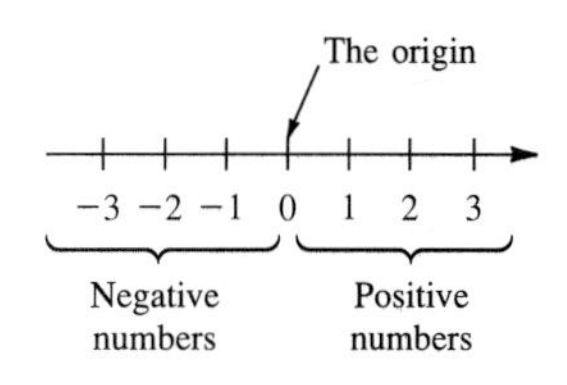

Negative numbers Numbers used to name points to the left of the origin on the number line.

Signed numbers A set containing both positive and negative numbers.

Integers The set consisting of the natural (or counting) numbers, their negatives, and zero.

The integers are

$\{\ldots, -3, -2, -1, 0, 1, 2, 3, \ldots\}$

Absolute Value The distance (on the number line) between the point named by a signed number and the origin.

$|7| = 7$
$|-10| = 10$

The absolute value of x is written $|x|$

Adding Signed Numbers [2.2]

To Add Signed Numbers

1. If two numbers have the same sign, add their absolute values. Give the sum the sign of the original numbers.
2. If two numbers have different signs, subtract their absolute values, the smaller from the larger. Give the sum the sign of the number with the larger absolute value.

$9 + 7 = 16$
$(-9) + (-7) = -16$

$15 + (-10) = 5$
$(-12) + 9 = -3$

Subtracting Signed Numbers [2.3]

To Subtract Signed Numbers

1. Rewrite the subtraction problem as an addition problem by (*a*) Changing the subtraction symbol to an addition symbol (*b*) Replacing the number being subtracted with its opposite
2. Add the resulting signed numbers as before.

$16 - 8 = 16 + (-8) = 8$

$8 - 15 = 8 + (-15) = -7$

$-9 - (-7) = -9 + 7 = -2$

Multiplying Signed Numbers [2.4]

To Multiply Signed Numbers Multiply the absolute values of the two numbers.

1. If the numbers have different signs, the product is negative.
2. If the numbers have the same sign, the product is positive.

$5(-7) = -35$

$(-10)(9) = -90$

$8 \cdot 7 = 56$

$(-9)(-8) = 72$

Dividing Signed Numbers [2.5]

To Divide Signed Numbers Divide the absolute values of the two numbers.

1. If the numbers have different signs, the quotient is negative.
2. If the numbers have the same sign, the quotient is positive.

$\frac{-32}{4} = -8$

$\frac{75}{-5} = -15$

$\frac{20}{5} = 4$

$\frac{-18}{-9} = 2$

Evaluating Algebraic Expressions [2.6]

To Evaluate an Expression

1. Replace each variable by the given number value.
2. Do the necessary arithmetic operations. (Be sure to follow the rules for the order of operations.)

Evaluate

$\frac{4a - b}{2c}$

if $a = -6$, $b = 8$, and $c = -4$.

$$\frac{4a - b}{2c} = \frac{4(-6) - 8}{2(-4)}$$

$$= \frac{-24 - 8}{-8} = \frac{-32}{-8} = 4$$

Summary Exercises Chapter 2

This supplementary exercise set will give you practice with each of the objectives of the chapter. Each exercise is keyed to the appropriate chapter section. The answers are provided in the instructor's manual. Your instructor will give you guidelines on how to best use these exercises in your instructional setting.

[2.1] Represent the integers on the number line shown.

1. 6, −18, −3, 2, 15, −9

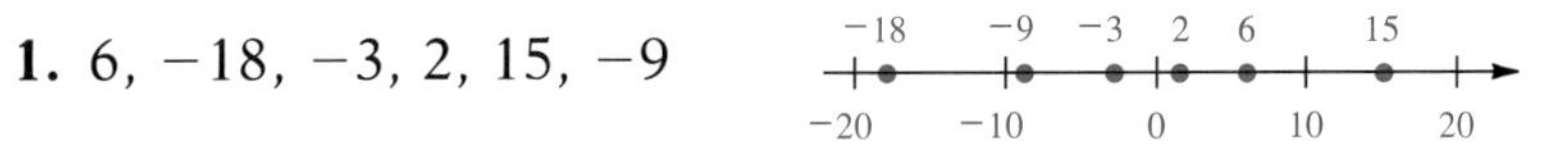

[2.1] Evaluate.

2. $|9|$
9

3. $|-9|$
9

4. $-|9|$
−9

5. $-|-9|$
−9

6. $|12 - 8|$
4

7. $|8| - |12|$
−4

8. $-|8 - 12|$
−4

9. $|-8| - |-12|$
−4

[2.2] Add.

10. $-3 + (-8)$
−11

11. $10 + (-4)$
6

12. $6 + (-6)$
0

13. $-9 + 6$
−3

14. $5 + (-12)$
−7

15. $-16 + (-16)$
−32

16. $-18 + 0$
−18

17. $\frac{3}{8} + \left(-\frac{11}{8}\right)$
−1

18. $5.7 + (-9.7)$
−4

19. $-18 + 7 + (-3)$
−14

[2.3] Subtract.

20. $8 - 13$
−5

21. $-7 - 10$
−17

22. $-2 - (-3)$
1

23. $10 - (-7)$
17

24. $-5 - (-1)$
−4

25. $-9 - (-9)$
0

26. $0 - (-2)$
2

27. $-\frac{5}{4} - \left(-\frac{17}{4}\right)$
3

28. $7.9 - (-8.1)$
16

[2.3] Perform the indicated operations.

29. $|8 - 12|$
4

30. $|8| - |12|$
−4

31. $|-8 - 12|$
20

32. $|-8| - |-12|$
−4

33. $-6 - (-2) + 3$
−1

34. $-5 - (5 - 8)$
−2

35. $7 - (3 - 7) + 4$
15

36. Subtract −7 from −8.
−1

37. Subtract −9 from the sum of 6 and −2.
13

[2.4] Multiply.

38. $(10)(-7)$
−70

39. $(-5)(6)$
−30

40. $(-8)(-5)$
40

41. $(-3)(-15)$
45

42. $(1)(-15)$
−15

43. $(0)(-8)$
0

44. $(-4)(-11)$
44

45. $\left(\frac{2}{3}\right)\left(-\frac{3}{2}\right)$
−1

46. $(-4)\left(\frac{3}{8}\right)$
$-\frac{3}{2}$

47. $\left(-\frac{5}{4}\right)(-1)$
$\frac{5}{4}$

48. $(-8)(-2)(5)$
80

49. $(-4)(-3)(2)$
24

50. $\left(\frac{2}{5}\right)(-10)\left(-\frac{5}{2}\right)$
10

51. $\left(\frac{4}{3}\right)(-6)\left(-\frac{3}{4}\right)$
6

[2.4] Perform the indicated operations.

52. $2(-4 + 3)$ −2

53. $(2)(-3) - (-5)(-3)$
−21

54. $(2 - 8)(2 + 8)$ −60

[2.5] Divide.

55. $\frac{80}{16}$ 5

56. $\frac{-63}{7}$ −9

57. $\frac{-81}{-9}$ 2

58. $\frac{0}{-5}$ 0

59. $\frac{32}{-8}$ −4

60. $\frac{-7}{0}$ Undefined

[2.5] Perform the indicated operations.

61. $\frac{-8 + 6}{-8 - (-10)}$ −1

62. $\frac{2(-3) - 1}{5 - (-2)}$ −1

63. $\frac{(-5)^2 - (-2)^2}{-5 - (-2)}$ −7

[2.6] Evaluate the expressions if $x = -3$, $y = 6$, $z = -4$, and $w = 2$.

64. $3x + w$
−7

65. $5y - 4z$
46

66. $x + y - 3z$
15

67. $5z^2$
80

68. $3x^2 - 2w^2$
19

69. $3x^3$
−81

70. $5(x^2 - w^2)$
25

71. $\frac{6z}{2w}$
−6

72. $\frac{2x - 4z}{y - z}$
1

73. $\frac{3x - y}{w - x}$
−3

74. $\frac{x(y^2 - z^2)}{(y + z)(y - z)}$
−3

75. $\frac{y(x - w)^2}{x^2 - 2xw + w^2}$
6

Self-Test for Chapter Two

The purpose of this self-test is to help you check your progress and to review for a chapter test in class. Allow yourself about an hour to take the test. When you are done, check your answers in the back of the book. If you missed any problems, go back and review the appropriate sections in the chapter and the exercises provided.

Represent the integers on the number line shown.

−17 −12 −7 4 5 18
−20 −10 0 10 20

1. 5, −12, 4, −7, 18, −17

Evaluate.

2. $|7|$ 7

3. $|-7|$ 7

4. $|18 - 7|$ 11

5. $|18| - |-7|$ 11

Add.

6. $-8 + (-5)$ −13

7. $6 + (-9)$ −3

8. $(-9) + (-12)$ −21

9. $-\frac{5}{3} + \frac{8}{3}$ 1

Subtract.

10. $9 - 15$ −6

11. $-9 - 15$ −24

12. $5 - (-4)$ 9

13. $-7 - (-7)$ 0

Multiply.

14. $(-8)(5)$ −40

15. $(-9)(-7)$ 63

16. $(4.5)(-6)$ −27

17. $(-2)(-3)(-4)$ −24

Divide.

18. $\dfrac{75}{-3}$ −25

19. $\dfrac{-27}{-9}$ 3

20. $\dfrac{-45}{9}$ −5

21. $\dfrac{9}{0}$ Undefined

Evaluate if $a = -2$, $b = 6$, and $c = -4$.

22. $4a - c$ −4

23. $5c^2$ 80

24. $6(2b - 3c)$ 144

25. $\dfrac{3a - 4b}{a + c}$ 5

Chapter Three

Equations and Inequalities

3.1 Equations—An Introduction

OBJECTIVE
To determine whether a given number is a solution for an equation.

In this chapter you will be working with one of the most important tools of mathematics, the equation. The ability to recognize and solve various types of equations is probably the most useful algebraic skill you will learn, and we will continue to build upon the methods of this chapter throughout the remainder of the text. To start, let's describe what we mean by an equation.

> An *equation* is a mathematical statement that two expressions are equal.

Some examples are

> $3 + 4 = 7$
> $x + 3 = 5$
> $P = 2L + 2W$

As you can see, an equals sign (=) separates the two sides of an equation.

$$\underbrace{x + 3}_{\text{Left side}} \underset{\text{Equals sign}}{=} \underbrace{5}_{\text{Right side}}$$

An equation like $x + 3 = 5$ is called a *conditional equation* because it can be either true or false depending on the value given to the variable.

An equation may be either true or false. For instance, $3 + 4 = 7$ is true because both sides name the same number. What about an equation like $x + 3 = 5$ that has a letter or variable on one side? There are many numbers that can replace x in the equation. However, in this case only one number will make the equation a true statement.

$$\text{If } x = \begin{cases} 1 & 1 + 3 = 5 \text{ is false} \\ 2 & 2 + 3 = 5 \text{ is true} \\ 3 & 3 + 3 = 5 \text{ is false} \end{cases}$$

The number 2 is called the *solution* (or *root*) of the equation $x + 3 = 5$ because substituting 2 for x gives a true statement.

A *solution* for an equation is any value for the variable that makes the equation a true statement.

Example 1

(*a*) Is 3 a solution for the equation $2x + 4 = 10$?

To find out, replace x with 3 and evaluate $2x + 4$ on the left.

Left side		Right side
$2 \cdot 3 + 4$	$\stackrel{?}{=}$	10
$6 + 4$	$\stackrel{?}{=}$	10
10	$=$	10

Since $10 = 10$ is a true statement, 3 is a solution of the equation.

(*b*) Is 5 a solution of the equation $3x - 2 = 2x + 1$?

To find out, replace x with 5 and evaluate each side separately.

Remember the rules for the order of operation. Multiply first; then add or subtract.

Left side		Right side
$3 \cdot 5 - 2$	$\stackrel{?}{=}$	$2 \cdot 5 + 1$
$15 - 2$	$\stackrel{?}{=}$	$10 + 1$
13	$\neq$	11

Since the two sides do not name the same number, we do not have a true statement and 5 is not a solution.

CHECK YOURSELF 1

For the equation

$$2x - 1 = x + 5$$

1. Is 4 a solution?
2. Is 6 a solution?

You may be wondering whether an equation can have more than one solution. It certainly can. For instance,

This is an example of a *quadratic equation.* We will consider methods of solution in Chapter 5 and then again in Chapter 10.

$$x^2 = 9$$

has two solutions. They are 3 and -3 because

$$3^2 = 9 \quad \text{and} \quad (-3)^2 = 9$$

In this chapter, however, we will always be working with *linear equations in one variable.* These are equations that can be put into the form

$ax + b = 0$

where the variable is x, a and b are real numbers, and a is not equal to 0. In a linear equation, the variable can appear only to the first power. No other power (x^2, x^3, etc.) can appear. Linear equations are also called *first-degree equations* for this reason.

> Linear equations in one variable that can be written in the form
>
> $ax + b = 0 \qquad a \neq 0$
>
> will have exactly one solution.

It is not difficult to find the solution for an equation such as $x + 3 = 8$ by guessing the answer to the following question:

What plus 3 is 8?

Here the answer to the question is 5, and that is also the solution for the equation. But for more complicated equations you are going to need something more than guesswork. So in Section 3.2 we will begin looking at a set of rules that will let you solve any linear equation in one variable.

CHECK YOURSELF ANSWERS

1. (1) 4 is not a solution; (2) 6 is a solution.

3.1 Exercises

Tell whether the number shown in parentheses is a solution for the given equation.

1. $x + 7 = 8$ (1) Yes

2. $x + 5 = 9$ (3) No

3. $x - 7 = 4$ (10) No

4. $x - 5 = 3$ (8) Yes

5. $5 - x = 2$ (4) No

6. $10 - x = 7$ (3) Yes

7. $4 - x = 6$ (-2) Yes

8. $5 - x = 6$ (-3) No

9. $2x + 3 = 9$ (5) No

10. $4x + 5 = 17$ (3) Yes

11. $2x - 1 = 3$ (4) No

12. $3x - 2 = 4$ (2) Yes

13. $3x - 5 = 16$ (7) Yes **14.** $5x - 3 = 21$ (5) No **15.** $3x - 8 = -12$ (-2) No

16. $7x + 21 = 0$ (-3) Yes **17.** $5 - 2x = 7$ (-1) Yes **18.** $4 - 5x = 9$ (-2) No

19. $12 + 2x = 0$ (-6) Yes **20.** $4x - 3 = -23$ (-5) Yes **21.** $4x - 5 = 2x + 3$ (4) Yes

22. $5x + 4 = 2x + 10$ (4) No **23.** $3x + 2 = x - 6$ (-5) No **24.** $6x + 3 = 2x - 9$ (-3) Yes

25. $x + 3 + 2x = 5 + x + 8$ (5) Yes **26.** $5x - 3 + 2x = 3 + x - 12$ (-2) No

27. $\frac{2}{3}x = 12$ (36) No **28.** $\frac{3}{4}x = 18$ (24) Yes

29. $\frac{3}{5}x + 5 = 11$ (10) Yes **30.** $\frac{2}{3}x + 8 = -12$ (-6) No

Skillscan (Section 1.4)

Simplify by combining like terms.

a. $x + 3x$
$4x$

b. $5x - 7x$
$-2x$

c. $5a - 3a$
$2a$

d. $3b + 7b$
$10b$

e. $-7x + 5x$
$-2x$

f. $9w - 9w$
0

g. $2y + 1 + 3y$
$5y + 1$

h. $5p - 3 + 2p$
$7p - 3$

Answers

1. Yes **3.** No **5.** No **7.** Yes **9.** No **11.** No **13.** Yes **15.** No **17.** Yes **19.** Yes **21.** Yes **23.** No **25.** Yes **27.** No **29.** Yes **a.** $4x$ **b.** $-2x$ **c.** $2a$ **d.** $10b$ **e.** $-2x$ **f.** 0 **g.** $5y + 1$ **h.** $7p - 3$

3.2 Solving Equations by Adding or Subtracting

OBJECTIVE

To use addition and/or subtraction to solve equations.

As we said in Section 3.1, guesswork is not a very good approach to solving more complicated equations. A better method is to transform the given equation to an *equivalent equation* whose solution can be found by inspection. Let's make a definition.

> Equations that have the same solution are called *equivalent equations.*

As examples,

$$2x + 3 = 5$$
$$2x = 2$$
$$x = 1$$

are all equivalent equations. They all have the same solution, 1. We say that a linear equation is *solved* when it is transformed to an equivalent equation of the form

$x = \square$

x is alone on the left side. The right side is some number.

Note: In some cases we'll write the equation in the form

$\square = x$

The number will be our solution when the equation has x isolated on the left or on the right.

Here is the first property you will need in solving equations.

> **THE ADDITION PROPERTY OF EQUALITY**
>
> If $\quad a = b$
>
> then $a + c = b + c$
>
> In words, adding the same quantity to both sides of an equation gives an equivalent equation.

Remember: An equation is a statement that the two sides are equal. Adding the same quantity on both sides does not change the equality or "balance."

Let's look at an example applying this property to solve an equation.

Example 1

Solve

$x - 3 = 9$

Remember that our goal is to isolate x on one side of the equation. Since the 3 is being subtracted from x, we can add 3 to remove it. We then use the addition property to add 3 to both sides of the equation.

$$\begin{array}{rcr} x - 3 & = & 9 \\ +3 & & +3 \\ \hline x & = & 12 \end{array}$$

Adding 3 "undoes" the subtraction and leaves x alone on the left.

Since 12 is the solution for the equivalent equation $x = 12$, it is the solution for our original equation.

To check, replace x with 12 in the original equation:

$x - 3 = 9$	The given equation
$12 - 3 = 9$	Substitute 12 for x
$9 = 9$	A true statement

Since we have a true statement, 12 is the solution.

CHECK YOURSELF 1

Solve and check:

$x - 5 = 4$

Example 2

Solve

$x + 5 = 9$

Note: Since subtraction can be defined in terms of addition,

$a - b = a + (-b)$

we can also use the addition property to *subtract* the same quantity from both sides of an equation.

In this case 5 is *added* to x on the left. We can use the addition property to subtract 5 from both sides to "undo" the addition and leave the variable x alone on one side of the equation.

$$\begin{array}{rcr} x + 5 & = & 9 \\ -5 & & -5 \\ \hline x & = & 4 \end{array}$$

The solution is 4. To check, replace x with 4:

$4 + 5 = 9$ (true)

CHECK YOURSELF 2

Solve and check:

$x + 6 = 13$

What if the equation has a variable term on both sides? You will have to use the addition property to add or subtract a term involving the variable to get the desired result.

Example 3

Solve

$5x = 4x + 7$

We will start by subtracting $4x$ from both sides of the equation. Do you see why? Remember that an equation is solved when we have an equivalent equation of the form

$x = \square$

$$\begin{array}{rcr} 5x & = & 4x + 7 \\ -4x & & -4x \\ \hline x & = & 7 \end{array}$$

← Subtracting $4x$ from both sides *removes* $4x$ from the right.

To check: Since 7 is a solution for the equivalent equation $x = 7$, it should be a solution for the original equation. To find out, replace x with 7:

$$5 \cdot 7 \stackrel{?}{=} 4 \cdot 7 + 7$$
$$35 \stackrel{?}{=} 28 + 7$$
$$35 = 35 \quad \text{(true)}$$

CHECK YOURSELF 3

Solve and check:

$7x = 6x + 3$

You may have to apply the addition property more than once to solve an equation. Look at the following example.

Example 4

Solve

$7x - 8 = 6x$

We want all variables on *one* side of the equation. If we choose the left, we can begin by subtracting $6x$ from both sides of the equation. This will remove $6x$ from the right:

$$\begin{array}{rcr} 7x - 8 & = & 6x \\ -6x & & -6x \\ \hline x - 8 & = & 0 \end{array}$$

Now we want the number on the right, so we add 8 to both sides. Adding 8 will isolate x:

$$\begin{array}{rcr} x - 8 & = & 0 \\ +8 & & +8 \\ \hline x & = & 8 \end{array}$$

The solution is 8. We'll leave it to you to check this result.

CHECK YOURSELF 4

Solve and check:

$9x + 3 = 8x$

Often an equation will have more than one term in the variable *and* more than one number. You will have to apply the addition property twice in solving these equations.

Example 5

(*a*) Solve

$5x - 7 = 4x + 3$

We would like the variable terms on the left, so we start by subtracting $4x$ to remove that term from the right side of the equation:

$$\begin{array}{rcr} 5x - 7 &=& 4x + 3 \\ -4x && -4x \\ \hline x - 7 &=& 3 \end{array}$$

Now, to get the numbers on the right, we add 7 to both sides to undo the subtraction on the left:

$$\begin{array}{rcr} x - 7 &=& 3 \\ +7 && +7 \\ \hline x &=& 10 \end{array}$$

The solution is 10. To check, replace x with 10 in the original equation:

$$5 \cdot 10 - 7 = 4 \cdot 10 + 3$$
$$43 = 43 \qquad \text{(true)}$$

Note: You could just as easily have added 7 to both sides and *then* subtracted $4x$. The result would be the same. In fact, some students prefer to combine the two steps.

(*b*) Solve

$7x + 3 = 6x - 2$

Start by subtracting $6x$ from both sides of the equation:

$$\begin{array}{rcr} 7x + 3 &=& 6x - 2 \\ -6x && -6x \\ \hline x + 3 &=& -2 \end{array}$$

Now subtract 3 from both sides. Do you see why?

$$\begin{array}{rcl} x + 3 &=& -2 \\ -3 && -3 \\ \hline x &=& -5 \end{array}$$

The solution is -5. Check this result yourself by replacing x with -5 in the original equation. Be careful of the rules for signs!

CHECK YOURSELF 5

Solve and check:

1. $4x - 5 = 3x + 2$
2. $6x + 2 = 5x - 4$

By *simplify* we mean to combine any like terms that may appear on one side of the equation.

In solving an equation, you should always simplify each side as much as possible before using the addition property.

Example 6

Solve

Like terms   Like terms

$$5 + 8x - 2 = 2x - 3 + 5x$$

Since like terms appear on each side of the equation, we start by combining the numbers on the left (5 and -2). Then we combine the like terms ($2x$ and $5x$) on the right. We then have

$$3 + 8x = 7x - 3$$

Now we can apply the addition property, as before:

$$\begin{array}{rcl} 3 + 8x &=& 7x - 3 \\ -7x &=& -7x \\ \hline 3 + x &=& -3 \\ -3 && -3 \\ \hline x &=& -6 \end{array}$$

The solution is -6. To check, always return to the original equation. That will catch any possible errors in simplifying. Replacing x with -6 gives

$$5 + 8(-6) - 2 \stackrel{?}{=} 2(-6) - 3 + 5(-6)$$
$$5 - 48 - 2 \stackrel{?}{=} -12 - 3 - 30$$
$$-45 = -45 \quad \text{(true)}$$

CHECK YOURSELF 6

Solve and check:

1. $3 + 6x + 4 = 8x - 3 - 3x$
2. $5x + 21 + 3x = 20 + 7x - 2$

We may have to apply some of the properties discussed in Section 1.2 in solving equations. Our final example illustrates the use of the distributive property to clear an equation of parentheses.

Example 7

Solve

$$2(3x + 4) = 5x - 6$$

Note: $2(3x + 4) = 2(3x) + 2(4) = 6x + 8$

Applying the distributive property on the left, we have

$$6x + 8 = 5x - 6$$

We can then proceed as before:

$$\begin{array}{rcr} 6x + 8 & = & 5x - 6 \\ -5x & & -5x \\ \hline x + 8 & = & -6 \\ -8 & & -8 \\ \hline x & = & -14 \end{array}$$

Subtract $5x$.

Subtract 8.

The solution is -14. We will leave the checking of this result to the reader. **Remember:** Always return to the original equation.

CHECK YOURSELF 7

Solve and check each of the following equations.

1. $4(5x - 2) = 19x + 4$
2. $3(5x + 1) = 2(7x - 3) - 4$

Note: Recall our comment that we could write an equation in the equivalent forms

$x = \square$ or $\square = x$

where $\square$ represents some number. Suppose we have an equation like

$$12 = x + 7$$

Subtracting 7, we will isolate x *on the right*:

$$\begin{array}{rcl} 12 &=& x + 7 \\ -7 & & -7 \\ \hline 5 &=& x \end{array}$$

and the solution is 5.

CHECK YOURSELF ANSWERS

1. 9 is the solution.
2. 7 is the solution.
3. 3 is the solution.
4. -3 is the solution.
5. (1) 7; (2) -6.
6. (1) -10; (2) -3.
7. (1) 12; (2) -13.

3.2 Exercises

Solve and check the following equations.

1. $x + 3 = 5$ 2

2. $x - 8 = 2$ 10

3. $x - 6 = 5$ 11

4. $x + 6 = 10$ 4

5. $x - 8 = -10$ -2

6. $x + 5 = 2$ -3

7. $x + 4 = -3$ -7

8. $x - 5 = -4$ 1

9. $11 = x + 5$ 6

10. $x + 7 = 0$ -7

11. $6 + x = 9$ 3

12. $5 = x - 8$ 13

13. $x - 5 = 0$ 5

14. $9 + x = 13$ 4

15. $3x = 2x + 4$ 4

16. $5x = 4x - 8$ -8

17. $8x = 7x - 10$ -10

18. $7x = 6x + 5$ 5

19. $6x + 3 = 5x$ -3

20. $12x - 6 = 11x$ 6

21. $8x - 4 = 7x$ 4

22. $9x - 7 = 8x$ 7

23. $2x + 3 = x + 5$ 2

24. $3x - 2 = 2x + 1$ 3

25. $5x - 7 = 4x - 3$ 4

26. $8x + 5 = 7x - 2$ -7

27. $7x - 2 = 6x + 4$ 6

28. $10x - 3 = 9x - 6$ -3 **29.** $2 + 5x = 5 + 4x$ 3 **30.** $3 + 6x = 2 + 5x$ -1

31. $3 + 7x = 6x - 8$ -11 **32.** $2 + 10x = 9x - 7$ -9

33. $2 + 4x + 1 = x + 9 + 2x$ 6 **34.** $5x - 2 + x = 3 + 5x + 6$ 11

35. $4x + 7 + 3x = 5x + 13 + x$ 6 **36.** $5x + 9 + 4x = 9 + 8x - 7$ -7

37. $3x - 5 + 2x - 7 + x = 5x + 2$ 14 **38.** $5x + 8 + 3x - x + 5 = 6x - 3$ -16

39. $3(2x + 5) = 5x - 3$ -18 **40.** $4(3x - 2) = 11x + 5$ 13

41. $3(7x + 2) = 5(4x + 1) + 17$ 16 **42.** $5(5x + 3) = 3(8x - 2) + 4$ -17

43. $6(6x - 1) - 5(7x + 3) = 3$ 24 **44.** $9(5x - 8) - 4(11x - 16) = 2$ 10

45. $\frac{5}{4}x - 1 = \frac{1}{4}x + 7$ 8 **46.** $\frac{7}{5}x + 3 = \frac{2}{5}x - 8$ -11

47. $\frac{9}{2}x - \frac{3}{4} = \frac{7}{2}x + \frac{5}{4}$ 2 **48.** $\frac{11}{3}x + \frac{1}{6} = \frac{8}{3}x + \frac{19}{6}$ 3

Translate the following statements to an algebraic equation. Let x represent the number in each case.

49. 3 more than a number is 7. $x + 3 = 7$

50. 5 less than a number is 12. $x - 5 = 12$

51. 7 less than 3 times a number is twice that same number. $3x - 7 = 2x$

52. 4 more than 5 times a number is 6 times that same number. $5x + 4 = 6x$

53. 2 times the sum of a number and 5 is 18 more than that same number. $2(x + 5) = x + 18$

54. 3 times the sum of a number and 7 is 4 times that same number. $3(x + 7) = 4x$

Skillscan (Section 2.4)
Multiply.

a. $\left(\frac{1}{3}\right)\left(3\right)$ **b.** $\left(-6\right)\left(-\frac{1}{6}\right)$ **c.** $\left(7\right)\left(\frac{1}{7}\right)$ **d.** $\left(-\frac{1}{4}\right)\left(-4\right)$

e. $\left(\frac{3}{5}\right)\left(\frac{5}{3}\right)$ **f.** $\left(\frac{7}{8}\right)\left(\frac{8}{7}\right)$ **g.** $\left(-\frac{4}{7}\right)\left(-\frac{7}{4}\right)$ **h.** $\left(-\frac{6}{11}\right)\left(-\frac{11}{6}\right)$

Answers
1. 2 **3.** 11 **5.** −2 **7.** −7 **9.** 6 **11.** 3 **13.** 5 **15.** 4 **17.** −10 **19.** −3 **21.** 4 **23.** 2 **25.** 4 **27.** 6 **29.** 3 **31.** −11 **33.** 6 **35.** 6 **37.** 14 **39.** −18 **41.** 16 **43.** 24 **45.** 8 **47.** 2 **49.** $x + 3 = 7$ **51.** $3x - 7 = 2x$ **53.** $2(x + 5) = x + 18$ **a.** 1 **b.** 1 **c.** 1 **d.** 1 **e.** 1 **f.** 1 **g.** 1 **h.** 1

3.3 Solving Equations by Multiplying or Dividing

OBJECTIVE
To use multiplication and/or division to solve equations.

Let's look at a different type of equation. For instance, what if we have an equation like

$6x = 18$

Adding or subtracting by the property of the last section won't help. We will need a second property for solving equations.

Again, as long as you do the *same* thing to *both* sides of the equation, the "balance" is maintained.

Do you see why the number cannot be 0? Multiplying by 0 gives 0 = 0. We have lost the variable!

THE MULTIPLICATION PROPERTY OF EQUALITY

If $a = b$, then $ac = bc$ where $c \neq 0$

In words, multiplying both sides of an equation by the same nonzero number gives an equivalent equation.

Let's work through some examples, using this second rule.

Example 1

Solve

$6x = 18$

Here the variable, x, is multiplied by 6 on the left. Apply the multiplication property to multiply both sides by $\frac{1}{6}$. Keep in mind that we want an equation of the form

$$x = \square$$

$\frac{1}{6}(6x) = \left(\frac{1}{6} \cdot 6\right)x$

$= 1 \cdot x$, or x

We then have x alone on the left, which is what we want.

$$\frac{1}{6}(6x) = \left(\frac{1}{6}\right)18$$

We can now simplify.

$$1 \cdot x = 3 \qquad \text{or} \qquad x = 3$$

The solution is 3. To check, replace x with 3:

$$6 \cdot 3 \stackrel{?}{=} 18$$
$$18 = 18 \qquad \text{(true)}$$

CHECK YOURSELF 1

Solve and check:

$$8x = 32$$

Our next example illustrates a slightly different approach to solving an equation by using the multiplication property.

Example 2

Since division is defined in terms of multiplication, we can also apply our multiplication property to divide both sides of an equation by the same nonzero number.

Solve

$$5x = -35$$

On the left x is multiplied by 5. *Divide* both sides by 5 to "undo" that multiplication:

$$\frac{5x}{5} = \frac{-35}{5}$$

$$x = -7$$

Note that the right side reduces to −7. Be careful with the rules for signs.

We will leave it to you to check the solution.

CHECK YOURSELF 2

Solve and check:

$7x = -42$

Example 3

Solve

$$-9x = 54$$

In this case x is multiplied by -9. So divide both sides by -9 to isolate x on the left:

$$\frac{-9x}{-9} = \frac{54}{-9}$$

$$x = -6$$

The solution is -6. To check:

$$(-9)(-6) \stackrel{?}{=} 54$$

$$54 = 54 \qquad \text{(true)}$$

CHECK YOURSELF 3

Solve and check:

$-10x = -60$

The following examples will illustrate the use of the multiplication property when fractions are involved in an equation.

Example 4

(*a*) Solve

$$\frac{x}{3} = 6$$

Here x is *divided* by 3 on the left. We will use multiplication to isolate x on the left:

$$3\left(\frac{x}{3}\right) = 3 \cdot 6$$

$$x = 18$$

This leaves x alone on the left because

$$3\left(\frac{x}{3}\right) = \frac{3}{1} \cdot \frac{x}{3} = \frac{x}{1} = x$$

To check:

$$\frac{18}{3} \stackrel{?}{=} 6$$

$$6 = 6 \qquad \text{(true)}$$

(*b*) Solve

$$\frac{x}{5} = -9$$

Since x is divided by 5 on the left, we multiply both sides by 5:

$$5\left(\frac{x}{5}\right) = 5(-9)$$

$$x = -45$$

The solution is -45. To check, we replace x with -45:

$$\frac{-45}{5} \stackrel{?}{=} -9$$

$$-9 = -9 \qquad \text{(true)}$$

The solution is verified.

CHECK YOURSELF 4

Solve and check:

1. $\frac{x}{7} = 3$ **2.** $\frac{x}{4} = -8$

Example 5

Solve

$$\frac{3}{5}x = 9$$

One approach is to multiply by 5 as the first step.

$$5\left(\frac{3}{5}x\right) = 5 \cdot 9$$

$$3x = 45$$

Now divide by 3.

$$\frac{3x}{3} = \frac{45}{3}$$

$$x = 15$$

To check:

$$\frac{3}{5} \cdot 15 \stackrel{?}{=} 9$$

$$9 = 9 \qquad \text{(true)}$$

A second method combines the multiplication and division steps and is generally a bit more efficient. We multiply by $\frac{5}{3}$.

Recall that $\frac{5}{3}$ is the *reciprocal* of $\frac{3}{5}$, and the product of a number and its reciprocal is just 1! So

$$\left(\frac{5}{3}\right)\left(\frac{3}{5}\right) = 1$$

$$\frac{5}{3}\left(\frac{3}{5}x\right) = \frac{5}{3} \cdot 9$$

So $x = 15$, as before.

CHECK YOURSELF 5

Solve and check:

$$\frac{2}{3}x = 18$$

Once again, you may have to simplify an equation before applying the methods of this section. Our final example illustrates.

Example 6

Solve and check:

$$3x + 5x = 40$$

Using the distributive property, we can combine the like terms on the left to write

$$8x = 40$$

We can now proceed as before.

Divide by 8.

$$\frac{8x}{8} = \frac{40}{8}$$

$$x = 5$$

The solution is 5. To check, we return to the original equation. Substituting 5 for x yields

A true statement.

$$3 \cdot 5 + 5 \cdot 5 \stackrel{?}{=} 40$$
$$15 + 25 \stackrel{?}{=} 40$$
$$40 = 40$$

The solution is verified.

CHECK YOURSELF 6

Solve and check:

$7x + 4x = -66$

CHECK YOURSELF ANSWERS

1. 4.
2. −6.
3. 6.
4. (1) 21; (2) −32.
5. 27.
6. −6.

3.3 Exercises

Solve for x and check your result.

1. $3x = 12$ 4 **2.** $4x = 20$ 5 **3.** $8x = 48$ 6

4. $5x = -35$ −7 **5.** $63 = 9x$ 7 **6.** $66 = 6x$ 11

7. $4x = -16$ −4 **8.** $-3x = 27$ −9 **9.** $-9x = 72$ −8

10. $10x = -100$ −10 **11.** $6x = -54$ −9 **12.** $-7x = 49$ −7

13. $-4x = -12$ 3 **14.** $52 = -4x$ −13 **15.** $-35 = 5x$ −7

16. $-5x = -25$ 5 **17.** $-8x = -72$ 9 **18.** $-10x = -60$ 6

19. $\frac{x}{2} = 4$ 8 **20.** $\frac{x}{3} = 2$ 6 **21.** $\frac{x}{5} = 3$ 15

22. $\frac{x}{8} = 5$ 40

23. $6 = \frac{x}{7}$ 42

24. $6 = \frac{x}{3}$ 18

25. $\frac{x}{4} = -5$ −20

26. $\frac{x}{5} = -7$ −35

27. $-\frac{x}{8} = 3$ −24

28. $-\frac{x}{3} = -4$ 12

29. $\frac{2}{3}x = 6$ 9

30. $\frac{4}{5}x = 8$ 10

31. $\frac{3}{4}x = -15$ −20

32. $\frac{7}{8}x = -21$ −24

33. $-\frac{2}{5}x = 10$ −25

34. $-\frac{5}{6}x = -15$ 18

35. $5x + 4x = 36$ 4

36. $8x - 3x = -50$ −10

37. $13x - 6x = -42$ −6

38. $9x + 3x = 60$ 5

39. $7x - 3x + 5x = 36$ 4

40. $9x + 3x - 4x = -48$ −6

Once again, certain equations involving decimal fractions can be solved by the methods of this section. For instance, to solve $2.3x = 6.9$ we simply use our multiplication property to divide both sides of the equation by 2.3. This will isolate x on the left as desired. Use this idea to solve the following equations.

41. $3.2x = 12.8$
4

42. $5.1x = -15.3$
−3

43. $-4.5x = 13.5$
−3

44. $-8.2x = -32.8$
4

45. $1.3x + 2.8x = 12.3$
3

46. $2.7x + 5.4x = -16.2$
−2

47. $9.3x - 6.2x = 12.4$
4

48. $12.5x - 7.2x = -21.2$
−4

Translate the following statements to an equation. Let x represent the number in each case.

49. 5 times a number is 40. $5x = 40$

50. Twice a number is 36. $2x = 36$

51. A number divided by 7 is equal to 6.
$\frac{x}{7} = 6$

52. A number divided by 5 is equal to −4.
$\frac{x}{5} = -4$

53. $\frac{1}{3}$ of a number is 8. $\frac{x}{3} = 8$

54. $\frac{1}{5}$ of a number is 10. $\frac{x}{5} = 10$

55. $\frac{3}{4}$ of a number is 18. $\frac{3}{4}x = 18$

56. $\frac{2}{7}$ of a number is 8. $\frac{2}{7}x = 8$

57. Twice a number, divided by 5, is 12.
$\frac{2x}{5} = 12$

58. 3 times a number, divided by 4, is 36.
$\frac{3x}{4} = 36$

Skillscan (Section 1.2)
Use the distributive property to remove the parentheses in the following expressions.

a. $2(x - 3)$
$2x - 6$

b. $3(a + 4)$
$3a + 12$

c. $5(2b + 1)$
$10b + 5$

d. $3(3p - 4)$
$9p - 12$

e. $7(3x - 4)$
$21x - 28$

f. $-4(5x + 4)$
$-20x - 16$

g. $-3(4x - 3)$
$-12x + 9$

h. $-5(3y - 2)$
$-15y + 10$

Answers
1. 4 **3.** 6 **5.** 7 **7.** −4 **9.** −8 **11.** −9 **13.** 3 **15.** −7 **17.** 9 **19.** 8 **21.** 15 **23.** 42 **25.** −20 **27.** −24 **29.** 9 **31.** −20 **33.** −25 **35.** 4 **37.** −6 **39.** 4 **41.** 4 **43.** −3 **45.** 3 **47.** 4 **49.** $5x = 40$ **51.** $\frac{x}{7} = 6$ **53.** $\frac{x}{3} = 8$ **55.** $\frac{3}{4}x = 18$ **57.** $\frac{2x}{5} = 12$ **a.** $2x - 6$ **b.** $3a + 12$ **c.** $10b + 5$ **d.** $9p - 12$ **e.** $21x - 28$ **f.** $-20x - 16$ **g.** $-12x + 9$ **h.** $-15y + 10$

3.4 Combining the Rules to Solve Equations

OBJECTIVE
To use both the addition/subtraction and the multiplication/division properties to solve equations.

In all our examples thus far, either the addition property or the multiplication property was used in solving an equation. Often, finding a solution will require the use of both properties.

Example 1

Solve

$4x - 5 = 7$

Here x is *multiplied* by 4. The result, $4x$, then has 5 subtracted from it on the left side of the equation. These two operations mean that both properties must be applied in solving the equation.

Since the variable term is already on the left, we start by adding 5 to both sides:

Note: From now on we'll show the steps of the solution in a *horizontal form,* which is what you will probably want to use in practice.

$$4x - 5 + 5 = 7 + 5 \qquad \text{or} \qquad 4x = 12$$

We now divide both sides by 4:

$$\frac{4x}{4} = \frac{12}{4}$$

$$x = 3$$

The solution is 3. To check, replace x with 3 in the original equation. Be careful to follow the rules for the order of operations.

$$4 \cdot 3 - 5 \stackrel{?}{=} 7$$
$$12 - 5 \stackrel{?}{=} 7$$
$$7 = 7 \qquad \text{(true)}$$

CHECK YOURSELF 1

Solve and check:

$$5x - 8 = 7$$

Example 2

Solve

$$3x + 8 = -4$$

Again, we want the numbers on the right, so in this case we want to subtract 8 from both sides. Do you see why?

$$3x + 8 = -4$$
$$3x + 8 - 8 = -4 - 8$$
$$3x = -12$$

Now divide both sides by 3 to isolate x on the left.

$$\frac{3x}{3} = \frac{-12}{3}$$

$$x = -4$$

The solution is -4. We'll leave the check of this result to the reader.

CHECK YOURSELF 2

Solve and check:

$6x + 9 = -15$

The variable may appear in any position in an equation. Just apply the rules carefully as you try to write an equivalent equation to find the solution. The following example illustrates.

Example 3

Solve

$3 - 2x = 9$

First subtract 3 from both sides.

$$\begin{aligned} 3 - 2x &= 9 \\ 3 - 3 - 2x &= 9 - 3 \\ -2x &= 6 \end{aligned}$$

Note: $\frac{-2}{-2} = 1$, so we divide by -2 to isolate x on the left.

Now divide both sides by -2. This will leave x alone on the left.

$$\frac{-2x}{-2} = \frac{6}{-2}$$

$$x = -3$$

The solution is -3. To check:

$$\begin{aligned} 3 - 2(-3) &\stackrel{?}{=} 9 \\ 3 + 6 &\stackrel{?}{=} 9 \\ 9 &= 9 \quad \text{(true)} \end{aligned}$$

CHECK YOURSELF 3

Solve and check:

$10 - 3x = 1$

You may also have to combine multiplication with addition or subtraction to solve an equation. Consider the following example.

Example 4

(*a*) Solve

$$\frac{x}{5} - 3 = 4$$

To get all numbers on the right, we first add 3 to both sides.

$$\frac{x}{5} - 3 + 3 = 4 + 3$$

$$\frac{x}{5} = 7$$

Now how can we isolate x on the left? Undo the division by multiplying both sides of the equation by 5.

$$5\left(\frac{x}{5}\right) = 5 \cdot 7$$

$$x = 35$$

The solution is 35. The check is the same as before. Just return to the original equation.

$$\frac{35}{5} - 3 = 4$$

$$7 - 3 = 4$$

$$4 = 4 \qquad \text{(true)}$$

(b) Solve

$$\frac{2}{3}x + 5 = 13$$

First subtract 5 from both sides.

$$\frac{2}{3}x + 5 - 5 = 13 - 5$$

$$\frac{2}{3}x = 8$$

Now multiply both sides by $\frac{3}{2}$, the reciprocal of $\frac{2}{3}$.

$$\left(\frac{3}{2}\right)\left(\frac{2}{3}x\right) = \left(\frac{3}{2}\right)8$$

or

$x = 12$

The solution is 12. We'll leave it to you to check this result.

CHECK YOURSELF 4

Solve and check:

1. $\frac{x}{6} + 5 = 3$ **2.** $\frac{3}{4}x - 8 = 10$

In Section 3.2 you learned how to solve certain equations when the variable appeared on both sides. The following examples will show you how to extend that work by using the multiplication property of equality.

Example 5

Solve

$6x - 4 = 3x - 2$

First add 4 to both sides. Remember, we want the number on the right. Adding 4 will undo the subtraction on the left.

$$\begin{aligned} 6x - 4 &= 3x - 2 \\ 6x - 4 + 4 &= 3x - 2 + 4 \\ 6x &= 3x + 2 \end{aligned}$$

Now subtract $3x$ so that the terms in x will be on the left.

$$\begin{aligned} 6x &= 3x + 2 \\ 6x - 3x &= 3x - 3x + 2 \\ 3x &= 2 \end{aligned}$$

Finally divide by 3.

$$\frac{3x}{3} = \frac{2}{3}$$

$$x = \frac{2}{3}$$

Check:

$$6\left(\frac{2}{3}\right) - 4 \stackrel{?}{=} 3\left(\frac{2}{3}\right) - 2$$

$$4 - 4 \stackrel{?}{=} 2 - 2$$

$$0 = 0 \qquad \text{(true)}$$

Note: As you know, the basic idea is to use our two properties to form an equivalent equation with the x term on the left and numbers on the right. Here we added 4 and then subtracted $3x$. You can do these steps in either order. Try it for yourself the other way. In either case, the multiplication property is then used as the *last step* in finding the solution.

CHECK YOURSELF 5

Solve and check:

$7x - 5 = 3x + 5$

Example 6

Solve $4x - 8 = 7x + 7$

$$4x - 8 = 7x + 7$$

$$4x - 8 + 8 = 7x + 7 + 8$$

Adding 8 will leave the numbers on the right.

$$4x = 7x + 15$$

$$4x - 7x = 7x - 7x + 15$$

Subtracting $7x$ will leave the variables on the left.

$$-3x = 15$$

$$\frac{-3x}{-3} = \frac{15}{-3}$$

Isolate x on the left by dividing by -3.

$$x = -5$$

We'll let you check this result.

To avoid the negative coefficient (-3) in Example 6, some students prefer a different approach. Let's return to our example.

Example 6 (An Alternative Method)

This time we'll work toward having the number on the *left* and the x term on the *right*, or

$\square = x$

$$4x - 8 = 7x + 7$$
$$4x - 8 - 7 = 7x + 7 - 7$$ Subtract 7 to write the numbers on the left
$$4x - 15 = 7x$$
$$4x - 4x - 15 = 7x - 4x$$ Subtract $4x$ to write the variables on the right.
$$-15 = 3x$$
$$\frac{-15}{3} = \frac{3x}{3}$$ Divide by 3 to isolate x on the right.
$$-5 = x$$

Since $-5 = x$ and $x = -5$ are equivalent equations, it really makes no difference; the solution is still -5! You can use whichever approach you prefer.

CHECK YOURSELF 6

Solve $5x + 3 = 9x - 21$ by finding equivalent equations of the form $x = \square$ and $\square = x$ to compare the two methods of finding the solution.

Example 7

Solve

$$7x - 3 + 5x + 4 = 6x + 25$$

We start the solution process by combining the like terms on the left. This gives

$$12x + 1 = 6x + 25$$
$$12x + 1 - 1 = 6x + 25 - 1$$ Subtract 1.
$$12x = 6x + 24$$
$$12x - 6x = 6x - 6x + 24$$ Subtract $6x$.
$$6x = 24$$
$$\frac{6x}{6} = \frac{24}{6}$$ Divide by 6.

or $$x = 4$$

The solution is 4. We leave the checking of this result to the reader.

CHECK YOURSELF 7

Solve and check:

$$9x - 6 - 3x + 1 = 2x + 15$$

It may also be necessary to remove grouping symbols in solving an equation. Our final example illustrates.

Example 8

Solve and check:

$5(x - 3) - 2x = x + 7$

As our first step we apply the distributive property to simplify on the left.

Note: $5(x - 3) = 5(x) - 5(3) = 5x - 15$

$5x - 15 - 2x = x + 7$

Combining like terms, we have

$3x - 15 = x + 7$

We then proceed as in previous examples.

Add 15.

$$3x - 15 + 15 = x + 7 + 15$$
$$3x = x + 22$$

Subtract x.

$$3x - x = x - x + 22$$

Divide by 2.

$$2x = 22$$
$$x = 11$$

The solution is 11. To check, substitute 11 for x in the original equation. Again note the use of our rules for the order of operations.

Simplify terms in the parentheses.

$$5(11 - 3) - 2 \cdot 11 \stackrel{?}{=} 11 + 7$$

Multiply.

$$5 \cdot 8 - 2 \cdot 11 \stackrel{?}{=} 11 + 7$$

Add and subtract.

$$40 - 22 \stackrel{?}{=} 11 + 7$$

A true statement.

$$18 = 18$$

CHECK YOURSELF 8

Solve and check.

$7(x + 5) - 3x = x - 7$

Such an outline of steps is sometimes called an *algorithm* for the process.

Let's summarize our work with an outline of the steps involved in solving linear equations.

SOLVING LINEAR EQUATIONS

STEP 1 Use the distributive property to remove any grouping symbols that appear. Then simplify by combining like terms on each side of the equation.

STEP 2 Add or subtract the same term on each side of the equation until the term involving the variable is on one side and a number is on the other.

STEP 3 Multiply or divide both sides of the equation by the same nonzero number so that the variable is alone on one side of the equation.

STEP 4 Check the solution in the original equation.

CHECK YOURSELF ANSWERS

1. 3.
2. -4.
3. 3.
4. (1) -12; (2) 24.
5. $\frac{5}{2}$.
6. 6.
7. 5.
8. -14.

3.4 Exercises

Solve for x and check your result.

1. $2x + 1 = 9$ 4

2. $3x - 1 = 17$ 6

3. $3x - 2 = 7$ 3

4. $5x + 3 = 23$ 4

5. $4x + 7 = 35$ 7

6. $7x - 8 = 13$ 3

7. $2x + 9 = 5$ -2

8. $6x + 25 = -5$ -5

9. $3x - 5 = -1$ $\frac{4}{3}$

10. $2x + 3 = -2$ $-\frac{5}{2}$

11. $3x - 2 = -26$ -8

12. $4x - 3 = -19$ -4

13. $2 - 5x = 12$ -2

14. $5 - 3x = -4$ 3

15. $7 - 6x = -11$ 3

16. $4 - 7x = 39$ -5

17. $\frac{x}{2} + 1 = 5$ 8

18. $\frac{x}{3} - 2 = 3$ 15

19. $\frac{x}{4} - 5 = 3$ 32

20. $\frac{x}{5} + 3 = 8$ 25

21. $\frac{2}{3}x + 5 = 17$ 18

22. $\frac{3}{4}x - 5 = 4$ 12

23. $\frac{4}{5}x - 3 = 13$ 20

24. $\frac{5}{7}x + 4 = 14$ 14

25. $4x = x + 9$ 3

26. $3x = 8 - x$ 2

27. $5x = 30 - x$ 5

28. $8x = 3x + 20$ 4

29. $8x = 4x - 3$ $-\frac{3}{4}$

30. $3x = 4 - 2x$ $\frac{4}{5}$

31. $5x - 2 = 2x + 19$ 7

32. $7x + 3 = 2x + 18$ 3

33. $9x + 2 = 3x + 38$ 6

34. $8x - 3 = 4x + 17$ 5

35. $4x - 8 = x - 14$ −2

36. $6x - 5 = 3x - 29$ −8

37. $5x + 7 = 2x - 3$ $-\frac{10}{3}$

38. $9x + 7 = 5x - 3$ $-\frac{5}{2}$

39. $7x - 3 = 9x + 5$ −4

40. $5x - 2 = 8x - 11$ 3

41. $5x + 4 = 7x - 8$ 6

42. $2x + 23 = 6x - 5$ 7

43. $3x - 5 = 6x - 10$ $\frac{5}{3}$

44. $2x + 9 = 4x - 2$ $\frac{11}{2}$

45. $2x - 3 + 5x = 7 + 4x + 2$ 4

46. $8x - 7 - 2x = 2 + 4x - 5$ 2

47. $6x + 7 - 4x = 8 + 7x - 26$ 5

48. $7x - 2 - 3x = 5 + 8x + 13$ −5

49. $9x - 2 + 7x + 13 = 10x - 13$ −4

50. $5x + 3 + 6x - 11 = 8x + 25$ 11

51. $8x - 7 + 5x - 10 = 10x - 12$ $\frac{5}{3}$

52. $10x - 9 + 2x - 3 = 8x - 18$ $-\frac{3}{2}$

53. $7(2x - 1) - 5x = x + 25$ 4

54. $9(3x + 2) - 10x = 12x - 7$ −5

55. $3x + 2(4x - 3) = 6x - 9$ $-\frac{3}{5}$

56. $7x + 3(2x + 5) = 10x + 17$ $\frac{2}{3}$

57. $\frac{8}{3}x - 3 = \frac{2}{3}x + 15$ 9

58. $\frac{12}{5}x + 7 = 31 - \frac{3}{5}x$ 8

59. $\frac{2}{5}x - 5 = \frac{12}{5}x + 8$ $-\frac{13}{2}$

60. $\frac{3}{7}x - 5 = \frac{24}{7}x + 7$ -4

61. $5.3x - 7 = 2.3x + 5$ 4

62. $9.8x + 2 = 3.8x + 20$ 3

Translate each of the following statements to an equation. Let x represent the number in each case.

63. 3 more than twice a number is 7. $2x + 3 = 7$

64. 5 less than 3 times a number is 25. $3x - 5 = 25$

65. 7 less than 4 times a number is 41. $4x - 7 = 41$

66. 10 more than twice a number is 44. $2x + 10 = 44$

67. 5 more than two-thirds of a number is 21. $\frac{2}{3}x + 5 = 21$

68. 3 less than three-fourths of a number is 24. $\frac{3}{4}x - 3 = 24$

69. 3 times a number is 12 more than that number. $3x = x + 12$

70. 5 times a number is 8 less than that number. $5x = x - 8$

Skillscan (Section 1.5)
Divide.

a. $\frac{3b}{3}$ b **b.** $\frac{5x}{5}$ x **c.** $\frac{4xy}{4x}$ y **d.** $\frac{6a^2b}{6a^2}$ b

e. $\frac{7mn^2}{7n^2}$ m **f.** $\frac{\pi ab}{\pi a}$ b **g.** $\frac{srt}{sr}$ t **h.** $\frac{x^2yz}{x^2z}$ y

Answers

1. 4 **3.** 3 **5.** 7 **7.** −2 **9.** $\frac{4}{3}$ **11.** −8 **13.** −2 **15.** 3 **17.** 8 **19.** 32 **21.** 18

23. 20 **25.** 3 **27.** 5 **29.** $-\frac{3}{4}$ **31.** 7 **33.** 6 **35.** -2 **37.** $-\frac{10}{3}$ **39.** -4 **41.** 6 **43.** $\frac{5}{3}$
45. 4 **47.** 5 **49.** -4 **51.** $\frac{5}{3}$ **53.** 4 **55.** $-\frac{3}{5}$ **57.** 9 **59.** $-\frac{13}{2}$ **61.** 4 **63.** $2x + 3 = 7$
65. $4x - 7 = 41$ **67.** $\frac{2}{3}x + 5 = 21$ **69.** $3x = x + 12$ **a.** b **b.** x **c.** y **d.** b **e.** m **f.** b
g. t **h.** y

3.5 Solving Literal Equations

OBJECTIVE
To be able to solve a literal equation for any one of its variables.

Formulas are extremely useful tools in any field in which mathematics is applied. Formulas are simply equations that express a relationship between more than one letter or variable. You are no doubt familiar with all kinds of examples, such as

$A = \frac{1}{2} bh$ The area of a triangle

$I = PRT$ Interest

$V = \pi r^2 h$ The volume of a cylinder

Actually a formula is really a *literal* equation. This is an equation that involves more than one letter or variable. For instance, our first formula or literal equation, $A = \frac{1}{2}bh$, involves the three letters A (for area), b (for base), and h (for height).

Unfortunately, formulas are not always given in the form needed to solve a particular problem. That's where algebra is needed to change the formula to a more useful equivalent equation, which is solved for a particular letter or variable. You will see that the steps used in the process will be very similar to those you saw earlier in solving linear equations. Let's consider an example.

Example 1

Suppose that we know the area A and the base b of a triangle and want to find its height h.

We are given

$$A = \frac{1}{2}b \cdot h$$

Our job is to find an equivalent equation with h, the unknown, by itself on one side. We call $\frac{1}{2}b$ the *coefficient* of h. We can remove the two *factors* of that coefficient, $\frac{1}{2}$ and b, separately.

Note:

$2\left(\frac{1}{2}bh\right)$

$= \left(2 \cdot \frac{1}{2}\right)(bh)$ Associative property

$= 1 \cdot bh$

$= bh$

$$2A = 2\left(\frac{1}{2}bh\right)$$ Multiply both sides by 2 to clear of fractions.

or

$$2A = bh$$

$$\frac{2A}{b} = \frac{bh}{b}$$ Divide by b to "isolate" h.

$$\frac{2A}{b} = h$$

or

$$h = \frac{2A}{b}$$ Reverse the sides to write h on the left.

We now have the height h in terms of the area A and the base b. This is called *solving the equation for h* and means that we are rewriting the formula as an equivalent equation of the form

$$h = \square$$

Here $\square$ means an expression containing all the numbers or letters *other than h.*

Fortunately, as we pointed out above, you have learned the methods needed to solve most literal equations or formulas for some specified variable. As Example 1 illustrates, the rules developed in Sections 3.2 and 3.3 can be used in exactly the same way as they were in solving equations with one variable.

CHECK YOURSELF 1

Solve $V = \frac{1}{3}B \cdot h$ for h.

Example 2

This is a formula for the circumference of a circle.

Solve $C = 2\pi r$ for r.

To isolate r, we must remove its coefficient, 2π. So to solve for r, let's divide both sides of the equation by 2π. Do you see that this will undo the multiplication on the right and leave r by itself?

$$\frac{C}{2\pi} = \frac{2\pi r}{2\pi}$$

$$\frac{C}{2\pi} = r$$

or

$$r = \frac{C}{2\pi}$$

Again we have switched the sides of the final equation so that the result has the form

$$r = \square$$

CHECK YOURSELF 2

Solve $I = Prt$ for t.

You may have to apply both the addition and the multiplication properties when solving a formula or literal equation for a specified variable. The following example illustrates this.

Example 3

This is a form for a linear equation in two variables. You will see this again in Chapter 7.

Solve $y = mx + b$ for x.

Remember that we want to end up with x alone on one side of the equation. Let's start by subtracting b from both sides to undo the addition on the right.

$$y = mx + b$$
$$y - b = mx + b - b$$
$$y - b = mx$$

If we now divide both sides by m, then x will be alone on the right-hand side.

$$\frac{y - b}{m} = \frac{mx}{m}$$
$$\frac{y - b}{m} = x$$

or

$$x = \frac{y - b}{m}$$

CHECK YOURSELF 3

Solve $v = v_0 + gt$ for t.

Let's summarize the steps illustrated by our examples.

SOLVING FORMULAS OR LITERAL EQUATIONS

STEP 1 If necessary, multiply both sides of the equation by the same term to clear of fractions.

STEP 2 Add or subtract the same term on both sides of the equation so that all terms involving the variable that you are solving for are on one side of the equation and all other terms are on the other side.

STEP 3 Divide both sides of the equation by the coefficient of the variable that you are solving for.

Let's look at one more example, using the above steps.

Example 4

This is a formula for the *amount* of money in an account after interest has been earned.

Solve $A = P + Prt$ for r.

$$A = P + Prt$$

$$A - P = P - P + Prt$$

$$A - P = Prt$$

Subtracting P from both sides will leave the term involving r alone on the right.

$$\frac{A - P}{Pt} = \frac{Prt}{Pt}$$

Dividing both sides by Pt will isolate r on the right.

$$\frac{A - P}{Pt} = r$$

or

$$r = \frac{A - P}{Pt}$$

CHECK YOURSELF 4

Solve $2x + 3y = 6$ for y.

Let's look at an application of solving a literal equation for a specified variable in our final example.

Example 5

Suppose that the amount in an account, 3 years after a principal of \$5000 was invested, is \$6050. What was the interest rate?

From our previous example,

$$A = P + Prt \qquad (1)$$

where A is the amount in the account, P is the principal, r is the interest rate, and t is the time that the money has been invested. By the result of Example 4 we have

$$r = \frac{A - P}{Pt} \qquad (2)$$

Do you see the advantage of having our equation solved for the desired variable?

and we can substitute the known values in Equation (2):

$$r = \frac{6050 - 5000}{(5000)(3)}$$

$$= \frac{1050}{15,000} = 0.07 = 7\%$$

The interest rate is 7 percent.

CHECK YOURSELF 5

Suppose that the amount in an account, 4 years after a principal of $3000 was invested, is $3720. What was the interest rate?

CHECK YOURSELF ANSWERS

1. $h = \frac{3V}{B}$.

2. $t = \frac{I}{Pr}$.

3. $t = \frac{v - v_0}{g}$.

4. $y = \frac{6 - 2x}{3}$.

5. 6 percent.

3.5 Exercises

Solve each literal equation for the indicated variable.

1. $p = 4s$ (for s) Perimeter of a square $\frac{p}{4}$

2. $V = Bh$ (for B) Volume of a prism $\frac{V}{h}$

3. $E = IR$ (for R) Electric circuits $\frac{E}{I}$

4. $I = Prt$ (for r) Simple interest $\frac{I}{Pt}$

5. $V = LWH$ (for H) Volume of a rectangular solid $\dfrac{V}{LW}$

6. $V = \pi r^2 h$ (for h) Volume of a cylinder $\dfrac{V}{\pi r^2}$

7. $A + B + C = 180$ (for B) Measure of angles in a triangle $180 - A - C$

8. $V - E + F = 2$ (for F) Euler's formula $2 - V + E$

9. $ax + b = 0$ (for x) Linear equation in one variable $-\dfrac{b}{a}$

10. $y = mx + b$ (for m) Point-slope form for a line $\dfrac{y - b}{x}$

11. $s = \dfrac{1}{2}gt^2$ (for g) Distance $\dfrac{2s}{t^2}$

12. $K = \dfrac{1}{2}mv^2$ (for m) Energy $\dfrac{2k}{v^2}$

13. $x + 3y = 6$ (for y) Linear equation $\dfrac{6 - x}{3}$

14. $3x + 4y = 12$ (for x) Linear equation $\dfrac{12 - 4y}{3}$

15. $P = 2L + 2W$ (for L) Perimeter of a rectangle $\dfrac{P - 2W}{2}$

16. $ax + by = c$ (for y) Linear equation in two variables $\dfrac{c - ax}{b}$

17. $V = \dfrac{KT}{P}$ (for T) Volume of a gas $\dfrac{PV}{K}$

18. $V = \dfrac{1}{3}\pi r^2 h$ (for h) Volume of a cone $\dfrac{3V}{\pi r^2}$

19. $x = \dfrac{a + b}{2}$ (for b) Average of two numbers $2x - a$

20. $D = \dfrac{C - s}{n}$ (for s) Depreciation $C - nD$

21. $F = \frac{9}{5}C + 32$ (for C) Celsius/Fahrenheit $\frac{5}{9}(F - 32)$

22. $A = P + Prt$ (for t) Amount at simple interest $\frac{A - P}{Pr}$

23. $S = 2\pi r^2 + 2\pi rh$ (for h) Total surface area of a cylinder $\frac{S - 2\pi r^2}{2\pi r}$

24. $A = \frac{1}{2}h(B + b)$ (for b) Area of a trapezoid $\frac{2A - hB}{h}$

25. A rectangular solid has a base with length 6 cm and width 4 cm. If the volume of the solid is 72 cm^3, find the height of the solid. See Exercise 5. 3 cm

26. A cylinder has a radius of 4 in. If the volume of the cylinder is 144π in^3, what is the height of the cylinder? See Exercise 6. 9 in

27. A principal of $2000 was invested in a savings account for 4 years. If the interest earned for that period was $480, what was the interest rate? See Exercise 4. 6%

28. If the perimeter of a rectangle is 60 ft and its width is 12 ft, find its length. See Exercise 15. 18 ft

29. The high temperature in New York for a particular day was reported at 77°F. How would that same temperature have been given in Celsius? See Exercise 21. 25°C

30. The area of a trapezoid is 36 in^2. If its height is 4 in and the length of one of the bases is 11 in, find the length of the other base. See Exercise 24. 7 in

We considered the notation for the four arithmetic operations in the programming language BASIC in Section 2.6. To review, those operations are indicated by the following:

Algebraic expression	BASIC expression
$a + b$	A + B
$a - b$	A − B
ab	A * B
$\frac{a}{b}$	A/B

Using the above information, write a BASIC expression for the given formula solved for the specified letter. *Hint:* In BASIC, the operations of multiplication and division are done *in order* from left to right. Then the operations of addition and subtraction are performed in the same manner. You may have to insert parentheses in the following answers to achieve your desired result.

31. $V = Bh$ (for h) V/B

32. $I = Prt$ (for t) I/(P * r)

33. $A = P + Prt$ (for t) (A − P)/(P * r)

34. $A = \frac{1}{2}h(B + b)$ (for b) (2 * A − h * B)/h

Skillscan (Section 2.1)

Locate each of the following numbers on the number line.

a. 4 **b.** −5 **c.** −3 **d.** 2

e. $-\frac{7}{2}$ **f.** $\frac{2}{3}$ **g.** 2.5 **h.** −1.1

Answers

1. $\frac{p}{4}$ **3.** $\frac{E}{I}$ **5.** $\frac{V}{LW}$ **7.** $180 - A - C$ **9.** $-\frac{b}{a}$ **11.** $\frac{2s}{t^2}$ **13.** $\frac{6 - x}{3}$ **15.** $\frac{P - 2W}{2}$ **17.** $\frac{PV}{K}$

19. $2x - a$ **21.** $\frac{5}{9}(F - 32)$ **23.** $\frac{S - 2\pi r^2}{2\pi r}$ **25.** 3 cm **27.** 6% **29.** 25°C **31.** h = V/B

33. t = (A − P)/(P * r)

a.–h.

3.6 Inequalities—An Introduction

OBJECTIVES

1. To understand the notation of inequalities
2. To graph the solution sets of inequalities

As pointed out in this chapter's introduction, an equation is just a statement that two expressions are equal. In algebra, an *inequality* is a statement that one expression is less than or greater than another. There are two new symbols used in writing inequalities. Their use is illustrated in the following example.

Example 1

The inequality symbols are less than (<) and greater than (>).

To help you remember, the "arrowhead" always points toward the smaller quantity.

$5 < 8$ is an inequality read "5 is less than 8."

$9 > 6$ is an inequality read "9 is greater than 6."

CHECK YOURSELF 1

Complete the statements, using the symbols < and >.

1. 12   8   **2.** 20   25

Just as was the case with equations, inequalities that involve variables may be either true or false depending on the value that we give to the variable. For instance, given the inequality

$x < 6$

$$\text{If } x = \begin{cases} 3 & 3 < 6 \text{ is true} \\ 5 & 5 < 6 \text{ is true} \\ -10 & -10 < 6 \text{ is true} \\ 8 & 8 < 6 \text{ is } \textit{false} \end{cases}$$

Therefore 3, 5, and -10 are some *solutions* for the original inequality $x < 6$. They make the inequality a true statement. You should see that 8 is *not* a solution. We call the set of all solutions the *solution set* of the inequality. Of course, there are many possible solutions.

Since there are so many solutions (an infinite number, in fact), we certainly do not want to try to list them all! A convenient way to show the solution set of an inequality is with the use of a number line.

Example 2

To graph the solution set for the inequality $x < 6$, we want to include all real numbers that are "less than" 6. This means all numbers *to the left* of 6 on the number line. We then start at 6 and draw an arrow extending left, as shown:

Note: The open circle at 6 means that we do not want to include 6 in the solution set (6 is not less than itself). The colored arrow shows all numbers in the solution set, with the arrowhead indicating that the solution set continues indefinitely to the left.

CHECK YOURSELF 2

Graph the solution set of $x < -2$.

The following example illustrates a graph when the "greater than" symbol is involved.

Example 3

Graph the solution set of $x > -3$. We start at -3 and draw an arrow extending to the right, as shown below.

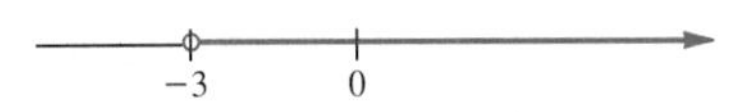

In this case we want all numbers greater than (or *to the right* of) -3.

CHECK YOURSELF 3

Graph the solution set of $x > 5$.

Two other symbols are used in writing inequalities. The inequality

$x \geq 5$

is really a combination of the two statements $x > 5$ and $x = 5$. It is read "x is greater than or equal to 5." The solution set will include 5 in this case.

Example 4

The solution set for $x \geq 5$ is graphed as follows.

Note: Here the closed circle means that we want to include 5 in the solution set.

We can also combine the "less than" symbol with equality.

Example 5

The inequality $x \le 2$ is read "x is less than or equal to 2." The graph of its solution set is

CHECK YOURSELF 4

Graph the solution sets.

1. $x \le -4$ **2.** $x \ge 3$

CHECK YOURSELF ANSWERS

1. (1) $12 > 8$; (2) $20 < 25$.

2. $x < -2$

3. $x > 5$

4. (1) $x \le -4$; (2) $x \ge 3$

3.6 Exercises

Complete the statements, using the symbol < or >.

1. 5 ______ 10
$5 < 10$

2. 9 ______ 8
$9 > 8$

3. 7 ______ −2
$7 > -2$

4. 0 ______ −5
$0 > -5$

5. 0 ______ 4
$0 < 4$

6. −10 ______ −5
$-10 < -5$

7. −3 ______ −8
$-3 > -8$

8. −8 ______ −12
$-8 > -12$

9. −7 ______ 0
$-7 < 0$

10. −8 ______ −9 $-8 > -9$

Write the inequalities in words.

11. $x < 3$
x is less than 3

12. $x \le -5$
x is less than or equal to −5

13. $x \ge -4$
x is greater than or equal to −4

14. $x < -2$
x is less than −2

15. $-5 \le x$
−5 is less than or equal to x

16. $2 < x$
2 is less than x

17. $x \ge 0$
x is greater than or equal to 0

18. $x \ge -7$
x is greater than or equal to −7

Graph the solution sets of the following inequalities.

19. $x > 1$

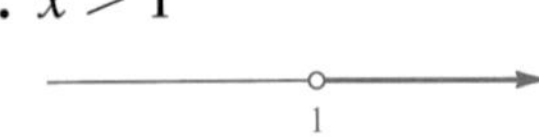

20. $x < -2$

−2

21. $x < 8$

8

22. $x > 3$

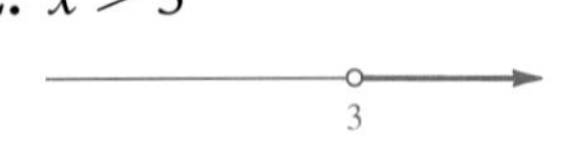

23. $x > -5$

−5

24. $x < -4$

−4

25. $x \geq 9$

26. $x \geq 0$

27. $x < 0$

0

28. $x \leq -3$

−3

29. $x \leq -10$

−10

30. $x \geq -8$

−8

Skillscan (Section 3.4)

Solve the following equations.

a. $3x + 4 = 19$
5

b. $2w - 5 = -13$
−4

c. $5a - 8 = 3a$
4

d. $9w + 14 = 7w$
−7

e. $8x - 5 = 5x - 23$
−6

f. $4y + 7 = 7y + 5$
$\frac{2}{3}$

g. $3(x - 8) = x - 2$
11

h. $\frac{2}{3}x - 4 = 8$
18

Answers

1. $5 < 10$ **3.** $7 > -2$ **5.** $0 < 4$ **7.** $-3 > -8$ **9.** $-7 < 0$ **11.** x is less than 3 **13.** x is greater than or equal to -4 **15.** -5 is less than or equal to x **17.** x is greater than or equal to 0

19. 1

21. 8

23. −5

25. 9

27. 0

29. −10

a. 5 **b.** −4 **c.** 4 **d.** −7 **e.** −6 **f.** $\frac{2}{3}$ **g.** 11 **h.** 18

3.7 Solving Linear Inequalities

OBJECTIVE

To be able to solve and graph the solution sets for linear inequalities in one variable.

You learned how to graph the solution sets of some simple inequalities, such as $x < 8$ or $x \geq 10$, in the last section. Now we will look at more complicated inequalities, such as

$$2x - 3 < x + 4$$

This is called a *linear inequality in one variable.* Only one variable is involved in the inequality, and it appears only to the first power. Fortunately, the methods used to solve this type of inequality are very similar to those we used earlier in this chapter to solve equations. Here is our first property.

Equivalent inequalities have exactly the same solution sets.

THE ADDITION PROPERTY OF INEQUALITY

If $a < b$, then $a + c < b + c$

In words, adding the same quantity to both sides of an inequality gives an *equivalent inequality.*

The following example illustrates the use of this property.

Example 1

Solve and graph the solution set for $x - 8 < 7$.

We will add 8 to both sides of the inequality by the addition property. Do you see why? The inequality is solved when the equivalent inequality has the form

$$x < \square \qquad \text{or} \qquad x > \square$$

That should look familiar!

$$x - 8 < 7$$
$$x - 8 + 8 < 7 + 8$$
$$x < 15$$

The graph of the solution set is

0 5 10 15 20

CHECK YOURSELF 1

Solve and graph the solution set for

$$x - 9 > -3$$

Example 2

Solve and graph the solution set for $3x > 5 + 2x$.

This time we want to subtract $2x$ from both sides to arrive at an equivalent inequality with the terms in x on the left.

As was the case with equations, the addition property also allows us to *subtract* the same quantity from both sides of an inequality.

$$3x > 5 + 2x$$
$$3x - 2x > 5 + 2x - 2x$$
$$x > 5$$

The graph of the solution set is

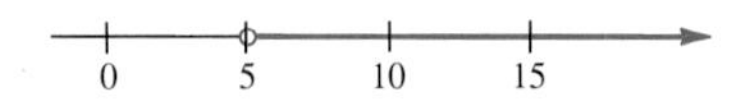

Example 3

Solve and graph the solution set for $4x - 2 \geq 3x + 5$.

First, we subtract $3x$ from both sides of the inequality.

$$4x - 2 \geq 3x + 5$$
$$4x - 3x - 2 \geq 3x - 3x + 5$$
$$x - 2 \geq 5$$

Now we add 2 to both sides.

$$x - 2 + 2 \geq 5 + 2$$
$$x \geq 7$$

The graph of the solution set is

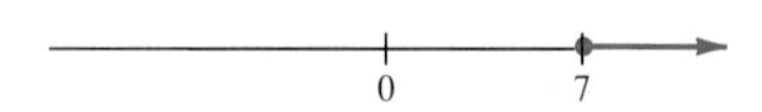

Note: We subtracted $3x$ and then added 2 to both sides. If these steps are done in the other order, the resulting inequality will be the same. You might want to verify that yourself.

CHECK YOURSELF 2

Solve and graph the solution set for

$7x - 8 \leq 6x + 2$

You will also need a rule for multiplying on both sides of an inequality. Here you'll have to be a bit careful. There is a difference between the multiplication property for inequalities and that for equations. Look at the following:

$2 < 7$ (a true inequality)

Let's multiply both sides by 3.

$$\begin{aligned} 2 &< 7 \\ 3 \cdot 2 &< 3 \cdot 7 \\ 6 &< 21 \quad \text{(a true inequality)} \end{aligned}$$

Now we multiply both sides by -3.

$$\begin{aligned} 2 &< 7 \\ (-3)(2) &< (-3)(7) \\ -6 &< -21 \quad \text{(\textit{not} a true inequality)} \end{aligned}$$

Let's try something different.

$$\begin{aligned} 2 &< 7 \\ (-3)(2) &> (-3)(7) \\ -6 &> -21 \quad \text{(this is now a true inequality)} \end{aligned}$$

Change the "sense" of the inequality: $<$ becomes $>$.

This suggests that multiplying both sides of an inequality by a negative number changes the "sense" of the inequality.

We can state the following general property.

THE MULTIPLICATION PROPERTY OF INEQUALITY

If $a < b$, then $ac < bc$ where $c > 0$
and $ac > bc$ where $c < 0$

In words, multiplying both sides of an inequality by the same *positive* number gives an equivalent inequality.

When you are multiplying both sides of an inequality by the same *negative* number, *reverse the sense* of the inequality to give an equivalent inequality.

Example 4

(*a*) Solve and graph the solution set for $5x < 30$.

If we multiply both sides of the inequality by $\frac{1}{5}$, the inequality will have the form

$x < \square$

$\frac{1}{5}(5x) = \frac{1}{5}(30)$

Simplifying, we have

$x < 6$

The graph of the solution set is

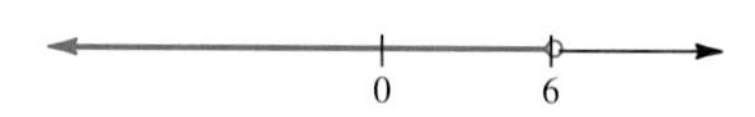

(*b*) Solve and graph the solution set for $-4x \geq 28$.

In this case we want to multiply both sides of the inequality by $-\frac{1}{4}$ to leave x alone on the left.

$\left(-\frac{1}{4}\right)(-4x) \leq \left(-\frac{1}{4}\right)(28)$

Reverse the sense of the inequality because you are multiplying by a negative number!

or $\quad x \leq -7$

The graph of the solution set is

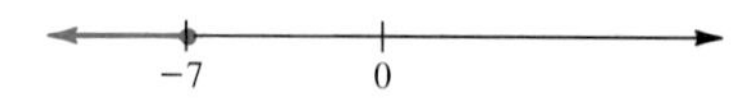

CHECK YOURSELF 3

Solve and graph the solution sets:

1. $7x > 35$ **2.** $-8x \leq 48$

The following example illustrates the use of the multiplication property when fractions are involved in an inequality.

Example 5

(*a*) Solve and graph the solution set for

$\frac{x}{4} > 3$

Here we multiply both sides of the inequality by 4. Do you see that this will isolate x on the left?

$$4\left(\frac{x}{4}\right) > 4(3)$$

or $x > 12$

The graph of the solution set is

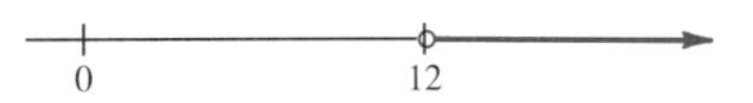

(*b*) Solve and graph the solution set for

$$-\frac{x}{6} \geq -3$$

In this case, we multiply both sides of the inequality by -6:

$$(-6)\left(-\frac{x}{6}\right) \leq (-6)(-3)$$

Note that we reverse the sense of the inequality since we are multiplying by a negative number.

and simplifying yields

$$x \leq 18$$

The graph of the solution set is

CHECK YOURSELF 4

Solve and graph the solution sets for the following inequalities.

1. $\frac{x}{5} \leq 4$ **2.** $-\frac{x}{3} < -7$

Often you will have to apply both properties of this section to solve an inequality. The following example illustrates.

Example 6

(*a*) Solve and graph the solution set for $5x - 3 < 2x$.

First, add 3 to both sides to undo the subtraction on the left.

$$5x - 3 < 2x$$
$$5x - 3 + 3 < 2x + 3$$
$$5x < 2x + 3$$

Now subtract $2x$, so that only the number remains on the right.

$$5x < 2x + 3$$
$$5x - 2x < 2x - 2x + 3$$
$$3x < 3$$

Note that the multiplication property also allows us to divide both sides by a nonzero number.

We have chosen here to *divide* both sides by 3.

$$\frac{3x}{3} < \frac{3}{3}$$
$$x < 1$$

The graph of the solution set is

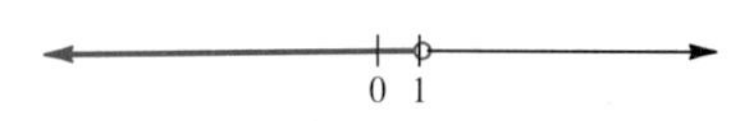

(*b*) Solve and graph the solution set for $2 - 5x < 7$.

$$2 - 5x < 7$$
$$2 - 2 - 5x < 7 - 2$$ First subtract 2.
$$-5x < 5$$
$$\frac{-5x}{-5} > \frac{5}{-5}$$ Divide by -5. Be sure to reverse the sense of the inequality.

or $\quad x > -1$

The graph is

−1 0

CHECK YOURSELF 5

Solve and graph the solution set:

1. $4x + 9 \geq x$ **2.** $5 - 6x < 41$

Example 7

Solve and graph the solution set for $5x - 5 \geq 3x + 4$.

$$5x - 5 \geq 3x + 4$$
$$5x - 5 + 5 \geq 3x + 4 + 5$$ First add 5.
$$5x \geq 3x + 9$$
$$5x - 3x \geq 3x - 3x + 9$$ Subtract $3x$.
$$2x \geq 9$$

$$\frac{2x}{2} \geq \frac{9}{2} \qquad \text{Divide by 2.}$$

$$x \geq \frac{9}{2}$$

The graph of the solution set is

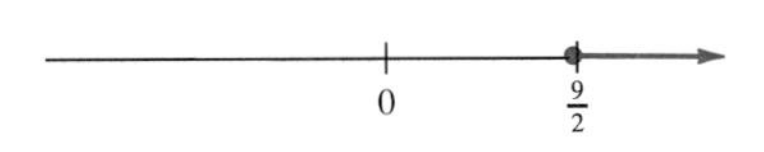

CHECK YOURSELF 6

Solve and graph the solution set for

$8x + 3 < 4x - 13$

Again, you should be especially careful when negative coefficients occur during the solution process for inequalities. Consider the following example.

Example 8

Solve and graph the solution set for $2x + 4 < 5x - 2$.

$$2x + 4 < 5x - 2$$

$$2x + 4 - 4 < 5x - 2 - 4 \qquad \text{Subtract 4.}$$

$$2x < 5x - 6$$

$$2x - 5x < 5x - 5x - 6 \qquad \text{Subtract } 5x.$$

$$-3x < -6$$

$$\frac{-3x}{-3} > \frac{-6}{-3} \qquad \text{Divide by } -3\text{, reverse the sense of the inequality.}$$

$$x > 2$$

The graph of the solution set is

CHECK YOURSELF 7

Solve and graph the solution set for

$5x + 12 \geq 10x - 8$

The solution of inequalities may also require the use of the distributive property to clear an inequality of parentheses. Our final example illustrates.

Example 9

Solve and graph the solution set for

$5(x - 2) \geq -8$

Applying the distributive property on the left yields

$5x - 10 \geq -8$

Solving as before yields

$$5x - 10 + 10 \geq -8 + 10 \quad \text{Add 10.}$$
$$5x \geq 2$$
or
$$x \geq \frac{2}{5} \quad \text{Divide by 5.}$$

The graph of the solution set is

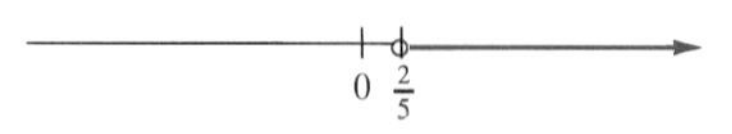

CHECK YOURSELF 8

Solve and graph the solution set for

$4(x + 3) < 9$

The following outline (or algorithm) will summarize our work in this section in solving linear inequalities.

> **SOLVING LINEAR INEQUALITIES**
>
> STEP 1 Remove any grouping symbols and combine any like terms appearing on either side of the inequality.
>
> STEP 2 Apply the addition property to write an equivalent inequality with the variable term on one side of the inequality and the number on the other.
>
> STEP 3 Apply the multiplication property to write an equivalent inequality with the variable isolated on one side of the inequality. Be sure to reverse the sense of the inequality if you multiply or divide by a negative number.
>
> STEP 4 Graph the solution derived in step 3 on a number line.

CHECK YOURSELF ANSWERS

1. $x > 6$

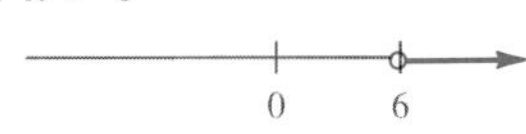

2. $x \leq 10$

3. (1) $x > 5$

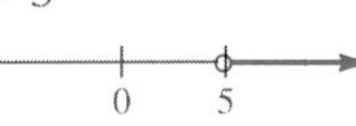

(2) $x \geq -6$

4. (1) $x \leq 20$

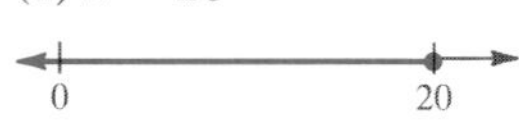

(2) $x > 21$

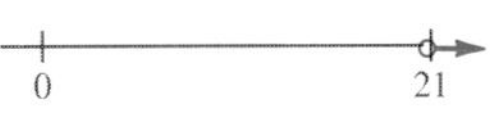

5. (1) $x \geq -3$

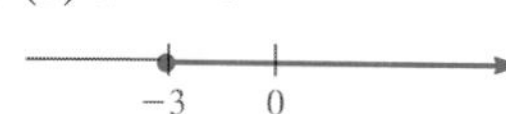

(2) $x > -6$

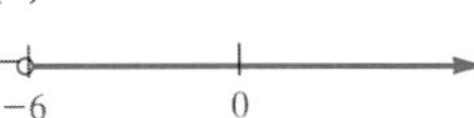

6. $x < -4$

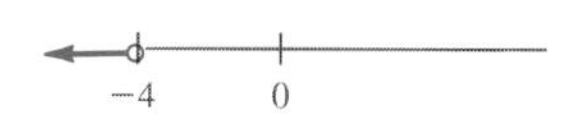

7. $x \leq 4$

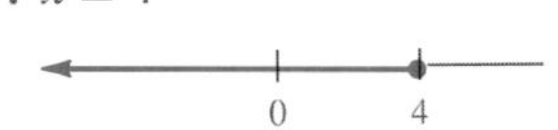

8. $x < -\frac{3}{4}$

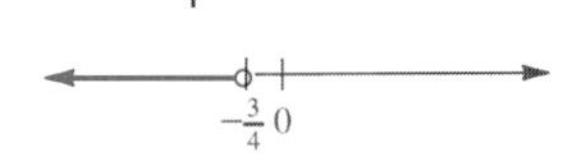

3.7 Exercises

Solve and graph the solution sets of the following inequalities.

1. $x - 5 < 8$ $x < 13$

2. $x + 3 \leq 2$ $x \leq -1$

3. $x + 9 \geq 11$ $x \geq 2$

4. $x - 5 > -8$ $x > -3$

5. $5x < 4x + 7$ $x < 7$

6. $3x \geq 2x - 4$ $x \geq -4$

7. $6x - 8 \leq 5x$ $x \leq 8$

8. $3x + 2 > 2x$ $x > -2$

9. $4x - 3 \geq 3x + 5$ $x \geq 8$

10. $5x + 2 \leq 4x - 6$ $x \leq -8$

11. $7x + 5 < 6x - 4$ $x < -9$

12. $8x - 7 > 7x + 3$ $x > 10$

13. $3x \leq 9$ $x \leq 3$

14. $5x > 20$ $x > 4$

15. $5x > -35$ $x > -7$

16. $7x \leq -21$ $x \leq -3$

17. $-6x \geq 18$ $x \leq -3$

18. $-9x < 45$ $x > -5$

19. $-10x < -60$ $x > 6$

20. $-12x \geq -48$ $x \leq 4$

21. $\frac{x}{4} > 5$ $x > 20$

22. $\frac{x}{3} \leq -3$ $x \leq -9$

23. $-\frac{x}{2} \geq -3$ $x \leq 6$

24. $-\frac{x}{5} < 4$ $x > -20$

25. $\frac{2x}{3} < 6$ $x < 9$

26. $\frac{3x}{4} \geq -9$ $x \geq -12$

27. $5x > 3x + 8$ $x > 4$

28. $4x \leq x - 9$ $x \leq -3$

29. $8x - 25 < 3x$ $x < 5$

30. $7x + 12 \geq x$ $x \geq -2$

31. $5x - 2 > 3x$ $x > 1$

32. $7x + 3 \geq 2x$ $x \geq -\frac{3}{5}$

33. $3 - 2x > 5$ $x < -1$

34. $5 - 3x \leq 17$ $x \geq -4$

35. $2x \geq 5x + 18$ $x \leq -6$

36. $3x < 7x - 28$ $x > 7$

37. $4x < 8x - 3$ $x > \frac{3}{4}$

38. $5x < 10x + 4$ $x > -\frac{4}{5}$

39. $5x - 3 \leq 3x + 15$ $x \leq 9$

40. $8x + 7 > 5x + 34$ $x > 9$

41. $9x + 7 > 2x - 28$ $x > -5$

42. $10x - 5 \leq 8x - 25$ $x \leq -10$

43. $7x - 5 < 3x + 2$ $x < \frac{7}{4}$

44. $5x - 2 \geq 2x - 7$ $x \geq -\frac{5}{3}$

45. $3x - 8 \leq 9x + 16$ $x \geq -4$

46. $2x + 7 > 7x - 18$ $x < 5$

47. $5x + 7 > 8x - 17$ $x < 8$

48. $4x - 3 \leq 9x + 27$ $x \geq -6$

49. $3x - 2 \leq 5x + 3$ $x \geq -\frac{5}{2}$

50. $2x + 3 > 8x - 2$ $x < \frac{5}{6}$

51. $3(x - 2) > 9$ $x > 5$

52. $5(x + 4) \leq -15$ $x \leq -7$

53. $4(x + 7) \leq 2x + 31$ $x \leq \frac{3}{2}$

54. $6(x - 5) > 3x - 26$ $x > \frac{4}{3}$

55. $2(x - 7) > 5x - 12$ $x < -\frac{2}{3}$

56. $3(x + 4) \leq 7x + 7$ $x \geq \frac{5}{4}$

Translate the following statements into inequalities. Let x represent the number in each case.

57. 5 more than a number is greater than 3. $x + 5 > 3$

58. 3 less than a number is less than or equal to 5. $x - 3 \leq 5$

59. 4 less than twice a number is less than or equal to 7. $2x - 4 \leq 7$

60. 10 more than a number is greater than negative 2. $x + 10 > -2$

61. 4 times a number, decreased by 15, is greater than that number. $4x - 15 > x$

62. 2 times a number, increased by 28, is less than or equal to 6 times that number.
$2x + 28 \leq 6x$

Skillscan (Section 1.1)
Write each statement, using symbols.

a. 7 more than x
$x + 7$

b. 5 less than a
$a - 5$

c. 4 times p
$4p$

d. 3 times w
$3w$

e. 3 more than twice s
$2s + 3$

f. 2 less than 5 times x
$5x - 2$

g. The sum of twice x and $x + 1$
$2x + x + 1$ or $3x + 1$

h. The sum of 3 times x and $x + 2$
$3x + x + 2$ or $4x + 2$

Answers

1. $x < 13$

13

3. $x \geq 2$

2

5. $x < 7$

7. $x \leq 8$

9. $x \geq 8$

8

11. $x < -9$

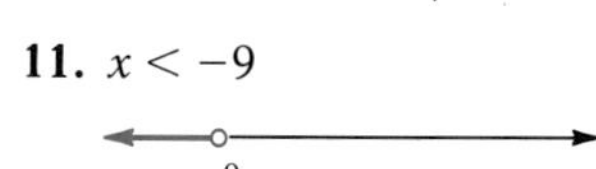

13. $x \leq 3$

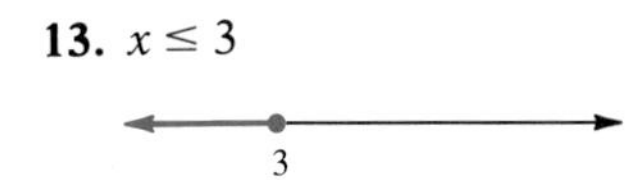

15. $x > -7$

−7

17. $x \leq -3$

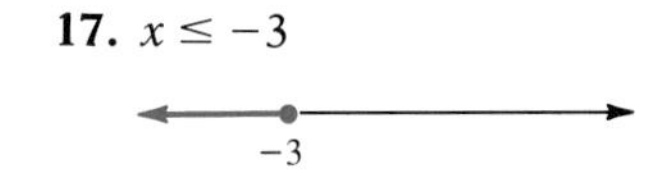

19. $x > 6$

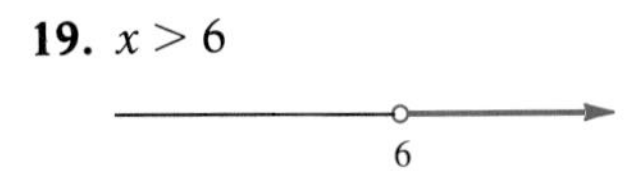

21. $x > 20$

20

23. $x \leq 6$

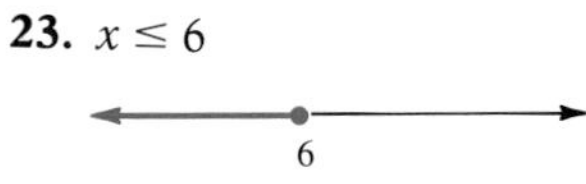

25. $x < 9$

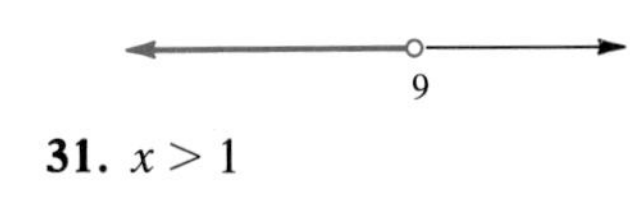

27. $x > 4$

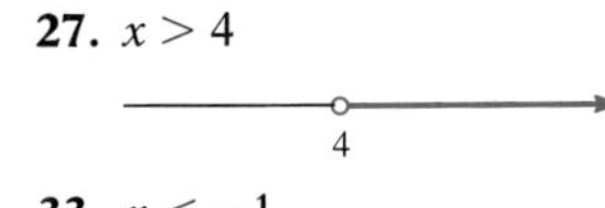

29. $x < 5$

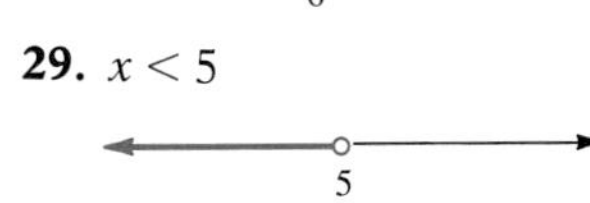

31. $x > 1$

1

33. $x < -1$

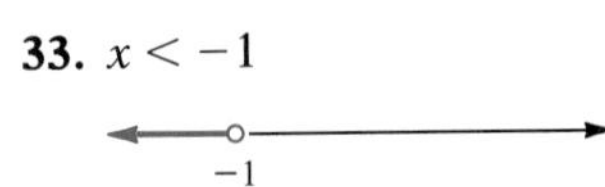

35. $x \leq -6$

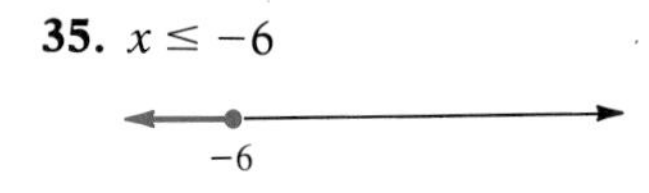

37. $x > \frac{3}{4}$

$\frac{3}{4}$

39. $x \leq 9$

9

41. $x > -5$

-5

43. $x < \frac{7}{4}$

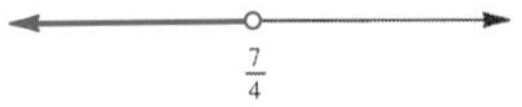

$\frac{7}{4}$

45. $x \geq -4$

-4

47. $x < 8$

8

49. $x \geq -\frac{5}{2}$

$-\frac{5}{2}$

51. $x > 5$

5

53. $x \leq \frac{3}{2}$

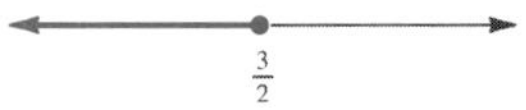

$\frac{3}{2}$

55. $x < -\frac{2}{3}$

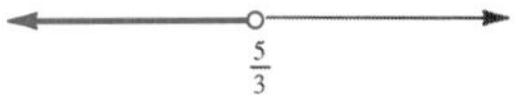

$\frac{5}{3}$

57. $x + 5 > 3$ **59.** $2x - 4 \leq 7$ **61.** $4x - 15 > x$ **a.** $x + 7$ **b.** $a - 5$

c. $4p$ **d.** $3w$ **e.** $2s + 3$ **f.** $5x - 2$ **g.** $2x + x + 1$ or $3x + 1$ **h.** $3x + x + 2$ or $4x + 2$

3.8 Applying Equations

OBJECTIVE
To use equations for the solution of word problems.

Earlier in this chapter you learned how to solve a variety of equations. The next step is to use this work in the solution of word problems. If you feel a bit uneasy about this subject, don't be too nervous. You have lots of company! To help you feel more comfortable when solving word problems, we are going to present a step-by-step approach that will (*with practice*) allow you to organize your work, and organization is the key to the solution of these problems.

TO SOLVE WORD PROBLEMS

STEP 1 Read the problem carefully. Then reread it to decide what you are asked to find.

STEP 2 Choose a letter to represent one of the unknowns in the problem. Then represent all the unknowns of the problem, with an expression using that same letter.

STEP 3 Translate the problem to the language of algebra to form an equation.

STEP 4 Solve the equation and answer the question of the original problem.

STEP 5 Verify your solution by returning to the original problem.

We discussed these translations in Section 1.1. You might find it helpful to review that section before going on.

The third step is usually the hardest part. We must translate words to the language of algebra. Before we look at a complete example, the following table may help you review that translation step.

TRANSLATING WORDS TO ALGEBRA

WORDS	ALGEBRA
The sum of x and y	$x + y$
3 plus a	$3 + a$ or $a + 3$
5 more than m	$m + 5$
b increased by 7	$b + 7$
The difference of x and y	$x - y$
4 less than a	$a - 4$
s decreased by 8	$s - 8$
The product of x and y	$x \cdot y$ or xy
5 times a	$5 \cdot a$ or $5a$
Twice m	$2m$
The quotient of x and y	$\frac{x}{y}$
a divided by 6	$\frac{a}{6}$
One-half of b	$\frac{b}{2}$ or $\frac{1}{2}b$

Now let's look at some typical examples of translating phrases to algebra.

Example 1

Translate each statement to an algebraic expression.

(*a*) The sum of a and 2 times b.

$a + 2b$

Sum (pointing to $+$); 2 times b (pointing to $2b$)

(*b*) 5 times m increased by 1.

$5m + 1$

5 times m (pointing to $5m$); Increased by 1 (pointing to $+ 1$)

(*c*) 5 less than 3 times x.

$3x - 5$

3 times x → $3x$; 5 less than → $- 5$

(*d*) The product of x and y, divided by 3.

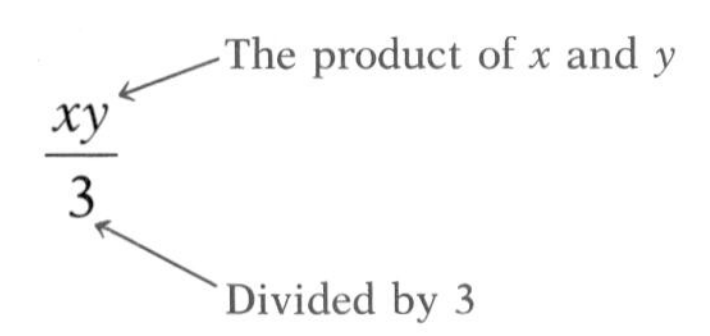

CHECK YOURSELF 1

Translate to algebra.

1. 2 more than twice x
2. 4 less than 5 times n
3. The product of twice a and b
4. The sum of s and t, divided by 5

Now let's work through a complete example, using our five-step approach.

Example 2

The sum of twice a number and 5 is 17. What is the number?

Step 1 *Read carefully*. You must find the unknown number.

Step 2 *Choose letters or variables*. Let x represent the unknown number.

Step 3 *Translate*.

The sum of → $+$

$2x + 5 = 17$

Twice x → $2x$; is → $=$

Step 4 *Solve*.

Subtract 5.

$$2x + 5 = 17$$
$$2x + 5 - 5 = 17 - 5$$
$$2x = 12$$

Divide by 2.

$$\frac{2x}{2} = \frac{12}{2}$$

$$x = 6$$

Always return to the *original problem* to check your result and *not* to the equation of step 3. This will prevent possible errors!

Step 5 *Check.* Is the sum of twice 6 and 5 equal to 17? Yes! (12 + 5 = 17) We have checked our solution.

CHECK YOURSELF 2

The sum of 3 times a number and 8 is 35. What is the number?

Consecutive integers are integers that follow one another, like 10, 11, and 12. To represent them in algebra:

If x is an integer, $x + 1$ is the next consecutive integer, $x + 2$ is the next, and so on.

We'll need this idea in the following example.

Example 3

The sum of two consecutive integers is 41. What are the two integers?

REMEMBER THE STEPS!

Read the problem carefully. What do you need to find?

Step 1 We want to find the two consecutive integers.

Assign letters to the unknown or unknowns.

Step 2 Let x be the first integer. Then $x + 1$ must be the next.

Write an equation.

Step 3

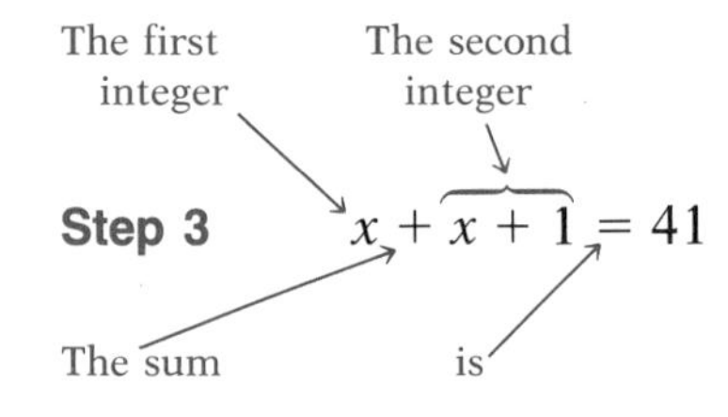

Solve the equation.

Step 4

$$\begin{aligned} x + x + 1 &= 41 \\ 2x + 1 &= 41 \\ 2x &= 40 \\ x &= 20 \end{aligned}$$

The first integer (x) is 20, and the next integer ($x + 1$) is 21.

Check.

Step 5 The sum of the two integers, 20 and 21, is 41.

CHECK YOURSELF 3

The sum of three consecutive integers is 51. What are the three integers?

Example 4

There were 55 more yes votes than no votes on an election measure. If 735 votes were cast in all, how many yes votes were there? How many no votes?

Step 1 We want to find the number of yes votes and the number of no votes.

Step 2 Let x be the number of no votes. Then

$$\underbrace{x + 55}$$

55 more than x

is the number of yes votes.

Step 3

$$x + \underbrace{x + 55} = 735$$

No votes Yes votes

Step 4

$$\begin{aligned} x + x + 55 &= 735 \\ 2x + 55 &= 735 \\ 2x &= 680 \\ x &= 340 \end{aligned}$$

$$\begin{aligned} \text{No votes } (x) &= 340 \\ \text{Yes votes } (x + 55) &= 395 \end{aligned}$$

Step 5 340 no votes plus 395 yes votes equals 735 total votes. The solution checks.

CHECK YOURSELF 4

Francine earns $120 per month more than Rob. If they earn a total of $2680 per month, what are their monthly salaries?

Similar methods will allow you to solve a variety of word problems. Look at the following example.

Example 5

Kevin is twice as old as Matt. Marcia, their sister, is 3 years older than Matt. If the sum of their ages is 31 years, find each of their ages.

Step 1 We want to find each age, so there are three unknowns.

There are other choices for x, but choosing the smallest quantity will usually give the easiest equation to write and solve.

Step 2 Let x be Matt's age.

Twice Matt's age

Then $2x$ is Kevin's age

3 more than Matt's age

and $x + 3$ is Marcia's age.

Step 3 Matt Kevin Marcia

$$x + 2x + x + 3 = 31$$

Sum of their ages

Step 4

$$x + 2x + x + 3 = 31$$
$$4x + 3 = 31$$
$$4x = 28$$
$$x = 7$$

Matt's age $(x) = 7$
Kevin's age $(2x) = 14$
Marcia's age $(x + 3) = 10$

Step 5 The sum of their ages $(7 + 14 + 10)$ is 31, and the solution is verified.

CHECK YOURSELF 5

Lucy jogged twice as many miles as Paul but 3 mi less than Isaac. If the three ran a total of 23 mi, how far did each person run?

Many word problems involve geometric figures and measurements. Let's look at an example.

Example 6

The perimeter of a rectangle is 46 cm. If the length is 3 cm more than the width, what are the dimensions of the rectangle?

Step 1 We want to find the dimensions (length and width) of the rectangle.

Step 2 Let x be the width. Then $\underbrace{x + 3}_{\text{3 more than } x}$ is the length.

Always draw a sketch at this point where figures are involved. It will help you form an equation in step 3.

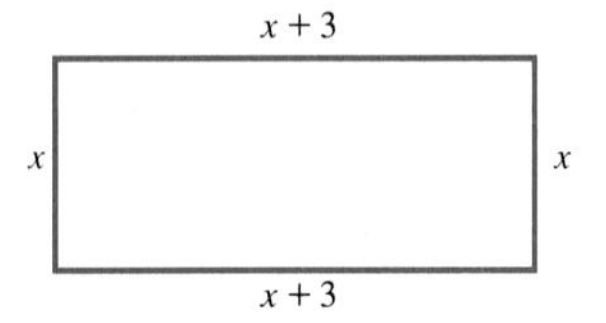

Step 3 The perimeter (the distance around) is 46 cm, so

$$\underset{\text{Width}}{x} + \underset{\text{Length}}{\underbrace{x + 3}} + \underset{\text{Width}}{x} + \underset{\text{Length}}{\underbrace{x + 3}} = 46$$

Step 4

$$\begin{aligned} x + x + 3 + x + x + 3 &= 46 \\ 4x + 6 &= 46 \\ 4x &= 40 \\ x &= 10 \\ \text{Width } (x) &= 10 \text{ cm} \\ \text{Length } (x + 3) &= 13 \text{ cm} \end{aligned}$$

Step 5 The perimeter is 10 + 13 + 10 + 13, or 46 cm. The solution is verified.

CHECK YOURSELF 6

The perimeter of a triangle is 55 in. If the length of the base is 5 in less than the length of the two equal legs, find the lengths of the legs and the base.

That is all we propose to do with word problems for the moment. Remember to use the five-step approach as you work on the problems for this section. We hope that you find it helpful.

CHECK YOURSELF ANSWERS

1. (1) $2x + 2$; (2) $5n - 4$; (3) $2ab$; (4) $\frac{s + t}{5}$.
2. The equation is $3x + 8 = 35$. The number is 9.
3. The equation is $x + x + 1 + x + 2 = 51$. The integers are 16, 17, and 18.

4. The equation is $x + x + 120 = 2680$. Rob's salary is $1280, and Francine's is $1400.
5. Paul: 4 mi; Lucy: 8 mi; Isaac: 11 mi.
6. Legs: 20 in; Base: 15 in.

3.8 Exercises

Solve the following word problems. Be sure to label the unknowns and to show the equation you use for the solution.

1. The sum of twice a number and 7 is 33. What is the number? 13

2. 3 times a number, increased by 8, is 50. Find the number. 14

3. 5 times a number, minus 12, is 78. Find the number. 18

4. 4 times a number, decreased by 20, is 44. What is the number? 16

5. The sum of two consecutive integers is 71. Find the two integers. 35, 36

6. The sum of two consecutive integers is 145. Find the two integers. 72, 73

7. The sum of three consecutive integers is 63. What are the three integers? 20, 21, 22

8. If the sum of three consecutive integers is 93, find the three integers. 30, 31, 32

9. The sum of two consecutive even integers is 66. What are the two integers? (*Hint:* Consecutive even integers such as 10, 12, and 14 can be represented by x, $x + 2$, $x + 4$, and so on.) 32, 34

10. If the sum of two consecutive even integers is 86, find the two integers. 42, 44

11. If the sum of two consecutive odd integers is 52, what are the two integers? (*Hint:* Consecutive odd integers such as 21, 23, and 25 can be represented by x, $x + 2$, $x + 4$, and so on.) 25, 27

12. The sum of two consecutive odd integers is 88. Find the two integers. 43, 45

13. The sum of three consecutive odd integers is 105. What are the three integers? 33, 35, 37

14. The sum of three consecutive even integers is 126. What are the three integers? 40, 42, 44

15. If the sum of four consecutive integers is 86, what are the four integers? 20, 21, 22, 23

16. The sum of four consecutive integers is 62. What are the four integers? 14, 15, 16, 17

17. 4 times an integer is 9 more than 3 times the next consecutive integer. What are the two integers? 12, 13

18. 4 times an integer is 30 less than 5 times the next consecutive even integer. Find the two integers. 20, 22

19. In an election, the winning candidate had 160 more votes than the loser. If the total number of votes cast was 3260, how many votes did each candidate receive? 1710, 1550

20. Jody earns $140 more per month than Frank. If their monthly salaries total $2760, what amount does each earn? Jody: $1450 Frank: $1310

21. A washer-dryer combination costs \$650. If the washer costs \$70 more than the dryer, what does each appliance cost? Washer: \$360 Dryer: \$290

22. Morgan has a board that is 98 in long. He wishes to cut the board into two pieces so that one piece will be 10 in longer than the other. What should be the length of each piece? 44 in, 54 in

23. Ken is 1 year less than twice as old as his sister. If the sum of their ages is 14 years, how old is Ken? 9

24. Diane is twice as old as her brother Dan. If the sum of their ages is 27 years, how old are Diane and her brother? Diane: 18 Dan: 9

25. José is 3 years less than 4 times as old as his daughter. If the sum of their ages is 37 years, how old is José? 29

26. Mrs. Jackson is 2 years more than 3 times as old as her son. If the difference between their ages is 22 years, how old is Mrs. Jackson? 32

27. On her vacation in Europe, Jovita's expenses for food and lodging were \$60 less than twice as much as her airfare. If she spent \$2400 in all, what was her airfare? \$820

28. Rachel earns \$6000 less than twice as much as Tom. If their two incomes total \$30,000, how much does each earn? Rachel: \$18,000 Tom: \$12,000

29. There are 99 students registered in three sections of algebra. There are twice as many students in the 10 o'clock section as in the 8 o'clock section and 7 more students at 12 o'clock than at 8 o'clock. How many students are in each section?
8:00: 23; 10:00: 46; 12:00: 30

30. The Randolphs used 12 more gallons (gal) of fuel oil in October than in September and twice as much oil in November as in September. If they used 132 gal for the 3 months, how much was used during each month? Sept.: 30 gal, Oct.: 42 gal, Nov.: 60 gal.

31. The length of a rectangle is 5 cm more than its width. The perimeter is 98 cm. What are the dimensions? (*Hint:* Remember to draw a sketch whenever geometric figures are involved in a word problem.) 22 cm, 27 cm

32. The length of a rectangle is 3 times its width. If the perimeter of the rectangle is 48 in, find the length and width of the rectangle. 6 in, 18 in

33. The length of a rectangle is 2 ft more than 3 times its width. If the perimeter is 68 ft, what are the length and width of the rectangle? 26 ft, 8 ft

34. The length of a rectangle is 3 cm more than twice its width. What are the dimensions of the rectangle if its perimeter is 48 cm? 7 cm, 17 cm

35. One side of a triangle is 4 meters (m) longer than the shortest side. The third side is twice the length of the shortest side. If the perimeter of the triangle is 44 m, find the lengths of the three sides of the triangle. 10 m, 14 m, 20 m

36. The equal legs of an isosceles triangle are each 4 ft more than twice the length of the base. If the perimeter of the triangle is 68 ft, find the lengths of the three sides of the triangle. 12 ft, 28 ft, 28 ft

Answers

1. 13 **3.** 18 **5.** 35, 36 **7.** 20, 21, 22 **9.** 32, 34 **11.** 25, 27 **13.** 33, 35, 37 **15.** 20, 21, 22, 23 **17.** 12, 13 **19.** 1710 votes, 1550 votes **21.** \$360, \$290 **23.** 9 **25.** 29 **27.** \$820 **29.** 23, 46, 30 **31.** 22 cm, 27 cm **33.** 26 ft, 8 ft **35.** 10 m, 14 m, 20 m

Summary

Algebraic Equations [3.1] to [3.4]

Equation A statement that two expressions are equal.

$$\underbrace{2x - 7}_{\text{Left side}} \overset{}{=} \underbrace{x + 3}_{\text{Right side}}$$

Left side, Equals sign, Right side

Solution A value for the variable that will make the equation a true statement.

4 is a solution for the equation

$3x - 5 = 7$

because

$3 \cdot 4 - 5 = 7$

$12 - 5 = 7$

$7 = 7$ (true)

Equivalent Equations Equations that have exactly the same solutions.

Writing Equivalent Equations There are two basic properties that will yield equivalent equations.

1. If $a = b$, then $a + c = b + c$.
   Adding (or subtracting) the same quantity on each side of an equation gives an equivalent equation.
2. If $a = b$, then $ac = bc$, $c \neq 0$.
   Multiplying (or dividing) both sides of an equation by the same number gives an equivalent equation.

$4x - 3 = 5$

$4x = 8$

$x = 2$

are all equivalent equations that have same solution, 2.

$$\begin{array}{rcr} x - 7 & = & 13 \\ +7 & & +7 \\ \hline x & = & 20 \end{array}$$

$x - 7 = 13$ and $x = 20$ are equivalent.

Solving Linear Equations We say that an equation is "solved" when we have an equivalent equation of the form

$$x = \square$$

Some number

$5x = 20$

$\frac{5x}{5} = \frac{20}{5}$

$x = 4$

$5x = 20$ and $x = 4$ are equivalent equations.

The steps of solving an equation are as follows:

1. Use the distributive property to remove any grouping symbols that appear. Then simplify by combining any like terms on each side of the equation.
2. Add or subtract the same term on both sides of the equation until the term involving the variable is on one side and a number is on the other.
3. Multiply or divide both sides of the equation by the same nonzero number so that the variable is alone on one side of the equation.
4. Check the solution in the original equation.

Solve:

$3(x - 2) + 4x = 3x + 14$

$3x - 6 + 4x = 3x + 14$

$$\begin{array}{rcr} 7x - 6 & = & 3x + 14 \\ +6 & & +6 \\ \hline 7x & = & 3x + 20 \\ -3x & & -3x \\ \hline 4x & = & 20 \end{array}$$

$\frac{4x}{4} = \frac{20}{4}$

$x = 5$

To check:

$7 \cdot 5 - 6 \stackrel{?}{=} 3 \cdot 5 + 14$

$35 - 6 \stackrel{?}{=} 15 + 14$

$29 = 29$ (true)

Literal Equations [3.5]

Literal Equation An equation that involves more than one letter or variable.

Solving Literal Equations

1. If necessary, multiply both sides of the equation by the same term to clear of fractions.
2. Add or subtract the same term on both sides of the equation so that all terms involving the variable you are solving for are on one side and all other terms are on the other side.
3. Divide both sides by any numbers or letters multiplying the variable that you are solving for.

Solve for b:

$a = \frac{2b + c}{3}$

$3 \cdot a = \left(\frac{2b + 3}{3}\right)3$

$$\begin{array}{rcl} 3a & = & 2b + c \\ -c & = & -c \\ \hline 3a - c & = & 2b \end{array}$$

$\frac{3a - c}{2} = b$

Inequalities [3.6] to [3.7]

Inequality A statement that one quantity is less than (or greater than) another. Two symbols are used:

$a < b$ — a is less than b

$a > b$ — a is greater than b

Graphing Inequalities To graph $x < a$,

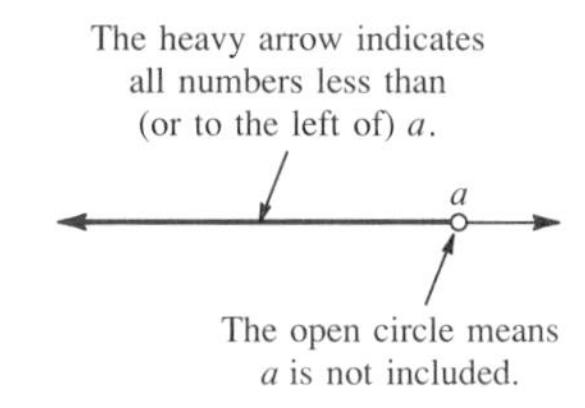

The heavy arrow indicates
all numbers less than
(or to the left of) a.
a
The open circle means
a is not included.

Two other symbols are $\leq$ and $\geq$.

To graph $x \geq b$,

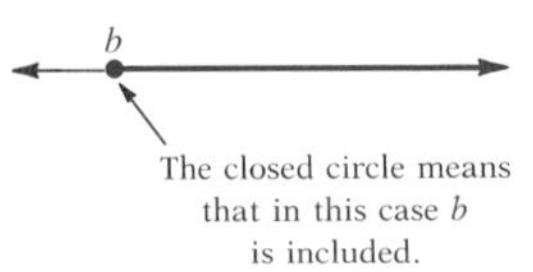

b
The closed circle means
that in this case b
is included.

Solving Inequalities An inequality is "solved" when it is in the form $x < \square$ or $x > \square$.

Proceed as in solving equations by using the following properties.

1. If $a < b$, then $a + c < b + c$.
   Adding (or subtracting) the same quantity to both sides of an inequality gives an equivalent inequality.
2. If $a < b$, then $ac < bc$, when $c > 0$; and $ac > bc$, when $c < 0$.
   Multiplying both sides of an inequality by the same *positive number* gives an equivalent inequality. When both sides of an inequality are multiplied by the same *negative number, reverse the sense* of the inequality to give an equivalent inequality.

$$\begin{array}{rcr} 2x - 3 & > & 5x + 6 \\ +3 & & +3 \\ \hline 2x & > & 5x + 9 \\ -5x & & -5x \\ \hline -3x & > & 9 \end{array}$$

$$\frac{-3x}{-3} < \frac{9}{-3}$$

$$x < -3$$

$-3 \qquad 0$

Applying Equations [3.8]

Using Equations to Solve Word Problems Use the following steps.

1. Read the problem carefully. Then reread it to decide what you are asked to find.
2. Choose a letter to represent one of the unknowns in the problem. Then represent all of the unknowns of the problem, with an expression using that same letter.
3. Translate the problem to the language of algebra to form an equation.
4. Solve the equation and answer the question of the original problem.
5. Verify your solution by returning to the original problem.

Summary Exercises Chapter 3

This summary exercise set is provided to give you practice with each of the objectives of the chapter. Each exercise is keyed to the appropriate chapter section. The answers are provided in the instructor's manual. Your instructor will give you guidelines on how to best use these exercises in your instructional setting.

[3.1] Tell whether the number shown in parentheses is a solution for the given equation.

1. $7x + 2 = 16$ (2) Yes

2. $5x - 8 = 3x + 2$ (4) No

3. $7x - 2 = 2x + 8$ (2) Yes

4. $4x + 3 = 2x - 11$ (−7) Yes

5. $x + 5 + 3x = 2 + x + 23$ (6) No

6. $\frac{2}{3}x - 2 = 10$ (21) No

[3.2] Solve the following equations and check your results.

7. $x + 5 = 7$ 2

8. $x - 9 = 3$ 12

9. $5x = 4x - 5$ −5

10. $3x - 9 = 2x$ 9

11. $5x - 3 = 4x + 2$ 5

12. $9x + 2 = 8x - 7$ −9

13. $7x - 5 = 6x - 4$ 1

14. $3 + 4x - 1 = x - 7 + 2x$ −9

15. $4(2x + 3) = 7x + 5$ −7

16. $5(5x - 3) = 6(4x + 1)$ 21

[3.3] Solve the following equations and check your results.

17. $5x = 35$ 7

18. $7x = -28$ −4

19. $-6x = 24$ −4

20. $-9x = -63$ 7

21. $\frac{x}{4} = 8$ 32

22. $-\frac{x}{5} = -3$ 15

23. $\frac{2}{3}x = 18$ 27

24. $\frac{3}{4}x = 24$ 32

[3.4] Solve the following equations and check your results.

25. $5x - 3 = 12$ 3

26. $4x + 3 = -13$ −4

27. $7x + 8 = 3x$ −2

28. $3 - 5x = -17$ 4

29. $3x - 7 = x$ $\frac{7}{2}$

30. $2 - 4x = 5$ $-\frac{3}{4}$

31. $\frac{x}{3} - 5 = 1$ 18

32. $\frac{3}{4}x - 2 = 7$ 12

33. $6x - 5 = 3x + 13$ 6

34. $3x + 7 = x - 9$ −8

35. $7x + 4 = 2x + 6$ $\frac{2}{5}$

36. $9x - 8 = 7x - 3$ $\frac{5}{2}$

37. $2x + 7 = 4x - 5$ 6

38. $3x - 15 = 7x - 10$ $-\frac{5}{4}$

39. $3x - 2 + 5x = 7 + 2x + 21$ 5

40. $8x + 3 - 2x + 5 = 3 - 4x$ $-\frac{1}{2}$

41. $5(3x - 1) - 6x = 3x - 2$ $\frac{1}{2}$

42. $5x + 2(3x - 4) = 14x + 7$ −5

43. $\frac{10}{3}x - 5 = \frac{4}{3}x + 7$ 6

44. $\frac{11}{4}x - 15 = 5 - \frac{5}{4}x$ 5

45. $3.7x + 8 = 1.7x + 16$ 4

46. $5.4x - 3 = 8.4x + 9$ −4

[3.5] Solve for the indicated variable.

47. $V = LWH$ (for L) $\frac{V}{WH}$

48. $P = 2L + 2W$ (for L) $\frac{P - 2W}{2}$

49. $ax + by = c$ (for y) $\frac{c - ax}{b}$

50. $A = \frac{1}{2}bh$ (for h) $\frac{2A}{b}$

51. $A = P + Prt$ (for t) $\frac{A - P}{Pr}$

52. $m = \frac{n - p}{q}$ (for n) $mq + p$

[3.6] Graph the solution sets.

53. $x > 5$

54. $x < -3$

55. $x \le -4$

56. $x \ge 9$

57. $x \ge -6$

58. $x < 0$

[3.7] Solve and graph the solution sets for the following inequalities.

59. $x - 4 \le 7$ $x \le 11$

60. $x + 3 > -2$ $x > -5$

61. $5x > 4x - 3$ $x > -3$

62. $4x \ge -12$ $x \ge -3$

63. $-12x < 36$ $x > -3$

64. $-\frac{x}{5} \ge 3$ $x \le -15$

65. $2x \le 8x - 3$ $x \ge \frac{1}{2}$

66. $2x + 3 \ge 9$ $x \ge 3$

67. $4 - 3x > 8$ $x < -\frac{4}{3}$

68. $5x - 2 \le 4x + 5$ $x \le 7$

69. $7x + 13 \ge 3x + 19$ $x \ge \frac{3}{2}$

70. $4x - 2 < 7x + 16$ $x > -6$

71. $5(x - 3) < 2x + 12$ $x < 9$

72. $4(x + 3) \ge x + 7$ $x \ge -\frac{5}{3}$

[3.8] Solve the following word problems. Be sure to label the unknowns and to show the equation you use for the solution.

73. The sum of 3 times a number and 7 is 25. What is the number? 6

74. 5 times a number, decreased by 8, is 32. Find the number. 8

75. If the sum of two consecutive integers is 85, find the two integers. 42, 43

76. The sum of three consecutive odd integers is 57. What are the three integers? 17, 19, 21

77. Rafael earns $35 more per week than Andrew. If their weekly salaries total $715, what amount does each earn? Rafael: $375, Andrew: $340

78. Larry is 2 years older than Susan, while Nathan is twice as old as Susan. If the sum of their ages is 30 years, find each of their ages.
Susan: 7 years, Larry: 9 years, Nathan: 14 years

79. The perimeter of a rectangle is 28 in. If the length is 1 in less than twice the width, what are the dimensions of the rectangle? 5 in, 9 in

80. One side of a triangle is 3 cm longer than the shortest side. The third side is twice the length of the shortest side. If the perimeter of the triangle is 35 cm, find the lengths of the three sides of the triangle. 8 cm, 11 cm, 16 cm

Self-Test for Chapter Three

The purpose of this self-test is to help you check your progress and to review for a chapter test in class. Allow yourself about an hour to take the test. When you are done, check your answers in the back of the book. If you missed any problems, be sure to go back and review the appropriate sections in the chapter and the exercises that are provided.

Tell whether the number shown in parentheses is a solution for the given equation.

1. $7x - 3 = 25$ (5) No

2. $8x - 3 = 5x + 9$ (4) Yes

Solve the following equations and check your results.

3. $x - 7 = 4$ 11

4. $7x - 12 = 6x$ 12

5. $9x - 2 = 8x + 5$ 7

Solve the following equations and check your results.

6. $7x = 49$ 7

7. $\frac{1}{4}x = -3$ -12

8. $\frac{4}{5}x = 20$ 25

Solve the following equations and check your results.

9. $7x - 5 = 16$ 3

10. $10 - 3x = -2$ 4

11. $7x - 3 = 4x - 5$ $-\frac{2}{3}$

12. $2x - 7 = 5x + 8$ -5

Solve for the indicated variable.

13. $C = 2\pi r$ (for r) $\frac{C}{2\pi}$

14. $V = \frac{1}{3}Bh$ (for h) $\frac{3V}{B}$

15. $3x + 2y = 6$ (for y) $\frac{6 - 3x}{2}$

Graph the solution sets.

16. $x \geq 9$

17. $x < -3$

Solve and graph the solution sets for the following inequalities.

18. $x - 5 \leq 9$ $\quad x \leq 14$

19. $5 - 3x > 17$ $\quad x < -4$

20. $5x + 13 \geq 2x + 17$ $\quad x \geq \frac{4}{3}$

21. $2x - 3 < 7x + 2$ $\quad x > -1$

Solve the following word problems. Be sure to show the equation you use for the solution.

22. 5 times a number, decreased by 7, is 28. What is the number? 7

23. The sum of three consecutive integers is 66. Find the three integers. 21, 22, 23

24. Jan is twice as old as Steve, while Rick is 5 years older than Jan. If the sum of their ages is 35 years, find each of their ages. Steve, 6; Jan, 12; Rick, 17

25. The perimeter of a rectangle is 62 in. If the length of the rectangle is 1 in more than twice its width, what are the dimensions of the rectangle? 10 in, 21 in

Cumulative Test for Chapters One to Three

This test is provided to help you in the process of reviewing the previous chapters. Answers are provided in the back of the book. If you missed any problems, be sure to go back and review the appropriate chapter sections.

Write, using symbols.

1. 3 times the sum of r and s

$3(r + s)$

2. The quotient when 5 less than x is divided by 3

$\frac{x - 5}{3}$

Identify the property that is illustrated by the following statements.

3. $7 + (5 + 2) = (7 + 5) + 2$

Associative property of addition

4. $2(3 + 5) = 2 \cdot 3 + 2 \cdot 5$

Distributive property

Write in exponential form.

5. $8 \cdot x \cdot x \cdot x \cdot y \cdot y$ $\quad 8x^3y^2$

Simplify the following expressions.

6. $7a^2b - 2a^2b$ $\quad 5a^2b$

7. $10a^2 + 5a + 2a^2 - 2a$ $\quad 12a^2 + 3a$

8. $3m^2n \cdot 5m^3n^2$ $\quad 15m^5n^3$

9. $\frac{25x^3y^4}{5x^2y^3}$ $\quad 5xy$

Evaluate the following expressions.

10. $2 \cdot 3^2 - 8 \cdot 2$

2

11. $5(7 - 3)^2$

80

12. $|12 - 5|$

7

13. $|12| - |5|$

7

Perform the indicated operations.

14. $(-7) + (-9)$ $\quad -16$

15. $\frac{17}{3} + \left(-\frac{5}{3}\right)$ $\quad 4$

16. $(-7)(-9)$ $\quad 63$

17. $(-3.2)(5)$ $\quad -16$

18. $\frac{-90}{5}$ $\quad -18$

19. $\frac{0}{-13}$ $\quad 0$

Evaluate each of the following expressions if $x = -2$, $y = 3$, and $z = 5$.

20. $3x - y$ $\quad -9$

21. $4x^2 - y$ $\quad 13$

22. $\frac{5z - 4x}{2y + z}$ $\quad 3$

Solve the following equations and check your results.

23. $9x - 5 = 8x$ 5

24. $-\frac{3}{4}x = 18$ -24

25. $6x - 8 = 2x - 3$ $\frac{5}{4}$

26. $2x + 3 = 7x + 5$ $-\frac{2}{5}$

27. $\frac{4}{3}x - 6 = 4 - \frac{2}{3}x$ 5

Solve the following equations for the indicated variable.

28. $I = Prt$ (for r) $\frac{I}{Pt}$

29. $A = \frac{1}{2}bh$ (for h) $\frac{2A}{b}$

30. $ax + by = c$ (for y) $\frac{c - ax}{b}$

Solve and graph the solution sets for the following inequalities.

31. $3x - 5 < 4$ $x < 3$

32. $7 - 2x \geq 10$ $x \leq -\frac{3}{2}$

33. $7x - 2 > 4x + 10$ $x > 4$

34. $2x + 5 \leq 8x - 3$ $x \geq \frac{4}{3}$

Solve the following word problems. Be sure to show the equation used for the solution.

35. If 4 times a number, decreased by 7, is 45, find that number. 13

36. The sum of two consecutive integers is 85. What are those two integers? 42, 43

37. If 3 times an integer is 12 more than the next consecutive odd integer, what is that integer? 7

38. Michelle earns \$120 more per week than David. If their weekly salaries total \$720, how much does Michelle earn? \$420

39. The length of a rectangle is 2 cm more than 3 times its width. If the perimeter of the rectangle is 44 cm, what are the dimensions of the rectangle? 5 cm, 17 cm

40. One side of a triangle is 5 in longer than the shortest side. The third side is twice the length of the shortest side. If the triangle perimeter is 37 in, find the length of each leg. 8 in, 13 in, 16 in

Chapter Four

Polynomials

4.1 Polynomials—An Introduction

OBJECTIVES

1. To identify polynomials, monomials, binomials, and trinomials
2. To find the degree of a polynomial
3. To write polynomials in descending form

Our work in this chapter deals with the most common kind of algebraic expression, a *polynomial.* To define a polynomial let's recall our earlier definition of the word *term.*

> A *term* is a number, or the product of a number and one or more variables, raised to a power.

A polynomial consists of specific kinds of terms in which the only allowable exponents are the whole numbers, 0, 1, 2, 3, . . . and so on. These terms are connected by addition or subtraction signs. In each term of a polynomial, the number is called the *numerical coefficient,* or more simply the *coefficient* of that term.

Example 1

(*a*) $5x + 3$ is a polynomial.

The terms are $5x$ and 3. The coefficients are 5 and 3.

Note: Each sign (+ or −) is attached to the term that follows that sign.

(*b*) $3x^2 - 2x + 5$ is also a polynomial. Its terms are $3x^2$, $-2x$, and 5. The coefficients are 3, −2, and 5.

(*c*) $5x^3 + 2 - \dfrac{3}{x}$ is *not* a polynomial because of the division by x in the third term.

CHECK YOURSELF 1

Which of the following are polynomials?

1. $5x^2$ **2.** $3y^3 - 2y + \dfrac{5}{y}$ **3.** $4x^2 - 2x + 3$

Certain polynomials occur very often and are given special names according to the number of terms that they have.

The prefix *mono* means 1.

A polynomial with one term is called a *monomial.*

The prefix *bi* means 2.

A polynomial with two terms is called a *binomial.*

The prefix *tri* means 3.

A polynomial with three terms is called a *trinomial.*

Example 1

Don't let this bother you! A term can be made up of several factors, but there is just one term.

(*a*) $3x^2y$ is a monomial. It has one term.

(*b*) $2x^3 + 5x$ is a binomial. It has two terms, $2x^3$ and $5x$.

(*c*) $5x^3 - 4x + 3$ is a trinomial. Its three terms are $5x^2$, $-4x$, and 3.

Note: There are no special names for polynomials with more than three terms.

CHECK YOURSELF 2

Classify as monomials, binomials, or trinomials.

1. $5x^4 - 2x^3$ **2.** $4x^7$ **3.** $2x^2 + 5x - 3$

Remember, in a polynomial the allowable exponents are the whole numbers 0, 1, 2, 3, and so on. The degree will be a whole number.

It is also useful to classify polynomials by their *degree.* The degree of a polynomial in one variable is the highest power appearing in any one term.

Example 3

The highest power

(*a*) $5x^3 - 3x^2 + 4x$ has degree 3.

The highest power

(*b*) $4x - 5x^4 + 3x^3 + 2$ has degree 4.

(*c*) $8x$ has degree 1.

(Because $8x = 8x^1$)

$x^0 = 1$ because of a definition in algebra. We will discuss this in Chapter 9.

(*d*) 7 has degree 0.

(Because $7 = 7 \cdot 1 = 7x^0$)

Note: Polynomials can have more than one variable, such as $4x^2y^3 + 5xy^2$. The degree is then the sum of the powers of the highest-degree term (here 2 + 3, or 5). In general, we will be working with polynomials in a single variable, such as x.

CHECK YOURSELF 3

Find the degree of each polynomial.

1. $6x^5 - 3x^3 - 2$ **2.** $5x$ **3.** $3x^3 + 2x^6 - 1$ **4.** 9

As you start working with operations on polynomials later in this chapter, it will be much easier if you get used to writing polynomials in *descending exponent* or *power form.* This simply means that the term with the highest exponent is written first, then the term with the next highest exponent, and so on.

Example 4

The exponents get smaller from left to right.

(*a*) $5x^7 - 3x^4 + 2x^2$ is in descending form.

(*b*) $4x^4 + 5x^6 - 3x^5$ is *not* in descending form. The polynomial should be written as

$$5x^6 - 3x^5 + 4x^4$$

Notice that the degree of the polynomial is the power of the *first* or *leading* term once the polynomial is arranged in descending form.

CHECK YOURSELF 4

Write the following polynomials in descending form.

1. $5x^4 - 4x^5 + 7$ **2.** $4x^3 + 9x^4 + 6x^8$

A polynomial may represent different numbers depending on the value given to the variable. Our final example illustrates.

Example 5

Given the polynomial

$$3x^3 - 2x^2 - 4x + 1$$

(*a*) Find the value of the polynomial when $x = 2$.

Substituting 2 for x, we have

Again note how the rules for the order of operations are applied. See Section 1.6 for a review if you would like.

$$\begin{aligned}&3(2)^3 - 2(2)^2 - 4(2) + 1\\ &= 3(8) - 2(4) - 4(2) + 1\\ &= 24 - 8 - 8 + 1\\ &= 9\end{aligned}$$

(*b*) Find the value of the polynomial when $x = -2$.

Now we substitute -2 for x.

Be particularly careful when dealing with powers of negative numbers!

$$\begin{aligned}&3(-2)^3 - 2(-2)^2 - 4(-2) + 1\\ &= 3(-8) - 2(4) - 4(-2) + 1\\ &= -24 - 8 + 8 + 1\\ &= -23\end{aligned}$$

CHECK YOURSELF 5

For the polynomial

$4x^3 - 3x^2 + 2x - 1$

find its value when

1. $x = 3$ **2.** $x = -3$

CHECK YOURSELF ANSWERS

1. (1) and (3) are polynomials.
2. (1) Binomial; (2) monomial; (3) trinomial.
3. (1) 5; (2) 1; (3) 6; (4) 0.
4. (1) $-4x^5 + 5x^4 + 7$; (2) $6x^8 + 9x^4 + 4x^3$.
5. (1) 86; (2) -142.

4.1 Exercises

Which of the following expressions are polynomials?

1. $5x^2$

Polynomial

2. $4x^2 - \dfrac{2}{x}$

Not a polynomial

3. $5x^2y - 2xy^2$

Polynomial

4. 3

Polynomial

5. -7

Polynomial

6. $4x^3 + x$

Polynomial

7. $\dfrac{3 + x}{x^2}$

Not a polynomial

8. $5a^2 - 2a + 7$

Polynomial

For each of the following polynomials, list the terms and the coefficients.

9. $2x^2 - 3x$ $2x^2, -3x; 2, -3$

10. $5x^3 + x$ $5x^3, x; 5, 1$

11. $4x^3 - 3x + 2$ $4x^3, -3x, 2; 4, -3, 2$

12. $7x^2$ $7x^2; 7$

Classify the following as monomials, binomials, or trinomials where possible.

13. $4x^2 - 2x$ Binomial

14. $3x^5$ Monomial

15. $6y^2 + 3y + 3$ Trinomial

16. $x^2 + 2xy + y^2$
Trinomial

17. $2x^4 - 3x^2 + 5x - 2$
Not classified

18. $x^4 + \frac{5}{x} + 7$
Not a polynomial

19. $5y^7$
Monomial

20. $3x^3 - 4x^2 + 5x - 3$
Not classified

21. $x^5 - \frac{3}{x^2}$
Not a polynomial

22. $4x^2 - 9$ Binomial

Arrange in descending form if necessary and give the degree of each polynomial.

23. $4x^5 - 3x^2$
$4x^5 - 3x^2; 5$

24. $5x^2 + 3x^3 + 4$
$3x^3 + 5x^2 + 4; 3$

25. $7x^7 - 5x^9 + 4x^3$
$-5x^9 + 7x^7 + 4x^3; 9$

26. $2 + x$
$x + 2; 1$

27. $4x$
$4x; 1$

28. $x^{17} - 3x^4$
$x^{17} - 3x^4; 17$

29. $5x^2 - 3x^5 + x^6 - 7$
$x^6 - 3x^5 + 5x^2 - 7; 6$

30. 5
$5; 0$

Find the values of the following polynomials for the given values of the variable.

31. $5x + 2$, $x = 1$ and $x = -1$ 7, −3

32. $3x - 3$, $x = 2$ and $x = -2$ 3, −9

33. $x^2 - x$, $x = 2$ and $x = -2$ 2, 6

34. $2x^2 + 5$, $x = 3$ and $x = -3$ 23, 23

35. $3x^2 + 4x - 2$, $x = 4$ and $x = -4$
62, 30

36. $2x^2 - 5x + 1$, $x = 2$ and $x = -2$
−1, 19

37. $-x^2 - 2x + 3$, $x = 1$ and $x = -3$
0, 0

38. $-x^2 - 5x - 6$, $x = -3$ and $x = -2$
0, 0

Indicate whether the following statements are always, sometimes, or never true.

39. A monomial is a polynomial.
Always

40. A binomial is a trinomial.
Never

41. The degree of a trinomial is 3.
Sometimes

42. A trinomial has three terms.
Always

43. A polynomial has four or more terms.
Sometimes

44. A binomial must have two coefficients.
Always

45. If x equals 0, the value of a polynomial in x equals 0.
Sometimes

46. The coefficient of the leading term in a polynomial is the largest coefficient of the polynomial.
Sometimes

Going Beyond

Capital letters such as P or Q are often used to name polynomials. For example, we might write $P(x) = 3x^3 - 5x^2 + 2$ where $P(x)$ is read "P of x." The notation permits a convenient shorthand. We write $P(2)$, read "P of 2," to indicate the value of the polynomial when $x = 2$. Here

$$\begin{aligned} P(2) &= 3(2)^3 - 5(2)^2 + 2 \\ &= 3 \cdot 8 - 5 \cdot 4 + 2 \\ &= 6 \end{aligned}$$

Use the information above in the following problems.

If $P(x) = x^3 - 2x^2 + 5$ and $Q(x) = 2x^2 + 3$, find

47. $P(1)$ 4

48. $P(-1)$ 2

49. $Q(2)$ 11

50. $Q(-2)$ 11

51. $P(3)$ 14

52. $Q(-3)$ 21

53. $P(0)$ 5

54. $Q(0)$ 3

55. $P(2) + Q(-1)$ 10

56. $P(-2) + Q(3)$ 10

Skillscan (Section 1.4)

Combine like terms where possible.

a. $8m + 7m$ $15m$

b. $9x - 5x$ $4x$

c. $9m^2 - 8m$ $9m^2 - 8m$

d. $8x^2 - 7x^2$ x^2

e. $5c^3 + 15c^3$ $20c^3$

f. $9s^3 + 8s^3$ $17s^3$

g. $8c^2 - 6c + 2c^2$
$10c^2 - 6c$

h. $8r^3 - 7r^2 + 5r^3$
$13r^3 - 7r^2$

Answers

1. Polynomial **3.** Polynomial **5.** Polynomial **7.** Not a polynomial **9.** $2x^2, -3x; 2, -3$ **11.** $4x^3, -3x^2, 2; 4, -3, 2$ **13.** Binomial **15.** Trinomial **17.** Not possible **19.** Monomial **21.** Not possible **23.** $4x^5 - 3x^2; 5$ **25.** $-5x^9 + 7x^7 + 4x^3; 9$ **27.** $4x; 1$ **29.** $x^6 - 3x^5 + 5x^2 - 7; 6$ **31.** 7, −3 **33.** 2, 6 **35.** 62, 30 **37.** 0, 0 **39.** Always **41.** Sometimes **43.** Sometimes **45.** Sometimes **47.** 4 **49.** 11 **51.** 14 **53.** 5 **55.** 10 **a.** $15m$ **b.** $4x$ **c.** $9m^2 - 8m$ **d.** x^2 **e.** $20c^3$ **f.** $17s^3$ **g.** $10c^2 - 6c$ **h.** $13r^3 - 7r^2$

4.2 Adding and Subtracting Polynomials

OBJECTIVES

1. To add polynomials
2. To subtract polynomials

Addition is always a matter of combining like quantities (two apples plus three apples, four books plus five books, and so on). If you keep that basic idea in mind, adding polynomials will be easy. It is just a matter of combining like terms. Suppose that you want to add

$$5x^2 + 3x + 4 \qquad \text{and} \qquad 4x^2 + 5x - 6$$

Parentheses are sometimes used to indicate addition, so for the sum of the polynomials, we can write

The + sign between the parentheses indicates the addition.

$$(5x^2 + 3x + 4) + (4x^2 + 5x - 6)$$

Now what about the parentheses? You can use the following rule.

> **REMOVING SIGNS OF GROUPING**
>
> If a plus sign (+) or no sign at all appears in front of parentheses, just remove the parentheses. No other changes are necessary.

Now let's return to the addition.

Just remove the parentheses. No other changes are necessary.

$$(5x^2 + 3x + 4) + (4x^2 + 5x - 6)$$
$$= 5x^2 + 3x + 4 + 4x^2 + 5x - 6$$

Like terms Like terms Like terms

Collect the like terms.

Note the use of the associative and commutative properties in reordering and regrouping.

$$= (5x^2 + 4x^2) + (3x + 5x) + (4 - 6)$$

Combine the like terms for the result:

$= 9x^2 + 8x - 2$

As should be clear, much of this work can be done mentally. You can then write the sum directly by locating the like terms and combining. The following examples illustrate.

Example 1

To add

$$(3x - 5) + (2x + 3)$$
$$= 3x - 5 + 2x + 3$$

Like terms Like terms

$$= 5x - 2$$

CHECK YOURSELF 1

Add $6x^2 + 2x$ and $4x^2 - 7x$.

Example 2

To add

$$(4a^2 - 7a + 5) + (3a^2 + 3a - 4)$$
$$= 4a^2 - 7a + 5 + 3a^2 + 3a - 4$$

Like terms Like terms Like terms

$$= 7a^2 - 4a + 1$$

CHECK YOURSELF 2

Add $5y^2 - 3y + 7$ and $3y^2 - 5y - 7$.

Example 3

To add

$$(2x^2 + 7x) + (4x - 6)$$
$$= 2x^2 + \underbrace{7x + 4x} - 6$$

These are the only like terms; $2x^2$ and -6 cannot be combined.

$$= 2x^2 + 11x - 6$$

CHECK YOURSELF 3

Add $5m^2 + 8$ and $8m^2 - 3m$.

As we mentioned in Section 4.1, writing polynomials in descending form usually makes the work easier. Look at the following example.

Example 4

To add

$3x - 2x^2 + 7$ and $5 + 4x^2 - 3x$

write the polynomials in descending form, then add.

$$(-2x^2 + 3x + 7) + (4x^2 - 3x + 5)$$
$$= 2x^2 + 12$$

CHECK YOURSELF 4

Add $8 - 5x^2 + 4x$ and $7x - 8 + 8x^2$.

Subtracting polynomials requires another rule for removing signs of grouping.

> **REMOVING SIGNS OF GROUPING**
>
> If a minus sign (−) appears in front of a set of parentheses, the parentheses can be removed by changing the sign of each term inside the parentheses.

The use of this rule is illustrated in our next example.

Example 5

Note: This uses the distributive property, since

$$-(2x + 3y) = (-1)(2x + 3y)$$
$$= -2x - 3y$$

(a) $-(2x + 3y) = -2x - 3y$

Change each sign to remove the parentheses.

(b) $a - (3b + c) = a - 3b - c$

Sign changes.

(c) $m - (5n - 3p) = m - 5n + 3p$

Sign changes.

(d) $2x - (-3y + z) = 2x + \underbrace{3y - z}_{\text{Sign changes.}}$

CHECK YOURSELF 5

Remove the parentheses.

1. $-(3m + 5n)$
2. $-(5w - 7z)$
3. $3r - (2s - 5t)$
4. $5a - (-3b - 2c)$

Subtracting polynomials is now a matter of using the previous rule to remove the parentheses and then combining the like terms. Consider the following example.

Example 6

(a) To subtract $5x - 3$ from $8x + 2$, write

Note: The expression after "from" is written first in the problem.

$$(8x + 2) - (5x - 3)$$
$$= 8x + 2 \underbrace{- 5x + 3}_{\text{Sign changes.}}$$
$$= 3x + 5$$

(b) To subtract $4x^2 - 8x + 3$ from $8x^2 + 5x - 3$, write

$$(8x^2 + 5x - 3) - (4x^2 - 8x + 3)$$
$$= 8x^2 + 5x - 3 \underbrace{- 4x^2 + 8x - 3}_{\text{Sign changes.}}$$
$$= 4x^2 + 13x - 6$$

CHECK YOURSELF 6

Subtract.

1. $7x + 3$ from $10x - 7$
2. $5x^2 - 3x + 2$ from $8x^2 - 3x - 6$

Again, writing all polynomials in descending form will make locating and combining like terms much easier. Look at the following example.

Example 7

(*a*) To subtract $4x^2 - 3x^3 + 5x$ from $8x^3 - 7x + 2x^2$, write

$$(8x^3 + 2x^2 - 7x) - (-3x^3 + 4x^2 + 5x)$$
$$= 8x^3 + 2x^2 - 7x + \underbrace{3x^3 - 4x^2 - 5x}_{\text{Sign changes.}}$$
$$= 11x^3 - 2x^2 - 12x$$

(*b*) To subtract $8x - 5$ from $-5x + 3x^2$, write

$$(3x^2 - 5x) - (8x - 5)$$
$$= 3x^2 \underbrace{- 5x - 8x}_{} + 5$$

Only the like terms can be combined.

$$= 3x^2 - 13x + 5$$

CHECK YOURSELF 7

Subtract.

1. $7x - 3x^2 + 5$ from $5 - 3x + 4x^2$
2. $3a - 2$ from $5a + 4a^2$

If you think back to addition and subtraction in arithmetic, you'll remember that the work was arranged vertically. That is, the numbers being added or subtracted were placed under one another so that each column represented the same place value. This meant that in adding or subtracting columns you were always dealing with "like quantities."

It is also possible to use a vertical method for adding or subtracting polynomials. Arrange the polynomials one under another, so that each column contains like terms. Then add or subtract in each column.

Example 8

Add $2x^2 - 5x$, $3x^2 + 2$, and $6x - 3$.

Write

Like terms

$$\begin{array}{rrr} 2x^2 & - 5x & \\ 3x^2 & & + 2 \\ & 6x & - 3 \\ \hline 5x^2 & + \ x & - 1 \end{array}$$

CHECK YOURSELF 8

Add $3x^2 + 5$, $x^2 - 4x$, and $6x + 7$.

Example 9

(*a*) Subtract $5x - 3$ from $8x - 7$.

Write

$$\begin{array}{rr} & 8x - 7 \\ (-) & \underline{5x - 3} \\ & 3x - 4 \end{array}$$

To subtract, mentally change each sign. Think of $-(5x - 3)$ or $-5x + 3$, than add.

(*b*) Subtract $5x^2 - 3x + 4$ from $8x^2 + 5x - 3$.

Write

$$\begin{array}{rr} & 8x^2 + 5x - 3 \\ (-) & \underline{5x^2 - 3x + 4} \\ & 3x^2 + 8x - 7 \end{array}$$

To subtract, mentally change each sign. Think of $-(5x^2 - 3x + 4)$ or $-5x^2 + 3x - 4$. Then add.

Subtracting in this form takes some practice. Take time to study the method carefully. You'll be using it in long division in Section 4.4.

CHECK YOURSELF 9

Subtract, using the vertical method.

1. $4x^2 - 3x$ from $8x^2 + 2x$ **2.** $8x^2 + 4x - 3$ from $9x^2 - 5x + 7$

CHECK YOURSELF ANSWERS

1. $10x^2 - 5x$. **2.** $8y^2 - 8y$. **3.** $13m^2 - 3m + 8$. **4.** $3x^2 + 11x$.
5. (1) $-3m - 5n$; (2) $-5w + 7z$; (3) $3r - 2s + 5t$; (4) $5a + 3b + 2c$.
6. (1) $3x - 10$; (2) $3x^2 - 8$. **7.** (1) $7x^2 - 10x$; (2) $4a^2 + 2a + 2$.
8. $4x^2 + 2x + 12$. **9.** (1) $4x^2 + 5x$; (2) $x^2 - 9x + 10$.

4.2 Exercises

Add.

1. $5a - 3$ and $4a + 7$ $9a + 4$

2. $7x + 8$ and $5x - 9$ $12x - 1$

3. $9b^2 - 12b$ and $4b^2 - 6b$ $13b^2 - 18b$

4. $3m^2 + 2m$ and $5m^2 - 7m$ $8m^2 - 5m$

5. $3x^2 - 2x$ and $-5x^2 + 2x$ $-2x^2$

6. $3p^2 + 5p$ and $-7p^2 - 5p$ $-4p^2$

7. $2x^2 + 5x - 3$ and $3x^2 - 7x + 4$
$5x^2 - 2x + 1$

8. $4d^2 - 8d + 7$ and $5d^2 - 6d - 9$
$9d^2 - 14d - 2$

9. $2b^2 + 8$ and $5b + 8$ $2b^2 + 5b + 16$

10. $4x - 3$ and $3x^2 - 9x$ $3x^2 - 5x - 3$

11. $8y^3 - 5y^2$ and $5y^2 - 2y$ $8y^3 - 2y$

12. $9x^4 - 2x^2$ and $2x^2 + 3$ $9x^4 + 3$

13. $a^2 - 3a^3$ and $2a^3 + 3a^2$ $-a^3 + 4a^2$

14. $7m^3 - 3m$ and $-5m - 2m^3$ $5m^3 - 8m$

15. $x^2 - 5 + 3x$ and $8 - 4x - 3x^2$
$-2x^2 - x + 3$

16. $b^3 - 5b + b^2$ and $4b^2 + 2b - 3b^3$
$-2b^3 + 5b^2 - 3b$

Remove the parentheses in each of the following expressions and simplify where possible.

17. $-(2a + 3b)$
$-2a - 3b$

18. $-(7x - 4y)$
$-7x + 4y$

19. $5a - (2b - 3c)$
$5a - 2b + 3c$

20. $7x - (4y + 3z)$
$7x - 4y - 3z$

21. $9r - (3r + 5s)$
$6r - 5s$

22. $10m - (3m - 2n)$
$7m + 2n$

23. $5p - (-3p + 2q)$
$8p - 2q$

24. $8d - (-7c - 2d)$
$7c + 10d$

Subtract.

25. $x + 4$ from $2x - 3$ $x - 7$

26. $x - 2$ from $3x + 5$ $2x + 7$

27. $2m^2 - m$ from $3m^2 - 4m$ $m^2 - 3m$

28. $5a^2 + 2a$ from $7a^2 - 3a$ $2a^2 - 5a$

29. $4y^2 + 3y$ from $2y^2 + 3y$ $-2y^2$

30. $7n^2 - 3n$ from $5n^2 - 3n$ $-2n^2$

31. $x^2 - 4x - 3$ from $3x^2 - 5x - 2$
$2x^2 - x + 1$

32. $3x^2 - 2x + 4$ from $5x^2 - 8x - 3$
$2x^2 - 6x - 7$

33. $3a + 7$ from $8a^2 - 9a$ $8a^2 - 12a - 7$

34. $3x^3 + x^2$ from $4x^3 - 5x$ $x^3 - x^2 - 5x$

35. $4b^2 - 3b$ from $5b - 2b^2$ $-6b^2 + 8b$

36. $7y - 3y^2$ from $3y^2 - 2y$ $6y^2 - 9y$

37. $x^2 - 5 - 8x$ from $3x^2 - 8x + 7$
$2x^2 + 12$

38. $4x - 2x^2 + 4x^3$ from $4x^3 + x - 3x^2$
$-x^2 - 3x$

Perform the indicated operations.

39. Subtract $3b + 2$ from the sum of $4b - 2$ and $5b + 3$. $6b - 1$

40. Subtract $5m - 7$ from the sum of $2m - 8$ and $9m - 2$. $6m - 3$

41. Subtract $3x^2 + 2x - 1$ from the sum of $x^2 + 5x - 2$ and $2x^2 + 7x - 8$. $10x - 9$

42. Subtract $4x^2 - 5x - 3$ from the sum of $x^2 - 3x - 7$ and $2x^2 - 2x + 9$. $-x^2 + 5$

43. Subtract $2x^2 - 3x$ from the sum of $4x^2 - 5$ and $2x - 7$. $2x^2 + 5x - 12$

44. Subtract $5a^2 - 3a$ from the sum of $3a - 3$ and $5a^2 + 5$. $6a + 2$

45. Subtract the sum of $3y^2 - 3y$ and $5y^2 + 3y$ from $2y^2 - 8y$. $-6y^2 - 8y$

46. Subtract the sum of $7r^3 - 4r^2$ and $-3r^3 + 4r^2$ from $2r^3 + 3r^2$. $-2r^3 + 3r^2$

Add, using the vertical method.

47. $w^2 + 5$, $2w - 3$, and $5w^2 - 4w$ $6w^2 - 2w + 2$

48. $x^2 - 3x - 2$, $5x - 2$, and $4x^2 + 7$ $5x^2 + 2x + 3$

49. $2x^2 + x - 3$, $3x^2 - x - 5$, and $4x^2 - x + 8$ $9x^2 - x$

50. $5x^2 + 2x - 4$, $x^2 - 2x - 3$, and $2x^2 - 4x - 3$ $8x^2 - 4x - 10$

Subtract, using the vertical method.

51. $3a^2 - 2a$ from $5a^2 + 3a$
$2a^2 + 5a$

52. $6r^3 + 4r^2$ from $4r^3 - 2r^2$
$-2r^3 - 6r^2$

53. $5x^2 - 6x + 7$ from $8x^2 - 5x + 7$
$3x^2 + x$

54. $8x^2 - 4x + 2$ from $9x^2 - 8x + 6$
$x^2 - 4x + 4$

55. $5x^2 - 3x$ from $8x^2 - 9$ $3x^2 + 3x - 9$

56. $7x^2 + 6x$ from $9x^2 - 3$ $2x^2 - 6x - 3$

Skillscan (Section 1.5)
Multiply.

a. $x^5 \cdot x^7$ x^{12} **b.** $y^8 \cdot y^{12}$ y^{20} **c.** $2a^3 \cdot a^4$ $2a^7$

d. $3m^5 \cdot m^2$ $3m^7$ **e.** $4r^5 \cdot 3r$ $12r^6$ **f.** $6w^2 \cdot 5w^3$ $30w^5$

g. $(-2x^2)(8x^7)$ $-16x^9$ **h.** $(-10a)(-3a^5)$ $30a^6$

Answers
1. $9a + 4$ **3.** $13b^2 - 18b$ **5.** $-2x^2$ **7.** $5x^2 - 2x + 1$ **9.** $2b^2 + 5b + 16$ **11.** $8y^3 - 2y$ **13.** $-a^3 + 4a^2$ **15.** $-2x^2 - x + 3$ **17.** $-2a - 3b$ **19.** $5a - 2b + 3c$ **21.** $6r - 5s$ **23.** $8p - 2q$ **25.** $x - 7$ **27.** $m^2 - 3m$ **29.** $-2y^2$ **31.** $2x^2 - x + 1$ **33.** $8a^2 - 12a - 7$ **35.** $-6b^2 + 8b$ **37.** $2x^2 + 12$ **39.** $6b - 1$ **41.** $10x - 9$ **43.** $2x^2 + 5x - 12$ **45.** $-6y^2 - 8y$ **47.** $6w^2 - 2w + 2$ **49.** $9x^2 - x$ **51.** $2a^2 + 5a$ **53.** $3x^2 + x$ **55.** $3x^2 + 3x - 9$ **a.** x^{12} **b.** y^{20} **c.** $2a^7$ **d.** $3m^7$ **e.** $12r^6$ **f.** $30w^5$ **g.** $-16x^9$ **h.** $30a^6$

4.3 Multiplying Polynomials

OBJECTIVES
1. To find the product of a monomial and any polynomial
2. To find the product of two polynomials

You have already had some experience in multiplying polynomials. In Section 1.5 we stated the first property of exponents and used that property to find the product of two monomials. Let's review briefly.

The first property for exponents:

$x^m \cdot x^n = x^{m+n}$

TO FIND THE PRODUCT OF MONOMIALS

STEP 1 Multiply the coefficients.

STEP 2 Use the first property of exponents to combine the variables.

Example 1

To multiply $3x^2y$ and $2x^3y^5$, write

$$(3x^2y)(2x^3y^5)$$
$$= (3 \cdot 2)(x^2 \cdot x^3)(y \cdot y^5)$$

Multiply the coefficients. Add the exponents.

$$= 6x^5y^6$$

CHECK YOURSELF 1

Multiply.

1. $(5a^2b)(3a^2b^4)$ **2.** $(-3xy)(4x^3y^5)$

The next step is to find the product of a monomial and a polynomial. The key to the multiplication is the distributive property, which we introduced in Section 1.2. That property leads us to the following rule for multiplication.

You might want to review that section now before going on.

Distributive property:
$a(b + c) = ab + ac$

TO MULTIPLY A POLYNOMIAL BY A MONOMIAL

Use the distributive property to multiply each term of the polynomial by the monomial.

Example 2

(*a*) To multiply $2x + 3$ by x:

Note: With practice you will do this step mentally.

$$x(2x + 3)$$
$$= x \cdot 2x + x \cdot 3$$
$$= 2x^2 + 3x$$

Multiply x by $2x$ and then by 3, the terms of the polynomial. We "distribute" the multiplication over the sum.

(*b*) To multiply $2a^3 + 4a$ by $3a^2$:

$$3a^2(2a^3 + 4a)$$
$$= 3a^2 \cdot 2a^3 + 3a^2 \cdot 4a$$
$$= 6a^5 + 12a^3$$

CHECK YOURSELF 2

Multiply.

1. $2y(y^2 + 3y)$ **2.** $3w^2(2w^3 + 5w)$

An extended distributive property will also let you extend the patterns of Example 2 to a polynomial with *any* number of terms, as well as to a polynomial with negative terms.

Example 3

(*a*) $3x(4x^3 + 5x^2 + 2)$
$$= 3x \cdot 4x^3 + 3x \cdot 5x^2 + 3x \cdot 2$$
$$= 12x^4 + 15x^3 + 6x$$

(*b*) $5y^2(2y^3 - 4)$
$= 5y^2 \cdot 2y^3 - (5y^2)(4)$
$= 10y^5 - 20y^2$

(*c*) $-5c(4c^2 - 8c)$
$= (-5c)(4c^2) - (-5c)(8c)$
$= -20c^3 + 40c^2$

(*d*) $3c^2d^2(7cd^2 - 5c^2d^3)$
$= 3c^2d^2 \cdot 7cd^2 - (3c^2d^2)(5c^2d^3)$
$= 21c^3d^4 - 15c^4d^5$

Again we have shown all the steps of the process. With practice you can write the product directly, and you should try to do so.

CHECK YOURSELF 3

Multiply.

1. $3(5a^2 + 2a + 7)$
2. $4x^2(8x^3 - 6)$
3. $-5m(8m^2 - 5m)$
4. $9a^2b(3a^3b - 6a^2b^4)$

We also use the distributive property in writing the product of two binomials. Consider the following example.

Example 4

(*a*) To multiply $x + 2$ by $x + 3$:

We can think of $x + 2$ as a single quantity and apply the distributive property.

Note that this ensures that each term, *x* and 2, of the first binomial is multiplied by each term, *x* and 3, of the second binomial.

$(x + 2)(x + 3)$
$= (x + 2)x + (x + 2)3$
$= x^2 + 2x + 3x + 6$
$= x^2 + 5x + 6$

Multiply $x + 2$ by x and then by 3.

(*b*) To multiply $a - 3$ by $a - 4$, think of $a - 3$ as a single quantity and distribute.

$(a - 3)(a - 4)$
$= (a - 3)a - (a - 3)(4)$
$= a^2 - 3a - (4a - 12)$
$= a^2 - 3a - 4a + 12$
$= a^2 - 7a + 12$

Note that the parentheses are needed here because a *negative sign* precedes the binomial.

CHECK YOURSELF 4

Multiply.

1. $(x + 4)(x + 5)$ **2.** $(y + 5)(y - 6)$

Fortunately, there is a pattern to this kind of multiplication that will allow you to write the product of the two binomials directly without having to go through all these steps. We call it the *FOIL* method of multiplying, and the reason for this name will be clear as we look at the process in more detail.

To multiply $(x + 2)(x + 3)$:

1. $(x + 2)(x + 3)$

$x \cdot x$ Find the product of the *first* terms of the factors.

Remember this by F!

2. $(x + 2)(x + 3)$

$x \cdot 3$ Find the product of the *outer* terms.

Remember this by O!

3. $(x + 2)(x + 3)$

$2 \cdot x$ Find the product of the *inner* terms.

Remember this by I!

4. $(x + 2)(x + 3)$

$2 \cdot 3$ Find the product of the *last* terms.

Remember this by L!

Combining the four steps, we have

$$(x + 2)(x + 3)$$
$$= x^2 + 3x + 2x + 6$$
$$= x^2 + 5x + 6$$

Of course these are the same four terms found in Example 4*a*.

With practice, the FOIL method will let you write the products quickly and easily. Consider the following examples that illustrate this approach.

Of course, it's called FOIL to give you an easy way of remembering the steps: *F*irst, *O*uter, *I*nner, and *L*ast.

Example 5

(*a*) $(x + 4)(x + 5)$

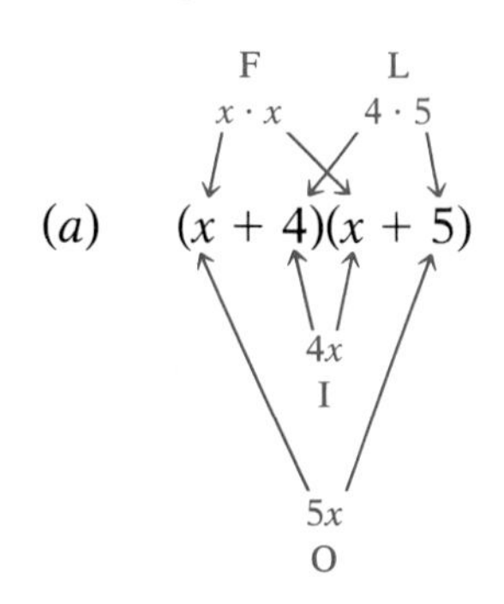

When possible, you should combine the outer and inner products mentally and write just the final product.

$$= x^2 + 5x + 4x + 20$$

F O I L

$$= x^2 + 9x + 20$$

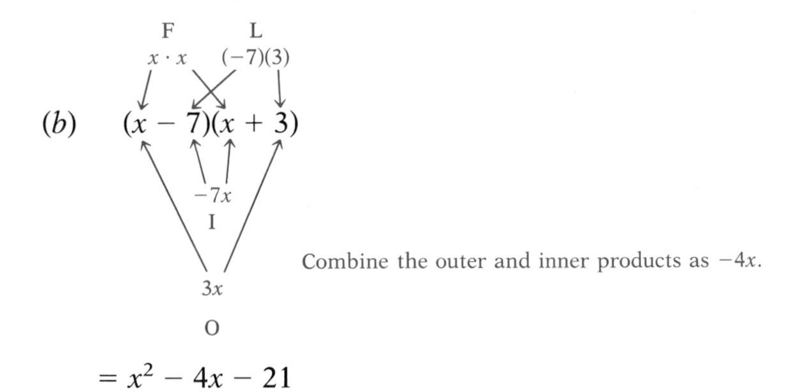

$$= x^2 - 4x - 21$$

CHECK YOURSELF 5

Multiply.

1. $(x + 6)(x + 7)$ **2.** $(x + 3)(x - 5)$ **3.** $(x - 2)(x - 8)$

Using the FOIL method, you can also find the product of binomials with coefficients other than 1 or with more than one variable.

Example 6

$12x^2$ -6

(*a*) $(4x - 3)(3x + 2)$

$-9x$ $8x$

Combine:
$-9x + 8x = -x$

$$= 12x^2 - x - 6$$

$6x^2$ $35y^2$

(*b*) $(3x - 5y)(2x - 7y)$

$-10xy$ $-21xy$

Combine:
$-10xy - 21xy = -31xy$

$$= 6x^2 - 31xy + 35y^2$$

The following rule will summarize our work in multiplying binomials.

TO MULTIPLY TWO BINOMIALS

STEP 1 The first term of the product of the binomials is the product of the first terms of the binomials (F).

STEP 2 The middle term of the product is the sum of the outer and inner products (OI).

STEP 3 The last term of the product is the product of the last terms of the binomials (L).

CHECK YOURSELF 6

Multiply.

1. $(5x + 2)(3x - 7)$ **2.** $(4a - 3b)(5a - 4b)$
3. $(3m + 5n)(2m + 3n)$

You should now be able to multiply any two binomials, using the FOIL method. But what if one of the factors has three or more terms? One method is the vertical format, shown in the following example.

Example 7

Multiply $x^2 - 5x + 8$ by $x + 3$.

Step 1

$$\begin{array}{r} x^2 - 5x + 8 \\ x + 3 \\ \hline 3x^2 - 15x + 24 \end{array}$$

Multiply each term of $x^2 - 5x + 8$ by 3.

Step 2

$$\begin{array}{rrrr} & x^2 & - 5x & + 8 \\ & & x & + 3 \\ \hline & 3x^2 & - 15x & + 24 \\ x^3 & - 5x^2 & + 8x & \\ \hline \end{array}$$

Now multiply each term by x.

Note that this line is shifted over so that the like terms are in the same columns.

Step 3

$$\begin{array}{rrrr} & x^2 & - 5x & + 8 \\ & & x & + 3 \\ \hline & 3x^2 & - 15x & + 24 \\ x^3 & - 5x^2 & + 8x & \\ \hline x^3 & - 2x^2 & - 7x & + 24 \end{array}$$

Now add to combine the like terms to write the product.

Note: Using this vertical method ensures that each term of one factor multiplies each term of the other. That's why it works!

CHECK YOURSELF 7

Multiply $2x^2 - 5x + 3$ by $3x + 4$.

Our final example illustrates the operation of multiplication when three factors are involved in a product.

Example 8

Multiply

$3x(x - 5)(x + 4)$

The best approach in this case is to apply the FOIL method to find the product of the two binomial factors. Here we have

$$3x(x - 5)(x + 4)$$
$$= 3x(x^2 - x - 20)$$

Note: It is best to use two steps here. Trying to do both operations at once often leads to errors.

Now apply the distributive property to multiply $x^2 - x - 20$ by $3x$.

$$= 3x^3 - 3x^2 - 60x$$

CHECK YOURSELF 8

Multiply:

$8m(m - 5)(m - 3)$

CHECK YOURSELF ANSWERS

1. (1) $15a^4b^5$; (2) $-12x^4y^6$.
2. (1) $2y^3 + 6y^2$; (2) $6w^5 + 15w^3$.
3. (1) $15a^2 + 6a + 21$; (2) $32x^5 - 24x^2$; (3) $-40m^3 + 25m^2$; (4) $27a^5b^2 - 54a^4b^5$.
4. (1) $x^2 + 9x + 20$; (2) $y^2 - y - 30$.
5. (1) $x^2 + 13x + 42$; (2) $x^2 - 2x - 15$; (3) $x^2 - 10x + 16$.
6. (1) $15x^2 - 29x - 14$; (2) $20a^2 - 31ab + 12b^2$; (3) $6m^2 + 19mn + 15n^2$.
7. $6x^3 - 7x^2 - 11x + 12$.
8. $8m^3 - 64m^2 + 120m$.

4.3 Exercises

Multiply.

1. $(5x^2)(3x^3)$ $15x^5$

2. $(7a^5)(4a^6)$ $28a^{11}$

3. $(-7b^4)(4b^6)$ $-28b^{10}$

4. $(7y^5)(-8y^5)$ $-56y^{10}$

5. $(-5p^5)(-8p^8)$ $40p^{13}$

6. $(-9m^9)(6m^6)$ $-54m^{15}$

7. $(4m^5)(-3m)$ $-12m^6$

8. $(-5r^7)(-3r)$ $15r^8$

9. $(4x^3y^2)(8x^2y)$ $32x^5y^3$

10. $(-3r^4s^2)(-7r^2s^5)$ $21r^6s^7$

11. $(-3m^5n^2)(2m^4n)$ $-6m^9n^3$

12. $(7a^3b^5)(-6a^4b)$ $-42a^7b^6$

Use the distributive property to multiply.

13. $4(3x + 7)$ $12x + 28$

14. $5(8b - 6)$ $40b - 30$

15. $5a(2a + 3)$ $10a^2 + 15a$

16. $4x(3x - 5)$ $12x^2 - 20x$

17. $3s^2(4s^2 - 7s)$ $12s^4 - 21s^3$

18. $9a^2(3a^3 + 5a)$ $27a^5 + 45a^3$

19. $2x(4x^2 - 2x + 1)$
$8x^3 - 4x^2 + 2x$

20. $5m(4m^3 - 3m^2 + 2)$
$20m^4 - 15m^3 + 10m$

21. $3xy(2x^2y + xy^2 + 5xy)$
$6x^3y^2 + 3x^2y^3 + 15x^2y^2$

22. $5ab^2(ab - 3a + 5b)$
$5a^2b^3 - 15a^2b^2 + 25ab^3$

23. $6m^2n(3m^2n - 2mn + mn^2)$
$18m^4n^2 - 12m^3n^2 + 6m^3n^3$

24. $8pq^2(2pq - 3p + 5q)$
$16p^2q^3 - 24p^2q^2 + 40pq^3$

Multiply.

25. $(x + 3)(x + 2)$
$x^2 + 5x + 6$

26. $(a - 3)(a - 7)$
$a^2 - 10a + 21$

27. $(m - 5)(m - 9)$
$m^2 - 14m + 45$

28. $(b + 7)(b + 5)$
$b^2 + 12b + 35$

29. $(p - 8)(p + 7)$
$p^2 - p - 56$

30. $(x - 10)(x + 9)$
$x^2 - x - 90$

31. $(w + 10)(w + 20)$
$w^2 + 30w + 200$

32. $(s - 12)(s - 8)$
$s^2 - 20s + 96$

33. $(3x - 5)(x - 8)$
$3x^2 - 29x + 40$

34. $(w + 5)(4w - 7)$
$4w^2 + 13w - 35$

35. $(2x - 3)(3x + 4)$
$6x^2 - x - 12$

36. $(5a + 1)(3a + 7)$
$15a^2 + 38a + 7$

37. $(3a - b)(4a - 9b)$
$12a^2 - 31ab + 9b^2$

38. $(7s - 3t)(3s + 8t)$
$21s^2 + 47st - 24t^2$

39. $(4p - 3q)(5p + 7q)$
$20p^2 + 13pq - 21q^2$

40. $(4x - 5y)(3x - 2y)$
$12x^2 - 23xy + 10y^2$

41. $(5x + 4y)(6x + 5y)$
$30x^2 + 49xy + 20y^2$

42. $(6x - 7y)(6x + 3y)$
$36x^2 - 24xy - 21y^2$

43. $(x + 5)^2$
$x^2 + 10x + 25$

44. $(y + 8)^2$
$y^2 + 16y + 64$

45. $(y - 9)^2$
$y^2 - 18y + 81$

46. $(2a + 3)^2$
$4a^2 + 12a + 9$

47. $(6m + n)^2$
$36m^2 + 12mn + n^2$

48. $(7b - c)^2$
$49b^2 - 14bc + c^2$

49. $(a - 5)(a + 5)$
$a^2 - 25$

50. $(x - 7)(x + 7)$
$x^2 - 49$

51. $(p - 3q)(p + 3q)$
$p^2 - 9q^2$

52. $(5x + y)(5x - y)$
$25x^2 - y^2$

53. $(7m + 4n)(7m - 4n)$
$49m^2 - 16n^2$

54. $(8a - 3b)(8a + 3b)$
$64a^2 - 9b^2$

Multiply, using the vertical method.

55. $(x + 2)(x^2 + 2x - 3)$
$x^3 + 4x^2 + x - 6$

56. $(a - 3)(a^2 + 4a + 1)$
$a^3 + a^2 - 11a - 3$

57. $(2m - 5)(2m^2 + 3m - 2)$
$4m^3 - 4m^2 - 19m + 10$

58. $(5p + 3)(p^2 + 4p + 1)$
$5p^3 + 23p^2 + 17p + 3$

59. $(3x + 4y)(x^2 + xy + 5y^2)$
$3x^3 + 7x^2y + 19xy^2 + 20y^3$

60. $(7a - 2b)(2a^2 - ab + 4b^2)$
$14a^3 - 11a^2b + 30ab^2 - 8b^3$

61. $(a^2 + 3ab - b^2)(a^2 - 5ab + b^2)$
$a^4 - 2a^3b - 15a^2b^2 + 8ab^3 - b^4$

62. $(m^2 - 5mn + 3n^2)(m^2 + 4mn - 2n^2)$
$m^4 - m^3n - 19m^2n^2 + 22mn^3 - 6n^4$

63. $(x - 2y)(x^2 + 2xy + 4y^2)$ $x^3 - 8y^3$

64. $(m + 3n)(m^2 - 3mn + 9n^2)$ $m^3 + 27n^3$

65. $(3a + b)(9a^2 - 3ab + b^2)$ $27a^3 + b^3$

66. $(4r - s)(16r^2 + 4rs + s^2)$ $64r^3 - s^3$

Multiply.

67. $2x(3x - 2)(4x + 1)$ $24x^3 - 10x^2 - 4x$

68. $3x(2x + 1)(2x - 1)$ $12x^3 - 3x$

69. $5a(4a - 3)(4a + 3)$
$80a^3 - 45a$

70. $6m(3m - 2)(3m - 7)$
$54m^3 - 162m^2 + 84m$

71. $3s(5s - 2)(4s - 1)$ $60s^3 - 39s^2 + 6s$

72. $7w(2w - 3)(2w + 3)$ $28w^3 - 63w$

73. $(x - 2)(x + 1)(x - 3)$ $x^3 - 4x^2 + x + 6$

74. $(y + 3)(y - 2)(y - 4)$ $y^3 - 3y^2 - 10y + 24$

75. $(a - 1)^3$ $a^3 - 3a^2 + 3a - 1$

76. $(x + 1)^3$ $x^3 + 3x^2 + 3x + 1$

Skillscan (Section 1.5)

a. $(3a)(3a)$ $9a^2$ **b.** $(3a)^2$ $9a^2$ **c.** $(5x)(5x)$ $25x^2$ **d.** $(5x)^2$ $25x^2$

e. $(-2w)(-2w)$
$4w^2$

f. $(-2w)^2$
$4w^2$

g. $(-4r)(-4r)$
$16r^2$

h. $(-4r)^2$
$16r^2$

Answers

1. $15x^5$ **3.** $-28b^{10}$ **5.** $40p^{13}$ **7.** $-12m^6$ **9.** $32x^5y^3$ **11.** $-6m^9n^3$ **13.** $12x + 28$ **15.** $10a^2 + 15a$ **17.** $12s^4 - 21s^3$ **19.** $8x^3 - 4x^2 + 2x$ **21.** $6x^3y^2 + 3x^2y^3 + 15x^2y^2$ **23.** $18m^4n^2 - 12m^3n^2 + 6m^3n^3$ **25.** $x^2 + 5x + 6$ **27.** $m^2 - 14m + 45$ **29.** $p^2 - p - 56$ **31.** $w^2 + 30w + 200$ **33.** $3x^2 - 29x + 40$ **35.** $6x^2 - x - 12$ **37.** $12a^2 - 31ab + 9b^2$ **39.** $20p^2 + 13pq - 21q^2$ **41.** $30x^2 + 49xy + 20y^2$ **43.** $x^2 + 10x + 25$ **45.** $y^2 - 18y + 81$ **47.** $36m^2 + 12mn + n^2$ **49.** $a^2 - 25$ **51.** $p^2 - 9q^2$ **53.** $49m^2 - 16n^2$ **55.** $x^3 + 4x^2 + x - 6$ **57.** $4m^3 - 4m^2 - 19m + 10$ **59.** $3x^3 + 7x^2y + 19xy^2 + 20y^3$ **61.** $a^4 - 2a^3b - 15a^2b^2 + 8ab^3 - b^4$ **63.** $x^3 - 8y^3$ **65.** $27a^3 + b^3$ **67.** $24x^3 - 10x^2 - 4x$ **69.** $80a^3 - 45a$ **71.** $60s^3 - 39s^2 + 6s$ **73.** $x^3 - 4x^2 + x + 6$ **75.** $a^3 - 3a^2 + 3a - 1$ **a.** $9a^2$ **b.** $9a^2$ **c.** $25x^2$ **d.** $25x^2$ **e.** $4w^2$ **f.** $4w^2$ **g.** $16r^2$ **h.** $16r^2$

4.4 Special Products

OBJECTIVES

1. To square a binomial
2. To find the product of two binomials that differ only in sign

Certain products occur frequently enough in algebra that it is worth learning special formulas for dealing with them. First, let's look at the product of two equal binomial factors. This is called the *square of a binomial.*

$$(x + y)^2 = (x + y)(x + y)$$
$$= x^2 + 2xy + y^2$$

$$(x - y)^2 = (x - y)(x - y)$$
$$= x^2 - 2xy + y^2$$

The patterns above lead us to the following rule.

TO SQUARE A BINOMIAL

STEP 1 The first term of the square is the square of the first term of the binomial.

STEP 2 The middle term is twice the product of the two terms of the binomial.

STEP 3 The last term is the square of the last term of the binomial.

Example 1

(*a*) $(x + 3)^2 = x^2 + \underbrace{2 \cdot x \cdot 3} + 3^2$

Square of first term — Twice the product of the two terms — Square of the last term

$= x^2 + 6x + 9$

(*b*) $(3a + 4b)^2 = (3a)^2 + 2(3a)(4b) + (4b)^2$
$= 9a^2 + 24ab + 16b^2$

(*c*) $(y - 5)^2 = y^2 + 2 \cdot y \cdot (-5) + (-5)^2$
$= y^2 - 10y + 25$

(*d*) $(5c - 3d)^2 = (5c)^2 + 2(5c)(-3d) + (-3d)^2$
$= 25c^2 - 30cd + 9d^2$

Again we have shown all the steps. With practice you can write just the square.

CHECK YOURSELF 1

Multiply.

1. $(2x + 1)^2$ **2.** $(4x - 3y)^2$

Be careful! A very common mistake in squaring binomials is to forget the middle term.

$(a + 3)^2$ is *not* equal to $a^2 + 9$

The correct square is

$(a + 3)^2 = a^2 + 6a + 9$

The middle term is twice the product of *a* and 3

CHECK YOURSELF 2

Multiply.

1. $(x + 5)^2$ **2.** $(3a + 2)^2$ **3.** $(y - 7)^2$ **4.** $(5x - 2y)^2$

A second special product will be very important in the next chapter, dealing with factoring. Suppose the form of a product is

$(x + y)(x - y)$

The two terms differ only in sign.

Let's see what happens when we multiply.

$$(x + y)(x - y) = x^2 \underbrace{- xy + xy}_{= 0} - y^2 = x^2 - y^2$$

Since the middle term becomes 0, we have the following rule.

> The product of two binomials that differ only in the sign between the terms is the square of the first term minus the square of the second.

Let's look at the application of this rule in our final example.

Example 2

(*a*) $(x + 5)(x - 5) = x^2 - 5^2$

Square of the first term — Square of the second term

$= x^2 - 25$

Note: $(2y)^2 = (2y)(2y) = 4y^2$

(*b*) $(x + 2y)(x - 2y) = x^2 - (2y)^2$

Square of the first term — Square of the second term

$= x^2 - 4y^2$

(*c*) $(3m + n)(3m - n) = 9m^2 - n^2$

(*d*) $(4a - 3b)(4a + 3b) = 16a^2 - 9b^2$

CHECK YOURSELF 3

Find the products.

1. $(a - 6)(a + 6)$
2. $(x - 3y)(x + 3y)$
3. $(5n + 2p)(5n - 2p)$
4. $(7b - 3c)(7b + 3c)$

CHECK YOURSELF ANSWERS

1. (1) $4x^2 + 4x + 1$; (2) $16x^2 - 24xy + 9y^2$.
2. (1) $x^2 + 10x + 25$; (2) $9a^2 + 12a + 4$; (3) $y^2 - 14y + 49$; (4) $25x^2 - 20xy + 4y^2$.
3. (1) $a^2 - 36$; (2) $x^2 - 9y^2$; (3) $25n^2 - 4p^2$; (4) $49b^2 - 9c^2$.

4.4 Exercises

Find the following binomial squares.

1. $(x + 4)^2$
   $x^2 + 8x + 16$

2. $(y + 8)^2$
   $y^2 + 16y + 64$

3. $(w - 7)^2$
   $w^2 - 14w + 49$

4. $(a - 9)^2$
   $a^2 - 18a + 81$

5. $(z + 12)^2$
   $z^2 + 24z + 144$

6. $(p - 20)^2$
   $p^2 - 40p + 400$

7. $(2a - 1)^2$
   $4a^2 - 4a + 1$

8. $(3x - 2)^2$
   $9x^2 - 12x + 4$

9. $(6m + 1)^2$
   $36m^2 + 12m + 1$

10. $(7b - 2)^2$
    $49b^2 - 28b + 4$

11. $(3x - y)^2$
    $9x^2 - 6xy + y^2$

12. $(5m + n)^2$
    $25m^2 + 10mn + n^2$

13. $(2r + 5s)^2$
    $4r^2 + 20rs + 25s^2$

14. $(3a - 4b)^2$
    $9a^2 - 24ab + 16b^2$

15. $(6a - 7b)^2$
    $36a^2 - 84ab + 49b^2$

16. $(5p + 4q)^2$
    $25p^2 + 40pq + 16q^2$

17. $\left(x + \frac{1}{2}\right)^2$
    $x^2 + x + \frac{1}{4}$

18. $\left(w - \frac{1}{4}\right)^2$
    $w^2 - \frac{1}{2}w + \frac{1}{16}$

Find the following products.

19. $(x - 5)(x + 5)$
    $x^2 - 25$

20. $(y + 7)(y - 7)$
    $y^2 - 49$

21. $(m + 12)(m - 12)$
    $m^2 - 144$

22. $(w - 10)(w + 10)$
    $w^2 - 100$

23. $\left(x - \frac{1}{2}\right)\left(x + \frac{1}{2}\right)$
    $x^2 - \frac{1}{4}$

24. $\left(x + \frac{2}{3}\right)\left(x - \frac{2}{3}\right)$
    $x^2 - \frac{4}{9}$

25. $(p - 0.4)(p + 0.4)$
    $p^2 - 0.16$

26. $(m - 0.6)(m + 0.6)$
    $m^2 - 0.36$

27. $(a - 3b)(a + 3b)$
    $a^2 - 9b^2$

28. $(p + 4q)(p - 4q)$
    $p^2 - 16q^2$

29. $(4r - s)(4r + s)$
    $16r^2 - s^2$

30. $(7x - y)(7x + y)$
    $49x^2 - y^2$

31. $(7w + 4z)(7w - 4z)$
    $49w^2 - 16z^2$

32. $(8c + 3d)(8c - 3d)$
    $64c^2 - 9d^2$

33. $(5x - 9y)(5x + 9y)$
    $25x^2 - 81y^2$

34. $(6s - 5t)(6s + 5t)$
    $36s^2 - 25t^2$

35. $x(x - 2)(x + 2)$
    $x^3 - 4x$

36. $a(a + 5)(a - 5)$
    $a^3 - 25a$

37. $2s(s - 3r)(s + 3r)$ $2s^3 - 18r^2s$

38. $5w(2w - z)(2w + z)$ $20w^3 - 5wz^2$

39. $4r(r + 5)^2$ $4r^3 + 40r^2 + 100r$

40. $6x(x - 3)^2$ $6x^3 - 36x^2 + 54x$

For the following problems, let x represent the number, then write and find the corresponding product.

41. The product of 6 more than a number and 6 less than that number $x^2 - 36$

42. The square of 5 more than a number $x^2 + 10x + 25$

43. The square of 4 less than a number $x^2 - 8x + 16$

44. The product of 5 less than a number and 5 more than that number $x^2 - 25$

Going Beyond

Note that $(28)(32) = (30 - 2)(30 + 2) = 900 - 4 = 896$. Use this pattern to find each of the following products.

45. (49)(51) 2499

46. (27)(33) 891

47. (34)(26) 884

48. (98)(102) 9996

49. (55)(65) 3575

50. (64)(56) 3584

Skillscan (Section 1.5)

Divide.

a. $\frac{2x^2}{2x}$ x

b. $\frac{3a^3}{3a}$ a^2

c. $\frac{6p^3}{2p^2}$ $3p$

d. $\frac{10m^4}{5m^2}$ $2m^2$

e. $\frac{20a^3}{5a^3}$ 4

f. $\frac{6x^2y}{3xy}$ $2x$

g. $\frac{12r^3s^2}{4rs}$ $3r^2s$

h. $\frac{49c^4d^6}{7cd^3}$ $7c^3d^3$

Answers

1. $x^2 + 8x + 16$ **3.** $w^2 - 14w + 49$ **5.** $z^2 + 24z + 144$ **7.** $4a^2 - 4a + 1$ **9.** $36m^2 + 12m + 1$ **11.** $9x^2 - 6xy + y^2$ **13.** $4r^2 + 20rs + 25s^2$ **15.** $36a^2 - 84ab + 49b^2$ **17.** $x^2 + x + \frac{1}{4}$ **19.** $x^2 - 25$ **21.** $m^2 - 144$ **23.** $x^2 - \frac{1}{4}$ **25.** $p^2 - 0.16$ **27.** $a^2 - 9b^2$ **29.** $16r^2 - s^2$ **31.** $49w^2 - 16z^2$ **33.** $25x^2 - 81y^2$ **35.** $x^3 - 4x$ **37.** $2s^3 - 18r^2s$ **39.** $4r^3 + 40r^2 + 100r$ **41.** $x^2 - 36$ **43.** $x^2 - 8x + 16$ **45.** 2499 **47.** 884 **49.** 3575 **a.** x **b.** a^2 **c.** $3p$ **d.** $2m^2$ **e.** 4 **f.** $2x$ **g.** $3r^2s$ **h.** $7c^3d^3$

4.5 Dividing Polynomials

OBJECTIVES

1. To divide a polynomial by a monomial
2. To divide a polynomial by another polynomial

In Section 1.5, we introduced the second property of exponents, which was used to divide one monomial by another monomial. Let's review that process.

The second property says: If x is not zero and $m > n$,

$$\frac{x^m}{x^n} = x^{m-n}$$

TO DIVIDE A MONOMIAL BY A MONOMIAL

STEP 1 Divide the coefficients.

STEP 2 Use the second property of exponents to combine the variables.

Example 1

(a) $\frac{8x^4}{2x^2} = 4x^{4-2}$ (Divide: $\frac{8}{2} = 4$; Subtract the exponents.)

$= 4x^2$

(b) $\frac{45a^5b^3}{9a^2b} = 5a^3b^2$

CHECK YOURSELF 1

Divide.

1. $\frac{16a^5}{8a^3}$ 2. $\frac{28m^4n^3}{7m^3n}$

Now let's look at how this can be extended to divide any polynomial by a monomial. Suppose that you want to divide $12a^3 + 8a^2$ by $4a$. Write

$$\frac{12a^3 + 8a^2}{4a}$$

This is the same as

Technically this step depends on the distributive law and the definition of division.

$$\frac{12a^3}{4a} + \frac{8a^2}{4a}$$

Now do each division.

$$= 3a^2 + 2a$$

The work above leads us to the following rule.

TO DIVIDE A POLYNOMIAL BY A MONOMIAL

Divide each term of the polynomial by the monomial. Then combine the results.

Example 2

Divide each term by 2.

$$(a)\quad \frac{4a^2 + 8}{2} = \frac{4a^2}{2} + \frac{8}{2}$$
$$= 2a^2 + 4$$

Divide each term by $6y$.

$$(b)\quad \frac{24y^3 - 18y^2}{6y} = \frac{24y^3}{6y} - \frac{18y^2}{6y}$$
$$= 4y^2 - 3y$$

Remember the rules for signs in division

$$(c)\quad \frac{15x^2 + 10x}{-5x} = \frac{15x^2}{-5x} + \frac{10x}{-5x}$$
$$= -3x - 2$$

With practice you can write just the quotient.

$$(d)\quad \frac{14x^4 + 28x^3 - 21x^2}{7x^2} = \frac{14x^4}{7x^2} + \frac{28x^3}{7x^2} - \frac{21x^2}{7x^2}$$
$$= 2x^2 + 4x - 3$$

(*e*) $$\frac{9a^3b^4 - 6a^2b^3 + 12ab^4}{3ab} = \frac{9a^3b^4}{3ab} - \frac{6a^2b^3}{3ab} + \frac{12ab^4}{3ab}$$
$$= 3a^2b^3 - 2ab^2 + 4b^3$$

CHECK YOURSELF 2

Divide.

1. $\dfrac{20y^3 - 15y^2}{5y}$ **2.** $\dfrac{8a^3 - 12a^2 + 4a}{-4a}$

3. $\dfrac{16m^4n^3 - 12m^3n^2 + 8mn}{4mn}$

We are now ready to look at dividing one polynomial by another polynomial (with more than one term). The process is very much like long division in arithmetic. The following example will illustrate.

Example 3

Divide $x^2 + 7x + 10$ by $x + 2$.

Divide x^2 by x to get x.

Step 1 $$x + 2\overline{)x^2 + 7x + 10} \quad \text{quotient: } x$$

Step 2 $$\begin{array}{r} x \\ x + 2\overline{)x^2 + 7x + 10} \\ \underline{x^2 + 2x} \end{array}$$

Multiply the divisor, $x + 2$, by x.

Remember: To subtract $x^2 + 2x$, mentally change each sign to $-x^2 - 2x$, and add. Take your time and be careful here. It's where most errors are made.

Step 3 $$\begin{array}{r} x \\ x + 2\overline{)x^2 + 7x + 10} \\ \underline{x^2 + 2x} \\ 5x + 10 \end{array}$$

Subtract and bring down 10.

Step 4

Divide $5x$ by x to get 5.

$$\begin{array}{r} x + 5 \\ x + 2\overline{)x^2 + 7x + 10} \\ \underline{x^2 + 2x} \\ 5x + 10 \end{array}$$

Note that we repeat the process until the degree of the remainder is less than that of the divisor or until there is no remainder.

Step 5

$$
\begin{array}{r}
x + 5 \\
x + 2\overline{)x^2 + 7x + 10} \\
\underline{x^2 + 2x} \quad\;\; \\
5x + 10 \\
\underline{5x + 10} \\
0
\end{array}
$$

Multiply $x + 2$ by 5 and then subtract.

The quotient is $x + 5$.

In the above example, we have shown all the steps separately to help you see the process. In practice your work will look like that illustrated in the following example.

Example 4

Divide $x^2 + x - 12$ by $x - 3$.

$$
\begin{array}{r}
x + 4 \\
x - 3\overline{)x^2 + x - 12} \\
\underline{x^2 - 3x} \quad\;\; \\
4x - 12 \\
\underline{4x - 12} \\
0
\end{array}
$$

THE STEPS

1. Divide x^2 by x to get x, the first term of the quotient.
2. Multiply $x - 3$ by x.
3. Subtract and bring down -12. Remember to mentally change the signs to $-x^2 + 3x$ and add.
4. Divide $4x$ by x to get 4, the second term of the quotient.
5. Multiply $x - 3$ by 4 and subtract.

The quotient is $x + 4$.

CHECK YOURSELF 3

Divide.

$(x^2 + 2x - 24) \div (x - 4)$

You may have a remainder in algebraic long division just as in arithmetic. Consider the following example.

Example 5

Divide $4x^2 - 8x + 11$ by $2x - 3$.

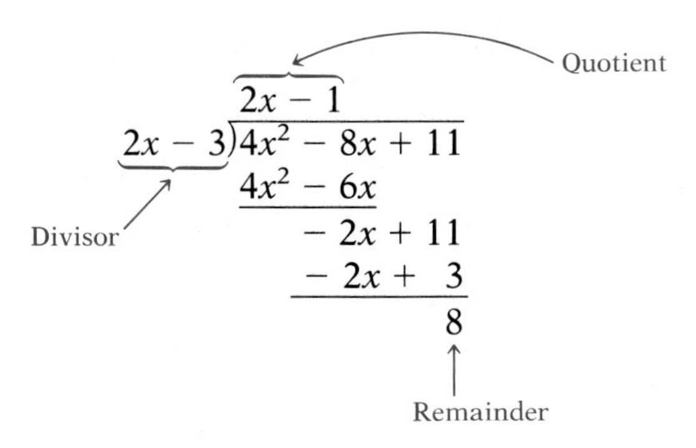

This result can be written as

$$\frac{4x^2 - 8x + 11}{2x - 3}$$

$$= \underbrace{2x - 1}_{\text{Quotient}} + \frac{8}{2x - 3}$$

Remainder (8); Divisor ($2x - 3$)

CHECK YOURSELF 4

Divide.

$(6x^2 - 7x + 15) \div (3x - 5)$

The division process shown in our previous examples can be extended to cases where the dividend has a higher degree. Note that the steps involved in the division process are exactly the same. The following example illustrates.

Example 6

Divide $6x^3 + x^2 - 4x - 5$ by $3x - 1$.

$$\begin{array}{r} 2x^2 + x - 1 \\ 3x - 1\overline{)6x^3 + x^2 - 4x - 5} \\ \underline{6x^3 - 2x^2} \\ 3x^2 - 4x \\ \underline{3x^2 - x} \\ -3x - 5 \\ \underline{-3x + 1} \\ -6 \end{array}$$

This result can be written as

$$\frac{6x^3 + x^2 - 4x - 5}{3x - 1} = 2x^2 + x - 1 + \frac{-6}{3x - 1}$$

CHECK YOURSELF 5

Divide $4x^3 - 2x^2 + 2x + 15$ by $2x + 3$.

Suppose that the dividend is "missing" a term in some power of the variable. You can use 0 as the coefficient for the missing terms. Consider the following example.

Example 7

Divide $x^3 - 2x^2 + 5$ by $x + 3$.

$$
\begin{array}{r}
x^2 - 5x + 15 \\
x + 3 \overline{) x^3 - 2x^2 + 0x + 5} \\
\underline{x^3 + 3x^2} \\
-5x^2 + 0x \\
\underline{-5x^2 - 15x} \\
15x + 5 \\
\underline{15x + 45} \\
-40
\end{array}
$$

Write $0x$ for the "missing" term in x.

This result can be written as

$$\frac{x^3 - 2x^2 + 5}{x + 3} = x^2 - 5x + 15 + \frac{-40}{x + 3}$$

CHECK YOURSELF 6

Divide.

$(4x^3 + x + 10) \div (2x - 1)$

You should always arrange the terms of the divisor and dividend in descending form before starting the long division process. This is illustrated in our final example.

Example 8

Divide $5x^2 - x + x^3 - 5$ by $-1 + x^2$.

Write the divisor as $x^2 - 1$ and the dividend as $x^3 + 5x^2 - x - 5$.

$$
\begin{array}{r}
x + 5 \\
x^2 - 1 \overline{) x^3 + 5x^2 - x - 5} \\
\underline{x^3 - x} \\
5x^2 - 5 \\
\underline{5x^2 - 5} \\
0
\end{array}
$$

Write the product of x and $x^2 - 1$, $x^3 - x$, so that like terms fall in the same columns.

CHECK YOURSELF 7

Divide:

$(5x^2 + 10 + 2x^3 + 4x) \div (2 + x^2)$

CHECK YOURSELF ANSWERS

1. (1) $2a^2$; (2) $4mn^2$.
2. (1) $4y^2 - 3y$; (2) $-2a^2 + 3a - 1$; (3) $4m^3n^2 - 3m^2n + 2$. **3.** $x + 6$.
4. $2x + 1 + \frac{20}{3x - 5}$. **5.** $2x^2 - 4x + 7 + \frac{-6}{2x + 3}$. **6.** $2x^2 + x + 1 + \frac{11}{2x - 1}$.
7. $2x + 5$.

4.5 Exercises

Divide.

1. $\frac{18x^6}{9x^2}$ $2x^4$

2. $\frac{20a^7}{5a^5}$ $4a^2$

3. $\frac{35m^3n^2}{7mn^2}$ $5m^2$

4. $\frac{42x^5y^2}{6x^3y}$ $7x^2y$

5. $\frac{3a + 6}{3}$ $a + 2$

6. $\frac{4x - 8}{4}$ $x - 2$

7. $\frac{9b^2 - 12}{3}$ $3b^2 - 4$

8. $\frac{10m^2 + 5m}{5}$ $2m^2 + m$

9. $\frac{16a^3 - 24a^2}{4a}$ $4a^2 - 6a$

10. $\frac{9x^3 + 12x^2}{3x}$ $3x^2 + 4x$

11. $\frac{12m^2 + 6m}{-3m}$ $-4m - 2$

12. $\frac{20b^3 - 25b^2}{-5b}$ $-4b^2 + 5b$

13. $\frac{18a^4 + 12a^3 - 6a^2}{6a}$ $3a^3 + 2a^2 - a$

14. $\frac{21x^5 - 28x^4 + 14x^3}{7x}$ $3x^4 - 4x^3 + 2x^2$

15. $\frac{20x^4y^2 - 15x^2y^3 + 10x^3y}{5x^2y}$ $4x^2y - 3y^2 + 2x$

16. $\frac{16m^3n^3 + 24m^2n^2 - 40mn^3}{8mn^2}$ $2m^2n + 3m - 5n$

Perform the indicated divisions.

17. $\frac{x^2 + 5x + 6}{x + 2}$ $x + 3$

18. $\frac{x^2 + 8x + 15}{x + 3}$ $x + 5$

19. $\frac{x^2 - x - 20}{x + 4}$ $x - 5$

20. $\frac{x^2 - 2x - 35}{x + 5}$ $x - 7$

21. $\frac{2x^2 + 5x - 3}{2x - 1}$ $x + 3$

22. $\frac{3x^2 + 20x - 32}{3x - 4}$ $x + 8$

23. $\frac{2x^2 - 3x - 5}{x - 3}$

$2x + 3 + \frac{4}{x - 3}$

24. $\frac{3x^2 + 17x - 12}{x + 6}$

$3x - 1 + \frac{-6}{x + 6}$

25. $\frac{4x^2 - 18x - 15}{x - 5}$

$4x + 2 + \frac{-5}{x - 5}$

26. $\frac{3x^2 - 18x - 32}{x - 8}$

$3x + 6 + \frac{16}{x - 8}$

27. $\frac{6x^2 - x - 10}{3x - 5}$

$2x + 3 + \frac{5}{3x - 5}$

28. $\frac{4x^2 + 6x - 25}{2x + 7}$

$2x - 4 + \frac{3}{2x + 7}$

29. $\frac{x^3 + x^2 - 4x - 4}{x + 2}$

$x^2 - x - 2$

30. $\frac{x^3 - 2x^2 + 4x - 21}{x - 3}$

$x^2 + x + 7$

31. $\frac{4x^3 + 7x^2 + 10x + 5}{4x - 1}$

$x^2 + 2x + 3 + \frac{8}{4x - 1}$

32. $\frac{2x^3 - 3x^2 + 4x + 4}{2x + 1}$

$x^2 - 2x + 3 + \frac{1}{2x + 1}$

33. $\frac{x^3 - x^2 + 5}{x - 2}$

$x^2 + x + 2 + \frac{9}{x - 2}$

34. $\frac{x^3 + 4x - 3}{x + 3}$

$x^2 - 3x + 13 + \frac{-42}{x + 3}$

35. $\frac{25x^3 + x}{5x - 2}$

$5x^2 + 2x + 1 + \frac{2}{5x - 2}$

36. $\frac{8x^3 - 6x^2 + 2x}{4x + 1}$

$2x^2 - 2x + 1 + \frac{-1}{4x + 1}$

37. $\frac{2x^2 - 8 - 3x + x^3}{x - 2}$

$x^2 + 4x + 5 + \frac{2}{x - 2}$

38. $\frac{x^2 - 18x + 2x^3 + 32}{x + 4}$

$2x^2 - 7x + 10 + \frac{-8}{x + 4}$

39. $\frac{x^4 - 1}{x - 1}$

$x^3 + x^2 + x + 1$

40. $\frac{x^4 + x^2 - 16}{x + 2}$

$x^3 - 2x^2 + 5x - 10 + \frac{4}{x + 2}$

41. $\frac{x^3 - 3x^2 - x + 3}{x^2 - 1}$

$x - 3$

42. $\frac{x^3 + 2x^2 + 3x + 6}{x^2 + 3}$

$x + 2$

43. $\frac{x^4 + 2x^2 - 2}{x^2 + 3}$

$x^2 - 1 + \frac{1}{x^2 + 3}$

44. $\frac{x^4 + x^2 - 5}{x^2 - 2}$ $x^2 + 3 + \frac{1}{x^2 - 2}$

Skillscan (Section 1.2)

Use the distributive property to simplify each expression.

a. $3(4a - 3)$
$12a - 9$

b. $5(3m + 7)$
$15m + 35$

c. $7(6x + 8)$
$42x + 56$

d. $6(4b - 7)$
$24b - 42$

e. $-3(2w + 5)$
$-6w - 15$

f. $-7(3x - 8)$
$-21x + 56$

g. $-5(-3y - 2)$
$15y + 10$

h. $-6(-4a + 2)$
$24a - 12$

Answers

1. $2x^4$ **3.** $5m^2$ **5.** $a + 2$ **7.** $3b^2 - 4$ **9.** $4a^2 - 6a$ **11.** $-4m - 2$ **13.** $3a^3 + 2a^2 - a$
15. $4x^2y - 3y^2 + 2x$ **17.** $x + 3$ **19.** $x - 5$ **21.** $x + 3$ **23.** $2x + 3 + \frac{4}{x - 3}$ **25.** $4x + 2 + \frac{-5}{x - 2}$

27. $2x + 3 + \frac{5}{3x - 5}$ **29.** $x^2 - x - 2$ **31.** $x^2 + 2x + 3 + \frac{8}{4x - 1}$ **33.** $x^2 + x + 2 + \frac{9}{x - 2}$

35. $5x^2 + 2x + 1 + \frac{2}{5x - 2}$ **37.** $x^2 + 4x + 5 + \frac{2}{x - 2}$ **39.** $x^3 + x^2 + x + 1$ **41.** $x - 3$

43. $x^2 - 1 + \frac{1}{x^2 + 3}$ **a.** $12a - 9$ **b.** $15m + 35$ **c.** $42x + 56$ **d.** $24b - 42$ **e.** $-6w - 15$

f. $-21x + 56$ **g.** $15y + 10$ **h.** $24a - 12$

4.6 More on Linear Equations

OBJECTIVE
To solve linear equations when signs of grouping are involved.

Our work earlier in this chapter, on removing signs of grouping, allows us to extend our work in solving equations where parentheses are involved. Recall that once the parentheses are removed, you can solve the equation by the methods learned in Chapter 3.

Let's start by reviewing an example similar to those we considered in Chapter 3.

Example 1

Solve $5(2x - 1) = 25$ for x. First, multiply on the left to remove the parentheses. The original equation is rewritten as

$$10x - 5 = 25$$

Now solve as before.

$$10x - 5 = 25$$

$$10x - 5 + 5 = 25 + 5 \quad \text{Add 5.}$$

$$10x = 30$$

$$\frac{10x}{10} = \frac{30}{10} \quad \text{Divide by 10.}$$

$$x = 3$$

The solution is 3. To check your answer, return to the *original equation*. Substitute 3 for x on the left and right. Then evaluate separately.

Again, returning to the *original equation* will catch any possible errors in the removal of the parentheses.

Left side	Right side
$5(2 \cdot 3 - 1) \stackrel{?}{=} 25$	
$5(6 - 1) \stackrel{?}{=} 25$	
$5 \cdot 5 \stackrel{?}{=} 25$	
$25 = 25$ (true)	

CHECK YOURSELF 1

Solve for x:

$8(3x + 5) = 16$

Example 2

Solve $8 - (3x + 1) = -8$.

First, remove the parentheses. Remember,

$$-(3x + 1) = -3x - 1$$

Change *both* signs.

The original equation then becomes

$$8 - 3x - 1 = -8$$

$$-3x + 7 = -8 \quad \text{Combine like terms.}$$

$$-3x + 7 - 7 = -8 - 7 \quad \text{Subtract 7.}$$

$$-3x = -15$$

$$x = 5 \quad \text{Divide by } -3.$$

The solution is 5. The check is left for the reader.

CHECK YOURSELF 2

Solve for x:

$7 - (4x - 3) = 22$

Our final example illustrates the solution process when more than one grouping symbol is involved in an equation.

Example 3

Solve $2(3x - 1) - 3(x + 5) = 4$.

$$2(3x - 1) - 3(x + 5) = 4$$

$$6x - 2 - 3x - 15 = 4 \quad \text{Use the distributive property to remove the parentheses.}$$

$$3x - 17 = 4 \quad \text{Combine like terms on the left.}$$

$$3x = 21 \quad \text{Add 17.}$$

$$x = 7 \quad \text{Divide by 3.}$$

The solution is 7.

Note how the rules for the order of operations are applied.

To check, return to the original equation to replace x with 7.

$$2(3 \cdot 7 - 1) - 3(7 + 5) \stackrel{?}{=} 4$$
$$2(21 - 1) - 3(7 + 5) \stackrel{?}{=} 4$$
$$2 \cdot 20 - 3 \cdot 12 \stackrel{?}{=} 4$$
$$40 - 36 \stackrel{?}{=} 4$$
$$4 = 4 \qquad \text{(a true statement)}$$

The solution is verified.

CHECK YOURSELF 3

Solve for x:

$5(2x + 4) = 7 - 3(1 - 2x)$

CHECK YOURSELF ANSWERS

1. -1.
2. -3.
3. -4.

4.6 Exercises

Solve each of the following equations for x and check your results.

1. $2(x - 3) = 8$ 7
2. $3(x + 4) = -6$ -6
3. $4(3x + 1) = 28$ 2
4. $5(2x - 3) = 65$ 8
5. $7(5x + 8) = -84$ -4
6. $6(3x + 2) = -60$ -4
7. $10 - (x - 2) = 15$ -3
8. $12 - (x + 3) = 3$ 6
9. $5 - (2x + 1) = 12$ -4
10. $9 - (3x - 2) = 2$ 3
11. $7 - (3x - 5) = 13$ $-\frac{1}{3}$
12. $5 - (4x + 3) = 4$ $-\frac{1}{2}$
13. $5x = 3(x - 6)$ -9
14. $5x = 2(x + 12)$ 8
15. $7(2x + 1) = 12x$ $-\frac{7}{2}$
16. $5(3x - 2) = 12x$ $\frac{10}{3}$
17. $4(7 - x) = 3x$ 4
18. $5(8 - x) = 3x$ 5
19. $2(2x - 1) = 3(x + 1)$ 5
20. $3(3x - 1) = 4(2x + 1)$ 7
21. $5(4x + 2) = 6(3x + 4)$ 7

22. $4(6x - 1) = 7(3x + 2)$ 6 **23.** $9(8x - 1) = 5(4x + 6)$ $\frac{3}{4}$ **24.** $7(3x + 11) = 9(3 - 6x)$ $-\frac{2}{3}$

25. $-4(2x - 1) + 3(3x + 1) = 9$ 2

26. $7(3x + 4) = 8(2x + 5) + 13$ 5

27. $5(2x - 1) - 3(x - 4) = 4(x + 4)$ 3

28. $2(x - 3) - 3(x + 5) = 3(x - 2) - 7$ −2

29. $3(3 - 4x) + 30 = 5x - 2(6x - 7)$ 5

30. $3x - 5(3x - 7) = 2(x + 9) + 45$ −2

Translate the following statements to equations. Let x represent the number in each case.

31. Twice the sum of a number and 4 is 20. $2(x + 4) = 20$

32. The sum of twice a number and 4 is 20. $2x + 4 = 20$

33. 3 times the difference of a number and 5 is 21. $3(x - 5) = 21$

34. The difference of 3 times a number and 5 is 21. $3x - 5 = 21$

35. The sum of twice an integer and 3 times the next consecutive integer is 48.
$2x + 3(x + 1) = 48$

36. The sum of 4 times an integer and twice the next consecutive odd integer is 46.
$4x + 2(x + 2) = 46$

Skillscan (Section 1.1)
Translate each of the following phrases into symbols.

a. 3 more than a $a + 3$

b. 2 less than x $x - 2$

c. 5 more than twice m $2m + 5$

d. 7 less than 4 times p $4p - 7$

e. 3 times r, decreased by 9 $3r - 9$

f. 5 times w, increased by 10 $5w + 10$

g. 12 less than 4 times q $4q - 12$

h. 9 more than twice c $2c + 9$

Answers

1. 7 **3.** 2 **5.** −4 **7.** −3 **9.** −4 **11.** $-\frac{1}{3}$ **13.** −9 **15.** $-\frac{7}{2}$ **17.** 4 **19.** 5 **21.** 7

23. $\frac{3}{4}$ **25.** 2 **27.** 3 **29.** 5 **31.** $2(x + 4) = 20$ **33.** $3(x - 5) = 21$ **35.** $2x + 3(x + 1) = 48$

a. $a + 3$ **b.** $x - 2$ **c.** $2m + 5$ **d.** $4p - 7$ **e.** $3r - 9$ **f.** $5w + 10$ **g.** $4q - 12$ **h.** $2c + 9$

4.7 More Applications

OBJECTIVES

1. To solve word problems involving numbers
2. To solve word problems involving geometric figures
3. To solve mixture problems
4. To solve motion problems

Many applications will lead to equations involving parentheses, and the methods of Section 4.6 will have to be applied during the solution process. Before we look at examples, let's review the five-step process for solving word problems that you studied earlier.

TO SOLVE WORD PROBLEMS

STEP 1 Read the problem carefully. Then reread it to decide what you are asked to find.

STEP 2 Choose a letter to represent the unknown or unknowns. Then represent all of the unknowns of the problem with an expression using the same letter.

STEP 3 Translate the problem to the language of algebra to form an equation.

STEP 4 Solve the equation and answer the question of the original problem.

STEP 5 Verify your solution by returning to the original problem.

These steps are illustrated in the following example.

Example 1

One number is 5 more than a second number. If 3 times the smaller number plus 4 times the larger is 104, find the two numbers.

Step 1 What are you asked to find? You must find the two numbers.

Step 2 Represent the unknowns. Let x be the smaller number. Then

$x + 5$ is the larger.

↑ "5 more than" x

Step 3 Write an equation.

Note that the parentheses are *essential* in writing the correct equation.

$3x + 4(x + 5) = 104$

3 times the smaller — Plus — 4 times the larger

Step 4 Solve the equation.

$$3x + 4(x + 5) = 104$$
$$3x + 4x + 20 = 104$$
$$7x + 20 = 104$$
$$7x = 84$$
$$x = 12$$

The smaller number (x) is 12 and the larger number ($x + 5$) is 17.

Step 5 Check the solution: 12 is the smaller number, and 17 is the larger number.

$3 \cdot 12 + 4 \cdot 17 = 104$ (true)

CHECK YOURSELF 1

One number is 4 more than another. If 6 times the smaller minus 4 times the larger is 4, what are the two numbers?

The solutions for many problems from geometry will also yield equations involving parentheses. Consider the following.

Example 2

The length of a rectangle is 1 cm less than 3 times the width. If the perimeter is 54 cm, find the dimensions of the rectangle.

Step 1 You want to find the dimensions (the width and length).

Step 2 Let x be the width.

Then $3x - 1$ is the length.

3 times the width — 1 less than

Whenever you are working on an application involving geometric figures, you should draw a sketch of the problem including the labels assigned in this step. In this example we would have

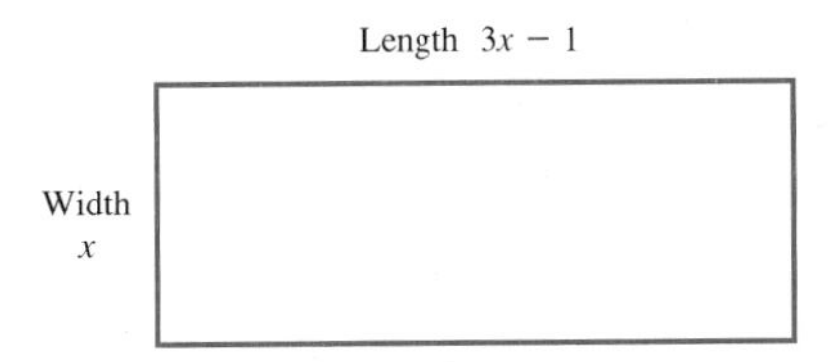

To write an equation, we'll use this formula for the perimeter of a rectangle:

$P = 2W + 2L$

So

$2x + 2(3x - 1) = 54$

Twice the width — Twice the length — Perimeter

Step 4 Solve the equation.

$$
\begin{aligned}
2x + 2(3x - 1) &= 54 \\
2x + 6x - 2 &= 54 \\
8x &= 56 \\
x &= 7
\end{aligned}
$$

Be sure to return to the original statement of the problem when checking your result.

The width x is 7 cm, and the length, $3x - 1$, is 20 cm. We will leave step 5, the check, to the reader.

CHECK YOURSELF 2

The length of a rectangle is 5 in more than twice the width. If the perimeter of the rectangle is 76 in, what are the dimensions of the rectangle?

Let's look at another group of applications called *coin problems*. There is one key idea in solving this type of word problem.

Suppose that you have 8 nickels. How much money do you have? To find the value of the coins, you must *multiply the number of coins by the value of one coin* (in this case 5¢). So if you have 8 nickels, you have $8 \cdot 5$¢, or 40¢. Let's see how this is used in the following example.

Example 3

Jacob has 4 more dimes than nickels. If the value of the coins is \$3.10, how many nickels and how many dimes does he have?

Step 1 You want to find the number of nickels and dimes.

Step 2 Let x be the number of nickels.

Then $x + 4$ is the number of dimes.

↑ 4 more than x

Step 3 Write an equation (in cents).

Note that we have chosen to write our expressions in terms of *cents* to avoid decimals.

The value of the nickels is $5x$.

↑ 5¢ times the number of nickels

The value of the dimes is $10(x + 4)$.

↑ 10¢ times the number of dimes

So

$$5x + 10(x + 4) = 310$$

↑ The total value is 310¢, or \$3.10.

Step 4 Solve the equation.

$$5x + 10(x + 4) = 310$$
$$5x + 10x + 40 = 310$$
$$15x + 40 = 310$$
$$15x = 270$$
$$x = 18 \quad \text{(nickels)}$$
$$\text{and} \quad x + 4 = 22 \quad \text{(dimes)}$$

Jacob has 18 nickels and 22 dimes.

Step 5 To check our result,

18 nickels have value	90¢		
22 dimes have value	220¢		
Total value	310¢	or	\$3.10

The solution is verified.

CHECK YOURSELF 3

Carlos has 3 more quarters than dimes. If the total value of the coins is \$3.90, how many of each type does he have?

Coin problems are actually a part of a larger group called *mixture problems*. Mixture problems give a relationship between groups of different kinds of objects. Look at the following example.

Example 4

Four hundred tickets were sold for a school play. General tickets were \$4, while student tickets were \$3. If the total ticket sales were \$1350, how many of each type of ticket were sold?

Step 1 You want to find the number of each type of ticket sold.

Step 2 Let x be the number of general tickets.

We subtract x, the number of general tickets, from 400, the total number of tickets, to find the number of student tickets.

Then $\underbrace{400 - x}$ student tickets were sold.

↑ 400 tickets were sold in all.

Step 3 The value of each kind of ticket is found in exactly the same way as in the coin problem of Example 3.

General tickets: $4x$ — \$4 for each of the x tickets.

Student tickets: $3(400 - x)$ — \$3 for each of the $400 - x$ tickets.

So to form an equation, we have

$$\underset{\text{Value of general tickets}}{4x} + \underset{\text{Value of student tickets}}{\underbrace{3(400 - x)}} = \underset{\text{Total value}}{1350}$$

Step 4

$$\begin{aligned} 4x + 3(400 - x) &= 1350 \\ 4x + 1200 - 3x &= 1350 \\ x + 1200 &= 1350 \\ x &= 150 \end{aligned}$$

So 150 general and 250 student tickets were sold. We will leave the check to you.

CHECK YOURSELF 4

Cheryl bought 25¢ stamps and 15¢ stamps at the post office. If she purchased 60 stamps at a cost of \$13, how many of each kind did she buy?

The last group of applications we will look at in this section involves *motion problems*. They involve a distance traveled, a rate or speed, and time. To solve motion problems, we need a relationship among these three quantities.

Suppose you travel at a rate of 50 miles per hour (mi/h) on a highway for 6 hours (h). How far (the distance) will you have gone? To find the distance, you multiply:

$$(50 \text{ mi/h})(6 \text{ h}) = 300 \text{ mi}$$

Speed or rate ↗ Time ↑ Distance ↑

In general, if r is a rate, t is the time, and d is the distance traveled,

Be careful to make your units consistent. If a rate is given in *miles per hour*, then the time must be given in *hours*, the distance in *miles*, and so on.

$$d = r \cdot t$$

This is the key relationship, and it will be used in all motion problems. Let's see how it is applied in the following example.

Example 5

On Friday morning Jane drove from her house to the beach in 4 h. In coming back on Sunday afternoon, heavy traffic slowed her speed by 10 mi/h, and the trip took 5 h. What was her average speed (rate) in each direction?

Step 1 We want the speed or rate in each direction.

Step 2 Let x be Jane's speed to the beach. Then $x - 10$ is her return speed.

It is always a good idea to sketch the given information in a motion problem. Here we would have

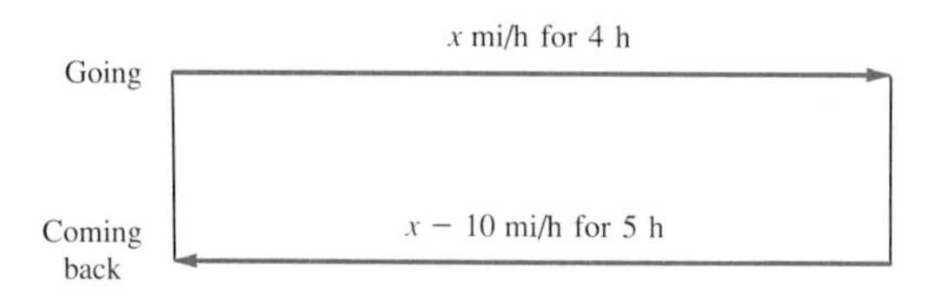

Step 3 Since we know that the distance is the same each way, we can write an equation using the fact that the product of the rate and the time each way must be the same.

So

$$4x = 5(x - 10)$$

Time · rate (going) ↗ Time · rate (coming back) ↑

Distance (going) = distance (coming back)

or

Time · rate = time · rate
(going) (coming back)

Step 4

$$4x = 5(x - 10)$$
$$4x = 5x - 50$$
$$-x = -50$$
$$x = 50 \quad \text{mi/h}$$

So Jane's rate going to the beach was 50 mi/h, and her rate coming back was 40 mi/h.

x was her rate going, $x - 10$ her rate coming back.

Step 5
To check, you should verify that the product of the time and the rate is the same in each direction.

CHECK YOURSELF 5

A plane made a flight (with the wind) between two towns in 2 h. Returning against the wind, the plane's speed was 60 mi/h slower, and the flight took 3 h. What was the plane's speed in each direction?

Example 6

Bruce leaves Las Vegas for Los Angeles at 10 A.M., driving at 50 mi/h. At 11 A.M. Don leaves Los Angeles for Las Vegas, driving at 55 mi/h along the same route. If the cities are 260 mi apart, at what time will they meet?

Step 1 Let's find the time that Bruce travels until they meet.

Step 2 Let x be Bruce's time.

Then $\underbrace{x - 1}$ is Don's time.

Don left 1 h later!

Again, you should draw a sketch of the given information.

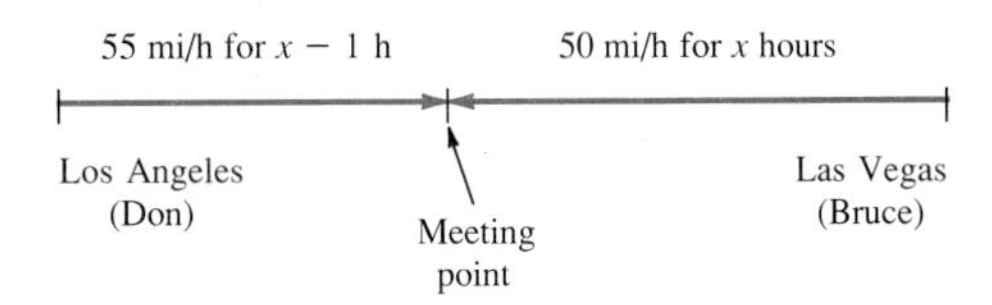

Step 3 To write an equation, we will again need the relationship $d = r \cdot t$. From this equation, we can write

$$\text{Bruce's distance} = 50x$$
$$\text{Don's distance} = 55(x - 1)$$

From the original problem, the sum of those distances is 260 mi, so

$$50x + 55(x - 1) = 260$$

Step 4

$$\begin{aligned} 50x + 55(x - 1) &= 260 \\ 50x + 55x - 55 &= 260 \\ 105x - 55 &= 260 \\ 105x &= 315 \\ x &= 3 \text{ h} \end{aligned}$$

Finally, since Bruce left at 10 A.M., the two will meet at 1 P.M. We will leave the check of this result to the reader.

CHECK YOURSELF 6

At noon a jogger leaves one point, running at 8 mi/h. One hour later a bicyclist leaves the same point, traveling at 20 mi/h in the opposite direction. At what time will they be 36 mi apart?

CHECK YOURSELF ANSWERS

1. The numbers are 10 and 14.
2. The width is 11; the length is 27.
3. 9 dimes and 12 quarters
4. 40 at 25¢ and 20 at 15¢
5. 180 mi/h with the wind and 120 mi/h against the wind.
6. At 2 P.M.

4.7 Exercises

Solve the following word problems. Be sure to show the equation you use for the solution.

1. One number is 8 more than another. If the sum of the smaller number and twice the larger number is 46, find the two numbers. 10, 18

2. One number is 3 less than another. If 4 times the smaller number minus 3 times the larger number is 4, find the two numbers. 13, 16

3. One number is 7 less than another. If 4 times the smaller number plus 2 times the larger number is 62, find the two numbers. 8, 15

4. One number is 10 more than another. If the sum of twice the smaller number and 3 times the larger number is 55, find the two numbers. 5, 15

5. Find two consecutive integers such that the sum of twice the first integer and 3 times the second integer is 28. (*Hint:* If x represents the first integer, $x + 1$ represents the next consecutive integer.) 5, 6

6. Find two consecutive odd integers such that 3 times the first integer is 5 more than twice the second. (*Hint:* If x represents the first integer, $x + 2$ represents the next consecutive odd integer.) 9, 11

7. The length of a rectangle is 1 in more than twice its width. If the perimeter of the rectangle is 74 in, find the dimensions of the rectangle. 12 in, 25 in

8. The length of a rectangle is 5 cm less than 3 times its width. If the perimeter of the rectangle is 46 cm, find the dimensions of the rectangle. 7 cm, 16 cm

9. The length of a rectangular garden is 4 m more than 3 times its width. The perimeter of the garden is 56 m. What are the dimensions of the garden? 6 m, 22 m

10. The length of a rectangular playing field is 5 ft less than twice its width. If the perimeter of the field is 230 ft, find the length and width of the field. 75 ft, 40 ft

11. The base of an isosceles triangle is 3 cm less than the length of the equal sides. If the perimeter of the triangle is 36 cm, find the length of each of the sides.
Legs, 13 cm; base, 10 cm

12. The length of one of the equal legs of an isosceles triangle is 3 in less than twice the length of the base. If the perimeter is 29 in, find the length of each of the sides.
Legs, 11 in; base, 7 in

13. Ken has \$1.35 in nickels and dimes. The number of dimes is one more than twice the number of nickels. How many of each kind of coin does he have? 5 nickels, 11 dimes

14. Marion has \$4.60 in quarters and dimes. If the number of quarters is 2 less than 3 times the number of dimes, how many quarters and how many dimes does she have?
6 dimes, 16 quarters

15. A charity collection jar has \$6.40 in nickels, dimes, and quarters. There are twice as many nickels as dimes and 4 more quarters than dimes. How many of each type of coin are in the jar? 24 nickels, 12 dimes, and 16 quarters

16. For his garage sale, Nick has \$36.00 in change in half-dollars, quarters, and dimes. The number of dimes is 8 more than the number of half-dollars, and the number of quarters is twice the number of half-dollars. Find the number of each type of coin.
40 dimes, 64 quarters, and 32 half-dollars

17. Tickets for a play cost \$8 for the main floor and \$6 in the balcony. If the total receipts from 500 tickets were \$3600, how many of each type of ticket were sold?
200 \$6 tickets, 300 \$8 tickets

18. Tickets for a basketball tournament were \$6 for students and \$9 for nonstudents. Total sales were \$10,500, and 250 more student tickets were sold than nonstudent tickets. How many of each type of ticket were sold? 850 student, 600 nonstudent

19. Todd bought 80 stamps at the post office in 25¢ and 20¢ denominations. If he paid \$18.50 for the stamps, how many of each denomination did he buy?
30 20¢ stamps, 50 25¢ stamps

20. A bank teller has a total of 125 \$10 bills and \$20 bills to start a day. If the value of the bills was \$1650, how many of each denomination did he have? 85 \$10 bills, 40 \$20 bills

21. Tickets for a train excursion were \$120 for a sleeping room, \$80 for a berth, and \$50 for a coach seat. The total ticket sales were \$8600. If there were 20 more berth tickets sold than sleeping room tickets, and 3 times as many coach tickets as sleeping room tickets, how many of each type of ticket were sold? 60 coach, 40 berth, and 20 sleeping room

22. Admission for a baseball game is \$6 for box seats, \$5 for the grandstand, and \$3 for the bleachers. The total receipts for one evening were \$9000. There were 100 more grandstand tickets sold than box seat tickets. Twice as many bleacher tickets were sold as box seat tickets. How many tickets of each type were sold?
500 box seats, 600 grandstand, and 1000 bleachers

23. Horace drove for 3 h to attend a meeting. On the return trip, his speed was 10 mi/h less, and the trip took 4 h. What was his speed each way? 40 mi/h, 30 mi/h

24. A bicyclist rode into the country for 5 h. In returning, her speed was 5 mi/h faster and the trip took 4 h. What was her speed each way? 20 mi/h, 25 mi/h

25. A car leaves a city and goes north at a rate of 50 mi/h at 2 P.M. One hour later a second car leaves, traveling south at a rate of 40 mi/h. At what time will the two cars be 320 mi apart? 6 P.M.

26. A passenger bus leaves a station at 1 P.M., traveling west at an average rate of 44 mi/h. One hour later a second bus leaves the same station, traveling east at a rate of 48 mi/h. At what time will the two buses be 274 mi apart? 4:30 P.M.

27. At 8:00 A.M., Catherine leaves on a trip at 45 mi/h. One hour later, Max decides to join her and leaves along the same route, traveling at 54 mi/h. When will Max catch up with Catherine? 2 P.M.

28. Martina leaves home, bicycling at a rate of 24 mi/h. Two hours later, John leaves, driving at a rate of 48 mi/h. How long will it take John to catch up with Martina? 4 h

29. Jean leaves Boston for Baltimore at 10 A.M., traveling at 45 mi/h. One hour later, Bill leaves Baltimore for Boston on the same route, traveling at 50 mi/h. If the two cities are 425 mi apart, when will Jean and Bill meet? 3 P.M.

30. A train leaves town A for town B, traveling at a rate of 35 mi/h. At the same time, a second train leaves town B for town A at 45 mi/h. If the two towns are 320 mi apart, how long will it take for the trains to meet? 4 h

Answers

1. 10, 18 **3.** 8, 15 **5.** 5, 6 **7.** 12 in, 25 in **9.** 6 m, 22 m **11.** Legs, 13 cm; base, 10 cm **13.** 5 nickels, 11 dimes **15.** 24 nickels, 12 dimes, and 16 quarters **17.** 200 $6 tickets, 300 $8 tickets **19.** 30 20¢ stamps, 50 25¢ stamps **21.** 60 coach, 40 berth, and 20 sleeping room **23.** 40 mi/h, 30 mi/h **25.** 6 P.M. **27.** 2 P.M. **29.** 3 P.M.

Summary

Polynomials [4.1]

Polynomial An algebraic expression made up of specific terms in which the only allowable exponents are whole numbers. These terms are connected by a + or a − sign. Each sign (+ or −) is attached to the term following that sign.

$4x^3 - 3x^2 + 5x$ is a polynomial. The terms of $4x^3 - 3x^2 + 5x$ are $4x^3$, $-3x^2$, and $5x$.

Term A *term* is a number, or the product of a number and one or more variables, raised to a power.

Coefficient In each term of a polynomial, the number is called the *numerical coefficient* or, more simply, the *coefficient* of that term.

The coefficients of $4x^3 - 3x^2 + 5x$ are 4, −3, and 5.

Types of Polynomials Polynomials can be classified according to the number of terms in the polynomial.

A *mono*mial has one term.

$2x^3$ is a monomial.

A *bi*nomial has two terms.

$3x^2 - 7x$ is a binomial.

A *tri*nomial has three terms.

$5x^5 - 5x^3 + 2$ is a trinomial.

Degree The highest power of the variable appearing in any one term.

The degree of $4x^5 - 5x^3 + 3x$ is 5.

Descending Form A polynomial is in descending form when it is written with the highest-degree term first, the next highest-degree term second, and so on.

$4x^5 - 5x^3 + 3x$ is written in descending form.

$5x^3 - 3x^4$ is *not* in descending form.

Adding and Subtracting Polynomials [4.2]

Removing Signs of Grouping

1. If a plus sign (+) or no sign at all appears in front of parentheses, just remove the parentheses. No other changes are necessary.

$(2x + 3) + (3x - 5)$
$= 2x + 3 + 3x - 5$

2. If a minus sign (−) appears in front of parentheses, the parentheses can be removed by changing the sign of each term inside the parentheses.

$-(2x^2 - 4x + 5)$
$= -2x^2 + 4x - 5$
↑ ↑ ↑
Sign changes

Adding Polynomials Remove the signs of grouping. Then collect and combine any like terms.

$(2x^2 + 3x) + (4x^2 - 5x)$
$= 2x^2 + 3x + 4x^2 - 5x$
$= 6x^2 - 2x$

Subtracting Polynomials Remove the signs of grouping by changing the signs of each term in the polynomial being subtracted. Then combine any like terms.

$(4x^2 - 5x) - (3x^2 - 2x)$
$= 4x^2 - 5x - 3x^2 + 2x$
$= x^2 - 3x$

Multiplying Polynomials [4.3]

To Multiply a Polynomial by a Monomial Multiply each term of the polynomial by the monomial, and add the results.

$3x^3(2x^2 - 5x)$
$= 3x^3 \cdot 2x^2 - 3x^3 \cdot 5x$
$= 6x^5 - 15x^4$

To Multiply a Binomial by a Binomial Use the FOIL method:

$(a + b)(c + d)$

F: $a \cdot c$ L: $b \cdot d$ I: $b \cdot c$ O: $a \cdot d$

$(2x - 3)(3x + 5)$
$= 6x^2 + 10x - 9x - 15$
F O I L
$= 6x^2 + x - 15$

To Multiply a Polynomial by a Polynomial Arrange the polynomials vertically. Multiply each term of the lower polynomial by each term of the upper polynomial, and add the results.

$$\begin{array}{r} x^2 - 3x + 5 \\ 2x - 3 \\ \hline -3x^2 + 9x - 15 \\ 2x^3 - 6x^2 + 10x \\ \hline 2x^3 - 9x^2 + 19x - 15 \end{array}$$

Special Products [4.4]

To Square a Binomial

1. The first term of the square is the square of the first term of the binomial.
2. The middle term is twice the product of the two terms of the binomial.
3. The last term is the square of the last term of the binomial.

$(2x - 5)^2$
$= 4x^2 + 2 \cdot 2x \cdot (-5) + 25$
$= 4x^2 - 20x + 25$

To Multiply Binomials That Differ Only in Sign The product is the square of the first term minus the square of the second term.

$(a + b)(a - b) = a^2 - b^2$

$(2x - 5y)(2x + 5y)$
$= (2x)^2 - (5y)^2$
$= 4x^2 - 25y^2$

Dividing Polynomials [4.5]

To Divide a Polynomial by a Monomial Divide each term of the polynomial by the monomial. Then combine the results.

$$\frac{9x^4 + 6x^3 - 15x^2}{3x}$$
$$= \frac{9x^4}{3x} + \frac{6x^3}{3x} - \frac{15x^2}{3x}$$
$$= 3x^3 + 2x^2 - 5x$$

To Divide a Polynomial by a Polynomial Use the long division method.

$$\begin{array}{r} x + 5 \\ x - 3\overline{)x^2 + 2x - 7} \\ \underline{x^2 - 3x} \\ 5x - 7 \\ \underline{5x - 15} \\ 8 \end{array}$$

Linear Equations and Applications [4.6] to [4.7]

Solving Linear Equations If parentheses (or other signs of grouping) are involved in an equation, use the distributive law to multiply, and remove the signs of grouping. Then solve the resulting equation by the methods of Chapter 3.

Solve:

$$5 - 2(3x - 1) = 31$$
$$5 - 6x + 2 = 31$$
$$-6x = 24$$
$$\frac{-6x}{-6} = \frac{24}{-6}$$
$$x = -4$$

Applying Linear Equations Use the same five-step process introduced in Section 3.8.

Summary Exercises Chapter 4

This summary exercise set is provided to give you practice with each of the objectives of the chapter. Each exercise is keyed to the appropriate chapter section. The answers are provided in the instructor's manual. Your instructor will give you guidelines on how to best use these exercises in your instructional setting.

[4.1] Classify the following polynomials as monomials, binomials, or trinomials, where possible.

1. $5x^3 - 2x^2$ Binomial

2. $7x^5$ Monomial

3. $4x^5 - 8x^3 + 5$ Trinomial

4. $x^3 + 2x^2 - 5x + 3$ Not classified

5. $9a - 18a^2$ Binomial

[4.1] Arrange in descending form, if necessary, and give the degree of each polynomial.

6. $5x^5 + 3x^2$
$5x^5 + 3x^2$, 5

7. $9x$
$9x$, 1

8. $6x^2 + 4x^4 + 6$
$4x^4 + 6x^2 + 6$, 4

9. $5 + x$
$x + 5$, 1

10. -8
-8, 0

11. $9x^4 - 3x + 7x^6$
$7x^6 + 9x^4 - 3x$, 6

[4.2] Add.

12. $9a^2 - 5a$ and $12a^2 + 3a$
$21a^2 - 2a$

13. $5x^2 + 3x - 5$ and $4x^2 - 6x - 2$
$9x^2 - 3x - 7$

14. $5y^3 - 3y^2$ and $4y + 3y^2$
$5y^3 + 4y$

[4.2] Subtract.

15. $4x^2 - 3x$ from $8x^2 + 5x$
$4x^2 + 8x$

16. $2x^2 - 5x - 7$ from $7x^2 - 2x + 3$
$5x^2 + 3x + 10$

17. $5x^2 + 3$ from $9x^2 - 4x$
$4x^2 - 4x - 3$

[4.2] Perform the indicated operations.

18. Subtract $5x - 3$ from the sum of $9x + 2$ and $-3x - 7$. $x - 2$

19. Subtract $5a^2 - 3a$ from the sum of $5a^2 + 2$ and $7a - 7$. $10a - 5$

20. Subtract the sum of $16w^2 - 3w$ and $8w + 2$ from $7w^2 - 5w + 2$. $-9w^2 - 10w$

[4.2] Add, using the vertical method.

21. $x^2 + 5x - 3$ and $2x^2 + 4x - 3$
$3x^2 + 9x - 6$

22. $9b^2 - 7$ and $8b + 5$
$9b^2 + 8b - 2$

23. $x^2 + 7$, $3x - 2$, and $4x^2 - 8x$
$5x^2 - 5x + 5$

[4.2] Subtract, using the vertical method.

24. $5x^2 - 3x + 2$ from $7x^2 - 5x - 7$
$2x^2 - 2x - 9$

25. $8m - 7$ from $9m^2 - 7$
$9m^2 - 8m$

[4.3] Multiply.

26. $(5a^3)(a^2)$
$5a^5$

27. $(2x^2)(3x^5)$
$6x^7$

28. $(-9p^3)(-6p^2)$
$54p^5$

29. $(3a^2b^3)(-7a^3b^4)$
$-21a^5b^7$

30. $5(3x - 8)$
$15x - 40$

31. $4a(3a + 7)$
$12a^2 + 28a$

32. $(-5rs)(2r^2s - 5rs)$
$-10r^3s^2 + 25r^2s^2$

33. $7mn(3m^2n - 2mn^2 + 5mn)$
$21m^3n^2 - 14m^2n^3 + 35m^2n^2$

34. $(x + 5)(x + 4)$
$x^2 + 9x + 20$

35. $(w - 9)(w - 10)$
$w^2 - 19w + 90$

36. $(a - 7b)(a + 7b)$
$a^2 - 49b^2$

37. $(p - 3q)^2$
$p^2 - 6pq + 9q^2$

38. $(a + 4b)(a + 3b)$
$a^2 + 7ab + 12b^2$

39. $(b - 8)(2b + 3)$
$2b^2 - 13b - 24$

40. $(3x - 5y)(2x - 3y)$
$6x^2 - 19xy + 15y^2$

41. $(5r + 7s)(3r - 9s)$
$15r^2 - 24rs - 63s^2$

42. $(y + 2)(y^2 - 2y + 3)$
$y^3 - y + 6$

43. $(b + 3)(b^2 - 5b - 7)$
$b^3 - 2b^2 - 22b - 21$

44. $(x - 2)(x^2 + 2x + 4)$
$x^3 - 8$

45. $(m^2 - 3)(m^2 + 7)$
$m^4 + 4m^2 - 21$

46. $2x(x + 5)(x - 6)$
$2x^3 - 2x^2 - 60x$

47. $3a(2a - 5b)(2a - 7b)$
$12a^3 - 72a^2b + 105ab^2$

Find the following products.

48. $(x + 7)^2$
$x^2 + 14x + 49$

49. $(a - 8)^2$
$a^2 - 16a + 64$

50. $(2w - 5)^2$
$4w^2 - 20w + 25$

51. $(3p + 4)^2$
$9p^2 + 24p + 16$

52. $(a + 7b)^2$
$a^2 + 14ab + 49b^2$

53. $(8x - 3y)^2$
$64x^2 - 48xy + 9y^2$

54. $(x - 5)(x + 5)$
$x^2 - 25$

55. $(y + 9)(y - 9)$
$y^2 - 81$

56. $(2m + 3)(2m - 3)$
$4m^2 - 9$

57. $(3r - 7)(3r + 7)$ $9r^2 - 49$

58. $(5r - 2s)(5r + 2s)$ $25r^2 - 4s^2$

59. $(7a + 3b)(7a - 3b)$ $49a^2 - 9b^2$

60. $2x(x - 5)^2$ $2x^3 - 20x^2 + 50x$

61. $3c(c + 5d)(c - 5d)$ $3c^3 - 75cd^2$

[4.5] Divide.

62. $\frac{9a^5}{3a^2}$ $3a^3$

63. $\frac{24m^4n^2}{6m^2n}$ $4m^2n$

64. $\frac{15a - 10}{5}$ $3a - 2$

65. $\frac{32a^3 + 24a}{8a}$ $4a^2 + 3$

66. $\frac{9r^2s^3 - 18r^3s^2}{-3rs^2}$ $-3rs + 6r^2$

67. $\frac{35x^3y^2 - 21x^2y^3 + 14x^3y}{7x^2y}$ $5xy - 3y^2 + 2x$

[4.5] Perform the indicated long division.

68. $\frac{x^2 - 2x - 15}{x + 3}$ $x - 5$

69. $\frac{2x^2 + 9x - 35}{2x - 5}$ $x + 7$

70. $\frac{x^2 - 8x + 17}{x - 5}$ $x - 3 + \frac{2}{x - 5}$

71. $\frac{6x^2 - x - 10}{3x + 4}$ $2x - 3 + \frac{2}{3x + 4}$

72. $\frac{6x^3 + 14x^2 - 2x - 6}{6x + 2}$ $x^2 + 2x - 1 + \frac{-4}{6x + 2}$

73. $\frac{4x^3 + x + 3}{2x - 1}$ $2x^2 + x + 1 + \frac{4}{2x - 1}$

74. $\frac{3x^2 + x^3 + 5 + 4x}{x + 2}$ $x^2 + x + 2 + \frac{1}{x + 2}$

75. $\frac{2x^4 - 2x^2 - 10}{x^2 - 3}$ $2x^2 + 4 + \frac{2}{x^2 - 3}$

[4.6] Solve the following equations for x.

76. $4(3x - 2) = 52$ 5

77. $8 - (4x + 1) = 15$ -2

78. $7 - (3x - 5) = 14$ $-\frac{2}{3}$

79. $9 - 3(x - 2) = 1 - x$ 7

80. $5(4x - 3) = 3(5x - 2) + 16$ 5

81. $5(8x + 4) - 2(4x - 2) = 48$ $\frac{3}{4}$

[4.7] Solve the following word problems.

82. One number is 6 more than another. If the sum of 3 times the smaller number and twice the larger number is 47, find the two numbers. 7, 13

83. Find two consecutive integers such that twice the first integer is 18 less than 3 times the second integer. 15, 16

84. Find three consecutive even integers such that the sum of the first integer and three times the second integer is 3 times the third integer. 6, 8, 10

85. The length of a rectangle is 5 in more than its width. If the perimeter of the rectangle is 50 in, find the dimensions of the rectangle. 10 in, 15 in

86. The length of a doubles tennis court is 6 ft more than twice its width. If the perimeter of the court is 228 ft, find the dimensions of the court. 36 ft, 78 ft

87. Dave has 33 coins, all nickels and dimes, with a value of $2.70. How many of each type of coin does he have? 12 nickels, 21 dimes

88. A cashier starts the day with $1690 in $5, $10, and $20 bills. If he has 4 more $10 bills than $5 bills and twice as many $20 bills as $5 bills, how many of each denomination does he have? 30 $5 bills, 34 $10 bills, 60 $20 bills

89. A boat makes a trip upriver against the current in 6 h. Coming back down the river, the boat can travel 6 mi/h faster and make the trip in 4 h. What is the speed of the boat in each direction? 12 mi/h, 18 mi/h

90. At 9 A.M. David left New Orleans for Tallahassee, averaging 47 mi/h. Two hours later, Gloria left Tallahassee for New Orleans along the same route, driving 5 mi/h faster than David. If the two cities are 391 mi apart, at what time did David and Gloria meet? 2 P.M.

Self-Test for Chapter Four

The purpose of this self-test is to help you check your progress and to review for a chapter test in class. Allow yourself about an hour to take the test. When you are done, check your answers in the back of the book. If you missed any problems, be sure to go back and review the appropriate sections in the chapter and do the exercises that are provided.

Classify the following polynomials as monomials, binomials, or trinomials.

1. $6x^2 + 7x$ Binomial

2. $5x^2 + 8x - 8$ Trinomial

Arrange in descending form, and give the coefficients and the degree of the polynomial.

3. $-3x^2 + 8x^4 - 7$ $8x^4 - 3x^2 - 7$; 8, −3, −7; 4

Add.

4. $3x^2 - 7x + 2$ and $7x^2 - 5x - 9$
$10x^2 - 12x - 7$

5. $7a^2 - 3a$ and $7a^3 + 4a^2$
$7a^3 + 11a^2 - 3a$

Subtract.

6. $5x^2 - 2x + 5$ from $8x^2 + 9x - 7$
$3x^2 + 11x - 12$

7. $2b^2 + 5$ from $3b^2 - 7b$
$b^2 - 7b - 5$

8. $5a^2 + a$ from the sum of $3a^2 - 5a$ and $9a^2 - 4a$ $7a^2 - 10a$

Add, using the vertical method.

9. $x^2 + 3$, $5x - 7$, and $3x^2 - 2$ $4x^2 + 5x - 6$

Subtract, using the vertical method.

10. $3x^2 - 5$ from $5x^2 - 7x$ $2x^2 - 7x + 5$

Multiply.

11. $5ab(3a^2b - 2ab + 4ab^2)$
$15a^3b^2 - 10a^2b^2 + 20a^2b^3$

12. $(x - 2)(3x + 7)$ $3x^2 + x - 14$

13. $(a - 7b)(a + 7b)$ $a^2 - 49b^2$

Multiply.

14. $(3m + 2n)^2$ $9m^2 + 12mn + 4n^2$

15. $(2x + y)(x^2 + 3xy - 2y^2)$
$2x^3 + 7x^2y - xy^2 - 2y^3$

Divide.

16. $\dfrac{14x^3y - 21xy^2}{7xy}$ $2x^2 - 3y$

17. $\dfrac{20c^3d - 30cd + 45c^2d^2}{5cd}$ $4c^2 - 6 + 9cd$

18. $(x^2 - 2x - 24) \div (x + 4)$ $x - 6$

19. $(2x^2 + x + 4) \div (2x - 3)$ $x + 2 + \dfrac{10}{2x - 3}$

20. $(6x^3 - 7x^2 + 3x + 9) \div (3x + 1)$ $2x^2 - 3x + 2 + \dfrac{7}{3x + 1}$

Solve the following equations for x.

21. $3(5x - 3) = 51$ 4

22. $7 - 2(3x - 4) = 5 + 11(6 + 2x)$ −2

Solve the following word problems.

23. One number is 5 more than another. If the sum of 4 times the smaller number and 3 times the larger is 57, find the two numbers. 6, 11

24. Sydney has 45 coins, all dimes and quarters, with a value of $8.25. How many dimes and quarters does she have? 20 dimes, 25 quarters

25. A plane makes a trip against a headwind in 9 h. Returning with the wind, the plane can travel 50 mi/h faster and makes the trip in 7 h. What is the speed of the plane in each direction? 175 mi/h, 225 mi/h

Chapter Five

Factoring

5.1 Factoring—An Introduction

OBJECTIVE
To factor a monomial from a polynomial.

In Chapter 4 you were given factors and asked to find a product. We are now going to reverse the process. You will be given a polynomial and asked to write the polynomial as a product of its factors. This is called *factoring*.

Let's start with an example from arithmetic. To *multiply* $5 \cdot 7$, you write

You find the product, 35, by multiplying.

$5 \cdot 7 = 35$

To *factor* 35, you would write

$35 = 5 \cdot 7$

Factoring is just the *reverse* of multiplication.
Now let's look at factoring in algebra. Up to now you have used the distributive property as

$$a(b + c) = ab + ac$$

For instance,

$$3(x + 5) = 3x + 15$$

3 and $x + 5$ are the factors of $3x + 15$.

To use the distributive property in factoring, we apply that property in the opposite fashion, as

$$ab + ac = a(b + c)$$

The property allows us to remove the common monomial factor a from $ab + ac$.

To use this in factoring, the first step is to see whether each term of the polynomial has a common monomial factor.

In our earlier example,

$$3x + 15 = 3 \cdot x + 3 \cdot 5$$

(Common factor: 3)

So, by the distributive property,

$$3x + 15 = 3(x + 5)$$

Again, factoring is just the reverse of multiplication.

To check this, multiply $3(x + 5)$.

The diagram below will relate the ideas of multiplication and factoring.

Multiplying →

$$3(x + 5) = 3x + 15$$

← Factoring

To factor a monomial from a polynomial, you can use these steps:

TO FACTOR A MONOMIAL FROM A POLYNOMIAL

STEP 1 Find the monomial with the largest common numerical coefficient and the largest power of each common variable. This is called the *greatest common factor* (GCF).

STEP 2 Indicate the GCF present in each term, then apply the distributive law.

STEP 3 Mentally check your factoring by multiplication.

The following examples illustrate the use of this rule for factoring polynomials.

Example 1

(*a*) Factor $8x^2 + 12x$.

The largest common numerical factor of 8 and 12 is 4, and x is the variable factor with the largest common power. So $4x$ is the GCF. Write

$$8x^2 + 12x = 4x \cdot 2x + 4x \cdot 3$$

(GCF: $4x$)

Now, by the distributive property, we have

$$8x^2 + 12x = 4x(2x + 3)$$

It is always a good idea to check your answer by multiplying to make sure that you get the original polynomial. Try it here. Multiply $4x$ by $2x + 3$.

(*b*) Factor $6a^4 - 18a^2$.

The GCF in this case is $6a^2$. Write

$$6a^4 - 18a^2 = \underbrace{6a^2}_{\text{GCF}} \cdot a^2 - \underbrace{6a^2}_{\text{GCF}} \cdot 3$$

Again, using the distributive property yields

$$6a^4 - 18a^2 = 6a^2(a^2 - 3)$$

You should check this by multiplying.

Note: In part *b* of the previous example, it is also true that

$$6a^4 - 18a^2 = 3a(2a^3 - 6a)$$

However, this is *not completely factored.* Do you see why? You want to find the common monomial factor with the *largest* possible coefficient and the *largest* exponent, in this case $6a^2$.

CHECK YOURSELF 1

Factor the following polynomials.

1. $5x + 20$ **2.** $6x^2 - 24x$ **3.** $10a^3 - 15a^2$

The process is exactly the same for polynomials with any number of terms. Consider the following examples.

Example 2

(*a*) Factor $5x^2 - 10x + 15$.

The GCF is 5.

$$5x^2 - 10x + 15 = \underbrace{5}_{\text{GCF}} \cdot x^2 - \underbrace{5}_{\text{GCF}} \cdot 2x + \underbrace{5}_{\text{GCF}} \cdot 3$$

$$= 5(x^2 - 2x + 3)$$

(*b*) Factor $6ab + 9ab^2 - 15a^2$.

The GCF is $3a$.

$$6ab + 9ab^2 - 15a^2 = 3a \cdot 2b + 3a \cdot 3b^2 - 3a \cdot 5a$$

GCF

$$= 3a(2b + 3b^2 - 5a)$$

(*c*) Factor $4a^4 + 12a^3 - 20a^2$.

The GCF is $4a^2$.

$$4a^4 + 12a^3 - 20a^2 = 4a^2 \cdot a^2 + 4a^2 \cdot 3a - 4a^2 \cdot 5$$

GCF

$$= 4a^2(a^2 + 3a - 5)$$

CHECK YOURSELF 2

Factor each of the following polynomials.

1. $8b^2 + 16b - 32$ **2.** $4xy - 8x^2y + 12x^3$
3. $7x^4 - 14x^3 + 21x^2$

With practice you should be able to factor polynomials like the ones we have been considering without having to write out the factors in each term. Try it in the following example.

Example 3

(*a*) Factor $6a^2b + 9ab^2 + 3ab$.

Mentally note that 3, a, and b are factors of each term, so

$$6a^2b + 9ab^2 + 3ab = 3ab(2a + 3b + 1)$$

(*b*) Factor $9x^4 - 27x^3 - 18x^2$.

Mentally note that 9 and x^2 are factors of each term.

Thus

$$9x^4 - 27x^3 - 18x^2 = 9x^2(x^2 - 3x - 2)$$

(*c*) Factor $3m^4 + 9m^3 - 12m^2 + 15m$.

Here 3 and m are factors of each term.

$$3m^4 + 9m^3 - 12m^2 + 15m = 3m(m^3 + 3m^2 - 4m + 5)$$

Again, you should check each of these examples by multiplying.

CHECK YOURSELF 3

Factor each of the following polynomials.

1. $5x^2y^2 - 10xy^2 + 15x^2y$
2. $8p^4 - 16p^3 - 40p^2$
3. $8a^4 - 16a^3 + 20a^2 - 24a$

Sometimes the GCF of an expression will be a binomial. Our final example illustrates.

Example 4

(*a*) Factor $x(x + y) + 3(x + y)$. Note that $x + y$ is a common binomial factor for each term. Removing that factor gives

$$x(x + y) + 3(x + y) = (x + y)(x + 3)$$

(*b*) Factor $a(a - b) - 2(a - b)$. Here $a - b$ is the common factor so

$$a(a - b) - 2(a - b) = (a - b)(a - 2).$$

CHECK YOURSELF 4

Factor each of the following.

1. $y(y - 1) - 7(y - 1)$
2. $a(a + b) + 3b(a + b)$

CHECK YOURSELF ANSWERS

1. (1) $5(x + 4)$; (2) $6x(x - 4)$; (3) $5a^2(2a - 3)$.
2. (1) $8(b^2 + 2b - 4)$; (2) $4x(y - 2xy + 3x^2)$; (3) $7x^2(x^2 - 2x + 3)$.
3. (1) $5xy(xy - 2y + 3x)$; (2) $8p^2(p^2 - 2p - 5)$; (3) $4a(2a^3 - 4a^2 + 5a - 6)$.
4. (1) $(y - 1)(y - 7)$; (2) $(a + b)(a + 3b)$.

5.1 Exercises

Find the greatest common factor for the following sets of terms.

1. 6, 8 2
2. 10, 125 5
3. 24, 16, 80 8
4. 121, 33, 66 11
5. x^2, x^5 x^2
6. y^7, y^9 y^7
7. a^3, a^6, a^9 a^3
8. b^4, b^6, b^8 b^4
9. $5x^4, 10x^5$ $5x^4$

10. $8y^9, 24y^3$ $8y^3$

11. $8a^4, 6a^6, 10a^{10}$ $2a^4$

12. $9b^3, 6b^5, 12b^4$ $3b^3$

13. $6xy^3, 9x^3y, 21x^2y^2$ $3xy$

14. $24a^2b^2, 12a^4b^4, 18a^5b^3$ $6a^2b^2$

15. $15a^2b, 5b^2c, 10b$ $5b$

16. $27x^2, 18x^2y^2, 6y^2$ 3

17. $15a^2bc^2, 9ab^2c^2, 6a^2b^2c^2$ $3abc^2$

18. $18x^3y^2z^3, 27x^4y^2z^3, 81xy^2z$ $9xy^2z$

19. $(x + y)^2, (x + y)^3$ $(x + y)^2$

20. $12(a + b)^4, 4(a + b)^3$ $4(a + b)^3$

Factor each of the following polynomials.

21. $8a + 4$ $4(2a + 1)$

22. $5x - 15$ $5(x - 3)$

23. $14m - 21n$ $7(2m - 3n)$

24. $6p + 12q$ $6(p + 2q)$

25. $6m^2 - 9m$ $3m(2m - 3)$

26. $18n^2 + 27n$ $9n(2n + 3)$

27. $10s^2 + 5s$ $5s(2s + 1)$

28. $12y^2 - 6y$ $6y(2y - 1)$

29. $12x^2 + 24x$ $12x(x + 2)$

30. $14b^2 - 28b$ $14b(b - 2)$

31. $15a^3 - 25a^2$ $5a^2(3a - 5)$

32. $36b^4 + 24b^2$ $12b^2(3b^2 + 2)$

33. $6pq + 18p^2q$ $6pq(1 + 3p)$

34. $8ab - 24ab^2$ $8ab(1 - 3b)$

35. $7m^3n - 21mn^3$ $7mn(m^2 - 3n^2)$

36. $36p^2q^2 - 9pq$ $9pq(4pq - 1)$

37. $6x^2 - 18x + 30$ $6(x^2 - 3x + 5)$

38. $7a^2 + 21a - 42$ $7(a^2 + 3a - 6)$

39. $3a^3 + 6a^2 - 12a$ $3a(a^2 + 2a - 4)$

40. $5x^3 - 15x^2 + 25x$ $5x(x^2 - 3x + 5)$

41. $4m + 8mn - 16mn^2$ $4m(1 + 2n - 4n^2)$

42. $9s - 12st + 15st^2$ $3s(3 - 4t + 5t^2)$

43. $7x^2y - 14xy + 28xy^2$ $7xy(x - 2 + 4y)$

44. $6a^2b - 18ab + 24ab^2$ $6ab(a - 3 + 4b)$

45. $10r^3s^2 + 25r^2s^2 - 15r^2s^3$
$5r^2s^2(2r + 5 - 3s)$

46. $28x^2y^3 - 35x^2y^2 + 42x^3y$
$7x^2y(4y^2 - 5y + 6x)$

47. $9a^5 - 15a^4 + 21a^3 - 27a$
$3a(3a^4 - 5a^3 + 7a^2 - 9)$

48. $8p^6 - 40p^4 + 24p^3 - 16p^2$
$8p^2(p^4 - 5p^2 + 3p - 2)$

49. $15m^3n^2 - 20m^2n + 35mn^3 - 10mn$
$5mn(3m^2n - 4m + 7n^2 - 2)$

50. $14ab^4 + 21a^2b^3 - 35a^3b^2 + 28ab^2$
$7ab^2(2b^2 + 3ab - 5a^2 + 4)$

51. $x(x - 2) + 3(x - 2)$
$(x - 2)(x + 3)$

52. $y(y + 5) - 3(y + 5)$
$(y + 5)(y - 3)$

53. $p(p - 2q) - q(p - 2q)$
$(p - 2q)(p - q)$

54. $2c(c + d) + 3d(c + d)$
$(c + d)(2c + 3d)$

55. $3(x + y)^2 + 9(x + y)$
$3(x + y)(x + y + 3)$

56. $6(a - b)^2 - 12(a - b)$
$6(a - b)(a - b - 2)$

57. The GCF of $2x - 6$ is 2. The GCF of $5x + 10$ is 5. Find the greatest common factor of the product $(2x - 6)(5x + 10)$. 10

58. The GCF of $3z + 12$ is 3. The GCF of $4z + 8$ is 4. Find the GCF of the following product: $(3z + 12)(4z + 8)$. 12

Skillscan (Section 4.4)
Multiply.

a. $(x - 1)(x + 1)$
$x^2 - 1$

b. $(a + 7)(a - 7)$
$a^2 - 49$

c. $(x - y)(x + y)$
$x^2 - y^2$

d. $(2x - 5)(2x + 5)$
$4x^2 - 25$

e. $(3a - b)(3a + b)$
$9a^2 - b^2$

f. $(5a - 4b)(5a + 4b)$
$25a^2 - 16b^2$

Answers

We provide the solutions for the odd-numbered exercises at the end of each exercise set. The solutions for the even-numbered exercises are found at the back of the book.

1. 2 **3.** 8 **5.** x^2 **7.** a^3 **9.** $5x^4$ **11.** $2a^4$ **13.** $3xy$ **15.** $5b$ **17.** $3abc^2$ **19.** $(x + y)^2$ **21.** $4(2a + 1)$ **23.** $7(2m - 3n)$ **25.** $3m(2m - 3)$ **27.** $5s(2s + 1)$ **29.** $12x(x + 2)$ **31.** $5a^2(3a - 5)$ **33.** $6pq(1 + 3p)$ **35.** $7mn(m^2 - 3n^2)$ **37.** $6(x^2 - 3x + 5)$ **39.** $3a(a^2 + 2a - 4)$ **41.** $4m(1 + 2n - 4n^2)$ **43.** $7xy(x - 2 + 4y)$ **45.** $5r^2s^2(2r + 5 - 3s)$ **47.** $3a(3a^4 - 5a^3 + 7a^2 - 9)$ **49.** $5mn(3m^2n - 4m + 7n^2 - 2)$ **51.** $(x - 2)(x + 3)$ **53.** $(p - 2q)(p - q)$ **55.** $3(x + y)(x + y + 3)$ **57.** 10 **a.** $x^2 - 1$ **b.** $a^2 - 49$ **c.** $x^2 - y^2$ **d.** $4x^2 - 25$ **e.** $9a^2 - b^2$ **f.** $25a^2 - 16b^2$

5.2 The Difference of Squares

OBJECTIVE
To factor a binomial that is a difference of two squares.

In Section 4.4 we dealt with some special products. Recall the following formula for the product of a sum and difference of two terms:

$$(a + b)(a - b) = a^2 - b^2$$

This also means that a binomial of the form $a^2 - b^2$ (called a *difference of two squares*) has as its factors $a + b$ and $a - b$.

To use this idea for factoring, we can write

$$a^2 - b^2 = (a + b)(a - b)$$

Note: To help you recognize a difference of squares, a "perfect square" term will have a coefficient that is a square (1, 4, 9, 16, 25, 36, etc.), and any variables will have exponents that are multiples of 2 (x^2, y^4, z^6, etc.).

Example 1

Factor $x^2 - 16$.

Think $x^2 - 4^2$

Since $x^2 - 16$ is a difference of squares, we have

$$x^2 - 16 = (x + 4)(x - 4)$$

You could also write $(x - 4)(x + 4)$. The order doesn't matter since multiplication is commutative.

CHECK YOURSELF 1

Factor $m^2 - 49$.

Example 2

Factor $4a^2 - 9$.

Think $(2a)^2 - 3^2$

So

$$4a^2 - 9 = (2a + 3)(2a - 3)$$

CHECK YOURSELF 2

Factor $9b^2 - 25$.

The process for factoring a difference of squares is identical when more than one variable is involved.

Example 3

Factor $25a^2 - 16b^4$.

$$\underset{(5a)^2}{\uparrow} \quad - \quad \underset{(4b^2)^2}{\uparrow}$$

So

$$25a^2 - 16b^4 = (5a + 4b^2)(5a - 4b^2)$$

CHECK YOURSELF 3

Factor $49c^4 - 9d^2$.

We will now consider an example that requires combining common-term factoring with the difference-of-squares technique. Note that the common factor is always removed as the *first step*.

Example 4

Factor $32x^2y - 18y^3$. Note that $2y$ is a common factor, so

Step 1
Remove the GCF.

$$32x^2y - 18y^3 = 2y(16x^2 - 9y^2)$$

Now the binomial in the parentheses is a difference of squares, and

Step 2
Factor the remaining binomial.

$$32x^2y - 18y^3 = 2y(4x + 3y)(4x - 3y)$$

CHECK YOURSELF 4

Factor $50a^3 - 8ab^2$.

CHECK YOURSELF ANSWERS

1. $(m + 7)(m - 7)$. **2.** $(3b + 5)(3b - 5)$. **3.** $(7c^2 + 3d)(7c^2 - 3d)$.
4. $2a(5a + 2b)(5a - 2b)$.

5.2 Exercises

For each of the following binomials, state whether the binomial is a difference of squares.

1. $2x^2 + y^2$ No

2. $2x^2 - y^2$ No

3. $9a^2 - 16b^2$ Yes

4. $4m^2 - 64n^2$ Yes

5. $16r^2 + 4$ No

6. $p^2 - 45$ No

7. $16a^2 - 12b^2$ No

8. $9a^2b^2 - 16c^2d^2$ Yes

9. $a^2b^2 - 25$ Yes

10. $4a^3 - b^3$ No

Factor the following binomials.

11. $m^2 - n^2$
$(m + n)(m - n)$

12. $r^2 - 9$
$(r + 3)(r - 3)$

13. $x^2 - 49$
$(x + 7)(x - 7)$

14. $c^2 - d^2$
$(c + d)(c - d)$

15. $25 - x^2$
$(5 + x)(5 - x)$

16. $64 - a^2$
$(8 + a)(8 - a)$

17. $4a^2 - 9$
$(2a + 3)(2a - 3)$

18. $p^2 - 25$
$(p + 5)(p - 5)$

19. $9w^2 - 25$
$(3w + 5)(3w - 5)$

20. $9x^2 - 64$
$(3x + 8)(3x - 8)$

21. $r^2 - 9s^2$
$(r + 3s)(r - 3s)$

22. $49x^2 - y^2$
$(7x + y)(7x - y)$

23. $9w^2 - 49z^2$
$(3w + 7z)(3w - 7z)$

24. $25x^2 - 81y^2$
$(5x + 9y)(5x - 9y)$

25. $16a^2 - 49b^2$
$(4a + 7b)(4a - 7b)$

26. $64m^2 - 9n^2$
$(8m + 3n)(8m - 3n)$

27. $x^4 - 36$
$(x^2 + 6)(x^2 - 6)$

28. $y^6 - 49$
$(y^3 + 7)(y^3 - 7)$

29. $x^2y^2 - 16$
$(xy + 4)(xy - 4)$

30. $m^2n^2 - 64$
$(mn + 8)(mn - 8)$

31. $25 - a^2b^2$
$(5 + ab)(5 - ab)$

32. $49 - w^2z^2$
$(7 + wz)(7 - wz)$

33. $r^4 - 4s^2$
$(r^2 + 2s)(r^2 - 2s)$

34. $p^2 - 9q^4$
$(p + 3q^2)(p - 3q^2)$

35. $81a^2 - 100b^6$
$(9a + 10b^3)(9a - 10b^3)$

36. $64x^4 - 25y^4$
$(8x^2 + 5y^2)(8x^2 - 5y^2)$

37. $18x^3 - 2xy^2$
$2x(3x + y)(3x - y)$

38. $50a^2b - 2b^3$
$2b(5a + b)(5a - b)$

39. $50m^3n - 18mn^3$
$2mn(5m + 3n)(5m - 3n)$

40. $32p^2q^2 - 8q^4$
$8q^2(2p + q)(2p - q)$

41. $48a^2b^2 - 27b^4$
$3b^2(4a + 3b)(4a - 3b)$

42. $20w^5 - 45w^3z^4$
$5w^3(2w + 3z^2)(2w - 3z^2)$

43. $x^2(x + y) - y^2(x + y)$
$(x + y)^2(x - y)$

44. $a^2(b - c) - 16b^2(b - c)$
$(b - c)(a + 4b)(a - 4b)$

45. $2m^2(m - 2n) - 18n^2(m - 2n)$
$2(m - 2n)(m + 3n)(m - 3n)$

46. $3a^3(2a + b) - 27ab^2(2a + b)$
$3a(2a + b)(a + 3b)(a - 3b)$

47. Find a value for k so that $kx^2 - 25$ will have the factors $2x + 5$ and $2x - 5$. 4

48. Find a value for k so that $9m^2 - kn^2$ will have the factors $3m + 7n$ and $3m - 7n$. 49

Skillscan (Section 4.3)
Multiply.

a. $(x - 1)(x + 2)$
$x^2 + x - 2$

b. $(a - 3)(a + 2)$
$a^2 - a - 6$

c. $(x + 4)(x + 6)$
$x^2 + 10x + 24$

d. $(w + 1)(w + 7)$
$w^2 + 8w + 7$

e. $(b + 1)(b + 3)$
$b^2 + 4b + 3$

f. $(a + 1)(a - 4)$
$a^2 - 3a - 4$

g. $(x - 1)(x - 1)$
$x^2 - 2x + 1$

h. $(p - 2)(p - 5)$
$p^2 - 7p + 10$

Answers

1. No **3.** Yes **5.** No **7.** No **9.** Yes **11.** $(m + n)(m - n)$ **13.** $(x + 7)(x - 7)$ **15.** $(5 + x)(5 - x)$ **17.** $(2a + 3)(2a - 3)$ **19.** $(3w + 5)(3w - 5)$ **21.** $(r + 3s)(r - 3s)$ **23.** $(3w + 7z)(3w - 7z)$ **25.** $(4a + 7b)(4a - 7b)$ **27.** $(x^2 + 6)(x^2 - 6)$ **29.** $(xy + 4)(xy - 4)$ **31.** $(5 + ab)(5 - ab)$ **33.** $(r^2 + 2s)(r^2 - 2s)$ **35.** $(9a + 10b^3)(9a - 10b^3)$ **37.** $2x(3x + y)(3x - y)$ **39.** $2mn(5m + 3n)(5m - 3n)$ **41.** $3b^2(4a + 3b)(4a - 3b)$ **43.** $(x + y)^2(x - y)$ **45.** $2(m - 2n)(m + 3n)(m - 3n)$ **47.** 4 **a.** $x^2 + x - 2$ **b.** $a^2 - a - 6$ **c.** $x^2 + 10x + 24$ **d.** $w^2 + 8w + 7$ **e.** $b^2 + 4b + 3$ **f.** $a^2 - 3a - 4$ **g.** $x^2 - 2x + 1$ **h.** $p^2 - 7p + 10$

5.3 Factoring Trinomials—Part 1

OBJECTIVE
To factor trinomials of the form

$x^2 + bx + c$

The process used to factor here is frequently called the *trial-and-error method.* You'll see the reason for the name as you work through this section.

You learned how to find the product of any two binomials by using the FOIL method in Section 4.3. Since factoring is the reverse of multiplication, we now want to use that pattern to find the factors of certain trinomials.

Recall that to multiply two binomials, we have

$$(x + 2)(x + 3) = x^2 + 5x + 6$$

The product of the first terms $(x \cdot x)$

The sum of the products of the outer and inner terms ($3x$ and $2x$)

The product of the last terms $(2 \cdot 3)$

Suppose now that you are given $x^2 + 5x + 6$ and want to find the factors. First, you know that the factors of a trinomial may be two binomials. So write

$$x^2 + 5x + 6 = (\qquad)(\qquad)$$

Since the first term of the trinomial is x^2, the first terms of the binomial factors must be x and x. We now have

$$x^2 + 5x + 6 = (x \qquad)(x \qquad)$$

The product of the last terms must be 6. Since 6 is positive, the factors must have *like* signs. Here are the possibilities:

$$\begin{aligned} 6 &= 1 \cdot 6 \\ &= 2 \cdot 3 \\ &= (-1)(-6) \\ &= (-2)(-3) \end{aligned}$$

This means that the possible factors are

$$(x + 1)(x + 6)$$
$$(x + 2)(x + 3)$$
$$(x - 1)(x - 6)$$
$$(x - 2)(x - 3)$$

How do we tell which is the correct pair? From the FOIL pattern we know that the sum of the outer and inner products must equal the middle term of the trinomial, in this case $5x$. This is the crucial step!

POSSIBLE FACTORS	MIDDLE TERMS	
$(x + 1)(x + 6)$	$7x$	
$(x + 2)(x + 3)$	$5x$	The correct middle term!
$(x - 1)(x - 6)$	$-7x$	
$(x - 2)(x - 3)$	$-5x$	

So we know that the correct factorization is

$$x^2 + 5x + 6 = (x + 2)(x + 3)$$

Are there any clues so far that will make this process quicker? Yes, there is an important one that you may have spotted. We started with a trinomial that had a positive middle term and a positive last term. The negative pairs of factors for 6 led to negative middle terms. So you don't need to bother with the negative factors if the middle term and the last term of the trinomial are both positive.

Example 1

(a) Factor $x^2 + 9x + 8$.

Since the middle term and the last term of the trinomial are both positive, consider only the positive factors of 8, i.e., $8 = 1 \cdot 8$ or $8 = 2 \cdot 4$.

If you are wondering why we didn't list $(x + 8)(x + 1)$ as a possibility, remember that multiplication is commutative. The order doesn't matter!

POSSIBLE FACTORS	MIDDLE TERMS
$(x + 1)(x + 8)$	$9x$
$(x + 2)(x + 4)$	$6x$

Since the first pair gives the correct middle term,

$$x^2 + 9x + 8 = (x + 1)(x + 8)$$

(b) Factor $x^2 + 12x + 20$.

The factors for 20 are

$$20 = 1 \cdot 20$$
$$= 2 \cdot 10$$
$$= 4 \cdot 5$$

POSSIBLE FACTORS	MIDDLE TERMS
$(x + 1)(x + 20)$	$21x$
$(x + 2)(x + 10)$	$12x$
$(x + 4)(x + 5)$	$9x$

so

$$x^2 + 12x + 20 = (x + 2)(x + 10)$$

CHECK YOURSELF 1

Factor.

1. $x^2 + 6x + 5$ **2.** $x^2 + 10x + 16$

Let's look at some examples in which the middle term of the trinomial is negative but the first and last terms are still positive. Consider

Positive Positive

$$x^2 - 11x + 18$$

Negative

Since we want a negative middle term ($-11x$), we use *two negative factors* for 18. Recall that the product of two negative numbers is positive.

Example 2

(a) Factor $x^2 - 11x + 18$.

The negative factors of 18 are

$$18 = (-1)(-18)$$
$$= (-2)(-9)$$
$$= (-3)(-6)$$

POSSIBLE FACTORS	MIDDLE TERMS
$(x - 1)(x - 18)$	$-19x$
$(x - 2)(x - 9)$	$-11x$
$(x - 3)(x - 6)$	$-9x$

So

$$x^2 - 11x + 18 = (x - 2)(x - 9)$$

(b) Factor $x^2 - 13x + 12$.

The negative factors of 12 are

$$12 = (-1)(-12)$$
$$= (-2)(-6)$$
$$= (-3)(-4)$$

POSSIBLE FACTORS	MIDDLE TERMS
$(x - 1)(x - 12)$	$-13x$
$(x - 2)(x - 6)$	$-8x$
$(x - 3)(x - 4)$	$-7x$

So

$$x^2 - 13x + 12 = (x - 1)(x - 12)$$

A few more clues: We have listed all the possible factors in the above examples. It really isn't necessary. Just work until you find the right pair. Also, with practice much of this work can be done mentally.

CHECK YOURSELF 2

Factor.

1. $x^2 - 10x + 9$ **2.** $x^2 - 10x + 21$

Let's look now at the process of factoring a trinomial whose last term is negative. For instance, to factor $x^2 + 2x - 15$, we can start as before:

$$x^2 + 2x - 15 = (x \quad ?)(x \quad ?)$$

Note that the product of the last terms must be negative (-15 here). So we must choose factors that have different signs.

What are our choices for the factors of -15?

$$
\begin{aligned}
-15 &= (1)(-15) \\
&= (-1)(15) \\
&= (3)(-5) \\
&= (-3)(5)
\end{aligned}
$$

This means that the possible factors and the resulting middle terms are

POSSIBLE FACTORS	MIDDLE TERMS
$(x + 1)(x - 15)$	$-14x$
$(x - 1)(x + 15)$	$14x$
$(x + 3)(x - 5)$	$-2x$
$(x - 3)(x + 5)$	$2x$

So

$$x^2 + 2x - 15 = (x - 3)(x + 5)$$

More clues: Some students prefer to look at the list of numerical factors rather than looking at the actual algebraic factors. Here you want the pair whose sum is 2, the coefficient of the middle term of the trinomial. That pair is -3 and 5, which leads us to the correct factors. Let's work through some further examples.

Example 3

(*a*) Factor $x^2 - 5x - 6$.

First, list the factors of -6. Of course, one will be positive, and one will be negative.

You may be able to pick the factors directly from this list. You want the pair whose sum is -5 (the coefficient of the middle term).

$$
\begin{aligned}
-6 &= (1)(-6) \\
&= (-1)(6) \\
&= (2)(-3) \\
&= (-2)(3)
\end{aligned}
$$

POSSIBLE FACTORS	MIDDLE TERMS
$(x + 1)(x - 6)$	$-5x$
$(x - 1)(x + 6)$	$5x$
$(x + 2)(x - 3)$	$-x$
$(x - 2)(x + 3)$	x

So

$$x^2 - 5x - 6 = (x + 1)(x - 6)$$

(b) Factor $x^2 + 8xy - 9y^2$.

The process is similar if two variables are involved in the trinomial you are to factor. Start with

$$x^2 + 8xy - 9y^2 = (x \quad ?)(x \quad ?).$$

The product of the last terms must be $-9y^2$.

$$\begin{aligned} -9y^2 &= (-y)(9y) \\ &= (y)(-9y) \\ &= (3y)(-3y) \end{aligned}$$

POSSIBLE FACTORS	MIDDLE TERMS
$(x - y)(x + 9y)$	$8xy$
$(x + y)(x - 9y)$	$-8xy$
$(x + 3y)(x - 3y)$	0

So

$$x^2 + 8xy - 9y^2 = (x - y)(x + 9y)$$

CHECK YOURSELF 3

Factor.

1. $x^2 + 7x - 30$ **2.** $x^2 - 3xy - 10y^2$

As was pointed out in the last section, any time that we have a common factor, that factor should be removed *before* we try any other factoring technique. Consider the following example.

Example 4

(a) Factor $3x^2 - 21x + 18$.

$$3x^2 - 21x + 18 = 3(x^2 - 7x + 6)$$ Remove the common factor of 3.

We now factor the remaining trinomial. For $x^2 - 7x + 6$:

POSSIBLE FACTORS	MIDDLE TERMS
$(x - 2)(x - 3)$	$-5x$
$(x - 1)(x - 6)$	$-7x$

The correct middle term

So

$$3x^2 - 21x + 18 = 3(x - 1)(x - 6)$$

Note: A common mistake is to forget to write the 3 that was factored out as the first step.

(b) Factor $2x^3 + 16x^2 - 40x$.

$2x^3 + 16x^2 - 40x = 2x(x^2 + 8x - 20)$ Remove the common factor of $2x$.

To factor the remaining trinomial, consider the following: For $x^2 + 8x - 20$ we have:

POSSIBLE FACTORS	MIDDLE TERMS	
$(x - 4)(x + 5)$	x	
$(x - 5)(x + 4)$	$-x$	
$(x - 10)(x + 2)$	$-8x$	
$(x - 2)(x + 10)$	$8x$	The correct middle term

so

$$2x^3 + 16x^2 - 40x = 2x(x - 2)(x + 10)$$

CHECK YOURSELF 4

Factor.

1. $3x^2 - 3x - 36$ **2.** $4x^3 + 24x^2 + 32x$

One further comment: Have you wondered if all trinomials are factorable? Look at the trinomial

$$x^2 + 2x + 6$$

The only possible factors are $(x + 1)(x + 6)$ and $(x + 2)(x + 3)$. Neither pair is correct (you should check the middle terms), and so this is an unfactorable trinomial over the integers. Of course, there are many others.

CHECK YOURSELF ANSWERS

1. (1) $(x + 1)(x + 5)$; (2) $(x + 2)(x + 8)$. **2.** (1) $(x - 9)(x - 1)$; (2) $(x - 3)(x - 7)$. **3.** (1) $(x + 10)(x - 3)$; (2) $(x + 2y)(x - 5y)$. **4.** (1) $3(x - 4)(x + 3)$; (2) $4x(x + 2)(x + 4)$.

5.3 Exercises

Complete the following statements.

1. $x^2 - 8x + 15 = (x - 3)(\quad)$ $x - 5$

2. $y^2 - 3y - 18 = (y - 6)(\quad)$ $y + 3$

3. $m^2 + 8m + 12 = (m + 2)(\quad)$ $m + 6$

4. $x^2 - 10x + 24 = (x - 6)(\quad)$ $x - 4$

5. $p^2 - 8p - 20 = (p + 2)(\quad)$
$p - 10$

6. $a^2 + 9a - 36 = (a + 12)(\quad)$
$a - 3$

7. $x^2 - 16x + 64 = (x - 8)(\quad)$
$x - 8$

8. $w^2 - 12w - 45 = (w + 3)(\quad)$
$w - 15$

9. $x^2 - 7xy + 10y^2 = (x - 2y)(\quad)$
$x - 5y$

10. $a^2 + 18ab + 81b^2 = (a + 9b)(\quad)$
$a + 9b$

Factor the following trinomials.

11. $x^2 + 8x + 15$
$(x + 3)(x + 5)$

12. $x^2 - 11x + 24$
$(x - 8)(x - 3)$

13. $x^2 - 11x + 28$
$(x - 4)(x - 7)$

14. $y^2 - y - 20$
$(y - 5)(y + 4)$

15. $s^2 + 13s + 30$
$(s + 10)(s + 3)$

16. $b^2 + 14b + 33$
$(b + 3)(b + 11)$

17. $a^2 - 2a - 48$
$(a - 8)(a + 6)$

18. $x^2 - 17x + 60$
$(x - 12)(x - 5)$

19. $x^2 - 8x + 7$
$(x - 1)(x - 7)$

20. $x^2 + 7x - 18$
$(x + 9)(x - 2)$

21. $m^2 + 3m - 28$
$(m + 7)(m - 4)$

22. $a^2 + 10a + 25$
$(a + 5)(a + 5)$

23. $x^2 - 6x - 40$
$(x - 10)(x + 4)$

24. $x^2 - 11x + 10$
$(x - 1)(x - 10)$

25. $x^2 - 14x + 49$
$(x - 7)(x - 7)$

26. $s^2 - 4s - 32$
$(s - 8)(s + 4)$

27. $p^2 - 10p - 24$
$(p - 12)(p + 2)$

28. $x^2 - 11x - 60$
$(x - 15)(x + 4)$

29. $x^2 + 5x - 66$
$(x + 11)(x - 6)$

30. $a^2 + 2a - 80$
$(a + 10)(a - 8)$

31. $c^2 + 19c + 60$
$(c + 4)(c + 15)$

32. $t^2 - 4t - 60$
$(t - 10)(t + 6)$

33. $n^2 + 5n - 50$
$(n + 10)(n - 5)$

34. $x^2 - 16x + 63$
$(x - 9)(x - 7)$

35. $x^2 + 7xy + 10y^2$
$(x + 2y)(x + 5y)$

36. $x^2 - 8xy + 12y^2$
$(x - 6y)(x - 2y)$

37. $a^2 - ab - 42b^2$
$(a - 7b)(a + 6b)$

38. $m^2 - 8mn + 16n^2$
$(m - 4n)(m - 4n)$

39. $x^2 - 13xy + 40y^2$
$(x - 5y)(x - 8y)$

40. $r^2 - 9rs - 36s^2$
$(r - 12s)(r + 3s)$

41. $9a^2 + 6ab + b^2$
$(3a + b)(3a + b)$

42. $x^2 + 3xy - 10y^2$
$(x + 5y)(x - 2y)$

43. $x^2 - 2xy - 8y^2$
$(x - 4y)(x + 2y)$

44. $u^2 + 6uv - 55v^2$
$(u + 11v)(u - 5v)$

45. $25m^2 + 10mn + n^2$
$(5m + n)(5m + n)$

46. $64m^2 - 16mn + n^2$
$(8m - n)(8m - n)$

Factor the following trinomials completely. Factor out the greatest common factor first.

47. $3a^2 - 3a - 126$
$3(a - 7)(a + 6)$

48. $2c^2 + 2c - 60$
$2(c + 6)(c - 5)$

49. $r^3 + 7r^2 - 18r$
$r(r + 9)(r - 2)$

50. $m^3 + 5m^2 - 14m$
$m(m + 7)(m - 2)$

51. $2x^3 - 20x^2 - 48x$
$2x(x - 12)(x + 2)$

52. $3p^3 + 48p^2 - 108p$
$3p(p + 18)(p - 2)$

53. $x^2y - 9xy^2 - 36y^3$
$y(x - 12y)(x + 3y)$

54. $4s^4 - 20s^3t - 96s^2t^2$
$4s^2(s - 8t)(s + 3t)$

55. $m^3 - 29m^2n + 120mn^2$
$m(m - 5n)(m - 24n)$

56. $2a^3 - 52a^2b + 96ab^2$
$2a(a - 2b)(a - 24b)$

Skillscan (Section 4.3)
Multiply.

a. $(2x - 1)(2x + 3)$
$4x^2 + 4x - 3$

b. $(3a - 1)(a + 4)$
$3a^2 + 11a - 4$

c. $(x - 4)(2x - 3)$
$2x^2 - 11x + 12$

d. $(2w - 11)(w + 2)$
$2w^2 - 7w - 22$

e. $(y + 5)(2y + 9)$
$2y^2 + 19y + 45$

f. $(2x + 1)(x - 12)$
$2x^2 - 23x - 12$

g. $(p + 9)(2p + 5)$
$2p^2 + 23p + 45$

h. $(3a - 5)(2a + 4)$
$6a^2 + 2a - 20$

Answers

1. $x - 5$ **3.** $m + 6$ **5.** $p - 10$ **7.** $x - 8$ **9.** $x - 5y$ **11.** $(x + 3)(x + 5)$ **13.** $(x - 4)(x - 7)$ **15.** $(s + 10)(s + 3)$ **17.** $(a - 8)(a + 6)$ **19.** $(x - 1)(x - 7)$ **21.** $(m + 7)(m - 4)$ **23.** $(x - 10)(x + 4)$ **25.** $(x - 7)(x - 7)$ **27.** $(p - 12)(p + 2)$ **29.** $(x + 11)(x - 6)$ **31.** $(c + 4)(c + 15)$ **33.** $(n + 10)(n - 5)$ **35.** $(x + 2y)(x + 5y)$ **37.** $(a - 7b)(a + 6b)$ **39.** $(x - 5y)(x - 8y)$ **41.** $(3a + b)(3a + b)$ **43.** $(x - 4y)(x + 2y)$ **45.** $(5m + n)(5m + n)$ **47.** $3(a - 7)(a + 6)$ **49.** $r(r + 9)(r - 2)$ **51.** $2x(x - 12)(x + 2)$ **53.** $y(x - 12y)(x + 3y)$ **55.** $m(m - 5n)(m - 24n)$ **a.** $4x^2 + 4x - 3$ **b.** $3a^2 + 11a - 4$ **c.** $2x^2 - 11x + 12$ **d.** $2w^2 - 7w - 22$ **e.** $2y^2 + 19y + 45$ **f.** $2x^2 - 23x - 12$ **g.** $2p^2 + 23p + 45$ **h.** $6a^2 + 2a - 20$

5.4 Factoring Trinomials—Part 2

1. To factor a trinomial of the form

   $ax^2 + bx + c$

2. To completely factor a trinomial

Factoring trinomials becomes a bit more time-consuming when the coefficient of the first term is not 1. Look at the following multiplication.

$$(5x + 2)(2x + 3) = 10x^2 + 19x + 6$$

Factors of $10x^2$ Factors of 6

Do you see the additional problem? We must consider all possible factors of the first coefficient (10 in the example) as well as those of the third term (6 in our example).

There is no easy way out! You need to form all possible combinations of factors and proceed by checking the middle term until the proper pair is found. If this seems a bit like guesswork, you're almost right. In fact some call this process factoring by "trial and error."

We can simplify the work a bit by reviewing the sign patterns found in Section 5.3.

SIGN PATTERNS FOR FACTORING TRINOMIALS

1. If all terms of a trinomial are positive, the signs between the terms in the binomial factors are both positive.
2. If the first and third terms of the trinomial are positive and the middle term is negative, the signs between the terms in the binomial factors are both negative.
3. If the third term of the trinomial is negative, the signs between the terms in the binomial factors are opposite (one is + and one is −).

Example 1

Factor $3x^2 + 14x + 15$.

First, list the possible factors of 3, the coefficient of the first term.

$3 = 1 \cdot 3$

Now list the factors of 15, the last term.

$15 = 1 \cdot 15$
$= 3 \cdot 5$

Because the signs of the trinomial are all positive, we know any factors will have the form

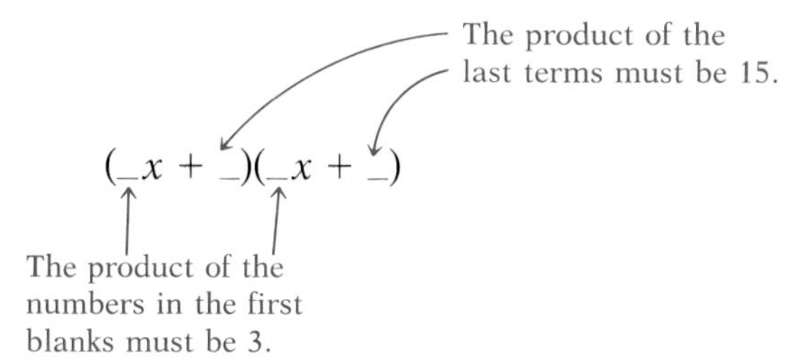

So the following represent the possible factors and the corresponding middle terms:

POSSIBLE FACTORS	MIDDLE TERMS	
$(x + 1)(3x + 15)$	$18x$	
$(x + 15)(3x + 1)$	$46x$	
$(3x + 3)(x + 5)$	$18x$	
$(3x + 5)(x + 3)$	$14x$	The correct middle term

So

$$3x^2 + 14x + 15 = (3x + 5)(x + 3)$$

CHECK YOURSELF 1

Factor.

1. $5x^2 + 14x + 8$ **2.** $3x^2 + 20x + 12$

Example 2

Factor $4x^2 - 11x + 6$.

Since only the middle term is negative, we know the factors have the form

$$(\_x - \_)(\_x - \_)$$

Both signs are negative.

Now look at the factors of the first coefficient and the last term.

$$4 = 1 \cdot 4 \qquad 6 = 1 \cdot 6$$
$$= 2 \cdot 2 \qquad = 2 \cdot 3$$

This gives us the possible factors:

POSSIBLE FACTORS	MIDDLE TERMS	
$(x - 1)(4x - 6)$	$-10x$	
$(x - 6)(4x - 1)$	$-25x$	
$(x - 2)(4x - 3)$	$-11x$	The correct middle term

Note that, in this example, we have *stopped the process* as soon as the correct pair of factors is found. That's exactly what you would do in practice.

So

$$4x^2 - 11x + 6 = (x - 2)(4x - 3)$$

CHECK YOURSELF 2

Factor.

1. $2x^2 - 9x + 9$ **2.** $6x^2 - 17x + 10$

Let's look now at an example of factoring a trinomial whose last term is negative.

Example 3

Factor $5x^2 + 6x - 8$.

Since the last term is negative, the factors will have the form

$(\_x + \_)(\_x - \_)$

Consider the factors of the first coefficient and the last term.

$5 = 1 \cdot 5 \qquad 8 = 1 \cdot 8$
$\qquad\qquad\qquad = 2 \cdot 4$

The possible factors are then

POSSIBLE FACTORS	MIDDLE TERMS
$(x + 1)(5x - 8)$	$-3x$
$(x + 8)(5x - 1)$	$39x$
$(5x + 1)(x - 8)$	$-39x$
$(5x + 8)(x - 1)$	$3x$
$(x + 2)(5x - 4)$	$6x$

Again we stop as soon as the correct pair of factors is found.

$5x^2 + 6x - 8 = (x + 2)(5x - 4)$

CHECK YOURSELF 3

Factor $4x^2 + 5x - 6$.

The process is similar if you want to factor a trinomial with more than one variable.

Example 4

Factor $6x^2 + 7xy - 10y^2$.

The form of the factors must be

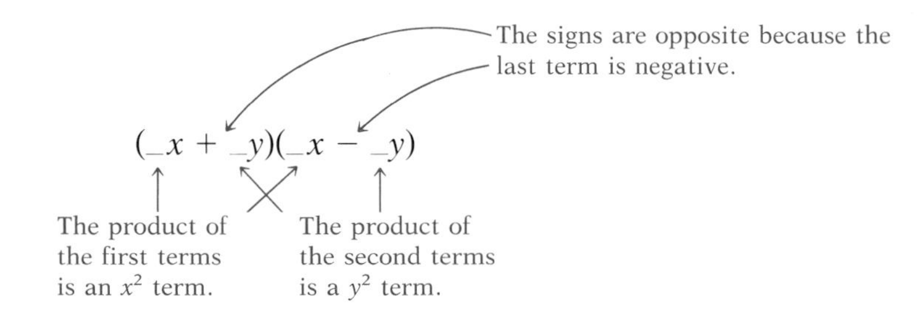
The signs are opposite because the last term is negative.
$(_x + _y)(_x - _y)$
The product of the first terms is an x^2 term.
The product of the second terms is a y^2 term.

Again look at the factors of the first and last coefficients.

$$6 = 1 \cdot 6 \qquad 10 = 1 \cdot 10$$
$$= 2 \cdot 3 \qquad\quad = 2 \cdot 5$$

POSSIBLE FACTORS	MIDDLE TERMS
$(x + y)(6x - 10y)$	$-4xy$
$(x + 10y)(6x - y)$	$59xy$
$(6x + y)(x - 10y)$	$-59xy$
$(6x + 10y)(x - y)$	$4xy$
$(x + 2y)(6x - 5y)$	$7xy$

Once more, we stop as soon as the correct factors are found.

$$6x^2 + 7xy - 10y^2 = (x + 2y)(6x - 5y)$$

CHECK YOURSELF 4

Factor $15x^2 - 4xy - 4y^2$.

Before we look at our next example, let's review one important point from Section 5.3. Recall that when you factor trinomials, you should not forget to look for a common factor as the first step. If there is a common factor, remove it and factor the remaining trinomial as before.

Example 5

Factor $18x^2 - 18x + 4$.

First look for a common factor in all three terms. Here that factor is 2, so write

$$18x^2 - 18x + 4 = 2(9x^2 - 9x + 2)$$

By our earlier methods, we can factor the remaining trinomial as

$$9x^2 - 9x + 2 = (3x - 1)(3x - 2)$$

So

$$18x^2 - 18x + 4 = 2(3x - 1)(3x - 2)$$

Don't forget the 2 that was factored out!

CHECK YOURSELF 5

Factor $16x^2 + 44x - 12$.

Example 6

Factor

$$6x^3 + 10x^2 - 4x$$

The common factor is $2x$.

So

$$6x^3 + 10x^2 - 4x = 2x(3x^2 + 5x - 2)$$

Factoring the remaining polynomial, we have

$$3x^2 + 5x - 2 = (3x - 1)(x + 2)$$

and

$$6x^3 + 10x^2 - 4x = 2x(3x - 1)(x + 2)$$

CHECK YOURSELF 6

Factor $6x^3 - 27x^2 + 30x$.

Strategies in Factoring

You have now had a chance to work with a variety of factoring techniques. Your success in factoring polynomials depends on your ability to recognize which factoring method should be applied in a particular problem.

We will close this section by presenting the following guidelines for applying the factoring methods you have studied in this chapter to a variety of polynomials.

FACTORING POLYNOMIALS

1. Look for a greatest common factor other than 1. If such a factor exists, factor out the GCF as the first step.
2. To continue the factoring process:
   a. If the polynomial that remains is a *binomial,* check to see if it is a difference of two squares. If it is, apply the difference-of-squares formula.
   b. If the polynomial that remains is a *trinomial,* try to factor the trinomial by the trial-and-error methods of Sections 5.3 and 5.4.

The following example illustrates the use of the strategy presented above.

Example 7

(*a*) Factor $5m^2n + 20n$.

First, we see that the GCF is $5n$. Removing that factor gives

$$5m^2n + 20n = 5n(m^2 + 4)$$

Since the binomial that remains is a *sum* of squares and is *not* factorable, we have completed the factorization.

(*b*) Factor $3x^3 - 48x$.

First, we see that the GCF is $3x$. Factoring out $3x$ yields

$$3x^3 - 48x = 3x(x^2 - 16)$$

Now we see that the binomial that remains, $x^2 - 16$, is a difference of squares. We then continue the factorization as

$$3x^3 - 48x^2 = 3x(x + 4)(x - 4)$$

(*c*) Factor $8r^2s + 20rs^2 - 12s^3$.

First, the GCF is $4s$, and we can write the original polynomial as

$$8r^2s + 20rs^2 - 12s^3 = 4s(2r^2 + 5rs - 3s^2)$$

Since the remaining polynomial is a trinomial, we can use the trial-and-error method to continue the factoring as

$$8r^2s + 20rs^2 - 12s^3 = 4s(2r - s)(r + 3s)$$

CHECK YOURSELF 7

Factor the following polynomials.

1. $8a^3 - 32ab^2$
2. $7x^3 + 7x^2y - 42xy^2$
3. $5m^4 + 15m^3 + 5m^2$

CHECK YOURSELF ANSWERS

1. (1) $(5x + 4)(x + 2)$; (2) $(3x + 2)(x + 6)$.
2. (1) $(2x - 3)(x - 3)$; (2) $(6x - 5)(x - 2)$.
3. $(4x - 3)(x + 2)$. 4. $(3x - 2y)(5x + 2y)$. 5. $4(4x - 1)(x + 3)$.
6. $3x(2x - 5)(x - 2)$. 7. (1) $8a(a + 2b)(a - 2b)$; (2) $7x(x + 3y)(x - 2y)$; (3) $5m^2(m^2 + 3m + 1)$.

5.4 Exercises

Complete the following statements.

1. $4x^2 - 4x - 3 = (2x + 1)(\quad)$
   $2x - 3$
2. $3w^2 + 11w - 4 = (w + 4)(\quad)$
   $3w - 1$
3. $6a^2 + 13a + 6 = (2a + 3)(\quad)$
   $3a + 2$
4. $25y^2 - 10y + 1 = (5y - 1)(\quad)$
   $5y - 1$
5. $15x^2 - 16x + 4 = (3x - 2)(\quad)$
   $5x - 2$
6. $6m^2 + 5m - 4 = (3m + 4)(\quad)$
   $2m - 1$
7. $16a^2 + 8ab + b^2 = (4a + b)(\quad)$
   $4a + b$
8. $6x^2 + 5xy - 4y^2 = (3x + 4y)(\quad)$
   $2x - y$
9. $4m^2 + 5mn - 6n^2 = (m + 2n)(\quad)$
   $4m - 3n$
10. $10p^2 - pq - 3q^2 = (5p - 3q)(\quad)$
    $2p + q$

Factor the following polynomials.

11. $3x^2 + 7x + 2$
    $(3x + 1)(x + 2)$
12. $5y^2 + 8y + 3$
    $(5y + 3)(y + 1)$
13. $2w^2 + 13w + 15$
    $(2w + 3)(w + 5)$
14. $3x^2 - 16x + 21$
    $(3x - 7)(x - 3)$
15. $5x^2 - 16x + 3$
    $(5x - 1)(x - 3)$
16. $2a^2 + 7a + 5$
    $(2a + 5)(a + 1)$
17. $4x^2 - 12x + 5$
    $(2x - 5)(2x - 1)$
18. $2x^2 + 11x + 12$
    $(2x + 3)(x + 4)$
19. $3x^2 - 5x - 2$
    $(3x + 1)(x - 2)$
20. $4m^2 - 23m + 15$
    $(4m - 3)(m - 5)$
21. $4p^2 + 19p - 5$
    $(4p - 1)(p + 5)$
22. $5x^2 - 36x + 7$
    $(5x - 1)(x - 7)$
23. $6x^2 + 19x + 10$
    $(3x + 2)(2x + 5)$
24. $6x^2 - 7x - 3$
    $(2x - 3)(3x + 1)$
25. $15x^2 + x - 6$
    $(5x - 3)(3x + 2)$

26. $12w^2 + 19w + 4$
$(4w + 1)(3w + 4)$

27. $6m^2 + 25m - 25$
$(6m - 5)(m + 5)$

28. $8x^2 - 6x - 9$
$(4x + 3)(2x - 3)$

29. $9x^2 - 12x + 4$
$(3x - 2)(3x - 2)$

30. $20x^2 - 23x + 6$
$(5x - 2)(4x - 3)$

31. $12x^2 - 8x - 15$
$(6x + 5)(2x - 3)$

32. $16a^2 + 40a + 25$
$(4a + 5)(4a + 5)$

33. $3y^2 + 7y - 6$
$(3y - 2)(y + 3)$

34. $12x^2 + 11x - 15$
$(3x + 5)(4x - 3)$

35. $8x^2 - 27x - 20$
$(8x + 5)(x - 4)$

36. $24v^2 + 5v - 36$
$(8v - 9)(3v + 4)$

37. $2x^2 + 3xy + y^2$
$(2x + y)(x + y)$

38. $3x^2 - 5xy + 2y^2$
$(3x - 2y)(x - y)$

39. $5a^2 - 8ab - 4b^2$
$(5a + 2b)(a - 2b)$

40. $5x^2 + 7xy - 6y^2$
$(5x - 3y)(x + 2y)$

41. $9x^2 + 4xy - 5y^2$
$(9x - 5y)(x + y)$

42. $16x^2 + 32xy + 15y^2$
$(4x + 3y)(4x + 5y)$

43. $6m^2 - 17mn + 12n^2$
$(3m - 4n)(2m - 3n)$

44. $15x^2 - xy - 6y^2$
$(5x + 3y)(3x - 2y)$

45. $36a^2 - 3ab - 5b^2$
$(12a - 5b)(3a + b)$

46. $3q^2 - 17qr - 6r^2$
$(3q + r)(q - 6r)$

Factor the following polynomials completely.

47. $4x^2 + 14x + 6$
$2(2x + 1)(x + 3)$

48. $9x^2 - 21x + 6$
$3(3x - 1)(x - 2)$

49. $20x^2 - 20x - 15$
$5(2x - 3)(2x + 1)$

50. $24x^2 - 18x - 6$
$6(4x + 1)(x - 1)$

51. $8m^2 + 12m + 4$
$4(2m + 1)(m + 1)$

52. $14x^2 - 20x + 6$
$2(7x - 3)(x - 1)$

53. $15r^2 - 21rs + 6s^2$
$3(5r - 2s)(r - s)$

54. $10x^2 + 5xy - 30y^2$
$5(2x - 3y)(x + 2y)$

55. $2x^3 - 2x^2 - 4x$
$2x(x - 2)(x + 1)$

56. $2y^3 + y^2 - 3y$
$y(2y + 3)(y - 1)$

57. $2y^4 + 5y^3 + 3y^2$
$y^2(2y + 3)(y + 1)$

58. $4z^3 - 18z^2 - 10z$
$2z(2z + 1)(z - 5)$

59. $36a^3 - 66a^2 + 18a$
$6a(3a - 1)(2a - 3)$

60. $20n^4 - 22n^3 - 12n^2$
$2n^2(2n - 3)(5n + 2)$

61. $9p^2 + 30pq + 21q^2$
$3(p + q)(3p + 7q)$

62. $12x^2 + 2xy - 24y^2$
$2(2x + 3y)(3x - 4y)$

63. $10(x + y)^2 - 11(x + y) - 6$
$(5x + 5y + 2)(2x + 2y - 3)$

64. $8(a - b)^2 + 14(a - b) - 15$
$(2a - 2b + 5)(4a - 4b - 3)$

Skillscan (Section 3.4)

Solve the following equations.

a. $x + 2 = 0$ -2

b. $x - 3 = 0$ 3

c. $2x - 3 = 0$ $\frac{3}{2}$

d. $3x - 2 = 0$ $\frac{2}{3}$

e. $6x - 5 = 0$ $\frac{5}{6}$

f. $3x + 2 = 0$ $-\frac{2}{3}$

g. $5x + 1 = 0$ $-\frac{1}{5}$

h. $8x - 3 = 0$ $\frac{3}{8}$

Answers

1. $2x - 3$ **3.** $3a + 2$ **5.** $5x - 2$ **7.** $4a + b$ **9.** $4m - 3n$ **11.** $(3x + 1)(x + 2)$ **13.** $(2w + 3)(w + 5)$ **15.** $(5x - 1)(x - 3)$ **17.** $(2x - 5)(2x - 1)$ **19.** $(3x + 1)(x - 2)$ **21.** $(4p - 1)(p + 5)$ **23.** $(3x + 2)(2x + 5)$ **25.** $(5x - 3)(3x + 2)$ **27.** $(6m - 5)(m + 5)$ **29.** $(3x - 2)(3x - 2)$ **31.** $(6x + 5)(2x - 3)$ **33.** $(3y - 2)(y + 3)$ **35.** $(8x + 5)(x - 4)$ **37.** $(2x + y)(x + y)$ **39.** $(5a + 2b)(a - 2b)$ **41.** $(9x - 5y)(x + y)$ **43.** $(3m - 4n)(2m - 3n)$ **45.** $(12a - 5b)(3a + b)$ **47.** $2(2x + 1)(x + 3)$ **49.** $5(2x - 3)(2x + 1)$ **51.** $4(2m + 1)(m + 1)$ **53.** $3(5r - 2s)(r - s)$ **55.** $2x(x - 2)(x + 1)$ **57.** $y^2(2y + 3)(y + 1)$ **59.** $6a(3a - 1)(2a - 3)$ **61.** $3(p + q)(3p + 7q)$ **63.** $(5x + 5y + 2)(2x + 2y - 3)$ **a.** -2 **b.** 3 **c.** $\frac{3}{2}$ **d.** $\frac{2}{3}$ **e.** $\frac{5}{6}$ **f.** $-\frac{2}{3}$ **g.** $-\frac{1}{5}$ **h.** $\frac{3}{8}$

5.5 Solving Equations by Factoring

OBJECTIVE

To solve quadratic equations by factoring.

There are many applications of our work with factoring. One important use of factoring is to solve certain types of equations. First we need to review an idea from arithmetic. If the product of two factors is 0, then one or both of the factors must be equal to 0. This is called the *zero-product principle*. In symbols,

> If $a \cdot b = 0$, then $a = 0$ or $b = 0$ or both.

Note: This rule only applies if a product is equal to zero. This will be very important later on.

Let's use this principle to solve an equation.

Example 1

Solve $(x - 3)(x + 2) = 0$.

Using the zero-product principle gives

If $\underbrace{(x - 3)}\underbrace{(x + 2)} = 0$, then

$x - 3 = 0$ or $x + 2 = 0$ Set each factor equal to 0.

$x = 3$ or $x = -2$ Solve each equation.

So 3 and -2 are the solutions for the equation.

CHECK YOURSELF 1

Solve $(x + 3)(x - 4) = 0$.

The left side of the equation of Example 1 was already in factored form. The next example illustrates a case in which factoring is necessary.

Example 2

Note: This equation is *not linear* because of the term in x^2. We will need different methods to solve this type of equation than those we saw earlier.

Solve $x^2 - 6x + 5 = 0$.

$$x^2 - 6x + 5 = 0$$
$$(x - 5)(x - 1) = 0 \quad \text{Factor on the left}$$

Again set each of the factors equal to zero.

$$x - 5 = 0 \quad \text{or} \quad x - 1 = 0$$
$$x = 5 \qquad\qquad x = 1$$

The solutions for the equation are 5 and 1.

We can check the solutions as before by substituting the two values back into the original equation.

Letting x be 5:	Letting x be 1:
$5^2 - 6 \cdot 5 + 5 = 0$	$1^2 - 6 \cdot 1 + 5 = 0$
$25 - 30 + 5 = 0$	$1 - 6 + 5 = 0$
$0 = 0$	$0 = 0$

Both solutions are verified.

CHECK YOURSELF 2

Solve $x^2 - 2x - 8 = 0$.

The equation of Example 2 has two solutions and has a special form.

Note: The equation is *set equal to 0* and the terms on the left are written in *descending form.*

> An equation with the form
>
> $ax^2 + bx + c = 0 \quad$ where $a \neq 0$
>
> is called a *quadratic equation in standard form.*

Quadratic equations must be in standard form before you factor, as the next example illustrates.

Example 3

Solve $2x^2 + 5x = 3$.

Be very careful! To use the zero-product principle, one side of the equation *must be zero*. So rewrite the equation in standard form by subtracting 3 from both sides. We now have

$$2x^2 + 5x - 3 = 0$$
$$(2x - 1)(x + 3) = 0 \quad \text{Again factor on the left.}$$

Set each of the factors equal to zero.

$$2x - 1 = 0 \quad \text{or} \quad x + 3 = 0$$
$$2x = 1 \qquad\qquad x = -3$$
$$x = \frac{1}{2}$$

So $\frac{1}{2}$ and -3 are the two solutions for the equation. We will leave the check of these solutions to you. Be sure to return to the original equation to verify these results.

CHECK YOURSELF 3

Solve.

1. $3x^2 - 5x = 2$ **2.** $2x^2 + 3 = 7x$

The following examples show how other factoring techniques are used in solving quadratic equations.

Example 4

(*a*) Solve $x^2 + 5x = 0$.

Note the common factor of x on the left. Factoring, we have

Note: Whenever x is a common factor of the quadratic member, you must set x equal to 0 to find the zero solution of the equation.

$$x(x + 5) = 0$$

Set each factor equal to 0.

$$x = 0 \quad \text{or} \quad x + 5 = 0$$
$$x = -5$$

The solutions are 0 and -5.

(*b*) Solve $x^2 = 4x$.

First, write the equation in standard form (set equal to 0).

$$x^2 - 4x = 0$$
$$x(x - 4) = 0 \quad \text{Factor on the left.}$$

$x = 0$ or $x - 4 = 0$
$x = 4$

The solutions are 0 and 4.

CHECK YOURSELF 4

Solve.

1. $x^2 + 8x = 0$ **2.** $x^2 = 9x$

The following example illustrates how the difference-of-squares factoring technique is applied in solving quadratic equations.

Example 5

Solve $x^2 = 9$.

Again, write the equation in standard form.

$$x^2 - 9 = 0$$

The left side is a difference of squares, so we have

$$(x + 3)(x - 3) = 0$$

So

$x + 3 = 0$ or $x - 3 = 0$
$x = -3$ $x = 3$

The solutions are -3 and 3.

CHECK YOURSELF 5

Solve $x^2 = 25$.

Example 6

Solve $3x^2 - 3x - 6 = 0$.

Note the common factor of 3 on the left. Write

$$3(x^2 - x - 2) = 0$$
$$3(x - 2)(x + 1) = 0$$

or, after dividing both sides by 3,

Note: On the right:

$\frac{0}{3} = 0$

$(x - 2)(x + 1) = 0$

So

$$x - 2 = 0 \quad \text{or} \quad x + 1 = 0$$
$$x = 2 \qquad\qquad x = -1$$

The solutions are 2 and -1.

CHECK YOURSELF 6

Solve $4x^2 + 14x = -6$.

The following rules summarize our work in solving quadratic equations by factoring.

TO SOLVE A QUADRATIC EQUATION BY FACTORING

STEP 1 Add or subtract the necessary terms on both sides of the equation so that the equation is in standard form (set equal to 0).

STEP 2 Factor the quadratic expression.

STEP 3 Set each factor equal to 0.

STEP 4 Solve the resulting equations to find the solutions.

STEP 5 Check each solution by substituting in the original equation.

Note: If the polynomial in step 2 is not factorable, you will have to use other methods for solving the equation. We will discuss other approaches in Chapter 10.

Strategies in Equation Solving

Keep in mind that the methods of this section dealt with solving quadratic equations. Earlier, in Chapter 3, we developed techniques to solve linear equations.

An important skill in equation solving is pattern recognition, that is, knowing which method to apply by recognizing the form of an equation. Our final example will help you in that process.

Example 7

For each of the following equations, determine whether the given equation is linear or quadratic. Then find the solution for each equation.

The equation is *linear* because it can be written in the form

$ax + b = 0$

where $a \neq 0$. The variable x can appear only to the first power.

(*a*) $2x + 1 = 0$

The equation $2x + 1 = 0$ is *linear*. To solve the equation, we want to isolate the variable x on the left.

$$2x = -1$$

$$x = -\frac{1}{2}$$

(*b*) $x(2x + 1) = 0$

This equation is *quadratic* because x appears to the second power.

Multiplying on the left we see that the equivalent equation, $2x^2 + x = 0$, is *quadratic*. The solutions, which can be found by the methods of this section are

$$x = 0 \qquad \text{or} \qquad x = -\frac{1}{2}$$

(*c*) $(3x + 1) + (x - 2) = 0$

Clearing parentheses and combining like terms, we see that the equivalent equation $4x - 1 = 0$ is *linear*. The solution in this case is $x = \frac{1}{4}$.

(*d*) $(3x + 1)(x - 2) = 26$

Multiplying on the left and writing the equation in standard form, we see that the equivalent equation, $3x^2 - 5x - 28 = 0$, is *quadratic*.

The solutions, which can be found by factoring, are $-\frac{7}{3}$ and 4.

CHECK YOURSELF 7

Determine whether each equation is linear or quadratic. Then find all solutions.

1. $3x - 5 = 0$
2. $(2x - 1)(3x + 4) = 10$
3. $(2x - 1) + (3x - 5) = 0$
4. $(x^2 + 2) + (2x^2 - 5) = 0$

CHECK YOURSELF ANSWERS

1. −3, 4. **2.** 4, −2. **3.** (1) 2, $-\frac{1}{3}$; (2) 3, $\frac{1}{2}$.

4. (1) 0, −8; (2) 0, 9. **5.** 5, −5. **6.** −3, $-\frac{1}{2}$.

7. (1) Linear, $\frac{5}{3}$; (2) quadratic, −2, $\frac{7}{6}$; (3) linear, $\frac{6}{5}$; (4) quadratic, −1, 1.

5.5 Exercises

Solve the following quadratic equations.

1. $(x-1)(x-2)=0$
$1, 2$

2. $(x-5)(x+3)=0$
$-3, 5$

3. $(2x+1)(x-5)=0$
$-\frac{1}{2}, 5$

4. $(3x-2)(x-5)=0$
$\frac{2}{3}, 5$

5. $x^2-2x-3=0$
$-1, 3$

6. $x^2+5x+4=0$
$-4, -1$

7. $x^2-7x+6=0$
$1, 6$

8. $x^2+3x-10=0$
$-5, 2$

9. $x^2+8x+15=0$
$-3, -5$

10. $x^2-3x-18=0$
$-3, 6$

11. $x^2+4x-21=0$
$-7, 3$

12. $x^2-12x+32=0$
$4, 8$

13. $x^2-6x=7$
$-1, 7$

14. $x^2+6x=-8$
$-2, -4$

15. $x^2+2x=15$
$-5, 3$

16. $x^2=7x-10$
$2, 5$

17. $2x^2+5x-3=0$
$-3, \frac{1}{2}$

18. $3x^2+7x+2=0$
$-\frac{1}{3}, -2$

19. $4x^2-24x+35=0$
$\frac{5}{2}, \frac{7}{2}$

20. $6x^2+11x-10=0$
$-\frac{5}{2}, \frac{2}{3}$

21. $4x^2+11x=-6$
$-\frac{3}{4}, -2$

22. $5x^2+2x=3$
$-1, \frac{3}{5}$

23. $5x^2+13x=6$
$-3, \frac{2}{5}$

24. $4x^2=13x+12$
$-\frac{3}{4}, 4$

25. $x^2-2x=0$
$0, 2$

26. $x^2+5x=0$
$0, -5$

27. $x^2=-8x$
$0, -8$

28. $x^2=7x$
$0, 7$

29. $5x^2-15x=0$
$0, 3$

30. $4x^2+20x=0$
$0, -5$

31. $x^2-25=0$
$-5, 5$

32. $x^2=49$
$-7, 7$

33. $x^2=81$
$-9, 9$

34. $x^2=64$
$-8, 8$

35. $2x^2-18=0$
$-3, 3$

36. $3x^2-75=0$
$-5, 5$

37. $3x^2+24x+45=0$
$-5, -3$

38. $4x^2-4x=24$
$-2, 3$

39. $8x^2+20x=12$
$-3, \frac{1}{2}$

40. $9x^2+42x=15$
$-5, \frac{1}{3}$

41. $(x-2)(x+1)=10$
$-3, 4$

42. $(x-4)(x+1)=14$
$-3, 6$

Translate each of the following statements to an equation. Let x represent the number in each case.

43. The square of a number minus 3 times that same number is 4. $x^2 - 3x = 4$

44. The sum of the square of a number and 5 times that same number is 14. $x^2 + 5x = 14$

45. If 4 times a number is added to the square of that same number, the sum is 12. $x^2 + 4x = 12$

46. If a number is subtracted from 3 times the square of that number, the difference is 2. $3x^2 - x = 2$

Simplify each equation, label the equation as linear or quadratic, and then find all solutions.

47. $3x + 4 = 2x$ Linear; -4

48. $3x + 4 = 2$ Linear; $-\frac{2}{3}$

49. $(2x + 5)(x + 1) = 0$ Quadratic; $-\frac{5}{2}, -1$

50. $(2x + 5) + (x + 1) = 0$ Linear; -2

51. $(2x - 1) - (x + 5) = 0$ Linear; 6

52. $(2x - 1)(x + 5) = 13$ Quadratic; $-6, \frac{3}{2}$

53. $x^2 + 4x - 5 = x^2$ Linear; $\frac{5}{4}$

54. $x^2 + 4x = 5$ Quadratic; $-5, 1$

55. $(x^2 + 2x - 3) + (x^2 + 10x - 29) = 0$ Quadratic; $-8, 2$

56. $x^2 + 2x - 3 = x^2 + 10x - 29$ Linear; $\frac{13}{4}$

57. $(x + 1)^2 = x^2 + 4$ Linear; $\frac{3}{2}$

58. $(2x + 1)^2 = 3x^2 + 13$ Quadratic; $-6, 2$

59. $x^2 - 3x - 5 = x + 7$ Quadratic; $-2, 6$

60. $x(x + 1) = (x + 2)^2$ Linear; $-\frac{4}{3}$

Skillscan (Section 3.8)

Solve each of the following word problems. Show the equation used for each solution.

a. One number is 5 more than another. If the sum of the two numbers is 19, find the two numbers. 7, 12

b. One number is 3 less than twice another. If the sum of the two numbers is 24, what are the two numbers? 9, 15

c. The sum of two consecutive even integers is 82. Find the two numbers. 40, 42

d. If the sum of three consecutive odd integers is 57, what are the two integers? 17, 19, 21

e. The length of a rectangle is 5 cm more than its width. If the perimeter of the rectangle is 42 cm, find the length and width of the rectangle. 13 cm, 8 cm

f. The length of a rectangle is 2 in more than 3 times its width. If the perimeter of the rectangle is 60 in, what are the dimensions of the rectangle? 7 in by 23 in

Answers

1. 1, 2 **3.** $-\frac{1}{2}, 5$ **5.** $-1, 3$ **7.** 1, 6 **9.** $-3, -5$ **11.** $-7, 3$ **13.** $-1, 7$ **15.** $-5, 3$ **17.** $-3, \frac{1}{2}$ **19.** $\frac{5}{2}, \frac{7}{2}$ **21.** $-\frac{3}{4}, -2$ **23.** $-3, \frac{2}{5}$ **25.** 0, 2 **27.** $0, -8$ **29.** 0, 3 **31.** $-5, 5$ **33.** $-9, 9$ **35.** $-3, 3$ **37.** $-5, -3$ **39.** $-3, \frac{1}{2}$ **41.** $-3, 4$ **43.** $x^2 - 3x = 4$ **45.** $x^2 + 4x = 12$ **47.** Linear; -4 **49.** Quadratic; $-\frac{5}{2}, -1$ **51.** Linear; 6 **53.** Linear; $\frac{5}{4}$ **55.** Quadratic; $-8, 2$ **57.** Linear; $\frac{3}{2}$ **59.** Quadratic; $-2, 6$ **a.** 7, 12 **b.** 9, 15 **c.** 40, 42 **d.** 17, 19, 21 **e.** 13 cm, 8 cm **f.** 7 in by 23 in

5.6 More Applications

OBJECTIVE
To solve word problems that lead to quadratic equations.

Certain types of word problems will lead to quadratic equations. You will have to use your work from Section 5.5 to solve them. Look at the following example.

Example 1

Remember our five-step process for solving word problems:

One number is 3 more than another. If their product is 54, find the two numbers.

1. Read the problem carefully. Then reread it to decide what you are asked to find.

Step 1 You are asked to find the two unknown numbers.

2. Choose a letter to represent the unknown or unknowns.

Step 2 Let x be the smaller number.

Then $\underbrace{x + 3}$ is the larger.

↑ "3 more than x"

3. Translate the problem to the language of algebra to form an equation.

Step 3

$$\underbrace{x(x + 3)} = 54$$

"Their product is 54."

4. Solve the equation.

Step 4

$x(x + 3) = 54$

$x^2 + 3x = 54$ Multiply on the left.

You should recognize that the equation is quadratic. It must be solved by the methods of Section 5.5.

$x^2 + 3x - 54 = 0$ Write the equation in standard form.

$(x + 9)(x - 6) = 0$ Factor.

$x + 9 = 0$ or $x - 6 = 0$

$x = -9$ $x = 6$

When word problems lead to quadratic equations, there may be two possible solutions. It is important to check, because one of the solutions may be eliminated by the conditions of the original problem. Here, both satisfy those conditions and so are valid solutions. The numbers are

Note that -9 and 6 are *not* the numbers asked for in the problem. The solutions are -9 and -6 *or* 6 and 9.

$x = -9 \leftarrow$ Smaller number $\rightarrow x = 6$

$x + 3 = -6 \leftarrow$ Larger number $\rightarrow x + 3 = 9$

5. Verify your solution by returning to the original problem.

Step 5 Since there are two separate solutions, both must be verified in the original problem. Since $(-9)(-6)$ and $6 \cdot 9$ are both 54, both pairs of numbers are solutions.

CHECK YOURSELF 1

The product of two consecutive integers is 90. What are the two integers?

Hint: Consecutive integers can be represented by x and $x + 1$.

Example 2

The sum of a number and its square is 30. What is the number?

Step 1 We want to find the unknown number.

Step 2 Let x be the number. Then the square of the number is x^2.

Step 3 Now write the equation.

$$\underbrace{x + x^2}_{\uparrow} = 30$$

The sum of the number and its square

Step 4 Writing the equation in standard form, we have

$$x^2 + x - 30 = 0$$

We factor and solve as before.

$$(x + 6)(x - 5) = 0$$

$$x + 6 = 0 \quad \text{or} \quad x - 5 = 0$$

$$x = -6 \qquad\qquad x = 5$$

Possible solutions are -6 and 5.

Step 5 Since we have two possible values, both must be checked.

$-6 + (-6)^2 = 30$	$5 + 5^2 = 30$
$-6 + 36 = 30$	$5 + 25 = 30$
$30 = 30$	$30 = 30$

Both solutions are verified.

CHECK YOURSELF 2

The square of an integer is 6 less than 7 times the integer. What is the integer?

Many problems involving geometric figures will also lead to quadratic equations that can be solved by factoring. The following example illustrates.

Example 3

A rectangle is 5 cm longer than it is wide. If the area of the rectangle is 84 cm^2, find the dimensions of the rectangle.

Step 1 You want to find the dimensions (the width and length) of the rectangle.

Step 2 Let x be the width of the rectangle; then $x + 5$ is the length. In solving geometric problems, always draw a sketch labeled with the variables assigned in this step. Here we would have

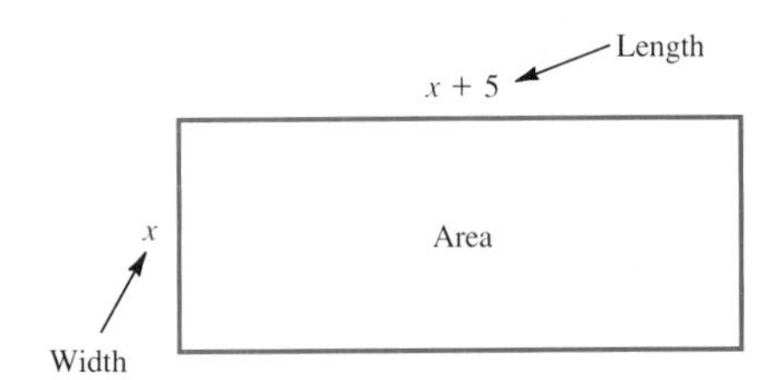

Step 3 Since the area of a rectangle is the product of its width and its length, we have

$$x(x + 5) = 84$$

Width ↗ Length ↑ Area ↖

Step 4 Multiply and write the equation in standard form.

$$x^2 + 5x - 84 = 0$$
$$(x + 12)(x - 7) = 0$$
$$x + 12 = 0 \quad \text{or} \quad x - 7 = 0$$
$$x = -12 \qquad\qquad x = 7$$

↑ This is not a possible solution—the width can't be negative!

The only possible solution is 7 cm. The width of the rectangle is 7 cm and the length is 12 cm.

Step 5 If the width is 7 cm and the length is 12 cm, the area is 84 cm^2. We have verified the solution.

CHECK YOURSELF 3

The length of a rectangle is 3 in more than twice its width. If the area of the rectangle is 90 in^2, what are its dimensions?

CHECK YOURSELF ANSWERS

1. 9 and 10 *or* −10 and −9. **2.** 1 or 6. **3.** Width, 6 in; length, 15 in.

5.6 Exercises

Solve the following word problems. Be sure to show the equations used for the solutions.

1. One integer is 5 less than another. If the product of the numbers is 66, find the two numbers. 6, 11 or −6, −11

2. One integer is 7 more than another. If the product of the numbers is 60, find the two numbers. 5, 12 or -5, -12

3. One integer is 1 less than twice another. If the product of the integers is 120, what are the two integers? 8, 15

4. One integer is 2 more than 3 times another. If the product of the integers is 56, find the two integers. 4, 14

5. The product of two consecutive integers is 132. Find the two integers. 11, 12 or -11, -12

6. If the product of two consecutive positive even integers is 120, find the two integers. 10, 12

7. The sum of an integer and its square is 72. What is the integer? -9 or 8

8. The square of an integer is 56 more than that integer. What is the integer? -7 or 8

9. The square of an integer is 20 more than 8 times that integer. What is the integer? -2 or 10

10. An integer is added to 3 times its square. The sum is 52. What is the integer? 4

11. If 2 times the square of an integer is increased by that integer, the sum is 55. Find the integer. 5

12. If the square of an integer is 21 less than 10 times that integer, what is the integer? 3 or 7

13. One positive integer is 1 more than twice another. If the difference of their squares is 65, what are the two integers? 4, 9

14. One positive integer is 3 less than twice another. If the difference of their squares is 24, find the two integers. 5, 7

15. The sum of the squares of two consecutive positive even integers is 100. Find the two integers. 6, 8

16. If the product of two consecutive positive odd integers is 63, find the two integers. 7, 9

17. The sum of the squares of two consecutive integers is 61. Find the integers. $-6, -5$ or 5, 6

18. The sum of the squares of two consecutive positive even integers is 164. Find the numbers. 8, 10

19. The sum of the squares of three consecutive positive odd integers is 83. What are the integers? 3, 5, 7

20. The sum of the squares of three consecutive integers is 50. Find the three integers. $-5, -4, -3$ or 3, 4, 5

21. Twice the square of a positive integer is 12 more than 5 times that integer. Find the integer. 4

22. Find an integer such that 10 more than the square of the integer is 40 more than the integer. -5 or 6

23. The length of a rectangle is 8 ft longer than its width. If the area of the rectangle is 65 ft^2, what are the dimensions of the rectangle? 5 ft by 13 ft

24. The length of a rectangle is 4 cm longer than its width. If the area of the rectangle is 140 cm^2, find the length and width of the rectangle. 10 cm, 14 cm

25. The width of a rectangle is 3 ft less than its length. If the area of the rectangle is 70 ft^2, what are the dimensions of the rectangle? 7 ft by 10 ft

26. The length of a rectangle is 5 cm more than its width. If the area of the rectangle is 150 cm^2, find the dimensions of the rectangle. 10 cm by 15 cm

27. The length of a rectangle is 1 in longer than twice its width. If the area of the rectangle is 105 in^2, what are the dimensions of the rectangle? 7 in by 15 in

28. The length of a rectangle is 2 cm less than 3 times its width. If the area of the rectangle is 40 cm^2, find the dimensions of the rectangle. 4 cm by 10 cm

29. The length of a rectangle is 2 cm more than 3 times its width. If the area of the rectangle is 85 cm^2, find the dimensions of the rectangle. 5 cm by 17 cm

30. If the length of a rectangle is 3 ft less than twice its width, and the area of the rectangle is 54 ft^2, what are the dimensions of the rectangle? 6 ft by 9 ft

31. If the sides of a square are increased by 3 in, the area is increased by 39 in^2. What were the dimensions of the original square? 5 in by 5 in

32. If the sides of a square are decreased by 2 cm, the area is decreased by 36 cm^2. What were the dimensions of the original square? 10 cm by 10 cm

33. The length of a rectangle is 3 in more than its width. If each dimension is increased by 2 in, the area of the rectangle is increased by 26 in^2. Find the dimensions of the original rectangle. 4 in by 7 in

34. The length of a rectangle is 4 ft more than its width. If each dimension is increased by 3 ft, the area of the rectangle is increased by 51 ft^2. Find the dimensions of the original rectangle. 5 ft by 9 ft

Skillscan (Section 3.5)

Solve each of the following equations for the indicated variable.

a. $2x - 3y = 12$ (for y) $\frac{2x - 12}{3}$

b. $V - E + F = 2$ (For E) $V + F - 2$

c. $S = 2\pi rh$ (for h) $\frac{S}{2\pi r}$

d. $s = \frac{1}{2}gt^2$ (for g) $\frac{2s}{t^2}$

e. $S = b^2 + \frac{1}{2}hb$ (for h) $\quad \frac{2S - 2b^2}{b}$

f. $S = \pi r^2 + 2\pi rh$ (for h) $\quad \frac{S - \pi r^2}{2\pi r}$

Answers

1. 6, 11 or −6, −11 **3.** 8, 15 **5.** 11, 12 or −11, −12 **7.** −9 or 8 **9.** −2 or 10 **11.** 5 **13.** 4, 9 **15.** 6, 8 **17.** −6, −5 or 5, 6 **19.** 3, 5, 7 **21.** 4 **23.** 5 ft by 13 ft **25.** 7 ft by 10 ft **27.** 7 in by 15 in **29.** 5 cm by 17 cm **31.** 5 in by 5 in **33.** 4 in by 7 in a. $\frac{2x - 12}{3}$ b. $V + F - 2$ c. $\frac{S}{2\pi r}$ d. $\frac{2s}{t^2}$ e. $\frac{2S - 2b^2}{b}$ f. $\frac{s - \pi r^2}{2\pi r}$

5.7 More on Literal Equations

OBJECTIVES

1. To solve literal equations involving parentheses
2. To solve literal equations by factoring

Back in Section 3.5 you dealt with literal equations, that is, equations that contained more than one letter or variable. Remember that the idea was to solve the literal equations for some particular letter. Often you will have to use the distributive property in order to remove parentheses in solving for that particular letter. Look at the following example.

Example 1

Solve $a(m + n) = c$ for m.

First, use the distributive property to remove the parentheses on the left.

$$a(m + n) = c$$

$$am + an = c$$

Remember that we want all terms involving m on one side and all other terms on the opposite side. So subtract an from both sides of the equation.

$$am + an - an = c - an$$

or $\qquad am = c - an$

Now divide both sides by a to isolate m.

$$\frac{am}{a} = \frac{c - an}{a}$$

or

$$m = \frac{c - an}{a}$$

CHECK YOURSELF 1

Solve $x(a - b) = c$ for a.

Example 2

Solve

You may recognize this as the formula for the area of a trapezoid.

$$A = \frac{h}{2}(B + b)$$

for B.

In this case the best first step is to multiply both sides of the equation by 2 to clear of fractions.

$$2A = 2 \cdot \frac{h}{2}(B + b)$$

or

$$2A = h(B + b)$$

Now remove the parentheses.

$$2A = hB + hb$$

Subtract hb from both sides.

$$2A - hb = hB$$

Remember that we want to isolate B on one side. It makes no difference whether B is on the left or the right.

Divide by h to isolate B.

$$\frac{2A - hb}{h} = \frac{hB}{h}$$

$$\frac{2A - hb}{h} = B$$

or

$$B = \frac{2A - hb}{h}$$

CHECK YOURSELF 2

Solve $S_n = \frac{n}{2}(a_1 + a_n)$ for a_n.

Solving literal equations may also require that you factor to isolate the specified letter. This will be the case whenever the letter that you are solving for appears in more than one term.

Example 3

Solve $ax = bx + ab$ for x.

First, subtract bx from both sides.

$$\underbrace{ax - bx}_{\uparrow} = ab$$

All terms with x are now on the left.

Since x appears in more than one term, factor.

$$x(a - b) = ab$$

Now divide both sides by $a - b$ to isolate x.

$$\frac{x(a - b)}{a - b} = \frac{ab}{a - b}$$

or

$$x = \frac{ab}{a - b}$$

CHECK YOURSELF 3

Solve $my = mn - ny$ for y.

The following summarizes our work in solving literal equations.

TO SOLVE A LITERAL EQUATION

STEP 1 If necessary, multiply both sides of the equation by the same term to clear of fractions.

STEP 2 Remove any parentheses by multiplying.

STEP 3 Add or subtract the same terms on both sides so that all terms involving the variable you are solving for are on one side and all other terms are on the opposite side of the equation.

STEP 4 Factor if the variable you are solving for appears in more than one term.

STEP 5 Divide both sides of the equation by the same expression to isolate the specified variable.

CHECK YOURSELF ANSWERS

1. $a = \frac{c + bx}{x}$. **2.** $a_n = \frac{2S_n - na_1}{n}$. **3.** $y = \frac{mn}{m + n}$.

5.7 Exercises

Solve each of the following equations for the indicated variable.

1. $a(x - y) = 3$ for x
$\frac{3 + ay}{a}$

2. $m(n + p) = 5$ for p
$\frac{5 - mn}{m}$

3. $P = 2(L + W)$ for L
$\frac{P - 2W}{2}$

4. $r = 3(s - t)$ for s
$\frac{r + 3t}{3}$

5. $C = \frac{5}{9}(F - 32)$ for F
$\frac{9C + 160}{5}$

6. $s = \frac{2}{3}(a + b)$ for b
$\frac{3}{2}s - a$

7. $ax + bx = c$ for x
$\frac{c}{a + b}$

8. $my - ny = p$ for y
$\frac{p}{m - n}$

9. $am = bm + n$ for m
$\frac{n}{a - b}$

10. $pq = p - 2q$ for q
$\frac{p}{p + 2}$

11. $m(a - b) = ab$ for b
$\frac{am}{a + m}$

12. $s(t + v) = tv$ for t
$\frac{-sv}{s - v}$

13. $m = \frac{a - b}{b}$ for b
$\frac{a}{m + 1}$

14. $s = \frac{c + d}{d}$ for d
$\frac{c}{s - 1}$

15. $t = a + (n - 1)d$ for n
$\frac{t - a + d}{d}$

16. $L = a(1 + ct)$ for c
$\frac{L - a}{at}$

17. $S = C - rC$ for C
$\frac{S}{1 - r}$

18. $S = C + rC$ for C
$\frac{S}{1 + r}$

19. $A = \frac{1}{2}h(B + b)$ for b
$\frac{2A - hB}{h}$

20. $A = P(1 + rt)$ for t
$\frac{A - P}{Pr}$

Answers

1. $x = \frac{3 + ay}{a}$ **3.** $L = \frac{P - 2W}{2}$ **5.** $F = \frac{9C + 160}{5}$ **7.** $x = \frac{c}{a + b}$ **9.** $m = \frac{n}{a - b}$ **11.** $b = \frac{am}{a + m}$
13. $b = \frac{a}{m + 1}$ **15.** $n = \frac{t - a + d}{d}$ **17.** $C = \frac{S}{1 - r}$ **19.** $b = \frac{2A - hB}{h}$

Summary Chapter 5

Common-Term Factoring [5.1]

Common Monomial Factor A single term that is a factor of every term of the polynomial. The greatest common monomial factor (GCF) has the largest possible numerical coefficient and the largest possible exponents.

$4x^2$ is the greatest common monomial factor of $8x^4 - 12x^3 + 16x^2$.

Factoring a Monomial from a Polynomial

1. Determine the greatest common monomial factor.
2. Apply the distributive law in the form

$$ab + ac = a(b + c)$$

↑ The greatest common monomial factor

$8x^4 - 12x^3 + 16x^2$
$= 4x^2(2x^2 - 3x + 4)$

The Difference of Squares [5.2]

To Factor a Difference of Squares Use the following form:

$$a^2 - b^2 = (a + b)(a - b)$$

To factor: $16x^2 - 25y^2$:

Think: $(4x)^2 - (5y)^2$

so

$16x^2 - 25y^2$
$= (4x + 5y)(4x - 5y)$

Factoring Trinomials [5.3] to [5.4]

Forming Possible Binomial Factors The product of the first terms of the binomial factors must equal the first term of the trinomial.

For $6x^2 - 19x + 10$ possible first terms are

$(6x \quad)(x \quad)$

or

$(3x \quad)(2x \quad)$

The product of the last terms of the binomial factors must equal the last term of the trinomial.

Possible last terms are

$(\quad 10)(\quad 1)$

or

$(\quad 5)(\quad 2)$

The signs in the binomial factors will follow the patterns developed in Section 5.3.

The signs in the factors are

$(\quad - \quad)(\quad - \quad)$

To Factor a Trinomial List all possible pairs of binomial factors. Find the correct pair of factors by calculating the sum of the inner and outer products. That sum must equal the middle term of the trinomial.

The possible binomial factors are

$(6x - 10)(x - 1)$
$(6x - 5)(x - 2)$
$(6x - 1)(x - 10)$
$(6x - 2)(x - 5)$
$(3x - 10)(2x - 1)$
$(3x - 1)(2x - 10)$
$(3x - 5)(2x - 2)$
$(3x - 2)(2x - 5)$ The correct factors

Solving Equations by Factoring [5.5]

To Solve a Quadratic Equation by Factoring

1. Add or subtract the necessary terms on both sides of the equation so that the equation is in standard form (set equal to 0).
2. Factor the quadratic expression.
3. Set each factor equal to 0.
4. Solve the resulting equations to find the solutions.
5. Check each solution by substituting in the original equation.

To solve:

$$x^2 + 7x = 30$$
$$x^2 + 7x - 30 = 0$$
$$(x + 10)(x - 3) = 0$$
$$x + 10 = 0 \text{ or } x - 3 = 0$$

$x = -10$ and $x = 3$ are solutions.

Applications of Quadratic Equations [5.6]

Solving Word Problems Certain types of word problems will lead to quadratic equations. Use the five-step process developed in Section 3.8.

In step 5 of the process (verify your solution), it is possible that one of the solutions resulting from the quadratic equation used will not satisfy the conditions of the original problem. Be sure to consider the possibility that one of the solutions must be discarded.

Solving Literal Equations [5.7]

1. If necessary, multiply both sides of the equation by the same term to clear fractions.
2. Remove any parentheses by multiplying.
3. Add or subtract the same terms on both sides so that all terms involving the variable you are solving for are on one side and all other terms are on the opposite side of the equation.
4. Factor if the variable you are solving for appears in more than one term.
5. Divide both sides of the equation by the same expression to isolate the specified variable.

To solve:

$$x = \frac{a + b}{a} \quad \text{for } a$$
$$ax = a + b$$
$$ax - a = b$$
$$a(x - 1) = b$$
$$\frac{a(x - 1)}{x - 1} = \frac{b}{x - 1}$$
$$a = \frac{b}{x - 1}$$

Summary Exercises Chapter 5

This summary exercise set is provided to give you practice with each of the objectives of the chapter. Each exercise is keyed to the appropriate chapter section. The answers are provided in the instructor's manual. Your instructor will give you guidelines on how to best use these exercises in your instructional setting.

[5.1] Factor each of the following polynomials.

1. $18a + 24$
$6(3a + 4)$

2. $9m^2 - 21m$
$3m(3m - 7)$

3. $24s^2t - 16s^2$
$8s^2(3t - 2)$

4. $18a^2b + 36ab^2$
$18ab(a + 2b)$

5. $35s^3 - 28s^2$
$7s^2(5s - 4)$

6. $3x^3 - 6x^2 + 15x$
$3x(x^2 - 2x + 5)$

7. $18m^2n^2 - 27m^2n + 45m^2n^3$
$9m^2n(2n - 3 + 5n^2)$

8. $121x^8y^3 + 77x^6y^3$
$11x^6y^3(11x^2 + 7)$

9. $8a^2b + 24ab - 16ab^2$
$8ab(a + 3 - 2b)$

10. $3x^2y - 6xy^3 + 9x^3y - 12xy^2$
$3xy(x - 2y^2 + 3x^2 - 4y)$

11. $x(2x - y) + y(2x - y)$
$(2x - y)(x + y)$

12. $5(w - 3z) - w(w - 3z)$
$(w - 3z)(5 - w)$

[5.2] Factor each of the following binomials completely.

13. $p^2 - 49$
$(p + 7)(p - 7)$

14. $25a^2 - 16$
$(5a + 4)(5a - 4)$

15. $m^2 - 9n^2$
$(m + 3n)(m - 3n)$

16. $16r^2 - 49s^2$
$(4r + 7s)(4r - 7s)$

17. $25 - z^2$
$(5 + z)(5 - z)$

18. $a^4 - 16b^2$
$(a^2 + 4b)(a^2 - 4b)$

19. $25a^2 - 36b^2$
$(5a + 6b)(5a - 6b)$

20. $x^6 - 4y^2$
$(x^3 + 2y)(x^3 - 2y)$

21. $3w^3 - 12wz^2$
$3w(w + 2z)(w - 2z)$

22. $16a^4 - 49b^2$
$(4a^2 + 7b)(4a^2 - 7b)$

23. $2m^2 - 72n^4$
$2(m + 6n^2)(m - 6n^2)$

24. $3w^3z - 12wz^3$
$3wz(w + 2z)(w - 2z)$

[5.3] Factor each of the following trinomials completely.

25. $x^2 + 9x + 20$
$(x + 4)(x + 5)$

26. $x^2 - 10x + 24$
$(x - 4)(x - 6)$

27. $a^2 - a - 12$
$(a - 4)(a + 3)$

28. $w^2 - 13w + 40$
$(w - 8)(w - 5)$

29. $x^2 + 12x + 36$
$(x + 6)(x + 6)$

30. $r^2 - 9r - 36$
$(r - 12)(r + 3)$

31. $b^2 - 4bc - 21c^2$
$(b - 7c)(b + 3c)$

32. $m^2n + 4mn - 32n$
$n(m + 8)(m - 4)$

33. $m^3 + 2m^2 - 35m$
$m(m + 7)(m - 5)$

34. $2x^2 - 2x - 40$
$2(x - 5)(x + 4)$

35. $3y^3 - 48y^2 + 189y$
$3y(y - 7)(y - 9)$

36. $3b^3 - 15b^2 - 42b$
$3b(b - 7)(b + 2)$

[5.4] Factor each of the following trinomials completely.

37. $3x^2 + 8x + 5$
$(3x + 5)(x + 1)$

38. $5w^2 + 13w - 6$
$(5w - 2)(w + 3)$

39. $2b^2 - 9b + 9$
$(2b - 3)(b - 3)$

40. $8x^2 + 2x - 3$
$(4x + 3)(2x - 1)$

41. $10x^2 - 11x + 3$
$(5x - 3)(2x - 1)$

42. $4a^2 + 7a - 15$
$(4a - 5)(a + 3)$

43. $9y^2 - 3yz - 20z^2$
$(3y - 5z)(3y + 4z)$

44. $8x^2 + 14xy - 15y^2$
$(2x + 5y)(4x - 3y)$

45. $8x^3 - 36x^2 - 20x$
$2x(4x + 2)(x - 5)$

46. $9x^2 - 15x - 6$
$3(3x + 1)(x - 2)$

47. $6x^3 - 3x^2 - 9x$
$3x(2x - 3)(x + 1)$

48. $5w^2 - 25wz + 30z^2$
$5(w - 2z)(w - 3z)$

[5.5] Solve each of the following quadratic equations.

49. $(x - 1)(2x + 3) = 0$
$1, -\frac{3}{2}$

50. $x^2 - 5x + 6 = 0$
$2, 3$

51. $x^2 - 10x = 0$
$0, 10$

52. $x^2 = 144$
$-12, 12$

53. $x^2 - 2x = 15$
$-3, 5$

54. $3x^2 - 5x - 2 = 0$
$-\frac{1}{3}, 2$

55. $4x^2 - 13x + 10 = 0$
$2, \frac{5}{4}$

56. $2x^2 - 3x = 5$
$-1, \frac{5}{2}$

57. $3x - 9x = 0$
$0, 3$

58. $x^2 - 25 = 0$
$-5, 5$

59. $2x^2 - 32 = 0$
$-4, 4$

60. $2x^2 - x - 3 = 0$
$-1, \frac{3}{2}$

61. $4x + 10x = 6$
$-3, \frac{1}{2}$

62. $3x^2 - 12x + 9 = 0$
$(1, 3)$

63. $4x^2 + 24x = -32$
$-4, -2$

[5.6] Solve each of the following applications. Be sure to show the equation used for the solution.

64. One integer is 8 less than another. If the product of the two integers is 84, what are the two integers? 6, 14 or −6, −14

65. Twice the square of a positive integer is 10 more than 8 times that integer. What is the integer? 5

66. The length of a rectangle is 3 cm less than twice its width, and the area of that rectangle is 35 cm^2. Find the length and width of the rectangle. 7 cm, 5 cm

67. The sides of a square are increased by 3 ft, and this increases the area of the square by 33 ft^2. What was the length of a side of the original square? 4 ft

[5.7] Solve each equation for the indicated variable.

68. $A = P(1 + rt)$ for t
$\frac{A - P}{Pr}$

69. $A = \frac{1}{2}h(B + b)$ for B
$\frac{2A - hb}{h}$

70. $p = \frac{x - y}{y}$ for y
$\frac{x}{p + 1}$

Self-Test for Chapter Five

The purpose of this self-test is to help you check your progress and to review for a chapter test in class. Allow yourself about an hour to take the test. When you are done, check your answers in the back of the book. If you missed any problems, be sure to go back and review the appropriate sections in the chapter and the exercises that are provided.

Factor each of the following polynomials.

1. $12b + 18$ $\quad 6(2b + 3)$

2. $9p^3 - 12p^2$ $\quad 3p^2(3p - 4)$

3. $5x^2 - 10x + 20$ $\quad 5(x^2 - 2x + 4)$

4. $6a^2b - 18ab + 12ab^2$ $\quad 6ab(a - 3 + 2b)$

Factor each of the following polynomials completely.

5. $a^2 - 25$ $\quad (a + 5)(a - 5)$

6. $64m^2 - n^2$ $\quad (8m + n)(8m - n)$

7. $49x^2 - 16y^2$ $\quad (7x + 4y)(7x - 4y)$

8. $32a^2b - 50b^3$ $\quad 2b(4a + 5b)(4a - 5b)$

Factor each of the following polynomials completely.

9. $a^2 - 5a - 14$ $\quad (a - 7)(a + 2)$

10. $b^2 + 8b + 15$ $\quad (b + 3)(b + 5)$

11. $x^2 - 11x + 28$ $\quad (x - 4)(x - 7)$

12. $y^2 + 12yz + 20z^2$ $\quad (y + 10z)(y + 2z)$

Factor each of the following polynomials completely.

13. $2x^2 + 15x - 8$ $\quad (2x - 1)(x + 8)$

14. $3w^2 + 10w + 7$ $\quad (3w + 7)(w + 1)$

15. $8x^2 - 2xy - 3y^2$ $\quad (4x - 3y)(2x + y)$

16. $6x^3 + 3x^2 - 30x$ $\quad 3x(2x + 5)(x - 2)$

Solve each of the following equations for x.

17. $x^2 - 8x + 15 = 0$ $\quad 3, 5$

18. $x^2 - 3x = 4$ $\quad -1, 4$

19. $3x^2 + x - 2 = 0$ $\quad -1, \frac{2}{3}$

20. $4x^2 - 12x = 0$ $\quad 0, 3$

21. $4x^2 + 26x = 14$ $\quad -7, \frac{1}{2}$

Solve the following word problems.

22. One integer is 3 less than twice another. If the product of the integers is 35, what are the two integers? 5, 7

23. The length of a rectangle is 2 cm more than 3 times its width. If the area of the rectangle is 33 cm^2, what are the dimensions of the rectangle? 3 cm by 11 cm

Solve for the variable listed after each of the following equations.

24. $P = 2(L + W)$ for W $\quad \frac{P - 2L}{2}$

25. $a = \frac{b + c}{c}$ for c $\quad \frac{b}{a - 1}$

Chapter Six

Algebraic Fractions

6.1 Algebraic Fractions—An Introduction

OBJECTIVES

1. To review the language of fractions.
2. To determine the excluded values for the variables of an algebraic fraction.

In arithmetic you learned about fractions, or rational numbers. Recall that a rational number is the quotient of two integers, $\frac{p}{q}$, where q is not equal to 0. For example, $\frac{2}{3}$, $-\frac{4}{5}$, $\frac{12}{7}$, and $\frac{5}{1}$ are all rational numbers. We now want to extend the idea of fractions to algebra. All that you learned in arithmetic about fractions will be very helpful here.

Algebraic fractions are also called *rational expressions*. Note the similarity to rational numbers.

Let's start by defining what is meant by an *algebraic fraction*. It is an expression of the form

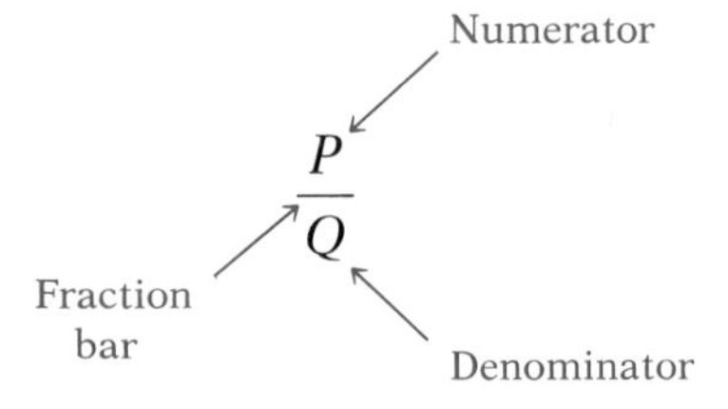

Recall that a divisor (the denominator here) cannot have the value 0, or else the division is undefined. See Section 2.5 for details.

where P and Q are polynomials and Q cannot have the value 0.

The condition that Q, the polynomial in the denominator, cannot be 0 means that certain values for the variable will have to be excluded.

Example 1

In the algebraic fractions:

(*a*) $\frac{x}{5}$, x can have any value.

(*b*) $\frac{3}{x}$, x cannot equal 0.

Note: If $x = 0$, then $\frac{3}{x}$ is undefined; 0 is the excluded value.

(*c*) $\frac{5}{x - 2}$, x cannot equal 2.

To see this, set $x - 2 = 0$ so $x = 2$ gives a zero denominator.

Note: If $x = 2$, then $\frac{5}{x - 2} = \frac{5}{2 - 2} = \frac{5}{0}$, which is undefined.

Again, if

$x + 3 = 0$
$x = -3$

or if

$x - 4 = 0$
$x = 4$

both values give a 0 denominator.

(*d*) $\frac{4}{(x + 3)(x - 4)}$, x cannot equal -3 or 4.

Note: If $x = -3$, then

$$\frac{4}{(x + 3)(x - 4)} = \frac{4}{(-3 + 3)(-3 - 4)} = \frac{4}{0 \cdot 7} = \frac{4}{0}$$

which is undefined. Try $x = 4$; you will see that the same thing happens.

Both values will make the denominator 0 and must be excluded.

CHECK YOURSELF 1

What values for x, if any, must be excluded?

1. $\frac{x}{7}$ **2.** $\frac{5}{x}$ **3.** $\frac{7}{x - 5}$ **4.** $\frac{2x}{(x - 6)(x + 1)}$

In some cases you will have to factor the denominator to see the restrictions on the values for the variable.

Example 2

What values for x must be excluded in the following fraction?

$$\frac{3}{x^2 - 6x - 16}$$

Factoring the denominator, we have

$$\frac{3}{(x - 8)(x + 2)}$$

Letting $x - 8 = 0$ or $x + 2 = 0$, we see that 8 and -2 cannot be values for x. Both values will make the denominator 0.

CHECK YOURSELF 2

What values for x must be excluded on the following fraction?

$$\frac{5}{x^2 - 3x - 10}$$

CHECK YOURSELF ANSWERS

1. (1) None; (2) 0; (3) 5; (4) 6, −1.
2. 5 and −2.

6.1 Exercises

What values for x, if any, must be excluded in each of the following algebraic fractions?

1. $\frac{x}{9}$ None

2. $\frac{7}{x}$ 0

3. $\frac{12}{x}$ 0

4. $\frac{x}{6}$ None

5. $\frac{3}{x - 2}$ 2

6. $\frac{x - 1}{5}$ None

7. $\frac{-5}{x + 4}$ −4

8. $\frac{4}{x + 3}$ −3

9. $\frac{x - 5}{2}$ None

10. $\frac{x - 1}{x - 5}$ 5

11. $\frac{3x}{(x + 1)(x - 2)}$ −1, 2

12. $\frac{5x}{(x - 3)(x + 7)}$ −7, 3

13. $\frac{x - 1}{(2x - 1)(x + 3)}$ $-3, \frac{1}{2}$

14. $\frac{x + 3}{(3x + 1)(x - 2)}$ $-\frac{1}{3}, 2$

15. $\frac{3}{x^2 - 4}$ −2, 2

16. $\frac{4x}{x^2 - 2x - 3}$ −1, 3

17. $\frac{x + 1}{x^2 - 6x + 5}$ 1, 5

18. $\frac{2x - 3}{x^2 - 25}$ −5, 5

19. $\frac{2x - 1}{3x^2 + x - 2}$ $-1, \frac{2}{3}$

20. $\frac{3x + 1}{4x^2 - 11x + 6}$ $\frac{3}{4}, 2$

Skillscan (Section 1.5)

Divide.

a. $\frac{12a}{3}$ $4a$

b. $\frac{35w}{7}$ $5w$

c. $\frac{15c^5}{3c^2}$ $5c^3$

d. $\frac{-48x^5}{3x^3}$ $-16x^2$

e. $\dfrac{56m^3}{-8m}$ $-7m^2$ **f.** $\dfrac{-72p^5}{-9p^2}$ $8p^3$ **g.** $\dfrac{100r^3s^4}{4r^2s^3}$ $25rs$ **h.** $\dfrac{-84x^4y^6}{7x^2y^2}$ $-12x^2y^4$

Answers

1. None **3.** 0 **5.** 2 **7.** −4 **9.** None **11.** −1, 2 **13.** $-3, \frac{1}{2}$ **15.** −2, 2 **17.** 1, 5 **19.** $-1, \frac{2}{3}$ **a.** $4a$ **b.** $5w$ **c.** $5c^3$ **d.** $-16x^2$ **e.** $-7m^2$ **f.** $8p^3$ **g.** $25rs$ **h.** $-12x^2y^4$

6.2 Writing Algebraic Fractions in Simplest Form

OBJECTIVE
To write algebraic fractions in simplest form.

As we pointed out in the last section, much of our work with algebraic fractions will be similar to your work in arithmetic. For instance, in algebra, as in arithmetic, many fractions name the same number. You will remember from arithmetic that

$$\frac{1}{4} = \frac{1 \cdot 2}{4 \cdot 2} = \frac{2}{8}$$

or

$$\frac{1}{4} = \frac{1 \cdot 3}{4 \cdot 3} = \frac{3}{12}$$

So $\frac{1}{4}, \frac{2}{8}$, and $\frac{3}{12}$ all name the same number. They are called *equivalent fractions*. These examples illustrate what is called the *fundamental principle of fractions*. In algebra it becomes

> For polynomials P, Q, and R,
>
> $$\frac{P}{Q} = \frac{PR}{QR} \quad \text{where } Q \neq 0 \text{ and } R \neq 0$$

This principal allows us to multiply or divide the numerator and denominator of a fraction by the same nonzero polynomial. The result will be an expression that is equivalent to the original one.

Our objective in this section is to simplify algebraic fractions using the fundamental principle. In algebra, as in arithmetic, to write a fraction in simplest form you divide the numerator and denominator of the fraction by all common factors. The numerator and denominator of the resulting fraction will have no common factors other than 1, and the fraction is then in *simplest form*. The following rule summarizes this procedure.

Note that step 2 uses the fact that $\frac{R}{R} = 1$, if $R \neq 0$.

TO WRITE ALGEBRAIC FRACTIONS IN SIMPLEST FORM

STEP 1 Factor the numerator and denominator.

STEP 2 Divide the numerator and denominator by all common factors.

STEP 3 The resulting fraction will be in lowest terms.

Example 1

(*a*) Write $\frac{18}{30}$ in simplest form.

$$\frac{18}{30} = \frac{2 \cdot 3 \cdot 3}{2 \cdot 3 \cdot 5} = \frac{\not{2} \cdot \not{3} \cdot 3}{\not{2} \cdot \not{3} \cdot 5} = \frac{3}{5}$$

Divide by the common factors. The slash lines indicate that we have divided the numerator and denominator by 2 and by 3.

(*b*) Write $\frac{4x^3}{6x}$ in simplest form.

$$\frac{4x^3}{6x} = \frac{\not{2} \cdot 2 \cdot \not{x} \cdot x \cdot x}{\not{2} \cdot 3 \cdot \not{x}} = \frac{2x^2}{3}$$

(*c*) Write $\frac{15x^3y^2}{20xy^4}$ in simplest form.

$$\frac{15x^3y^2}{20xy^4} = \frac{3 \cdot \not{5} \cdot \not{x} \cdot x \cdot x \cdot \not{y} \cdot \not{y}}{4 \cdot \not{5} \cdot \not{x} \cdot \not{y} \cdot \not{y} \cdot y \cdot y} = \frac{3x^2}{4y^2}$$

(*d*) Write $\frac{3a^2b}{9a^3b^2}$ in simplest form.

$$\frac{3a^2b}{9a^3b^2} = \frac{\not{3} \cdot \not{a} \cdot \not{a} \cdot \not{b}}{\not{3} \cdot 3 \cdot \not{a} \cdot \not{a} \cdot a \cdot \not{b} \cdot b} = \frac{1}{3ab}$$

CHECK YOURSELF 1

Write each fraction in simplest form.

1. $\frac{30}{66}$ **2.** $\frac{5x^4}{15x}$ **3.** $\frac{12xy^4}{18x^3y^2}$ **4.** $\frac{5m^2n}{10m^3n^3}$

In fact you will see most of the methods of this chapter center around our factoring work of the last chapter.

In simplifying arithmetic fractions, common factors are generally easy to recognize. With algebraic fractions, the factoring techniques you studied in Chapter 5 will have to be used as the *first step* in determining those factors.

Example 2

Write each fraction in simplest form.

(a) $\dfrac{2x-4}{x^2-4} = \dfrac{2(x-2)}{(x+2)(x-2)}$ Factor the numerator and denominator.

$= \dfrac{2\cancel{(x-2)}}{(x+2)\cancel{(x-2)}}$ Divide by the common factor $x - 2$. The lines drawn indicate that we have divided by that common factor.

$= \dfrac{2}{x+2}$

(b) $\dfrac{3x^2-3}{x^2-2x-3} = \dfrac{3(x-1)\cancel{(x+1)}}{(x-3)\cancel{(x+1)}}$

$= \dfrac{3x-3}{x-3}$

(c) $\dfrac{2x^2+x-6}{2x^2-x-3} = \dfrac{(x+2)\cancel{(2x-3)}}{(x+1)\cancel{(2x-3)}}$

$= \dfrac{x+2}{x+1}$

Be Careful! The expression $\dfrac{x+2}{x+1}$ is already in simplest form. Students are often tempted to divide as follows:

$\dfrac{\cancel{x}+2}{\cancel{x}+1}$ is *not equal* to $\dfrac{2}{1}$

Pick any value, other than 0, for x and substitute. You will quickly see that

$\dfrac{x+2}{x+1} \neq \dfrac{2}{1}$

The x's are terms in the numerator and denominator. They *cannot* be divided out. Only factors can be divided. The fraction

$\dfrac{x+2}{x+1}$

is in its simplest form.

CHECK YOURSELF 2

Write each fraction in simplest form.

1. $\dfrac{5x-15}{x^2-9}$ **2.** $\dfrac{a^2-5a+6}{3a^2-6a}$

3. $\dfrac{3x^2 + 14x - 5}{3x^2 + 2x - 1}$ **4.** $\dfrac{5p - 15}{p^2 - 4}$

Remember the rules for signs in division. The quotient of a positive number and a negative number is always negative. So there are three equivalent ways to write such a quotient. For instance,

$$\frac{-2}{3} = \frac{2}{-3} = -\frac{2}{3}$$

Note: $\dfrac{-2}{3}$, with the negative sign in the numerator, is the standard way to write the quotient.

The quotient of two positive numbers or two negative numbers is always positive. For example,

$$\frac{-2}{-3} = \frac{2}{3}$$

Example 3

Write each fraction in simplest form.

(*a*) $\dfrac{6x^2}{-3xy} = \dfrac{2 \cdot \cancel{3} \cdot \cancel{x} \cdot x}{(-1) \cdot \cancel{3} \cdot \cancel{x} \cdot y} = \dfrac{-2x}{y}$

(*b*) $\dfrac{-5a^2b}{-10b^2} = \dfrac{\cancel{(-1)} \cdot \cancel{5} \cdot a \cdot a \cdot \cancel{b}}{\cancel{(-1)} \cdot 2 \cdot \cancel{5} \cdot \cancel{b} \cdot b} = \dfrac{a^2}{2b}$

CHECK YOURSELF 3

Write each fraction in simplest form.

1. $\dfrac{8x^3y}{-4xy^2}$ **2.** $\dfrac{-16a^4b^2}{-12a^2b^5}$

Reducing certain algebraic fractions will be easier with the following result.

First, verify for yourself that

$$5 - 8 = -(8 - 5)$$

In general, it is true that

$$a - b = -(b - a)$$

or, by dividing both sides of the equation by $b - a$,

$$\frac{a - b}{b - a} = \frac{-(b - a)}{b - a}$$

So dividing by $b - a$ on the right, we have

$$\frac{a - b}{b - a} = -1$$

Let's look at some applications of that result.

Example 4

Write each fraction in simplest form.

(a) $\dfrac{2x - 4}{4 - x^2} = \dfrac{2(x - 2)}{(2 + x)(2 - x)}$ ⟵ This is equal to −1.

$$= \frac{2(-1)}{2 + x} = \frac{-2}{2 + x}$$

(b) $\dfrac{9 - x^2}{x^2 + 2x - 15} = \dfrac{(3 + x)(3 - x)}{(x + 5)(x - 3)}$ ⟵ This is equal to −1.

$$= \frac{(3 + x)(-1)}{x + 5}$$

$$= \frac{-x - 3}{x + 5}$$

CHECK YOURSELF 4

Write each fraction in simplest form.

1. $\dfrac{3x - 9}{9 - x^2}$ **2.** $\dfrac{x^2 - 6x - 27}{81 - x^2}$

CHECK YOURSELF ANSWERS

1. (1) $\dfrac{5}{11}$; (2) $\dfrac{x^3}{3}$; (3) $\dfrac{2y^2}{3x^2}$; (4) $\dfrac{1}{2mn^2}$.

2. (1) $\dfrac{5}{x + 3}$; (2) $\dfrac{a - 3}{3a}$; (3) $\dfrac{x + 5}{x + 1}$; (4) $\dfrac{5p - 15}{p^2 - 4}$.

3. (1) $\dfrac{-2x^2}{y}$; (2) $\dfrac{4a^2}{3b^3}$.

4. (1) $\dfrac{-3}{x + 3}$; (2) $\dfrac{-x - 3}{x + 9}$.

6.2 Exercises

Write each fraction in simplest form.

1. $\frac{14}{21}$ $\frac{2}{3}$

2. $\frac{35}{40}$ $\frac{7}{8}$

3. $\frac{40}{90}$ $\frac{4}{9}$

4. $\frac{45}{75}$ $\frac{3}{5}$

5. $\frac{4x^5}{6x^2}$ $\frac{2x^3}{3}$

6. $\frac{10x^2}{15x^4}$ $\frac{2}{3x^2}$

7. $\frac{9x^3}{27x^6}$ $\frac{1}{3x^3}$

8. $\frac{25w^6}{20w^2}$ $\frac{5w^4}{4}$

9. $\frac{10a^2b^5}{25ab^2}$ $\frac{2ab^3}{5}$

10. $\frac{18x^4y^3}{24x^2y^3}$ $\frac{3x^2}{4}$

11. $\frac{42x^3y}{14xy^3}$ $\frac{3x^2}{y^2}$

12. $\frac{18pq}{45p^2q^2}$ $\frac{2}{5pq}$

13. $\frac{-4m^3n}{6mn^2}$ $\frac{-2m^2}{3n}$

14. $\frac{-15x^3y^3}{-20xy^4}$ $\frac{3x^2}{4y}$

15. $\frac{-8ab^3}{-16a^3b}$ $\frac{b^2}{2a^2}$

16. $\frac{14x^2y}{-21xy^4}$ $\frac{-2x}{3y^3}$

17. $\frac{8r^2s^3t}{16rs^4t^3}$ $\frac{r}{2st^2}$

18. $\frac{-10a^3b^2c^3}{15ab^4c}$ $\frac{-2a^2c^2}{b^2}$

19. $\frac{3x+18}{5x+30}$ $\frac{3}{5}$

20. $\frac{4x-28}{5x-35}$ $\frac{4}{5}$

21. $\frac{3x-6}{5x-15}$ $\frac{3x-6}{5x-15}$

22. $\frac{x^2-25}{3x-15}$ $\frac{x+5}{3}$

23. $\frac{6a-24}{a^2-16}$ $\frac{6}{a+4}$

24. $\frac{5x-5}{x^2-4}$ $\frac{5x-5}{x^2-4}$

25. $\frac{x^2+2x+1}{5x+5}$ $\frac{x+1}{5}$

26. $\frac{4w^2-8w}{w^2+w-6}$ $\frac{4w}{w+3}$

27. $\frac{x^2-5x-14}{x^2-49}$ $\frac{x+2}{x+7}$

28. $\frac{y^2-9}{y^2+8y+15}$ $\frac{y-3}{y+5}$

29. $\frac{2m^2+3m-5}{2m^2+11m+15}$ $\frac{m-1}{m+3}$

30. $\frac{6x^2-x-2}{3x^2-5x+2}$ $\frac{2x+1}{x-1}$

31. $\frac{p^2+2pq-15q^2}{p^2-25q^2}$ $\frac{p-3q}{p-5q}$

32. $\frac{4r^2-25s^2}{2r^2+3rs-20s^2}$ $\frac{2r+5s}{r+4s}$

33. $\frac{2x-10}{25-x^2}$ $\frac{-2}{x+5}$

34. $\frac{3a-12}{16-a^2}$ $\frac{-3}{a+4}$

35. $\frac{25-a^2}{a^2+a-30}$ $\frac{-a-5}{a+6}$

36. $\frac{2x^2-7x+3}{9-x^2}$ $\frac{-2x+1}{x+3}$

37. $\dfrac{x^2 + xy - 6y^2}{4y^2 - x^2}$ $\quad \dfrac{-x - 3y}{2y + x}$

38. $\dfrac{16z^2 - w^2}{2w^2 - 5wz - 12z^2}$ $\quad \dfrac{-w - 4z}{2w + 3z}$

Skillscan (Appendix A)

Perform the indicated operations.

a. $\dfrac{2}{3} \cdot \dfrac{4}{5}$ $\quad \dfrac{8}{15}$

b. $\dfrac{5}{6} \cdot \dfrac{4}{11}$ $\quad \dfrac{10}{33}$

c. $\dfrac{4}{7} \div \dfrac{8}{5}$ $\quad \dfrac{5}{14}$

d. $\dfrac{1}{6} \div \dfrac{7}{9}$ $\quad \dfrac{3}{14}$

e. $\dfrac{5}{8} \cdot \dfrac{16}{15}$ $\quad \dfrac{2}{3}$

f. $\dfrac{15}{21} \div \dfrac{10}{7}$ $\quad \dfrac{1}{2}$

g. $\dfrac{15}{8} \cdot \dfrac{24}{25}$ $\quad \dfrac{9}{5}$

h. $\dfrac{28}{16} \div \dfrac{21}{20}$ $\quad \dfrac{5}{3}$

Answers

1. $\dfrac{2}{3}$ **3.** $\dfrac{4}{9}$ **5.** $\dfrac{2x^3}{3}$ **7.** $\dfrac{1}{3x^3}$ **9.** $\dfrac{2ab^3}{5}$ **11.** $\dfrac{3x^2}{y^2}$ **13.** $\dfrac{-2m^2}{3n}$ **15.** $\dfrac{b^2}{2a^2}$ **17.** $\dfrac{r}{2st^2}$ **19.** $\dfrac{3}{5}$ **21.** $\dfrac{3x - 6}{5x - 15}$ **23.** $\dfrac{6}{a + 4}$ **25.** $\dfrac{x + 1}{5}$ **27.** $\dfrac{x + 2}{x + 7}$ **29.** $\dfrac{m - 1}{m + 3}$ **31.** $\dfrac{p - 3q}{p - 5q}$ **33.** $\dfrac{-2}{x + 5}$ **35.** $\dfrac{-a - 5}{a + 6}$ **37.** $\dfrac{-x - 3y}{2y + x}$ **a.** $\dfrac{8}{15}$ **b.** $\dfrac{10}{33}$ **c.** $\dfrac{5}{14}$ **d.** $\dfrac{3}{14}$ **e.** $\dfrac{2}{3}$ **f.** $\dfrac{1}{2}$ **g.** $\dfrac{9}{5}$ **h.** $\dfrac{5}{3}$

6.3 Multiplying and Dividing Algebraic Fractions

OBJECTIVES

1. To write the product of algebraic fractions in simplest form.
2. To write the quotient of algebraic fractions in simplest form.

In arithmetic, you found the product of two fractions by multiplying the numerators and multiplying the denominators. For example,

$$\frac{2}{5} \cdot \frac{3}{7} = \frac{2 \cdot 3}{5 \cdot 7} = \frac{6}{35}$$

In symbols, we have

P, *Q*, *R*, and *S* again represent polynomials.

TO MULTIPLY FRACTIONS

$$\frac{P}{Q} \cdot \frac{R}{S} = \frac{PR}{QS} \quad \text{where } Q \neq 0 \text{ and } S \neq 0$$

It is easiest to divide numerator and denominator by any common factors *before* multiplying. Consider the following.

$$\frac{3}{8}\cdot\frac{4}{9}=\frac{3\cdot 4}{8\cdot 9}=\frac{1}{6}$$

Divide by the common factors of 3 and 4. The alternative is to multiply *first:*

$$\frac{3}{8}\cdot\frac{4}{9}=\frac{12}{72}$$

and then reduce to lowest terms

$$\frac{12}{72}=\frac{1}{6}$$

In algebra we multiply fractions in exactly the same way.

TO MULTIPLY ALGEBRAIC FRACTIONS

STEP 1 Factor the numerators and denominators.

STEP 2 Divide the numerator and denominator by any common factors.

STEP 3 Write the product of the remaining factors in the numerator over the product of the remaining factors in the denominator.

The following example illustrates.

Example 1

(a) $$\frac{2x^3}{5y^2}\cdot\frac{10y}{3x^2}=\frac{2x^3\cdot 10y}{5y^2\cdot 3x^2}=\frac{4x}{3y}$$

Divide by the common factors of 5, x^2, and y.

(b) $$\frac{x}{x^2-3x}\cdot\frac{6x-18}{9x}=\frac{x}{x(x-3)}\cdot\frac{6(x-3)}{9x}$$

Factor

$$=\frac{x\cdot 6\cancel{(x-3)}}{x\cancel{(x-3)}\cdot 9x}$$

$$=\frac{2}{3x}$$

Divide by the common factors of 3, x, and $x-3$.

(c) $$\frac{4}{x^2-2x}\cdot\frac{10-5x}{8}=\frac{4}{x(x-2)}\cdot\frac{5(2-x)}{8}$$

$$=\frac{4\cdot 5\cancel{(2-x)}}{x\cancel{(x-2)}\cdot 8}=\frac{-5}{2x}$$

Note: $\frac{2-x}{x-2}=-1$

(d) $$\frac{x^2-2x-8}{3x^2}\cdot\frac{6x}{3x-12}=\frac{\cancel{(x-4)}(x+2)}{3x^2}\cdot\frac{6x}{3\cancel{(x-4)}}$$

$$=\frac{2(x+2)}{3x}$$

Divide by the common factors of $x-4$, x, and 3.

$$(e)\ \frac{x^2 - y^2}{5x - 5y} \cdot \frac{10xy}{x^2 + 2xy + y^2} = \frac{\cancel{(x - y)}\cancel{(x + y)}}{5\cancel{(x - y)}} \cdot \frac{10xy}{\cancel{(x + y)}(x + y)}$$

$$= \frac{2xy}{x + y}$$

CHECK YOURSELF 1

Multiply.

1. $\frac{3x^2}{5y^2} \cdot \frac{10y^5}{15x^3}$
2. $\frac{5x + 15}{x} \cdot \frac{2x^2}{x^2 + 3x}$
3. $\frac{x}{2x - 6} \cdot \frac{3x - x^2}{2}$
4. $\frac{3x - 15}{6x^2} \cdot \frac{2x}{x^2 - 25}$
5. $\frac{x^2 - 5x - 14}{4x^2} \cdot \frac{8x}{x^2 - 49}$

You can also use your experience from arithmetic in dividing fractions. Recall that to divide fractions, we *invert the divisor* (the *second* fraction) and multiply.

Division of algebraic fractions is defined in exactly the same way.

Once more P, Q, R, and S are polynomials.

TO DIVIDE FRACTIONS

$$\frac{P}{Q} \div \frac{R}{S} = \frac{P}{Q} \cdot \frac{S}{R} = \frac{PS}{QR}$$

where $Q \neq 0$, $R \neq 0$, and $S \neq 0$.

The following example illustrates the use of this rule.

Example 2

$$(a)\ \frac{6}{x^2} \div \frac{9}{x^3} = \frac{6}{x^2} \cdot \frac{x^3}{9}$$

Invert the divisor and multiply.

$$= \frac{\cancel{6}}{\cancel{x^2}} \cdot \frac{\cancel{x^3}}{\cancel{9}}$$

No simplification can be done until the divisor is inverted. Then divide by the common factors of 3 and x^2.

$$= \frac{2x}{3}$$

$$(b)\ \frac{3x^2y}{8xy^3} \div \frac{9x^3}{4y^4} = \frac{3x^2y}{8xy^3} \cdot \frac{4y^4}{9x^3}$$

$$= \frac{y^2}{6x^2}$$

(c) $\frac{2x+4y}{9x-18y} \div \frac{4x+8y}{3x-6y} = \frac{2x+4y}{9x-18y} \cdot \frac{3x-6y}{4x+8y}$

$= \frac{2(x+2y)}{9(x-2y)} \cdot \frac{3(x-2y)}{4(x+2y)}$

$= \frac{1}{6}$

(d) $\frac{x^2-x-6}{2x-6} \div \frac{x^2-4}{4x^2} = \frac{x^2-x-6}{2x-6} \cdot \frac{4x^2}{x^2-4}$

$= \frac{(x-3)(x+2)}{2(x-3)} \cdot \frac{4x^2}{(x+2)(x-2)}$

$= \frac{2x^2}{x-2}$

CHECK YOURSELF 2

Divide.

1. $\frac{4}{x^5} \div \frac{12}{x^3}$ **2.** $\frac{5xy^2}{7x^3y} \div \frac{10y^2}{14x^3}$ **3.** $\frac{3x-9y}{2x+10y} \div \frac{x^2-3xy}{4x+20y}$

4. $\frac{x^2-9}{4x} \div \frac{x^2-2x-15}{2x-10}$

Before concluding this section, let's review why the invert-and-multiply rule works for dividing fractions. We will use an example from arithmetic for the explanation. Suppose that we want to divide as follows:

$$\frac{3}{5} \div \frac{2}{3} \qquad (1)$$

We can write

$$\underbrace{\frac{3}{5} \div \frac{2}{3}}_{(1)} = \frac{\frac{3}{5}}{\frac{2}{3}} = \frac{\frac{3}{5} \cdot \frac{3}{2}}{\frac{2}{3} \cdot \frac{3}{2}}$$

We are multiplying by 1.

Interpret the division as a fraction.

$$= \frac{\frac{3}{5} \cdot \frac{3}{2}}{1}$$

$\frac{2}{3} \cdot \frac{3}{2} = 1$

$$= \underbrace{\frac{3}{5} \cdot \frac{3}{2}}_{(2)}$$

We then have

(1) $\frac{3}{5} \div \frac{2}{3} = \frac{3}{5} \cdot \frac{3}{2}$ (2)

Comparing expressions (1) and (2), you should see the rule for dividing fractions. Invert the *second* fraction and multiply.

CHECK YOURSELF ANSWERS

1. (1) $\frac{2y^3}{5x}$; (2) 10; (3) $-\frac{x^2}{4}$; (4) $\frac{1}{x(x+5)}$; (5) $\frac{2(x+2)}{x(x+7)}$.
2. (1) $\frac{1}{3x^2}$; (2) $\frac{x}{y}$; (3) $\frac{6}{x}$; (4) $\frac{x-3}{2x}$.

6.3 Exercises

Multiply.

1. $\frac{2}{5} \cdot \frac{10}{18}$ $\frac{2}{9}$

2. $\frac{7}{12} \cdot \frac{3}{28}$ $\frac{1}{16}$

3. $\frac{x}{3} \cdot \frac{y}{4}$ $\frac{xy}{12}$

4. $\frac{w}{7} \cdot \frac{5}{4}$ $\frac{5w}{28}$

5. $\frac{3a}{2} \cdot \frac{4}{a^2}$ $\frac{6}{a}$

6. $\frac{5x^3}{3x} \cdot \frac{9}{20x}$ $\frac{3x}{4}$

7. $\frac{3x^3y}{10xy^3} \cdot \frac{5xy^2}{9xy^2}$ $\frac{x^2}{6y^2}$

8. $\frac{8xy^5}{5x^3y^2} \cdot \frac{15y^2}{16xy^3}$ $\frac{3y^2}{2x^3}$

9. $\frac{-4ab^2}{15a^3} \cdot \frac{25ab}{-16b^3}$ $\frac{5}{12a}$

10. $\frac{-3m^3n}{10mn^3} \cdot \frac{5mn^2}{-9mn^3}$ $\frac{m^2}{6n^3}$

11. $\frac{3x}{2x-6} \cdot \frac{x^2-3x}{6}$ $\frac{x^2}{4}$

12. $\frac{x^2+5x}{3x^2} \cdot \frac{10x}{5x+25}$ $\frac{2}{3}$

13. $\frac{x^2-3x-10}{5x} \cdot \frac{15x^2}{3x-15}$ $x(x+2)$

14. $\frac{p^2-8p}{4p} \cdot \frac{12p^2}{p^2-64}$ $\frac{3p^2}{p+8}$

15. $\frac{a^2-81}{a^2+9a} \cdot \frac{5a^2}{a^2-7a-18}$ $\frac{5a}{a+2}$

16. $\dfrac{m^2 - 4m - 21}{3m^2} \cdot \dfrac{m^2 + 7m}{m^2 - 49}$ $\quad \dfrac{m + 3}{3m}$

17. $\dfrac{2x^2 - x - 3}{3x^2 + 7x + 4} \cdot \dfrac{3x^2 - 11x - 20}{4x^2 - 9}$ $\quad \dfrac{x - 5}{2x + 3}$

18. $\dfrac{4r^2 - 1}{2r^2 - 9r - 5} \cdot \dfrac{3r^2 - 13r - 10}{9r^2 - 4}$ $\quad \dfrac{2r - 1}{3r - 2}$

19. $\dfrac{a^2 + ab}{2a^2 - ab - 3b^2} \cdot \dfrac{4a^2 - 9b^2}{5a^2 - 4ab}$ $\quad \dfrac{2a + 3b}{5a - 4b}$

20. $\dfrac{x^2 - 4y^2}{x^2 - xy - 6y^2} \cdot \dfrac{7x^2 - 21xy}{5x - 10y}$ $\quad \dfrac{7x}{5}$

21. $\dfrac{x^2 + 5x}{3x - 6} \cdot \dfrac{x^2 - 4}{3x^2 + 15x} \cdot \dfrac{6x}{x^2 + 6x + 8}$ $\quad \dfrac{2x}{3(x + 4)}$

22. $\dfrac{m^2 - n^2}{m^2 - mn} \cdot \dfrac{6m}{2m^2 + mn - n^2} \cdot \dfrac{8m - 4n}{12m^2 + 12mn}$

$\dfrac{2}{m(m + n)}$

23. $\dfrac{2x - 6}{x^2 + 2x} \cdot \dfrac{3x}{3 - x}$

$\dfrac{-6}{x + 2}$

24. $\dfrac{3x - 15}{x^2 + 3x} \cdot \dfrac{4x}{5 - x}$ $\quad \dfrac{-12}{x + 3}$

Divide.

25. $\dfrac{5}{8} \div \dfrac{15}{16}$

$\dfrac{2}{3}$

26. $\dfrac{4}{9} \div \dfrac{12}{18}$

$\dfrac{2}{3}$

27. $\dfrac{5}{x^2} \div \dfrac{10}{x}$

$\dfrac{1}{2x}$

28. $\dfrac{w^2}{3} \div \dfrac{w}{9}$

$3w$

29. $\dfrac{4x^2y^2}{9x^3} \div \dfrac{8y^2}{27xy}$

$\dfrac{3y}{2}$

30. $\dfrac{8x^3y}{27xy^3} \div \dfrac{16x^3y}{45y}$

$\dfrac{5}{6xy^2}$

31. $\dfrac{3x + 6}{8} \div \dfrac{5x + 10}{6}$

$\dfrac{9}{20}$

32. $\dfrac{x^2 - 2x}{4x} \div \dfrac{6x - 12}{8}$

$\dfrac{1}{3}$

33. $\dfrac{4a - 12}{5a + 15} \div \dfrac{8a^2}{a^2 + 3a}$

$\dfrac{a - 3}{10a}$

34. $\dfrac{6p - 18}{9p} \div \dfrac{3p - 9}{p^2 + 2p}$ $\quad \dfrac{2(p + 2)}{9}$

35. $\dfrac{x^2 + 2x - 8}{9x^2} \div \dfrac{x^2 - 16}{3x - 12}$ $\quad \dfrac{x - 2}{3x^2}$

36. $\dfrac{16x}{4x^2 - 16} \div \dfrac{4x - 24}{x^2 - 4x - 12}$ $\quad \dfrac{x}{x - 2}$

37. $\dfrac{x^2 - 9}{2x^2 - 6x} \div \dfrac{2x^2 + 5x - 3}{4x^2 - 1}$ $\quad \dfrac{2x + 1}{2x}$

38. $\dfrac{2m^2 - 5m - 7}{4m^2 - 9} \div \dfrac{5m^2 + 5m}{2m^2 + 3m}$ $\dfrac{2m - 7}{5(2m - 3)}$

39. $\dfrac{a^2 - 9b^2}{4a^2 + 12ab} \div \dfrac{a^2 - ab - 6b^2}{12ab}$ $\dfrac{3b}{a + 2b}$

40. $\dfrac{r^2 + 2rs - 15s^2}{r^3 + 5r^2s} \div \dfrac{r^2 - 9s^2}{5r^3}$ $\dfrac{5r}{r + 3s}$

41. $\dfrac{x^2 - 16y^2}{3x^2 - 12xy} \div (x^2 + 4xy)$ $\dfrac{1}{3x^2}$

42. $\dfrac{p^2 - 4pq - 21q^2}{4p - 28q} \div (2p^2 + 6pq)$ $\dfrac{1}{8p}$

43. $\dfrac{x - 7}{2x + 6} \div \dfrac{21 - 3x}{x^2 + 3x}$ $\dfrac{-x}{6}$

44. $\dfrac{x - 4}{x^2 + 2x} \div \dfrac{16 - 4x}{3x + 6}$ $\dfrac{-3}{4x}$

Perform the indicated operations.

45. $\dfrac{x^2 - 2x - 8}{2x - 8} \cdot \dfrac{x^2 + 5x}{x^2 + 5x + 6} \div \dfrac{x^2 + 2x - 15}{x^2 - 9}$ $\dfrac{x}{2}$

46. $\dfrac{14x - 7}{x^2 + 3x - 4} \cdot \dfrac{x^2 + 6x + 8}{2x^2 + 5x - 3} \div \dfrac{x^2 + 2x}{x^2 + 2x - 3}$ $\dfrac{7}{x}$

Skillscan (Appendix A.1)

Perform the indicated operations.

a. $\dfrac{3}{10} + \dfrac{4}{10}$ $\dfrac{7}{10}$

b. $\dfrac{5}{8} - \dfrac{4}{8}$ $\dfrac{1}{8}$

c. $\dfrac{5}{12} - \dfrac{1}{12}$ $\dfrac{1}{3}$

d. $\dfrac{7}{16} + \dfrac{3}{16}$ $\dfrac{5}{8}$

e. $\dfrac{7}{20} + \dfrac{9}{20}$ $\dfrac{4}{5}$

f. $\dfrac{13}{8} - \dfrac{5}{8}$ 1

g. $\dfrac{11}{6} - \dfrac{2}{6}$ $\dfrac{3}{2}$

h. $\dfrac{5}{9} + \dfrac{7}{9}$ $\dfrac{4}{3}$

Answers

1. $\dfrac{2}{9}$ 3. $\dfrac{xy}{12}$ 5. $\dfrac{6}{a}$ 7. $\dfrac{x^2}{6y^2}$ 9. $\dfrac{5}{12a}$ 11. $\dfrac{x^2}{4}$ 13. $x(x + 2)$ 15. $\dfrac{5a}{a + 2}$ 17. $\dfrac{x - 5}{2x + 3}$ 19. $\dfrac{2a + 3b}{5a - 4b}$ 21. $\dfrac{2x}{3(x + 4)}$ 23. $\dfrac{-6}{x + 2}$ 25. $\dfrac{2}{3}$ 27. $\dfrac{1}{2x}$ 29. $\dfrac{3y}{2}$ 31. $\dfrac{9}{20}$ 33. $\dfrac{a - 3}{10a}$ 35. $\dfrac{x - 2}{3x^2}$ 37. $\dfrac{2x + 1}{2x}$ 39. $\dfrac{3b}{a + 2b}$ 41. $\dfrac{1}{3x^2}$ 43. $\dfrac{-x}{6}$ 45. $\dfrac{x}{2}$ a. $\dfrac{7}{10}$ b. $\dfrac{1}{8}$ c. $\dfrac{1}{3}$ d. $\dfrac{5}{8}$ e. $\dfrac{4}{5}$ f. 1 g. $\dfrac{3}{2}$ h. $\dfrac{4}{3}$

6.4 Adding and Subtracting Like Fractions

OBJECTIVE
To write the sum or difference of like fractions as a single fraction in lowest terms.

You probably remember from arithmetic that *like fractions* are fractions that have the same denominator. The same is true in algebra.

$\frac{2}{5}, \frac{12}{5},$ and $\frac{4}{5}$ are like fractions.

$\frac{x}{3}, \frac{y}{3},$ and $\frac{z-5}{3}$ are like fractions.

$\frac{3x}{2}, \frac{x}{4},$ and $\frac{3x}{8}$ are unlike fractions.

The fractions have different denominators.

$\frac{3}{x}, \frac{2}{x^2},$ and $\frac{x+1}{x^3}$ are unlike fractions.

Adding or subtracting like fractions in algebra is straightforward (just as it was in arithmetic). You can use the following steps.

TO ADD OR SUBTRACT LIKE FRACTIONS

STEP 1 Add or subtract the numerators.

STEP 2 Write the sum or difference over the common denominator.

STEP 3 Write the resulting fraction in simplest form.

Example 1

Add the numerators.

$$(a)\ \frac{2}{9} + \frac{4}{9} = \frac{2+4}{9} = \frac{6}{9}$$

$$= \frac{2}{3}$$

Simplify the sum.

$$(b)\ \frac{2x}{15} + \frac{x}{15} = \frac{2x+x}{15}$$

$$= \frac{3x}{15} = \frac{x}{5}$$

Subtract the numerators.

$$(c)\ \frac{5y}{6} - \frac{y}{6} = \frac{5y - y}{6}$$

$$= \frac{4y}{6} = \frac{2y}{3}$$

Simplify.

$$(d)\ \frac{3}{x} + \frac{5}{x} = \frac{3+5}{x} = \frac{8}{x}$$

$$(e)\ \frac{9}{a^2} - \frac{7}{a^2} = \frac{9-7}{a^2} = \frac{2}{a^2}$$

CHECK YOURSELF 1

Add or subtract as indicated.

1. $\frac{3a}{10} + \frac{2a}{10}$ **2.** $\frac{7b}{8} - \frac{3b}{8}$ **3.** $\frac{4}{x} + \frac{3}{x}$

If polynomials are involved in the numerators or denominators, the process is exactly the same.

Example 2

$$(a)\ \frac{5}{x+3} + \frac{2}{x+3} = \frac{5+2}{x+3} = \frac{7}{x+3}$$

$$(b)\ \frac{4x}{x-4} - \frac{16}{x-4} = \frac{4x-16}{x-4}$$

Factor and simplify.

$$= \frac{4\cancel{(x-4)}}{\cancel{x-4}} = 4$$

$$(c)\ \frac{a-b}{3} + \frac{2a+b}{3} = \frac{(a-b)+(2a+b)}{3}$$

$$= \frac{3a}{3} = a$$

Be sure to enclose the second numerator in parentheses!

$$(d)\ \frac{3x+y}{2x} - \frac{x-3y}{2x} = \frac{(3x+y)-(x-3y)}{2x}$$

Change both signs.

$$= \frac{3x + y - x + 3y}{2x}$$

$$= \frac{2x + 4y}{2x}$$

$$= \frac{2(x + 2y)}{2x}$$ Factor and divide by the common factor of 2.

$$= \frac{x + 2y}{x}$$

CHECK YOURSELF 2

Add or subtract as indicated.

1. $\frac{4}{x - 5} - \frac{2}{x - 5}$ 2. $\frac{3x}{x + 3} + \frac{9}{x + 3}$ 3. $\frac{5x - y}{3y} - \frac{2x - 4y}{3y}$

CHECK YOURSELF ANSWERS

1. (1) $\frac{a}{2}$; (2) $\frac{b}{2}$; (3) $\frac{7}{x}$. **2.** (1) $\frac{2}{x - 5}$; (2) 3; (3) $\frac{x + y}{y}$.

6.4 Exercises

Add or subtract as indicated. Express your results in simplest form.

1. $\frac{8}{15} + \frac{2}{15}$ $\frac{2}{3}$

2. $\frac{7}{12} - \frac{5}{12}$ $\frac{1}{6}$

3. $\frac{11}{20} - \frac{6}{20}$ $\frac{1}{4}$

4. $\frac{5}{9} + \frac{7}{9}$ $\frac{4}{3}$

5. $\frac{x}{8} + \frac{3x}{8}$ $\frac{x}{2}$

6. $\frac{5y}{16} + \frac{7y}{16}$ $\frac{3y}{4}$

7. $\frac{7a}{10} - \frac{3a}{10}$ $\frac{2a}{5}$

8. $\frac{5x}{12} - \frac{x}{12}$ $\frac{x}{3}$

9. $\frac{5}{x} + \frac{3}{x}$ $\frac{8}{x}$

10. $\frac{9}{y} - \frac{3}{y}$ $\frac{6}{y}$

11. $\frac{8}{w} - \frac{2}{w}$ $\frac{6}{w}$

12. $\frac{7}{z} + \frac{9}{z}$ $\frac{16}{z}$

13. $\frac{6}{x-3} + \frac{2}{x-3}$ $\frac{8}{x-3}$

14. $\frac{5}{x+2} - \frac{3}{x+2}$ $\frac{2}{x+2}$

15. $\frac{2x}{x-2} - \frac{4}{x-2}$ 2

16. $\frac{7w}{w+3} + \frac{21}{w+3}$ 7

17. $\frac{8p}{p+4} + \frac{32}{p+4}$ 8

18. $\frac{5a}{a-3} - \frac{15}{a-3}$ 5

19. $\frac{x^2}{x+4} + \frac{3x-4}{x+4}$ $x-1$

20. $\frac{x^2}{x-3} - \frac{9}{x-3}$ $x+3$

21. $\frac{m^2}{m-5} - \frac{25}{m-5}$ $m+5$

22. $\frac{s^2}{s+3} + \frac{2s-3}{s+3}$ $s-1$

23. $\frac{a-1}{3} + \frac{2a-5}{3}$ $a-2$

24. $\frac{y+2}{5} + \frac{4y+8}{5}$ $y+2$

25. $\frac{3x-1}{4} - \frac{x+7}{4}$ $\frac{x-4}{2}$

26. $\frac{4x+2}{3} - \frac{x-1}{3}$ $x+1$

27. $\frac{3m-2}{5m} + \frac{2m+12}{5m}$ $\frac{m+2}{m}$

28. $\frac{5x-2y}{3y} - \frac{2x+y}{3y}$ $\frac{x-y}{y}$

29. $\frac{4w-7}{w-5} - \frac{2w+3}{w-5}$ 2

30. $\frac{3b-8}{b-6} + \frac{b-16}{b-6}$ 4

31. $\frac{x-7}{x^2-x-6} + \frac{2x-2}{x^2-x-6}$ $\frac{3}{x+2}$

32. $\frac{5a-12}{a^2-8a+15} - \frac{3a-2}{a^2-8a+15}$ $\frac{2}{a-3}$

Skillscan (Appendix 1)

a. $\frac{3}{4} + \frac{1}{2}$ $\frac{5}{4}$

b. $\frac{5}{6} - \frac{2}{3}$ $\frac{1}{6}$

c. $\frac{7}{10} - \frac{3}{5}$ $\frac{1}{10}$

d. $\frac{5}{8} + \frac{3}{4}$ $\frac{11}{8}$

e. $\frac{5}{6} + \frac{3}{8}$ $\frac{29}{24}$

f. $\frac{7}{8} - \frac{3}{5}$ $\frac{11}{40}$

g. $\frac{9}{10} - \frac{2}{15}$ $\frac{23}{30}$

h. $\frac{5}{12} + \frac{7}{18}$ $\frac{29}{36}$

Answers

1. $\frac{2}{3}$ **3.** $\frac{1}{4}$ **5.** $\frac{x}{2}$ **7.** $\frac{2a}{5}$ **9.** $\frac{8}{x}$ **11.** $\frac{6}{w}$ **13.** $\frac{8}{x-3}$ **15.** 2 **17.** 8 **19.** $x-1$ **21.** $m+5$ **23.** $a-2$ **25.** $\frac{x-4}{2}$ **27.** $\frac{m+2}{m}$ **29.** 2 **31.** $\frac{3}{x+2}$ **a.** $\frac{5}{4}$ **b.** $\frac{1}{6}$ **c.** $\frac{1}{10}$ **d.** $\frac{11}{8}$ **e.** $\frac{29}{24}$ **f.** $\frac{11}{40}$ **g.** $\frac{23}{30}$ **h.** $\frac{29}{36}$

6.5 Adding and Subtracting Unlike Fractions

OBJECTIVE
To write the sum or difference of unlike fractions as a single fraction in simplest form.

Adding or subtracting *unlike fractions* (fractions that do not have the same denominator) requires a bit more work than adding or subtracting the like fractions of the previous section. When the denominators are not the same, we must use the idea of the *lowest common denominator* (LCD). Each fraction is "built up" to an equivalent fraction having the LCD as a denominator. You can then add or subtract as before.

Let's review with an example from arithmetic.

Example 1

Add $\frac{5}{9}+\frac{1}{6}$.

Step 1 Find the LCD. Factor each denominator.

$9 = 3 \cdot 3$ ⟵ 3 appears twice.

$6 = 2 \cdot 3$

To form the LCD, include each factor the greatest number of times it appears in any single denominator. Use one 2, since 2 appears only once in the factorization of 6. Use two 3s since 3 appears twice in the factorization of 9. Thus the LCD for the fractions is $2 \cdot 3 \cdot 3 = 18$.

Step 2 "Build up" each fraction to an equivalent fraction with the LCD as the denominator. Do this by multiplying the numerator and denominator of the given fractions by the same number.

Do you see that this uses the fundamental principle in the form

$$\frac{P}{Q}=\frac{PR}{QR}$$

$$\frac{5}{9}=\frac{5\cdot 2}{9\cdot 2}=\frac{10}{18}$$

$$\frac{1}{6}=\frac{1\cdot 3}{6\cdot 3}=\frac{3}{18}$$

Step 3 Add the fractions.

$$\frac{5}{9} + \frac{1}{6} = \frac{10}{18} + \frac{3}{18} = \frac{13}{18}$$

$\frac{13}{18}$ is in simplest form, and so we are done!

CHECK YOURSELF 1

Add.

1. $\frac{1}{6} + \frac{3}{8}$ **2.** $\frac{3}{10} + \frac{4}{15}$

The process is exactly the same in algebra. We can summarize the steps with the following rule:

TO ADD OR SUBTRACT UNLIKE FRACTIONS

STEP 1 Find the lowest common denominator.

STEP 2 Convert each fraction to an equivalent fraction with the LCD as a denominator.

STEP 3 Add or subtract the like fractions formed in step 2.

STEP 4 Write the sum or difference in simplest form.

Example 2

(*a*) Add $\frac{3}{2x} + \frac{4}{x^2}$.

Step 1 Factor the denominators.

$$2x = 2 \cdot x$$
$$x^2 = x \cdot x$$

The LCD must contain the factors 2 and x. The factor x must appear *twice* because it appears twice as a factor in the second denominator.

The LCD is $2 \cdot x \cdot x$, or $2x^2$.

Step 2

$$\frac{3}{2x} = \frac{3 \cdot x}{2x \cdot x} = \frac{3x}{2x^2}$$

$$\frac{4}{x^2} = \frac{4 \cdot 2}{x^2 \cdot 2} = \frac{8}{2x^2}$$

Step 3

$$\frac{3}{2x} + \frac{4}{x^2} = \frac{3x}{2x^2} + \frac{8}{2x^2} = \frac{3x + 8}{2x^2}$$

The sum is in simplest form.

(*b*) Subtract $\frac{4}{3x^2} - \frac{3}{2x^3}$.

Step 1 Factor the denominators.

$$3x^2 = 3 \cdot x \cdot x$$
$$2x^3 = 2 \cdot x \cdot x \cdot x$$

The factor x must appear *three* times. Do you see why?

The LCD must contain the factors 2, 3, and x. The LCD is

$2 \cdot 3 \cdot x \cdot x \cdot x$, or $6x^3$

Step 2

$$\frac{4}{3x^2} = \frac{4 \cdot 2x}{3x^2 \cdot 2x} = \frac{8x}{6x^3}$$

$$\frac{3}{2x^3} = \frac{3 \cdot 3}{2x^3 \cdot 3} = \frac{9}{6x^3}$$

Step 3

$$\frac{4}{3x^2} - \frac{3}{2x^3} = \frac{8x}{6x^3} - \frac{9}{6x^3} = \frac{8x - 9}{6x^3}$$

The difference is in simplest form.

CHECK YOURSELF 2

Add or subtract as indicated.

1. $\frac{5}{x^2} + \frac{3}{x^3}$ **2.** $\frac{3}{5x} - \frac{1}{4x^2}$

Example 3

Add $\frac{2}{3x^2y} + \frac{3}{4x^3}$.

Step 1 Factor the denominators.

$$3x^2y = 3 \cdot x \cdot x \cdot y$$
$$4x^3 = 2 \cdot 2 \cdot x \cdot x \cdot x$$

The LCD is $12x^3y$. Do you see why?

Step 2

$$\frac{2}{3x^2y} = \frac{2 \cdot 4x}{3x^2y \cdot 4x} = \frac{8x}{12x^3y}$$

$$\frac{3}{4x^3} = \frac{3 \cdot 3y}{4x^3 \cdot 3y} = \frac{9y}{12x^3y}$$

Step 3

$$\frac{2}{3x^2y} + \frac{3}{4x^3} = \frac{8x}{12x^3y} + \frac{9y}{12x^3y}$$
$$= \frac{8x + 9y}{12x^3y}$$

CHECK YOURSELF 3

Add as indicated.

$$\frac{2}{3x^2y} + \frac{1}{6xy^2}$$

Example 4

(*a*) Add $\dfrac{5}{x} + \dfrac{2}{x - 1}$

Step 1 The LCD must have factors of x and $x - 1$. The LCD is $x(x - 1)$.

Step 2

$$\frac{5}{x} = \frac{5(x - 1)}{x(x - 1)}$$

$$\frac{2}{x - 1} = \frac{2x}{x(x - 1)}$$

Step 3

$$\frac{5}{x} + \frac{2}{x-1} = \frac{5(x-1)}{x(x-1)} + \frac{2x}{x(x-1)}$$

$$= \frac{5x - 5 + 2x}{x(x-1)}$$

$$= \frac{7x - 5}{x(x-1)}$$

(*b*) Subtract $\frac{3}{x-2} - \frac{4}{x+2}$

Step 1 The LCD must have factors of $x - 2$ and $x + 2$. The LCD is $(x - 2)(x + 2)$.

Step 2

Multiply numerator and denominator by $x + 2$.

$$\frac{3}{x-2} = \frac{3(x+2)}{(x-2)(x+2)}$$

Multiply numerator and denominator by $x - 2$.

$$\frac{4}{x+2} = \frac{4(x-2)}{(x+2)(x-2)}$$

Step 3

$$\frac{3}{x-2} - \frac{4}{x+2} = \frac{3(x+2) - 4(x-2)}{(x+2)(x-2)}$$

Note the sign changes.

$$= \frac{3x + 6 - 4x + 8}{(x+2)(x-2)}$$

$$= \frac{-x + 14}{(x+2)(x-2)}$$

CHECK YOURSELF 4

Add or subtract as indicated.

1. $\frac{3}{x+2} + \frac{5}{x}$ **2.** $\frac{4}{x+3} - \frac{2}{x-3}$

The following examples will show how factoring must sometimes be used in forming the LCD.

Example 5

(*a*) Add $\frac{3}{2x-2} + \frac{5}{3x-3}$.

Step 1 Factor the denominators.

$2x - 2 = 2(x - 1)$
$3x - 3 = 3(x - 1)$

The LCD must have factors of 2, 3, and $x - 1$. The LCD is $2 \cdot 3(x - 1)$ or $6(x - 1)$.

Step 2

$$\frac{3}{2x-2} = \frac{3}{2(x-1)} = \frac{3 \cdot 3}{2(x-1) \cdot 3} = \frac{9}{6(x-1)}$$

$$\frac{5}{3x-3} = \frac{5}{3(x-1)} = \frac{5 \cdot 2}{3(x-1) \cdot 2} = \frac{10}{6(x-1)}$$

Step 3

$$\frac{3}{2x-2} + \frac{5}{3x-3} = \frac{9}{6(x-1)} + \frac{10}{6(x-1)}$$
$$= \frac{9+10}{6(x-1)}$$
$$= \frac{19}{6(x-1)}$$

(*b*) Subtract $\frac{3}{2x-4} - \frac{6}{x^2-4}$.

Step 1 Factor the denominators.

$2x - 4 = 2(x - 2)$
$x^2 - 4 = (x + 2)(x - 2)$

The LCD must have factors of 2, $x - 2$, and $x + 2$. The LCD is $2(x - 2)(x + 2)$.

Step 2

Multiply numerator and denominator by $x + 2$.

$$\frac{3}{2x-4} = \frac{3}{2(x-2)} = \frac{3(x+2)}{2(x-2)(x+2)}$$

Multiply numerator and denominator by $x - 2$.

$$\frac{6}{x^2-4} = \frac{6}{(x+2)(x-2)} = \frac{6 \cdot 2}{2(x+2)(x-2)} = \frac{12}{2(x+2)(x-2)}$$

Step 3

$$\frac{3}{2x - 4} - \frac{6}{x^2 - 4} = \frac{3(x + 2) - 12}{2(x - 2)(x + 2)}$$

Remove the parentheses and combine like terms in the numerator.

$$= \frac{3x + 6 - 12}{2(x - 2)(x + 2)}$$

$$= \frac{3x - 6}{2(x - 2)(x + 2)}$$

Step 4 Simplify the difference.

Factor the numerator and divide by the common factor $x - 2$.

$$\frac{3x - 6}{2(x - 2)(x + 2)} = \frac{3\cancel{(x - 2)}}{2\cancel{(x - 2)}(x + 2)} = \frac{3}{2(x + 2)}$$

(c) Subtract $\frac{5}{x^2 - 1} - \frac{2}{x^2 + 2x + 1}$.

Step 1 Factor the denominators.

$$x^2 - 1 = (x - 1)(x + 1)$$
$$x^2 + 2x + 1 = (x + 1)(x + 1)$$

The LCD is $(x - 1)\underbrace{(x + 1)(x + 1)}$.

Two factors are needed.

Step 2

$$\frac{5}{(x - 1)(x + 1)} = \frac{5(x + 1)}{(x - 1)(x + 1)(x + 1)}$$

$$\frac{2}{(x + 1)(x + 1)} = \frac{2(x - 1)}{(x + 1)(x + 1)(x - 1)}$$

Step 3

$$\frac{5}{x^2 - 1} - \frac{2}{x^2 + 2x + 1} = \frac{5(x + 1) - 2(x - 1)}{(x - 1)(x + 1)(x + 1)}$$

Remove the parentheses and simplify in the numerator.

$$= \frac{5x + 5 - 2x + 2}{(x - 1)(x + 1)(x + 1)}$$

$$= \frac{3x + 7}{(x - 1)(x + 1)(x + 1)}$$

CHECK YOURSELF 5

Add or subtract as indicated.

1. $\frac{5}{2x + 2} + \frac{1}{5x + 5}$ 2. $\frac{3}{x^2 - 9} - \frac{1}{2x - 6}$

3. $\frac{4}{x^2 - x - 2} - \frac{3}{x^2 + 4x + 3}$

Recall from Section 6.2 that

$a - b = -(b - a)$

Let's see how this can be used in adding or subtracting algebraic fractions.

Example 6

Add $\frac{4}{x - 5} + \frac{2}{5 - x}$

Rather than try a denominator of $(x - 5)(5 - x)$, let's simplify first.

Replace $5 - x$ with $-(x - 5)$.

$$\frac{4}{x - 5} + \frac{2}{5 - x} = \frac{4}{x - 5} + \frac{2}{-(x - 5)}$$

We now use the fact that $\frac{a}{-b} = -\frac{a}{b}$

$$= \frac{4}{x - 5} - \frac{2}{x - 5}$$

The LCD is now $x - 5$, and we can combine the fractions as

$$= \frac{4 - 2}{x - 5}$$

$$= \frac{2}{x - 5}$$

CHECK YOURSELF 6

Subtract.

$\frac{3}{x - 3} - \frac{1}{3 - x}$

CHECK YOURSELF ANSWERS

1. (1) $\frac{13}{24}$; (2) $\frac{17}{30}$. 2. (1) $\frac{5x + 3}{x^3}$; (2) $\frac{12x - 5}{20x^2}$. 3. $\frac{4y + x}{6x^2y^2}$.

4. (1) $\frac{8x + 10}{x(x + 2)}$; (2) $\frac{2x - 18}{(x + 3)(x - 3)}$.

5. (1) $\frac{27}{10(x + 1)}$; (2) $\frac{-1}{2(x + 3)}$; (3) $\frac{x + 18}{(x + 1)(x - 2)(x + 3)}$.

6. $\frac{4}{x - 3}$.

6.5 Exercises

Add or subtract as indicated. Express your result in simplest form.

1. $\frac{5}{8} + \frac{3}{5}$ $\frac{49}{40}$

2. $\frac{7}{9} - \frac{1}{6}$ $\frac{11}{18}$

3. $\frac{11}{15} - \frac{3}{10}$ $\frac{13}{30}$

4. $\frac{7}{8} + \frac{5}{6}$ $\frac{41}{24}$

5. $\frac{y}{4} + \frac{3y}{5}$ $\frac{17y}{20}$

6. $\frac{5x}{6} - \frac{2x}{3}$ $\frac{x}{6}$

7. $\frac{7a}{3} - \frac{a}{7}$ $\frac{46a}{21}$

8. $\frac{3m}{4} + \frac{m}{9}$ $\frac{31m}{36}$

9. $\frac{3}{x} - \frac{4}{5}$ $\frac{15 - 4x}{5x}$

10. $\frac{5}{x} + \frac{2}{3}$ $\frac{15 + 2x}{3x}$

11. $\frac{5}{a} + \frac{a}{5}$ $\frac{25 + a^2}{5a}$

12. $\frac{y}{3} - \frac{3}{y}$ $\frac{y^2 - 9}{3y}$

13. $\frac{5}{m} + \frac{3}{m^2}$ $\frac{5m + 3}{m^2}$

14. $\frac{4}{x^2} - \frac{3}{x}$ $\frac{4 - 3x}{x^2}$

15. $\frac{1}{x^2} - \frac{2}{3x}$ $\frac{3 - 2x}{3x^2}$

16. $\frac{5}{2w} + \frac{7}{w^3}$ $\frac{5w^2 + 14}{2w^3}$

17. $\frac{3}{5s} + \frac{2}{s^2}$ $\frac{3s + 10}{5s^2}$

18. $\frac{7}{a^2} - \frac{3}{7a}$ $\frac{49 - 3a}{7a^2}$

19. $\frac{3}{4b^2} + \frac{5}{3b^3}$ $\frac{9b + 20}{12b^3}$

20. $\frac{4}{5x^3} - \frac{3}{2x^2}$ $\frac{8 - 15x}{10x^2}$

21. $\frac{x}{x + 2} + \frac{2}{5}$ $\frac{7x + 4}{5(x + 2)}$

22. $\frac{3}{4} - \frac{a}{a - 1}$ $\frac{-a - 3}{4(a - 1)}$

23. $\frac{y}{y - 4} - \frac{3}{4}$ $\frac{y + 12}{4(y - 4)}$

24. $\frac{m}{m + 3} + \frac{2}{3}$ $\frac{5m + 6}{3(m + 3)}$

25. $\frac{4}{x} + \frac{3}{x + 1}$ $\frac{7x + 4}{x(x + 1)}$

26. $\frac{2}{x} - \frac{1}{x - 2}$ $\frac{x - 4}{x(x - 2)}$

27. $\frac{5}{a - 1} - \frac{2}{a}$ $\frac{3a + 2}{a(a - 1)}$

28. $\frac{4}{x + 2} + \frac{3}{x}$ $\frac{7x + 6}{x(x + 2)}$

29. $\dfrac{4}{3s-2}+\dfrac{2}{3s}$ $\quad \dfrac{18s-4}{3s(3s-2)}$

30. $\dfrac{5}{2a-1}-\dfrac{3}{2a}$ $\quad \dfrac{4a+3}{2a(2a-1)}$

31. $\dfrac{2}{x+1}+\dfrac{3}{x+2}$ $\quad \dfrac{5x+7}{(x+1)(x+2)}$

32. $\dfrac{4}{y-1}+\dfrac{2}{y+3}$ $\quad \dfrac{6y+10}{(y-1)(y+3)}$

33. $\dfrac{5}{x-3}-\dfrac{1}{x+1}$ $\quad \dfrac{4x+8}{(x-3)(x+1)}$

34. $\dfrac{4}{x+5}-\dfrac{3}{x-1}$ $\quad \dfrac{x-19}{(x+5)(x-1)}$

35. $\dfrac{2}{a-2}+\dfrac{5}{3a-6}$ $\quad \dfrac{11}{3(a-2)}$

36. $\dfrac{3}{a+4}-\dfrac{2}{5a+20}$ $\quad \dfrac{13}{5(a+4)}$

37. $\dfrac{x}{x+3}-\dfrac{2}{3x+9}$ $\quad \dfrac{3x-2}{3(x+3)}$

38. $\dfrac{x}{x-4}+\dfrac{5}{2x-8}$ $\quad \dfrac{2x+5}{2(x-4)}$

39. $\dfrac{2}{3y+3}+\dfrac{1}{2y+2}$ $\quad \dfrac{7}{6(y+1)}$

40. $\dfrac{4}{5m-5}-\dfrac{3}{2m-2}$ $\quad \dfrac{-7}{10(m-1)}$

41. $\dfrac{5}{4w-8}-\dfrac{1}{3w-6}$ $\quad \dfrac{11}{12(w-2)}$

42. $\dfrac{2}{5x+5}+\dfrac{5}{4x+4}$ $\quad \dfrac{33}{20(x+1)}$

43. $\dfrac{5}{3a+6}-\dfrac{a}{5a+10}$ $\quad \dfrac{25-3a}{15(a+2)}$

44. $\dfrac{4}{3b-9}+\dfrac{3b}{2b-6}$ $\quad \dfrac{8+9b}{6(b-3)}$

45. $\dfrac{y-1}{y+1}-\dfrac{y}{2y+2}$ $\quad \dfrac{y-2}{2(y+1)}$

46. $\dfrac{x+3}{x-3}-\dfrac{x}{2x-6}$ $\quad \dfrac{x+6}{2(x-3)}$

47. $\dfrac{4}{x^2-4}+\dfrac{2}{x+2}$ $\quad \dfrac{2x}{(x-2)(x+2)}$

48. $\dfrac{3}{x-2}+\dfrac{2}{x^2-x-2}$ $\quad \dfrac{3x+5}{(x-2)(x+1)}$

49. $\dfrac{4x}{x^2-3x+2}-\dfrac{1}{x-2}$ $\quad \dfrac{3x+1}{(x-1)(x-2)}$

50. $\dfrac{w}{w^2-1}-\dfrac{2}{w+1}$ $\quad \dfrac{-w+2}{(w-1)(w+1)}$

51. $\dfrac{2x}{x^2-7x+12}+\dfrac{6}{x-3}$ $\quad \dfrac{8}{x-4}$

52. $\dfrac{15a}{a^2-a-6}-\dfrac{6}{a+2}$ $\quad \dfrac{9}{a-3}$

53. $\dfrac{3}{2w-2}+\dfrac{1}{3w+3}$ $\quad \dfrac{11w+7}{6(w-1)(w+1)}$

54. $\dfrac{2}{5x-10}-\dfrac{3}{2x+4}$ $\quad \dfrac{-11x+38}{10(x-2)(x+2)}$

55. $\dfrac{4}{3a-9}-\dfrac{3}{2a+6}$ $\quad \dfrac{-a+51}{6(a-3)(a+3)}$

56. $\dfrac{2}{3b-6}+\dfrac{3}{4b+8}$

$\dfrac{17b-2}{12(b-2)(b+2)}$

57. $\dfrac{5}{x^2-16}-\dfrac{3}{x^2-x-12}$

$\dfrac{2x+3}{(x+4)(x-4)(x+3)}$

58. $\dfrac{3}{x^2+4x+3}-\dfrac{1}{x^2-9}$

$\dfrac{2x-10}{(x+1)(x-3)(x+3)}$

59. $\dfrac{2}{y^2+y-6}+\dfrac{3y}{y^2-2y-15}$

$\dfrac{3y^2-4y-10}{(y+3)(y-2)(y-5)}$

60. $\dfrac{2a}{a^2-a-12}-\dfrac{3}{a^2-2a-8}$

$\dfrac{2a^2+a-9}{(a-4)(a+3)(a+2)}$

61. $\dfrac{6x}{x^2-9}-\dfrac{5x}{x^2+x-6}$

$\dfrac{x}{(x-3)(x-2)}$

62. $\dfrac{4y}{y^2+6y+5}+\dfrac{2y}{y^2-1}$

$\dfrac{6y}{(y+5)(y-1)}$

63. $\dfrac{3}{a-7}+\dfrac{2}{7-a}$

$\dfrac{1}{a-7}$

64. $\dfrac{5}{x-5}-\dfrac{3}{5-x}$

$\dfrac{8}{x-5}$

65. $\dfrac{2x}{2x-3}-\dfrac{1}{3-2x}$

$\dfrac{2x+1}{2x-3}$

66. $\dfrac{9m}{3m-1}+\dfrac{3}{1-3m}$

3

67. $\dfrac{1}{a-3}-\dfrac{1}{a+3}+\dfrac{2a}{a^2-9}$

$\dfrac{2}{a-3}$

68. $\dfrac{1}{p+1}+\dfrac{1}{p-3}-\dfrac{4}{p^2-2p-3}$

$\dfrac{2}{p+1}$

Skillscan (Section 6.3)

Multiply.

a. $\dfrac{3}{4}\cdot 8$

6

b. $\dfrac{7}{10}\cdot 20$

14

c. $\dfrac{4}{x^2}\cdot x^2$

4

d. $\dfrac{9}{w^2}\cdot w^3$

$9w$

e. $\dfrac{1}{xy}\cdot xy^2$

y

f. $\dfrac{2}{a^2}\cdot a^2b^2$

$2b^2$

g. $\dfrac{3}{pq}\cdot p^2q^2$

$3pq$

h. $\dfrac{2}{a^2b}\cdot a^2b^2$

$2b$

Answers

1. $\dfrac{49}{40}$ 3. $\dfrac{13}{30}$ 5. $\dfrac{17y}{20}$ 7. $\dfrac{46a}{21}$ 9. $\dfrac{15-4x}{5x}$ 11. $\dfrac{25+a^2}{5a}$ 13. $\dfrac{5m+3}{m^2}$ 15. $\dfrac{3-2x}{3x^2}$ 17. $\dfrac{3s+10}{5s^2}$ 19. $\dfrac{9b+20}{12b^3}$ 21. $\dfrac{7x+4}{5(x+2)}$ 23. $\dfrac{y+12}{4(y-4)}$ 25. $\dfrac{7x+4}{x(x+1)}$ 27. $\dfrac{3a+2}{a(a-1)}$ 29. $\dfrac{18s-4}{3s(3s-2)}$ 31. $\dfrac{5x+7}{(x+1)(x+2)}$ 33. $\dfrac{4x+8}{(x-3)(x+1)}$ 35. $\dfrac{11}{3(a-2)}$ 37. $\dfrac{3x-2}{3(x+3)}$ 39. $\dfrac{7}{6(y+1)}$

41. $\frac{11}{12(w-2)}$ **43.** $\frac{25-3a}{15(a+2)}$ **45.** $\frac{y-2}{2(y+1)}$ **47.** $\frac{2x}{(x-2)(x+2)}$ **49.** $\frac{3x+1}{(x-2)(x-1)}$ **51.** $\frac{8}{x-4}$
53. $\frac{11w+7}{6(w-1)(w+1)}$ **55.** $\frac{-a+51}{6(a-3)(a+3)}$ **57.** $\frac{2x+3}{(x+4)(x-4)(x+3)}$ **59.** $\frac{3y^2-4y-10}{(y+3)(y-2)(y-5)}$
61. $\frac{x}{(x-3)(x-2)}$ **63.** $\frac{1}{x-7}$ **65.** $\frac{2x+1}{2x-3}$ **67.** $\frac{2}{a-3}$ **a.** 6 **b.** 14 **c.** 4 **d.** $9w$ **e.** y
f. $2b^2$ **g.** $3pq$ **h.** $2b$

6.6 Complex Fractions

OBJECTIVE
To simplify complex fractions.

A fraction that has a fraction in its numerator, in its denominator, or in both is called a *complex fraction*.

$$\frac{\frac{5}{6}}{\frac{3}{4}} \quad \frac{\frac{4}{x}}{\frac{3}{x^2}} \quad \text{and} \quad \frac{\frac{a+2}{3}}{\frac{a-2}{5}}$$

are all complex fractions.

There are two methods for simplifying a complex fraction. To develop the first, remember that we can always multiply the numerator and the denominator of a fraction by the same nonzero term.

$$\frac{P}{Q} = \frac{P \cdot R}{Q \cdot R} \qquad \text{where } Q \neq 0 \text{ and } R \neq 0$$

For our first approach to simplifying a complex fraction, multiply the numerator and denominator by the LCD of all fractions that appear within the complex fraction.

Example 1

Simplify $\frac{\frac{3}{4}}{\frac{5}{8}}$.

The LCD of $\frac{3}{4}$ and $\frac{5}{8}$ is 8. So multiply the numerator and denominator by 8.

$$\frac{\frac{3}{4}}{\frac{5}{8}} = \frac{\frac{3}{4} \cdot 8}{\frac{5}{8} \cdot 8} = \frac{3 \cdot 2}{5 \cdot 1} = \frac{6}{5}$$

CHECK YOURSELF 1

Simplify.

1. $\dfrac{\frac{4}{7}}{\frac{3}{7}}$ **2.** $\dfrac{\frac{3}{8}}{\frac{5}{6}}$

The same method can be used to simplify a complex fraction when variables are involved in the expression. Consider the following example.

Example 2

Simplify $\dfrac{\frac{5}{x}}{\frac{10}{x^2}}$.

The LCD of $\frac{5}{x}$ and $\frac{10}{x^2}$ is x^2, so multiply the numerator and denominator by x^2.

$$\frac{\frac{5}{x}}{\frac{10}{x^2}} = \frac{\left(\frac{5}{x}\right)x^2}{\left(\frac{10}{x^2}\right)x^2} = \frac{5x}{10} = \frac{x}{2}$$

CHECK YOURSELF 2

Simplify.

1. $\dfrac{\frac{6}{x^3}}{\frac{9}{x^2}}$ **2.** $\dfrac{\frac{m^4}{15}}{\frac{m^3}{20}}$

We may also have a sum or a difference in the numerator or denominator of a complex fraction. The simplification steps are exactly the same. Consider the following example.

Example 3

Simplify $\dfrac{1 + \dfrac{x}{y}}{1 - \dfrac{x}{y}}$.

The LCD of 1, $\frac{x}{y}$, 1, and $\frac{x}{y}$ is y, so multiply the numerator and denominator by y.

Note the use of the distributive property to multiply *each term* in the numerator and in the denominator by y.

$$\frac{1 + \dfrac{x}{y}}{1 - \dfrac{x}{y}} = \frac{\left(1 + \dfrac{x}{y}\right)y}{\left(1 - \dfrac{x}{y}\right)y} = \frac{y + \dfrac{x}{y} \cdot y}{y - \dfrac{x}{y} \cdot y}$$

$$= \frac{y + x}{y - x}$$

CHECK YOURSELF 3

Simplify.

$$\frac{\dfrac{x}{y} - 2}{\dfrac{x}{y} + 2}$$

Our second method for simplifying complex fractions uses the fact that

To divide by a fraction we invert the divisor (it *follows* the division sign) and multiply.

$$\frac{\dfrac{P}{Q}}{\dfrac{R}{S}} = \frac{P}{Q} \div \frac{R}{S} = \frac{P}{Q} \cdot \frac{S}{R}$$

To use this method, we must write the numerator and denominator of the complex fraction as single fractions. We can then divide the numerator by the denominator as before.

Example 4

We have written the numerator and denominator as single fractions. Be sure you see how this is done.

$$\frac{1 + \frac{x}{y}}{1 - \frac{x}{y}} = \frac{\frac{y + x}{y}}{\frac{y - x}{y}}$$

$$= \frac{y + x}{y} \div \frac{y - x}{y}$$

$$= \frac{y + x}{y} \cdot \frac{y}{y - x}$$

$$= \frac{y + x}{y - x}$$

Of course the answer is the same as that which we obtained in Example 3. You can use whichever method you find easier in a particular problem.

CHECK YOURSELF 4

Using the second method, simplify $\dfrac{\frac{x}{y} - 2}{\frac{x}{y} + 2}$.

Let's look at one more example of simplifying a complex fraction by the second method.

Example 5

Simplify $\dfrac{4 - \frac{y^2}{x^2}}{2 + \frac{y}{x}}$.

In this approach we must first work *separately* in the numerator and denominator to form single fractions.

$$\frac{4 - \frac{y^2}{x^2}}{2 + \frac{y}{x}} = \frac{\frac{4x^2 - y^2}{x^2}}{\frac{2x + y}{x}}$$

$$= \frac{4x^2 - y^2}{x^2} \cdot \frac{x}{2x + y}$$

Invert the divisor (the denominator) and multiply.

$$= \frac{(2x - y)\cancel{(2x + y)}}{x^2} \cdot \frac{x}{\cancel{2x + y}}$$

Factor and divide by the common factors of $2x + y$ and x.

$$= \frac{2x - y}{x}$$

CHECK YOURSELF 5

Simplify $\dfrac{\frac{a^2}{b^2} - 1}{\frac{a}{b} + 1}$.

The following algorithm will summarize our work with the two methods of simplifying complex fractions.

SIMPLIFYING COMPLEX FRACTIONS

METHOD 1

1. Multiply the numerator and denominator of the complex fraction by the LCD of all the fractions that appear within the complex fraction.
2. Simplify the resulting fraction, writing the result in simplest form.

METHOD 2

1. Write the numerator and denominator of the complex fraction as single fractions, if necessary.
2. Invert the denominator and multiply as before, writing the result in simplest form.

CHECK YOURSELF ANSWERS

1. (1) $\frac{4}{3}$; (2) $\frac{9}{20}$. **2.** (1) $\frac{2}{3x}$; (2) $\frac{4m}{3}$. **3.** $\frac{x - 2y}{x + 2y}$.
4. $\frac{x - 2y}{x + 2y}$. **5.** $\frac{a - b}{b}$.

6.6 Exercises

Simplify each complex fraction.

1. $\dfrac{\frac{2}{3}}{\frac{6}{8}}$ $\quad \frac{8}{9}$

2. $\dfrac{\frac{5}{6}}{\frac{10}{15}}$ $\quad \frac{5}{4}$

3. $\dfrac{1+\frac{1}{2}}{2+\frac{1}{4}}$ $\quad \frac{2}{3}$

4. $\dfrac{1+\frac{3}{4}}{2-\frac{1}{8}}$ $\quad \frac{14}{15}$

5. $\dfrac{2+\frac{1}{3}}{3-\frac{1}{5}}$ $\quad \frac{5}{6}$

6. $\dfrac{2+\frac{3}{5}}{1+\frac{3}{10}}$ $\quad 2$

7. $\dfrac{\frac{2}{3}+\frac{1}{2}}{\frac{3}{4}-\frac{1}{3}}$ $\quad \frac{14}{5}$

8. $\dfrac{\frac{3}{4}+\frac{1}{2}}{\frac{7}{8}-\frac{1}{4}}$ $\quad 2$

9. $\dfrac{\frac{x}{8}}{\frac{x^2}{4}}$ $\quad \frac{1}{2x}$

10. $\dfrac{\frac{m^2}{10}}{\frac{m^3}{15}}$ $\quad \frac{3}{2m}$

11. $\dfrac{\frac{3}{a}}{\frac{2}{a^2}}$ $\quad \frac{3a}{2}$

12. $\dfrac{\frac{6}{x^2}}{\frac{9}{x^3}}$ $\quad \frac{2x}{3}$

13. $\dfrac{\frac{y+1}{y}}{\frac{y-1}{2y}}$ $\quad \frac{2y+2}{y-1}$

14. $\dfrac{\frac{w+3}{4w}}{\frac{w-3}{2w}}$ $\quad \frac{w+3}{2w-6}$

15. $\dfrac{2-\frac{1}{x}}{2+\frac{1}{x}}$ $\quad \frac{2x-1}{2x+1}$

16. $\dfrac{3+\frac{1}{a}}{3-\frac{1}{a}}$ $\quad \frac{3a+1}{3a-1}$

17. $\dfrac{3-\frac{x}{y}}{\frac{6}{y}}$ $\quad \frac{3y-x}{6}$

18. $\dfrac{2+\frac{x}{y}}{\frac{4}{y}}$ $\quad \frac{2y+x}{4}$

19. $\dfrac{2+\frac{p}{q}}{1+\frac{p}{q}}$ $\quad \frac{2q+p}{q+p}$

20. $\dfrac{\frac{m}{n}-3}{\frac{m}{n}+3}$ $\quad \frac{m-3n}{m+3n}$

21. $\dfrac{a^2-1}{1-\frac{1}{a}}$ $\quad a(a+1)$

22. $\dfrac{1+\frac{1}{2x}}{4x^2-1}$ $\quad \frac{1}{2x(2x-1)}$

23. $\dfrac{\frac{x^2}{y^2}-1}{\frac{x}{y}+1}$ $\quad \frac{x-y}{y}$

24. $\dfrac{\frac{a}{b}+2}{\frac{a^2}{b^2}-4}$ $\quad \frac{b}{a-2b}$

25. $\dfrac{1 + \dfrac{3}{x} - \dfrac{4}{x^2}}{1 + \dfrac{2}{x} - \dfrac{3}{x^2}}$ $\quad \dfrac{x + 4}{x + 3}$

26. $\dfrac{1 - \dfrac{2}{r} - \dfrac{8}{r^2}}{1 - \dfrac{1}{r} - \dfrac{6}{r^2}}$ $\quad \dfrac{r - 4}{r - 3}$

27. $\dfrac{\dfrac{1}{x} + \dfrac{1}{y}}{\dfrac{2}{x} - \dfrac{2}{y}}$ $\quad \dfrac{y + x}{2(y - x)}$

28. $\dfrac{\dfrac{3}{a} - \dfrac{3}{b}}{\dfrac{1}{a} + \dfrac{1}{b}}$ $\quad \dfrac{3(b - a)}{a + b}$

29. $\dfrac{\dfrac{2}{x} - \dfrac{1}{xy}}{\dfrac{1}{xy} + \dfrac{2}{y}}$ $\quad \dfrac{2y - 1}{1 + 2x}$

30. $\dfrac{\dfrac{1}{xy} + \dfrac{2}{x}}{\dfrac{3}{y} - \dfrac{1}{xy}}$ $\quad \dfrac{1 + 2y}{3x - 1}$

31. $\dfrac{\dfrac{x^2}{y} + 2x + y}{\dfrac{1}{y^2} - \dfrac{1}{x^2}}$ $\quad \dfrac{x^2y(x + y)}{x - y}$

32. $\dfrac{\dfrac{x}{y} + 1 - \dfrac{2y}{x}}{\dfrac{1}{y^2} - \dfrac{4}{x^2}}$ $\quad \dfrac{xy(x - y)}{x - 2y}$

33. $\dfrac{\dfrac{2}{x - 1} + 1}{1 - \dfrac{3}{x - 1}}$ $\quad \dfrac{x + 1}{x - 4}$

34. $\dfrac{\dfrac{3}{a + 2} - 1}{1 + \dfrac{2}{a + 2}}$ $\quad \dfrac{1 - a}{a + 4}$

35. $\dfrac{1 - \dfrac{1}{y - 1}}{y - \dfrac{8}{y + 2}}$ $\quad \dfrac{y + 2}{(y - 1)(y + 4)}$

36. $\dfrac{1 + \dfrac{1}{x + 2}}{x - \dfrac{18}{x - 3}}$ $\quad \dfrac{x - 3}{(x + 2)(x - 6)}$

37. $1 + \dfrac{1}{1 + \dfrac{1}{x}}$ $\quad \dfrac{2x + 1}{x + 1}$

38. $1 + \dfrac{1}{1 - \dfrac{1}{y}}$ $\quad \dfrac{2y - 1}{y - 1}$

Skillscan (Section 3.4)

Solve each of the following equations.

a. $x + 8 = 10$ $\quad 2$

b. $5x - 4 = 2$ $\quad \dfrac{6}{5}$

c. $3x + 8 = 4$ $\quad -\dfrac{4}{3}$

d. $3(x - 2) - 4 = 5$ $\quad 5$

e. $4(2x + 1) - 3 = -23$ $\quad -3$

f. $4(2x - 5) - 3(3x + 1) = -8$ $\quad -15$

Answers

1. $\dfrac{8}{9}$ 3. $\dfrac{2}{3}$ 5. $\dfrac{5}{6}$ 7. $\dfrac{14}{5}$ 9. $\dfrac{1}{2x}$ 11. $\dfrac{3a}{2}$ 13. $\dfrac{2y + 2}{y - 1}$ 15. $\dfrac{2x - 1}{2x + 1}$ 17. $\dfrac{3y - x}{6}$

19. $\frac{2q+p}{q+p}$ 21. $a(a+1)$ 23. $\frac{x-y}{y}$ 25. $\frac{x+4}{x+3}$ 27. $\frac{y+x}{2(y-x)}$ 29. $\frac{2y-1}{1+2x}$ 31. $\frac{x^2y(x+y)}{x-y}$
33. $\frac{x+1}{x-4}$ 35. $\frac{y+2}{(y-1)(y+4)}$ 37. $\frac{2x+1}{x+1}$ a. 2 b. $\frac{6}{5}$ c. $-\frac{4}{3}$ d. 5 e. -3 f. -15

6.7 Equations Involving Fractions

OBJECTIVE
To solve fractional equations.

In Chapter 3 you learned how to solve a variety of equations. We now want to extend that work to the solution of *fractional equations*. These are equations that involve algebraic fractions as one or more of their terms.

The resulting equation *will* be equivalent unless a solution results that makes a denominator in the original equation 0. More about this later!

To solve a fractional equation, we multiply each term of the equation by the LCD of any fractions that appear. The resulting equation should be equivalent to the original equation and be cleared of all fractions.

Example 1

Solve

Note: This equation has three terms, $\frac{x}{2}$, $-\frac{1}{3}$, and $\frac{2x+3}{6}$.

$$\frac{x}{2} - \frac{1}{3} = \frac{2x+3}{6} \tag{1}$$

The LCD for $\frac{x}{2}$, $\frac{1}{3}$, and $\frac{2x+3}{6}$ is 6. Multiply *each* term by 6.

$$6 \cdot \frac{x}{2} - 6 \cdot \frac{1}{3} = 6\left(\frac{2x+3}{6}\right)$$

or

By the multiplication property of equality, this equation is equivalent to the original equation, labeled (1).

$$3x - 2 = 2x + 3 \tag{2}$$

Solving as before, we have

$$x = 5$$

To check, substitute 5 for x in the *original* equation.

$$\frac{5}{2} - \frac{1}{3} \stackrel{?}{=} \frac{2 \cdot 5 + 3}{6}$$

$$\frac{13}{6} = \frac{13}{6} \qquad \text{(true)}$$

Be Careful! Many students have difficulty because they don't distinguish between adding or subtracting *expressions* (as we did in Sections 6.4 and 6.5) and solving equations (illustrated in the above example). In the expression

$$\frac{x+1}{2} + \frac{x}{3}$$

we want to add the two fractions to form a single fraction. In the equation

$$\frac{x+1}{2} = \frac{x}{3} + 1$$

we want to solve for x.

CHECK YOURSELF 1

Solve and check:

$$\frac{x}{4} - \frac{1}{6} = \frac{4x-5}{12}$$

The steps of the solution illustrated in Example 1 are summarized in the following rule.

TO SOLVE A FRACTIONAL EQUATION

STEP 1 Remove the fractions in the equation by multiplying each term by the LCD of all the fractions that appear.

STEP 2 Solve the equation resulting from step 1 as before.

STEP 3 Check your solution in the *original equation*.

The equation that is formed in Step 2 can be solved by the methods of Sections 3.4 and 5.5.

We can also solve fractional equations with variables in the denominator using the above algorithm. The following example illustrates.

Example 2

Solve

$$\frac{7}{4x} - \frac{3}{x^2} = \frac{1}{2x^2}$$

The LCD of the three terms in the equation is $4x^2$, and so we multiply each term by $4x^2$.

$$4x^2 \cdot \frac{7}{4x} - 4x^2 \cdot \frac{3}{x^2} = 4x^2 \cdot \frac{1}{2x^2}$$

Simplifying, we have

$$7x - 12 = 2$$

$$7x = 14$$

$$x = 2$$

We'll leave the check to you. Be sure to return to the original equation.

CHECK YOURSELF 2

Solve and check:

$$\frac{5}{2x} - \frac{4}{x^2} = \frac{7}{2x^2}$$

The process of solving fractional equations is exactly the same when binomials are involved in the denominators.

Example 3

(*a*) Solve

$$\frac{x}{x-3} - 2 = \frac{1}{x-3}$$

The LCD is $x - 3$, and so we multiply each term by $x - 3$.

$$\cancel{(x-3)} \cdot \frac{x}{\cancel{x-3}} - 2(x-3) = \cancel{(x-3)} \cdot \frac{1}{\cancel{x-3}}$$

Simplifying, we have

$$x - 2(x - 3) = 1$$

Be careful of the signs!

$$x - 2x + 6 = 1$$
$$-x = -5$$
$$x = 5$$

To check, substitute 5 for x in the original equation.

$$\frac{5}{5-3} - 2 \stackrel{?}{=} \frac{1}{5-3}$$
$$\frac{5}{2} - 2 \stackrel{?}{=} \frac{1}{2}$$
$$\frac{1}{2} = \frac{1}{2} \qquad \text{(true)}$$

(*b*) Solve

$$\frac{3}{x-3} - \frac{7}{x+3} = \frac{2}{x^2-9}$$

Recall that

$x^2 - 9 = (x - 3)(x + 3)$.

In factored form, the three denominators are $x - 3$, $x + 3$, and $(x + 3)(x - 3)$. This means that the LCD is $(x + 3)(x - 3)$, and so we multiply:

$$(x+3)\cancel{(x-3)}\left(\frac{3}{\cancel{x-3}}\right) - \cancel{(x+3)}(x-3)\left(\frac{7}{\cancel{x+3}}\right) = \cancel{(x+3)(x-3)}\left(\frac{2}{\cancel{x^2-9}}\right)$$

Simplifying, we have

$$3(x + 3) - 7(x - 3) = 2$$
$$3x + 9 - 7x + 21 = 2$$
$$-4x + 30 = 2$$
$$-4x = -28$$
$$x = 7$$

CHECK YOURSELF 3

Solve and check.

1. $\dfrac{x}{x-5} - 2 = \dfrac{2}{x-5}$ **2.** $\dfrac{4}{x-4} - \dfrac{3}{x+1} = \dfrac{5}{x^2 - 3x - 4}$

You should be aware of one further problem in dealing with fractional equations. The following example shows that possibility.

Example 4

Solve

$$\frac{x}{x-2} - 7 = \frac{2}{x-2}$$

The LCD is $x - 2$, and so we multiply each term by $x - 2$.

$$\cancel{(x-2)}\left(\frac{x}{\cancel{x-2}}\right) - 7(x-2) = \cancel{(x-2)}\left(\frac{2}{\cancel{x-2}}\right)$$

Simplifying, we have

$$\begin{aligned} x - 7x + 14 &= 2 \\ -6x &= -12 \\ x &= 2 \end{aligned}$$

Now, when we try to check our result we have

$$\frac{2}{2-2} - 7 = \frac{2}{2-2} \quad \text{or} \quad \frac{2}{0} - 7 = \frac{2}{0}$$

These terms are undefined.

What went wrong? Remember that two of the terms in our original equation were $\dfrac{x}{x-2}$ and $\dfrac{2}{x-2}$. The variable x cannot have the value 2 because 2 is an excluded value (it makes the denominator 0). So our original equation has *no solution*.

CHECK YOURSELF 4

Solve if possible.

$$\frac{x}{x+3} - 6 = \frac{-3}{x+3}$$

Equations involving fractions may also lead to quadratic equations, as the following example illustrates.

Example 5

Solve

$$\frac{x}{x-4} = \frac{15}{x-3} - \frac{2x}{x^2 - 7x + 12}$$

The LCD is $(x - 4)(x - 3)$, or $x^2 - 7x + 12$. Multiply each term by $(x - 4)(x - 3)$.

$$\frac{x}{(x-4)}(x-4)(x-3) = \frac{15}{(x-3)}(x-4)(x-3) - \frac{2x}{(x-4)(x-3)}(x-4)(x-3)$$

Simplifying, we have

$$x(x - 3) = 15(x - 4) - 2x$$

Multiply to clear of parentheses:

$$x^2 - 3x = 15x - 60 - 2x$$

In standard form, the equation is

Note that this equation is *quadratic*. It can be solved by the methods of Section 5.5.

$$x^2 - 16x + 60 = 0$$

or

$$(x - 6)(x - 10) = 0$$

Setting the factors to 0, we have

$$x - 6 = 0 \quad \text{or} \quad x - 10 = 0$$
$$x = 6 \qquad\qquad x = 10$$

So $x = 6$ and $x = 10$ are possible solutions. We will leave the check of *each* solution to you.

CHECK YOURSELF 5

Solve and check:

$$\frac{3x}{x+2} - \frac{2}{x+3} = \frac{36}{x^2 + 5x + 6}$$

Strategies in Equation Solving

As the examples of this section have illustrated, *whenever* an equation involves algebraic fractions, the *first step* of the solution is to clear the equation of fractions by multiplication.

The following algorithm will summarize our work in solving equations that involve algebraic fractions.

SOLVING EQUATIONS INVOLVING FRACTIONS

1. Remove the fractions appearing in the equation by multiplying each term by the LCD of all the fractions that appear.
2. Solve the equation resulting from step 1. If the equation is linear, use the methods of Section 3.4 for the solution. If the equation is quadratic, use the methods of Section 5.5.
3. Check all solutions by substitution in the *original equation*. Be sure to discard any *extraneous* solutions, that is, solutions that would result in a zero denominator in the original equation.

CHECK YOURSELF ANSWERS

1. 3. **2.** 3. **3.** (1) 8; (2) -11. **4.** No solution **5.** $x = -5$ or $x = \frac{8}{3}$.

6.7 Exercises

Solve each of the following equations for x.

1. $\frac{x}{2} + 3 = 6$ 6

2. $\frac{x}{3} - 2 = 1$ 9

3. $\frac{x}{2} - \frac{x}{3} = 2$ 12

4. $\frac{x}{6} - \frac{x}{8} = 1$ 24

5. $\frac{x}{5} - \frac{1}{3} = \frac{x - 7}{3}$ 15

6. $\frac{x}{6} + \frac{3}{4} = \frac{x - 1}{4}$ 12

7. $\frac{x}{4} - \frac{1}{5} = \frac{4x + 3}{20}$ 7

8. $\frac{x}{12} - \frac{1}{6} = \frac{2x - 7}{12}$ 5

9. $\frac{3}{x} + 2 = \frac{7}{x}$ 2

10. $\frac{4}{x} - 3 = \frac{16}{x}$ -4

11. $\frac{4}{x} + \frac{3}{4} = \frac{10}{x}$ 8

12. $\frac{3}{x} = \frac{5}{3} - \frac{7}{x}$ 6

13. $\frac{5}{2x} - \frac{1}{x} = \frac{9}{2x^2}$ 3

14. $\frac{4}{3x} + \frac{1}{x} = \frac{14}{3x^2}$ 2

15. $\frac{2}{x-3} + 1 = \frac{7}{x-3}$ 8

16. $\frac{x}{x+1} + 2 = \frac{14}{x+1}$ 4

17. $\frac{12}{x+3} = \frac{x}{x+3} + 2$ 2

18. $\frac{5}{x-3} + 3 = \frac{x}{x-3}$ 2

19. $\frac{3}{x-5} + 4 = \frac{2x+5}{x-5}$ 11

20. $\frac{24}{x+5} - 2 = \frac{x+2}{x+5}$ 4

21. $\frac{2}{x+3} + \frac{1}{2} = \frac{x+6}{x+3}$ −5

22. $\frac{6}{x-5} - \frac{2}{3} = \frac{x-9}{x-5}$ 11

23. $\frac{x}{3x+12} + \frac{x-1}{x+4} = \frac{5}{3}$ −23

24. $\frac{x}{4x-12} - \frac{x-4}{x-3} = \frac{1}{8}$ 5

25. $\frac{x}{x-3} - 2 = \frac{3}{x-3}$ No solution

26. $\frac{x}{x-5} + 2 = \frac{5}{x-5}$ No solution

27. $\frac{x-1}{x+3} - \frac{x-3}{x} = \frac{3}{x^2+3x}$ 6

28. $\frac{x}{x-2} - \frac{x+1}{x} = \frac{8}{x^2-2x}$ 6

29. $\frac{1}{x-2} - \frac{2}{x+2} = \frac{2}{x^2-4}$ 4

30. $\frac{1}{x+4} + \frac{1}{x-4} = \frac{12}{x^2-16}$ 6

31. $\frac{5}{x-4} = \frac{1}{x+2} - \frac{2}{x^2-2x-8}$ −4

32. $\frac{11}{x+2} = \frac{5}{x^2-x-6} + \frac{1}{x-3}$ 4

33. $\frac{3}{x-1} - \frac{1}{x+9} = \frac{18}{x^2+8x-9}$ −5

34. $\frac{2}{x+2} = \frac{3}{x+6} + \frac{9}{x^2+8x+12}$ −3

35. $\frac{3}{x+3} + \frac{25}{x^2+x-6} = \frac{5}{x-2}$ No solution

36. $\frac{5}{x+6} + \frac{2}{x^2+7x+6} = \frac{3}{x+1}$ $\frac{11}{2}$

37. $\frac{7}{x-5} - \frac{3}{x+5} = \frac{40}{x^2-25}$ $-\frac{5}{2}$

38. $\frac{3}{x-3} - \frac{18}{x^2-9} = \frac{5}{x+3}$ No solution

39. $\frac{2x}{x-3} + \frac{2}{x-5} = \frac{3x}{x^2-8x+15}$ $-\frac{1}{2}, 6$

40. $\frac{x}{x-4} = \frac{5x}{x^2-x-12} - \frac{3}{x+3}$ $-4, 3$

41. $\frac{2x}{x+2} = \frac{5}{x^2-x-6} - \frac{1}{x-3}$ $-\frac{1}{2}$

42. $\frac{3x}{x-1} = \frac{2}{x-2} - \frac{2}{x^2-3x+2}$ $\frac{2}{3}$

43. $\frac{7}{x-2} + \frac{16}{x+3} = 3$ $-\frac{1}{3}, 7$

44. $\frac{5}{x-2} + \frac{6}{x+2} = 2$ $-\frac{1}{2}, 6$

45. $\frac{11}{x-3} - 1 = \frac{10}{x+3}$ $-8, 9$

46. $\frac{17}{x-4} - 2 = \frac{10}{x+2}$ $-\frac{9}{2}, 10$

Skillscan (Section 4.6)

Write the equation necessary for the solution of each of the following problems. Then solve the equation.

a. One number is 5 less than another. If 3 times the first number is 9 more than the second number, find the two numbers. 7, 12

b. One number is 3 more than another. If twice the smaller number is 5 less than the larger number, what are the two numbers? −2, 1

c. The sum of an integer and 3 times the next consecutive integer is 47. What are the two integers? 11, 12

d. 5 times an integer is one more than 4 times the next consecutive odd integer. Find the two integers. 9, 11

e. Claudia rowed upstream for 3 h. While she was rowing downstream, her speed was 2 mi/h faster and the trip took 2 h. What was her speed each way? 4 mi/h, 6 mi/h

f. Joe left the city at 11 A.M., heading west at a rate of 55 mi/h. An hour later Jeanine headed east at a rate of 45 mi/h. At what time will they be 305 mi apart? 2:30 PM

Answers

1. 6 **3.** 12 **5.** 15 **7.** 7 **9.** 2 **11.** 8 **13.** 3 **15.** 8 **17.** 2 **19.** 11 **21.** -5 **23.** -23
25. No solution **27.** 6 **29.** 4 **31.** -4 **33.** -5 **35.** No solution **37.** $-\frac{5}{2}$ **39.** $-\frac{1}{2}, 6$
41. $-\frac{1}{2}$ **43.** $-\frac{1}{3}, 7$ **45.** $-8, 9$ **a.** 7, 12 **b.** $-2, 1$ **c.** 11, 12 **d.** 9, 11 **e.** 4 mi/h, 6 mi/h
f. 2:30 P.M.

6.8 More Applications

OBJECTIVE
To solve word problems that lead to fractional equations.

Many word problems will lead to fractional equations that must be solved by using the methods of the last section. The five steps in solving word problems are, of course, the same as you saw earlier.

Example 1

If one-third of a number is added to three-fourths of that same number, the sum is 26. Find the number.

Step 1 Read the problem carefully. You want to find the unknown number.

Step 2 Choose a letter to represent the unknown. Let x be the unknown number.

Step 3 Form an equation.

$$\frac{1}{3}x + \frac{3}{4}x = 26$$

One-third of the number (points to $\frac{1}{3}x$); Three-fourths of the number (points to $\frac{3}{4}x$)

Step 4 Solve the equation. Multiply each term of the equation by 12, the LCD.

$$12 \cdot \frac{1}{3}x + 12 \cdot \frac{3}{4}x = 12 \cdot 26$$

Simplifying yields

$$4x + 9x = 312$$
$$13x = 312$$
$$x = 24$$

So the number is 24.

Step 5 Check your solution by returning to the *original problem*. If the number is 24, we have

$$\frac{1}{3} \cdot 24 + \frac{3}{4} \cdot 24 = 8 + 18 = 26$$

and the solution is verified.

CHECK YOURSELF 1

The sum of two-fifths of a number and one-half of that number is 18. Find the number.

Example 2

One number is twice another number. If the sum of their reciprocals is $\frac{3}{10}$, what are the two numbers?

Step 1 You want to find the two numbers.

Step 2 Let x be one number.

Then $2x$ is the other number.

Twice the first

Step 3

$$\frac{1}{x} + \frac{1}{2x} = \frac{3}{10}$$

The reciprocal of the first number, x

The reciprocal of the second number, $2x$

Step 4 The LCD of the fractions is $10x$, and so we multiply by $10x$.

$$10x\left(\frac{1}{x}\right) + 10x\left(\frac{1}{2x}\right) = 10x\left(\frac{3}{10}\right)$$

Simplifying, we have

$$10 + 5 = 3x$$
$$15 = 3x$$
$$5 = x$$

x was one number and $2x$ was the other.

The numbers are 5 and 10.

Step 5
Again check the result by returning to the original problem. If the numbers are 5 and 10, we have

The sum of the reciprocals is $\frac{3}{10}$.

$$\frac{1}{5} + \frac{1}{10} = \frac{2+1}{10} = \frac{3}{10}$$

CHECK YOURSELF 2

One number is 3 times another. If the sum of their reciprocals is $\frac{2}{9}$, find the two numbers.

The solution of many motion problems will also involve fractional equations. Remember that the key equation for solving all motion problems relates the distance traveled, the speed or rate, and the time:

$$d = r \cdot t$$

Often we will use this equation in different forms by solving for r or for t. So

$$r = \frac{d}{t} \quad \text{or} \quad t = \frac{d}{r}$$

Example 3

Vince took 1 h longer to drive 180 mi than he did on a trip of 135 mi. If his speed was the same both times, how long did each trip take?

Step 1 You want to find the times taken for the 180-mi trip and for the 135-mi trip.

Note: It is often helpful to choose your variable to "suggest" the unknown quantity—here t for "time."

Step 2 Let t be the time for the 135-mi trip (in hours).

1 h longer

Then $t + 1$ is the time for the 180-mi trip.

It is often helpful to arrange the information in tabular form such as that shown below.

Remember that rate is distance divided by time. Our third column is formed by using that relationship.

	DISTANCE	TIME	RATE
135-mi trip	135	t	$\frac{135}{t}$
180-mi trip	180	$t + 1$	$\frac{180}{t+1}$

Step 3 To form an equation, remember that the speed (or rate) for each trip was the same. That is the *key* idea. You can equate the rates for the two trips that were found in step 2. From the third column of the above table, if the rates are equal, we can write

$$\frac{135}{t} = \frac{180}{t+1}$$

Step 4 To solve the above equation, multiply each term by $t(t + 1)$, the LCD of the fractions.

$$\cancel{t}(t+1)\left(\frac{135}{\cancel{t}}\right) = t\cancel{(t+1)}\left(\frac{180}{\cancel{t+1}}\right)$$

Simplifying, we have

$$\begin{aligned} 135(t + 1) &= 180t \\ 135t + 135 &= 180t \\ 135 &= 45t \\ t &= 3 \text{ h} \end{aligned}$$

The time for the 135-mi trip was 3 h and the time for the 180-mi trip was 4 h. We'll leave the check to you.

CHECK YOURSELF 3

Cynthia took 1 h longer to bicycle 60 mi than she did on a trip of 45 mi. If her speed was the same each time, find the time for each trip.

Example 4

A train makes a trip of 300 mi in the same time that a bus can travel 250 mi. If the speed of the train is 10 mi/h faster than the speed of the bus, find the speed of each.

Step 1 You want to find the speed of the train and of the bus.

Step 2 Let r be the speed (or rate) of the bus (in miles per hour).

Then $r + 10$ is the rate of the train.

10 mi/h faster

Again let's form a table of the information.

Remember that time is distance divided by rate. Here our third column is found by using that relationship.

	DISTANCE	RATE	TIME
Train	300	$r + 10$	$\frac{300}{r+10}$
Bus	250	r	$\frac{250}{r}$

Step 3 To form an equation, remember that the times for the train and bus are the same. We can equate the expressions for time found in step 2. Again, working from column 3, we have

$$\frac{250}{r} = \frac{300}{r + 10}$$

Step 4 We multiply each term by $r(r + 10)$, the LCD of the fractions.

$$\cancel{r}(r + 10)\left(\frac{250}{\cancel{r}}\right) = r\cancel{(r + 10)}\left(\frac{300}{\cancel{r + 10}}\right)$$

Simplifying, we have

$$\begin{aligned} 250(r + 10) &= 300r \\ 250r + 2500 &= 300r \\ 2500 &= 50r \\ r &= 50 \text{ mi/h} \end{aligned}$$

The rate of the bus is 50 mi/h, and the rate of the train is 60 mi/h. You can check this result.

CHECK YOURSELF 4

A car makes a trip of 280 mi in the same time that a truck travels

245 mi. If the speed of the truck is 5 mi/h slower than that of the car, find the speed of each.

A final group of applications involves fractions in decimal form. Mixture problems often use percentages, and those percentages can be written as decimals. The following example illustrates.

Example 5

A solution of antifreeze is 20 percent alcohol. How much pure alcohol must be added to 12 quarts (qt) of the solution to make a 40% solution?

Step 1 You want to find the number of quarts of pure alcohol that must be added.

Step 2 Let x be the number of quarts of pure alcohol to be added.

Step 3 To form our equation, note that the amount of alcohol present before mixing *must be the same* as the amount in the combined solution.

A picture will help.

12 qt 20% + x qt 100% = $12 + x$ qt 40%

So

$$\underbrace{12(0.20)}_{\text{The amount of alcohol in the first solution (20\% is 0.20)}} + \underbrace{x(1.00)}_{\text{The amount of pure alcohol ("pure" is 100\%, or 1.00)}} = \underbrace{(12 + x)(0.40)}_{\text{The amount of alcohol in the mixture}}$$

Step 4 Most students prefer to clear the decimals at this stage. It's easy here—multiplying by 100 will move the decimal point *two places to the right*. We then have

$$\begin{aligned} 12(20) + x(100) &= (12 + x)(40) \\ 240 + 100x &= 480 + 40x \\ 60x &= 240 \\ x &= 4 \text{ qt} \end{aligned}$$

CHECK YOURSELF 5

How much pure alcohol must be added to 500 cubic centimeters (cm^3) of a 40% alcohol mixture to make a solution that is 80% alcohol?

CHECK YOURSELF ANSWERS

1. The number is 20.
2. The numbers are 6 and 18.
3. 60-mi trip: 4 h; 45-mi trip: 3 h.
4. Car: 40 mi/h; truck: 35 mi/h.
5. 1000 cm^3.

6.8 Exercises

Solve the following word problems. Be sure to show the equation used for the solution.

1. If two-thirds of a number is added to one-half of that number, the sum is 35. Find the number. 30

2. If one-third of a number is subtracted from three-fourths of that number, the difference is 15. What is the number? 36

3. If one-fourth of a number is subtracted from two-fifths of a number, the difference is 3. Find that number. 20

4. If five-sixths of a number is added to one-fifth of the number, the sum is 31. What is the number? 30

5. If one-third of an integer is added to one-half of the next consecutive integer, the sum is 13. What are the two integers? 15, 16

6. If one-half of one integer is subtracted from three-fifths of the next consecutive integer, the difference is 3. What are the two integers? 24, 25

7. One number is twice another number. If the sum of their reciprocals is $\frac{1}{4}$, find the two numbers. 6, 12

8. One number is 3 times another. If the sum of their reciprocals is $\frac{1}{6}$, find the two numbers. 8, 24

9. One number is 4 times another. If the sum of their reciprocals is $\frac{5}{12}$, find the two numbers. 3, 12

10. One number is 3 times another. If the sum of their reciprocals is $\frac{4}{15}$, what are the two numbers? 5, 15

11. One number is 5 times another number. If the sum of their reciprocals is $\frac{6}{35}$, what are the two numbers? 7, 35

12. One number is 4 times another. The sum of their reciprocals is $\frac{5}{24}$. What are the two numbers? 6, 24

13. If the reciprocal of 5 times a number is subtracted from the reciprocal of that number, the result is $\frac{4}{25}$. What is the number? 5

14. If the reciprocal of a number is added to 4 times the reciprocal of that number, the result is $\frac{5}{9}$. Find that number. 9

15. Lee can ride his bicycle 50 mi in the same time it takes him to drive 125 mi. If his driving rate is 30 mi/h faster than his rate bicycling, find each rate. 20 mi/h, 50 mi/h

16. Tina can run 12 mi in the same time that it takes her to bicycle 72 mi. If her bicycling rate is 20 mi/h faster than her running rate, find each rate. 4 mi/h, 24 mi/h

17. An express bus can travel 275 mi in the same time that it takes a local bus to travel 225 mi. If the rate of the express bus is 10 mi/h faster than that of the local bus, find the rate for each bus. Express 55 mi/h, local 45 mi/h

18. A light plane took 1 h longer to travel 450 mi on the first portion of a trip than it took to fly 300 mi on the second. If the speed was the same for each portion, what was the flying time for each part of the trip? 3 h, 2 h

19. A passenger train can travel 325 mi in the same time a freight train takes to travel 200 mi. If the speed of the passenger train is 25 mi/h faster than the speed of the freight, find the speed of each. Freight 40 mi/h, passenger 65 mi/h

20. A small business jet took 1 h longer to fly 810 mi on the first part of a flight than to fly 540 mi on the second portion. If the jet's rate was the same for each leg of the flight, what was the flying time for each leg? 3 h, 2 h

21. Charles took 2 h longer to drive 240 mi on the first day of a vacation trip than to drive 144 mi on the second day. If his average driving rate was the same on both days, what was his driving time for each of the days? 5 h, 3 h

22. Julie took 2 h longer to drive 360 mi on the first day of a trip than she took to drive 270 mi on the second day. If her speed was the same both days, what was the driving time each day? 8 h, 6 h

23. An airplane took 3 h longer to fly 1200 mi than it took for a flight of 480 mi. If the plane's rate was the same on each trip, what was the time of each flight? 5 h, 2 h

24. A train travels 80 mi in the same time that a light plane can fly 280 mi. If the speed of the plane is 100 mi/h faster than that of the train, find each of the rates. 40 mi/h, 140 mi/h

25. Jan and Ed took a canoeing trip, traveling 6 mi upstream against a 2-mi/h current. They then returned to the same point downstream. If their entire trip took 4 h, how fast can they paddle in still water? *Hint:* If r is their rate (in miles per hour) in still water, their rate upstream is $r - 2$. 4 mi/h

26. A plane flies 720 mi against a steady 30-mi/h headwind and then returns to the same point with the wind. If the entire trip takes 10 h, what is the plane's speed in still air? 150 mi/h

27. How much pure alcohol must be added to 40 ounces (oz) of a 25% solution to produce a mixture that is 40% alcohol? 10 oz

28. How many centiliters (cL) of pure acid must be added to 200 cL of a 40% acid solution to produce a 50% solution? 40 cL

29. How many centiliters of a 25% acid solution must be added to 100 cL of a 4% acid solution to make a 10% acid solution? 40 cL

30. A storeroom has an 8% acid solution and a 30% solution. How many milliliters (mL) of the 30% solution should be added to 500 mL of the 8% solution to produce a solution that is 20% acid? 600 mL

Skillscan (Section 3.3)
Solve the following equations.

a. $\frac{x}{4} = 7$ 28 **b.** $\frac{x}{8} = 9$ 72 **c.** $\frac{x}{5} = \frac{8}{10}$ 4 **d.** $\frac{x}{2} = \frac{3}{4}$ $\frac{3}{2}$

e. $\frac{x}{9} = \frac{7}{3}$ 21 **f.** $\frac{x}{6} = \frac{28}{8}$ 21 **g.** $\frac{2x}{6} = \frac{4}{3}$ 4 **h.** $\frac{3x}{5} = \frac{9}{10}$ $\frac{3}{2}$

Answers

1. 30 **3.** 20 **5.** 15, 16 **7.** 6, 12 **9.** 3, 12 **11.** 7, 35 **13.** 5 **15.** 20 mi/h, 50 mi/h **17.** 45 mi/h, 55 mi/h **19.** 65 mi/h, 40 mi/h **21.** 5 h, 3 h **23.** 5 h, 2 h **25.** 4 mi/h **27.** 10 oz **29.** 40 cL **a.** 28 **b.** 72 **c.** 4 **d.** $\frac{3}{2}$ **e.** 21 **f.** 21 **g.** 4 **h.** $\frac{3}{2}$

6.9 Ratio and Proportion

OBJECTIVES

1. To solve a proportion for an unknown
2. To apply proportions to the solution of word problems

To begin this section, let's return to an equation that was developed in the previous section. In Example 3, we had to solve the equation

$$\frac{135}{t} = \frac{180}{t + 1}$$

Such an equation is said to be in *proportion form,* or more simply it is called a *proportion*. This type of equation occurs often enough in algebra that it is worth developing some special methods for its solution. First, we will need some definitions.

A *ratio* is a means of comparing two quantities. A ratio can be written as a fraction. For instance, the ratio of 2 to 3 can be written as $\frac{2}{3}$.

A statement that two ratios are equal is called a *proportion*. A proportion has the form

$$\frac{a}{b} = \frac{c}{d}$$

In the proportion above, a and d are called the *extremes* of the proportion, and b and c are called the *means*.

A useful property of proportions is easily developed. If

$$\frac{a}{b} = \frac{c}{d}$$

and we multiply both sides by $b \cdot d$, then

$$\left(\frac{a}{b}\right)bd = \left(\frac{c}{d}\right)bd \qquad \text{or} \qquad ad = bc$$

In words,

> In any proportion, the product of the extremes (ad) is equal to the product of the means (bc).

Since a proportion is a special kind of fractional equation, the above rule gives us an alternative approach to solving equations that are in the proportion form.

Example 1

Solve the equations for x.

The extremes are x and 15. The means are 5 and 12.

(a) $\dfrac{x}{5} = \dfrac{12}{15}$

Set the product of the extremes equal to the product of the means.

$$15x = 5 \cdot 12$$
$$15x = 60$$
$$x = 4$$

Our solution is 4. You can check as before, by substituting in the original proportion.

(b) $\dfrac{x + 3}{10} = \dfrac{x}{7}$

Set the product of the extremes equal to the product of the means.

$$7(x + 3) = 10x$$
$$7x + 21 = 10x$$
$$21 = 3x$$
$$7 = x$$

We will leave the checking of this result to the reader.

CHECK YOURSELF 1

Solve for x.

1. $\dfrac{x}{8} = \dfrac{3}{4}$ **2.** $\dfrac{x - 1}{9} = \dfrac{x + 2}{12}$

There are many types of applications that lead to proportions in their solution. Typically these applications will involve a common ratio, such as miles to gallons or miles to hours, and they can be solved with three basic steps.

TO SOLVE AN APPLICATION USING PROPORTIONS

STEP 1 Assign a variable to represent the unknown quantity.

STEP 2 Write a proportion using the known and unknown quantities. Be sure each ratio involves the same units.

STEP 3 Solve the proportion written in step 2 for the unknown quantity.

The following examples illustrate.

Example 2

A car uses 3 gallons (gal) of gas to travel 105 mi. At that mileage rate, how many gallons will be used on a trip of 385 mi?

Step 1 Assign a variable to represent the unknown quantity. Let x be the number of gallons of gas that will be used on the 385-mi trip.

Step 2 Write a proportion. Note that the ratio of miles to gallons must stay the same.

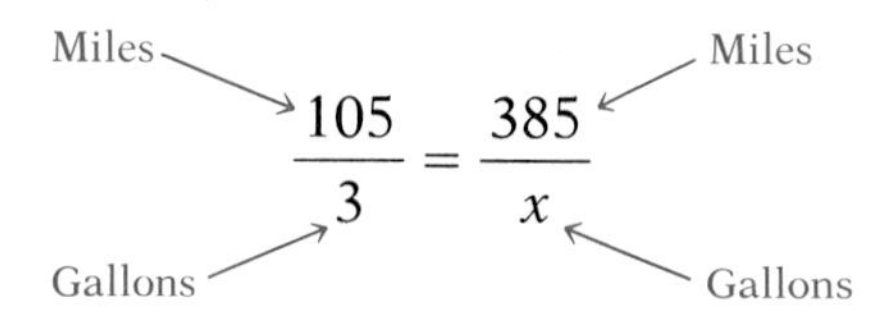

$$\frac{105}{3} = \frac{385}{x}$$

Step 3 Solve the proportion. The product of the extremes is equal to the product of the means.

$$105x = 3 \cdot 385$$
$$105x = 1155$$
$$\frac{105x}{105} = \frac{1155}{105}$$
$$x = 11 \text{ gal}$$

11 gal of gas will be used for the 385-mi trip.

To verify your solution, you can return to the original problem and check that the two ratios involved are the equivalent.

CHECK YOURSELF 2

A car uses 8 liters (L) of gasoline in traveling 100 kilometers (km). At that rate, how many liters of gas will be used on a trip of 250 km?

Example 3

A piece of wire 60 in long is to be cut into two pieces whose lengths have the ratio 5 to 7. Find the length of each piece.

Step 1 Let x represent the length of the shorter piece. Then $60 - x$ is the length of the longer piece.

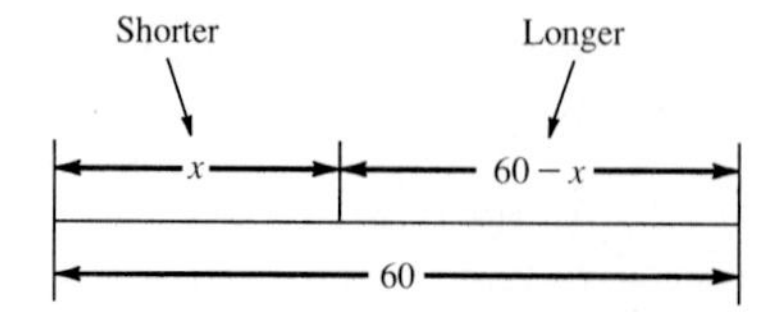

Step 2 The two pieces have the ratio $\frac{5}{7}$, so

On the left and right, we have the ratio of the length of the shorter piece to that of the longer piece.

$$\frac{x}{60 - x} = \frac{5}{7}$$

Step 3 Solving as before, we get

$$\begin{aligned} 7x &= (60 - x)5 \\ 7x &= 300 - 5x \\ 12x &= 300 \\ x &= 25 \qquad \text{(shorter piece)} \\ 60 - x &= 35 \qquad \text{(longer piece)} \end{aligned}$$

CHECK YOURSELF 3

A board 21 ft long is to be cut into two pieces so that the ratio of their lengths is 3 to 4. Find the lengths of the two pieces.

Example 4

Sam and Jill agree to divide a commission of \$1650 in the ratio of 5 to 6. What amount of money should each receive?

Step 1 Let x represent Sam's amount. Then $1650 - x$ will represent Jill's amount.

Step 2 The amounts have the ratio $\frac{5}{6}$, so

$$\frac{x}{1650 - x} = \frac{5}{6}$$

Step 3

$$\begin{aligned} 6x &= (1650 - x)5 \\ 6x &= 8250 - 5x \\ 11x &= 8250 \\ x &= 750 \qquad \text{(Sam's amount)} \\ 1650 - x &= 900 \qquad \text{(Jill's amount)} \end{aligned}$$

CHECK YOURSELF 4

Joe and Jason agree to divide an inheritance of \$8250 in the ratio of 4 to 7. What amount of money will each receive?

CHECK YOURSELF ANSWERS

1. (1) 6; (2) 10. **2.** 20 L. **3.** 9 ft, 12 ft. **4.** Joe $3000; Jason $5250.

6.9 Exercises

Solve each of the following equations for x.

1. $\frac{x}{7} = \frac{8}{14}$ 4

2. $\frac{3}{x} = \frac{9}{15}$ 5

3. $\frac{5}{8} = \frac{20}{x}$ 32

4. $\frac{x}{10} = \frac{9}{30}$ 3

5. $\frac{x+1}{5} = \frac{20}{25}$ 3

6. $\frac{2}{5} = \frac{x-2}{20}$ 10

7. $\frac{3}{5} = \frac{x-1}{20}$ 13

8. $\frac{5}{x-3} = \frac{15}{21}$ 10

9. $\frac{x}{6} = \frac{x+5}{16}$ 3

10. $\frac{x-2}{x+2} = \frac{12}{20}$ 8

11. $\frac{x}{x-3} = \frac{10}{7}$ 10

12. $\frac{x}{8} = \frac{x+3}{16}$ 3

13. $\frac{2}{x-1} = \frac{6}{x+9}$ 6

14. $\frac{3}{x-3} = \frac{4}{x-5}$ −3

15. $\frac{1}{x-2} = \frac{7}{x^2-4}$ 5

16. $\frac{1}{x-2} = \frac{3}{x^2-4x+4}$ 5

17. A speed of 60 mi/h corresponds to 88 feet per second (ft/s). If a light plane's speed is 150 mi/h, what is its speed in feet per second? 220 ft/s

18. John completed 24 pages of his reading assignment in 60 minutes (min). If the assignment is 100 pages long, how long will the entire assignment take if John reads at the same rate? 250 min

19. A car uses 5 gal of gasoline on a trip of 160 mi. At the same mileage rate, how much gasoline will a 384-mi trip require? 12 gal

20. A car uses 12 L of gasoline in traveling 150 km. At that rate, how many liters of gasoline will be used on a trip of 400 km? 32 L

21. Susan earns $6500 commission in 20 weeks in her new sales position. At that rate, how much will she earn in 1 year (52 weeks)? $16,900

22. Kevin earned $165 interest for 1 year on an investment of $1500. At that same rate, what amount of interest would be earned by an investment of $2500? $275

23. A board 24 ft long is to be divided into two pieces whose lengths have the ratio 5 to 7. Find the length of each piece. 10 ft, 14 ft

24. A 90-ft piece of cable is to be cut into two pieces whose lengths have the ratio 4 to 5. Find the length of each piece. 40 ft, 50 ft

25. A brother and sister are to divide an inheritance of $12,000 in the ratio of 2 to 3. What amount will each receive? $4800, $7200

26. Carlos wants to invest a total of $5600 in two savings accounts in the ratio of 3 to 4. What amount should he invest in each account? $2400, $3200

Answers

1. 4 **3.** 32 **5.** 3 **7.** 13 **9.** 3 **11.** 10 **13.** 6 **15.** 5 **17.** 220 ft/s **19.** 12 gal
21. $16,900 **23.** 10 ft, 14 ft **25.** $4800, $7200

Summary

Algebraic Fractions [6.1]

Algebraic Fractions These have the form

$$\frac{P}{Q}$$

(Numerator: P; Fraction bar; Denominator: Q)

where P and Q are polynomials and Q cannot have the value 0.

$\frac{x^2 - 3x}{x - 2}$ is an algebraic fraction. The variable x cannot have the value 2.

Simplifying Algebraic Fractions [6.2]

Simplest Form A fraction is in simplest form if its numerator and denominator have no common factors other than 1.

$\frac{x+2}{x-1}$ is in simplest form.

To Write in Simplest Form

1. Factor the numerator and denominator.
2. Divide the numerator and denominator by all common factors.
3. The resulting fraction will be in simplest form.

$$\frac{x^2-4}{x^2-2x-8} = \frac{(x-2)(x+2)}{(x-4)(x+2)} = \frac{(x-2)\cancel{(x+2)}}{(x-4)\cancel{(x+2)}} = \frac{x-2}{x-4}$$

Multiplying and Dividing [6.3]

Multiplying Fractions

$$\frac{P}{Q} \cdot \frac{R}{S} = \frac{PR}{QS}$$

where $Q \neq 0$ and $S \neq 0$.

$$\frac{2}{3} \cdot \frac{4}{5} = \frac{2 \cdot 4}{3 \cdot 5} = \frac{8}{15}$$

Multiplying Algebraic Fractions

1. Factor the numerators and denominators.
2. Divide the numerator and denominator by any common factors.
3. Write the product of the remaining factors in the numerator over the product of the remaining factors in the denominator.

$$\frac{2x-4}{x^2-4} \cdot \frac{x^2+2x}{6x+18} = \frac{2(x-2)}{(x-2)(x+2)} \cdot \frac{x(x+2)}{6(x+3)} = \frac{2\cancel{(x-2)}}{\cancel{(x-2)}\cancel{(x+2)}} \cdot \frac{x\cancel{(x+2)}}{6(x+3)} = \frac{x}{3(x+3)}$$

Dividing Fractions

$$\frac{P}{Q} \div \frac{R}{S} = \frac{P}{Q} \cdot \frac{S}{R}$$

where $Q \neq 0$, $R \neq 0$, and $S \neq 0$. In words, invert the divisor (the second fraction) and multiply.

$$\frac{3x}{2x-6} \div \frac{9x^2}{x^2-9} = \frac{3x}{2x-6} \cdot \frac{x^2-9}{9x^2} = \frac{3x}{2\cancel{(x-3)}} \cdot \frac{(x+3)\cancel{(x-3)}}{9x^2} = \frac{x+3}{6x}$$

Adding and Subtracting [6.4] to [6.5]

Like Fractions

1. Add or subtract the numerators.
2. Write the sum or difference over the common denominator.
3. Write the resulting fraction in simplest form.

$$\frac{2x}{x^2+3x} + \frac{6}{x^2+3x} = \frac{2x+6}{x^2+3x} = \frac{2\cancel{(x+3)}}{x\cancel{(x+3)}} = \frac{2}{x}$$

The Lowest Common Denominator Finding the LCD:

1. Factor each denominator.
2. Write each factor the greatest number of times it appears in any single denominator.
3. The LCD is the product of the factors found in step 2.

For $\frac{2}{x^2 + 2x + 1}$ and $\frac{3}{x^2 + x}$

Factor:

$x^2 + 2x + 1 = (x + 1)(x + 1)$

$x^2 + x = x(x + 1)$

The LCD is $x(x + 1)(x + 1)$

Unlike Fractions To add or subtract unlike fractions:

1. Find the LCD.
2. Convert each fraction to an equivalent fraction with the LCD as a common denominator.
3. Add or subtract the like fractions formed.
4. Write the sum or difference in simplest form.

$$\frac{2}{x^2 + 2x + 1} - \frac{3}{x^2 + x}$$
$$= \frac{2x}{x(x + 1)(x + 1)} - \frac{3(x + 1)}{x(x + 1)(x + 1)}$$
$$= \frac{2x - 3x - 3}{x(x + 1)(x + 1)}$$
$$= \frac{-x - 3}{x(x + 1)(x + 1)}$$

Complex Fractions [6.6]

Complex Fractions Fractions that have a fraction in their numerator, their denominator, or both.

$\dfrac{\frac{x - 2}{x}}{\frac{x^2 - 4}{x^2}}$ is a complex fraction.

Simplifying To simplify a complex fraction: you can apply either of the following methods.

METHOD 1

1. Multiply the numerator and denominator by the LCD of all functions that appear within the complex fraction.
2. Simplify the resulting fraction, writing the result in simplest form, if necessary.

$$\frac{\left(\frac{x - 2}{x}\right)x^2}{\left(\frac{x^2 - 4}{x^2}\right)x^2}$$
$$= \frac{(x - 2)x}{x^2 - 4}$$
$$= \frac{x\cancel{(x - 2)}}{(x + 2)\cancel{(x - 2)}}$$
$$= \frac{x}{(x + 2)}$$

METHOD 2

1. Write the numerator and denominator as single fractions if necessary.
2. Invert the denominator and multiply as before, writing the result in simplest form.

$$\frac{1 - \frac{x}{y}}{1 - \frac{x^2}{y^2}}$$
$$= \frac{\frac{y - x}{y}}{\frac{y^2 - x^2}{y^2}}$$
$$= \frac{y - x}{y} \cdot \frac{y^2}{y^2 - x^2}$$
$$= \frac{\cancel{y - x}}{y} \cdot \frac{y^2}{\cancel{(y - x)}(y + x)}$$
$$= \frac{y}{y + x}$$

Fractional Equations [6.7]

To Solve

1. Remove the fractions in the equation by multiplying *each term* of the equation by the LCD of all the fractions that appear.
2. Solve the resulting equation as before.
3. Check your solution in the *original equation*.

$$\frac{2}{x-2} - \frac{3}{x+2} = \frac{2}{x^2-4}$$

Multiply by $(x - 2)(x + 2)$. We then have

$2(x + 2) - 3(x - 2) = 2$

Solving, we get

$x = 8$

Ratio and Proportion [6.9]

Ratio A means of comparing two quantities. A ratio can be written as a fraction.

$\frac{a}{b}$ is the ratio of a to b,

$\frac{2}{3}$ is the ratio of 2 to 3.

Proportion A statement that two ratios are equal. The form is

$$\frac{a}{b} = \frac{c}{d}$$

where a and d are the extremes and b and c are the means.

$$\frac{2}{3} = \frac{8}{12}$$

is a proportion; 2 and 12 are the extremes, 3 and 8 are the means.

The Proportion Rule If

$$\frac{a}{b} = \frac{c}{d} \quad \text{then } ad = bc$$

In words, the product of the extremes is equal to the product of the means. This rule can be applied in solving fractional equations that are in the proportion form.

Solve:

$$\frac{x-2}{3} = \frac{x+2}{7}$$

Set the product of the extremes equal to the product of the means.

$$7(x - 2) = 3(x + 2)$$
$$7x - 14 = 3x + 6$$
$$4x = 20$$
$$x = 5$$

Summary Exercises Chapter 6

This summary exercise set is provided to give you practice with each of the objectives of the chapter. Each exercise is keyed to the appropriate chapter section. The answers are provided in the instructor's manual. Your instructor will give you guidelines on how to best use these exercises in your instructional setting.

[6.1] What values for x, if any, must be excluded in the following algebraic fractions?

1. $\frac{x}{5}$ None

2. $\frac{3}{x - 4}$ 4

3. $\frac{2}{(x + 1)(x - 2)}$ −1, 2

4. $\frac{7}{x^2 - 16}$ −4, 4

5. $\frac{x - 1}{x^2 + 3x + 2}$ −1, −2

6. $\frac{2x + 3}{3x^2 + x - 2}$ $-1, \frac{2}{3}$

[6.2] Write each fraction in simplest form.

7. $\frac{6a^2}{9a^3}$ $\frac{2}{3a}$

8. $\frac{-12x^4y^3}{18x^2y^2}$ $\frac{-2x^2y}{3}$

9. $\frac{w^2 - 25}{2w - 8}$ $\frac{w^2 - 25}{2w - 8}$

10. $\frac{3x^2 + 11x - 4}{2x^2 + 11x + 12}$ $\frac{3x - 1}{2x + 3}$

11. $\frac{m^2 - 2m - 3}{9 - m^2}$ $\frac{-m - 1}{m + 3}$

12. $\frac{3c^2 - 2cd - d^2}{6c^2 + 2cd}$ $\frac{c - d}{2c}$

[6.3] Multiply or divide as indicated.

13. $\frac{6x}{5} \cdot \frac{10}{18x^2}$ $\frac{2}{3x}$

14. $\frac{-2a^2}{ab^3} \cdot \frac{3ab^2}{-4ab}$ $\frac{3a}{2b^2}$

15. $\frac{2x + 6}{x^2 - 9} \cdot \frac{x^2 - 3x}{4}$ $\frac{x}{2}$

16. $\frac{a^2 + 5a + 4}{2a^2 + 2a} \cdot \frac{a^2 - a - 12}{a^2 - 16}$ $\frac{a + 3}{2a}$

17. $\frac{3p}{5} \div \frac{9p^2}{10}$ $\frac{2}{3p}$

18. $\frac{8m^3}{5mn} \div \frac{12m^2n^2}{15mn^3}$ $2m$

19. $\frac{x^2 + 7x + 10}{x^2 + 5x} \div \frac{x^2 - 4}{2x^2 - 7x + 6}$ $\frac{2x - 3}{x}$

20. $\frac{2w^2 + 11w - 21}{w^2 - 49} \div (4w - 6)$ $\frac{1}{2(w - 7)}$

21. $\frac{a^2b + 2ab^2}{a^2 - 4b^2} \div \frac{4a^2b}{a^2 - ab - 2b^2}$ $\frac{a + b}{4a}$

22. $\frac{2x^2 + 6x}{4x} \cdot \frac{6x + 12}{x^2 + 2x - 3} \div \frac{x^2 - 4}{x^2 - 3x + 2}$ 3

[6.4] Add or subtract as indicated.

23. $\frac{x}{9} + \frac{2x}{9}$ $\quad \frac{x}{3}$

24. $\frac{7a}{15} - \frac{2a}{15}$ $\quad \frac{a}{3}$

25. $\frac{8}{x+2} + \frac{3}{x+2}$ $\quad \frac{11}{x+2}$

26. $\frac{y-2}{5} - \frac{2y+3}{5}$
$\frac{-y-5}{5}$

27. $\frac{7r-3s}{4r} + \frac{r-s}{4r}$ $\quad \frac{2r-s}{r}$

28. $\frac{x^2}{x-4} - \frac{16}{x-4}$
$x + 4$

29. $\frac{5w-6}{w-4} - \frac{3w+2}{w-4}$ $\quad 2$

30. $\frac{x+3}{x^2-2x-8} + \frac{2x+3}{x^2-2x-8}$ $\quad \frac{3}{x-4}$

[6.5] Add or subtract as indicated.

31. $\frac{5x}{6} + \frac{x}{3}$
$\frac{7x}{6}$

32. $\frac{3y}{10} - \frac{2y}{5}$
$\frac{-y}{10}$

33. $\frac{5}{2m} - \frac{3}{m^2}$
$\frac{5m-6}{2m^2}$

34. $\frac{x}{x-3} - \frac{2}{3}$
$\frac{x+6}{3(x-3)}$

35. $\frac{4}{x-3} - \frac{1}{x}$
$\frac{3x+3}{x(x-3)}$

36. $\frac{2}{s+5} + \frac{3}{s+1}$
$\frac{5s+17}{(s+5)(s+1)}$

37. $\frac{5}{w-5} - \frac{2}{w-3}$
$\frac{3w-5}{(w-5)(w-3)}$

38. $\frac{4x}{2x-1} + \frac{2}{1-2x}$
2

39. $\frac{2}{3x-3} - \frac{5}{2x-2}$
$\frac{-11}{6(x-1)}$

40. $\frac{4y}{y^2-8y+15} + \frac{6}{y-3}$
$\frac{10}{y-5}$

41. $\frac{3a}{a^2+5a+4} + \frac{2a}{a^2-1}$
$\frac{5a}{(a+4)(a-1)}$

42. $\frac{3x}{x^2+2x-8} - \frac{1}{x-2} + \frac{1}{x+4}$

$\frac{3}{x+4}$

[6.6] Simplify the complex fractions.

43. $\dfrac{\frac{x^2}{12}}{\frac{x^3}{8}}$ $\quad \dfrac{2}{3x}$

44. $\dfrac{3 + \frac{1}{a}}{3 - \frac{1}{a}}$ $\quad \dfrac{3a + 1}{3a - 1}$

45. $\dfrac{1 + \frac{x}{y}}{1 - \frac{x}{y}}$ $\quad \dfrac{y + x}{y - x}$

46. $\dfrac{1 + \frac{1}{p}}{p^2 - 1}$ $\quad \dfrac{1}{p(p - 1)}$

47. $\dfrac{\frac{1}{m} - \frac{1}{n}}{\frac{1}{m} + \frac{1}{n}}$ $\quad \dfrac{n - m}{n + m}$

48. $\dfrac{2 - \frac{x}{y}}{4 - \frac{x^2}{y^2}}$ $\quad \dfrac{y}{2y + x}$

49. $\dfrac{\frac{2}{a + 1} + 1}{1 - \frac{4}{a + 1}}$ $\quad \dfrac{a + 3}{a - 3}$

50. $\dfrac{\frac{a}{b} - 1 - \frac{2b}{a}}{\frac{1}{b^2} - \frac{1}{a^2}}$ $\quad \dfrac{ab(a - 2b)}{a - b}$

[6.7] Solve the following equations for x.

51. $\dfrac{x}{4} - \dfrac{x}{5} = 2$ $\quad 40$

52. $\dfrac{13}{4x} + \dfrac{3}{x^2} = \dfrac{5}{2x}$ $\quad -4$

53. $\dfrac{x}{x - 2} + 1 = \dfrac{x + 4}{x - 2}$ $\quad 6$

54. $\dfrac{x}{x - 4} - 3 = \dfrac{4}{x - 4}$ $\quad$ No solution

55. $\dfrac{x}{2x - 6} - \dfrac{x - 4}{x - 3} = \dfrac{1}{8}$ $\quad 7$

56. $\dfrac{7}{x} - \dfrac{1}{x - 3} = \dfrac{9}{x^2 - 3x}$ $\quad 5$

57. $\dfrac{x}{x - 5} = \dfrac{3x}{x^2 - 7x + 10} + \dfrac{8}{x - 2}$ $\quad 8$

58. $\dfrac{6}{x + 5} + 1 = \dfrac{3}{x - 5}$ $\quad -10, 7$

59. $\dfrac{24}{x + 2} - 2 = \dfrac{2}{x - 3}$ $\quad 4, 8$

[6.8] Solve the following applications.

60. If two-fifths of a number is added to one-half of that number, the sum is 27. Find the number. 30

61. One number is 3 times another. If the sum of their reciprocals is $\frac{1}{3}$, what are the two numbers? 4, 12

62. If the reciprocal of 4 times a number is subtracted from the reciprocal of that number, the result is $\frac{1}{8}$. What is the number? 6

63. Robert made a trip of 240 mi. Returning by a different route, he found that the distance was only 200 mi, but traffic slowed his speed by 8 mi/h. If the trip took the same time in both directions, what was Robert's rate each way? 48 mi/h, 40 mi/h

64. On the first day of a vacation trip, Jean drove 225 mi. On the second day it took her 1 h longer to drive 270 mi. If her average speed was the same both days, how long did she drive each day? 5 h, 6 h

65. A light plane flies 700 mi against a steady 20 mi/h headwind and then returns to the same point with the wind. If the entire trip took 12 h, what was the speed of the plane in still air? 120 mi/h

66. How much pure alcohol should be added to 300 milliliters (mL) of a 30% solution to obtain a 40% solution? 50 mL

67. A chemist has a 10% acid solution and a 40% solution. How much of the 40% solution should be added to 300 mL of the 10% solution to produce a mixture with a concentration of 20%? 150 mL

[6.9] Solve the following proportion problems.

68. $\frac{x-3}{8} = \frac{x-2}{10}$ 7

69. $\frac{1}{x-3} = \frac{7}{x^2-x-6}$ 5

70. Melina wants to invest a total of $10,800 in two types of savings accounts. If she wants the ratio of the amounts deposited in the two accounts to be 4 to 5, what amount should she invest in each account? $4800, $6000

Self-Test for Chapter Six

The purpose of this self-test is to help you check your progress and to review for a chapter test in class. Allow yourself about an hour to take the test. When you are done, check your answers in the back of the book. If you missed any problems, be sure to go back and review the appropriate sections in the chapter and the exercises that are provided.

What values for x, if any, must be excluded in the following algebraic fractions?

1. $\frac{8}{x - 4}$ 4

2. $\frac{3}{x^2 - 9}$ −3, 3

Write each fraction in simplest form.

3. $\frac{-21x^5y^3}{28xy^5}$ $\frac{-3x^4}{4y^2}$

4. $\frac{4a - 24}{a^2 - 6a}$ $\frac{4}{a}$

5. $\frac{3x^2 + x - 2}{3x^2 - 8x + 4}$ $\frac{x + 1}{x - 2}$

Multiply or divide as indicated.

6. $\frac{3pq^2}{5pq^3} \cdot \frac{20p^2q}{21q}$ $\frac{4p^2}{7q}$

7. $\frac{x^2 - 3x}{5x^2} \cdot \frac{10x}{x^2 - 4x + 3}$ $\frac{2}{x - 1}$

8. $\frac{2x^2}{3xy} \div \frac{8x^2y}{9xy}$ $\frac{3}{4y}$

9. $\frac{3m - 9}{m^2 - 2m} \div \frac{m^2 - m - 6}{m^2 - 4}$ $\frac{3}{m}$

Add or subtract as indicated.

10. $\frac{3a}{8} + \frac{5a}{8}$ a

11. $\frac{2x}{x + 3} + \frac{6}{x + 3}$ 2

12. $\frac{7x - 3}{x - 2} - \frac{2x + 7}{x - 2}$ 5

13. $\frac{x}{3} + \frac{4x}{5}$ $\frac{17x}{50}$

14. $\frac{3}{s} - \frac{2}{s^2}$ $\frac{3s - 2}{s^2}$

15. $\dfrac{5}{x-2} - \dfrac{1}{x+3}$ $\quad \dfrac{4x+17}{(x-2)(x+3)}$

16. $\dfrac{6}{w-2} + \dfrac{9w}{w^2-7w+10}$ $\quad \dfrac{15}{w-5}$

Simplify the complex fractions.

17. $\dfrac{\dfrac{x^2}{18}}{\dfrac{x^3}{12}}$ $\quad \dfrac{2}{3x}$

18. $\dfrac{2-\dfrac{m}{n}}{4-\dfrac{m^2}{n^2}}$ $\quad \dfrac{n}{2n+m}$

Solve the following equations for x.

19. $\dfrac{x}{3} - \dfrac{x}{4} = 3$ $\quad 36$

20. $\dfrac{x}{x+3} + 1 = \dfrac{3x-6}{x+3}$ $\quad 9$

21. $\dfrac{5}{x} - \dfrac{x-3}{x+2} = \dfrac{22}{x^2+2x}$ $\quad 2, 6$

Solve the following applications.

22. One number is 3 times another. If the sum of their reciprocals is $\dfrac{1}{3}$, find the two numbers.
4, 12

23. Mark drove 250 mi to visit Sandra. Returning by a shorter route, he found that the trip was only 225 mi, but traffic slowed his speed by 5 mi/h. If the two trips took exactly the same time, what was his rate each way? 50 mi/h, 45 mi/h

Solve the following proportion problems.

24. $\dfrac{x-1}{5} = \dfrac{x+2}{8}$ $\quad 6$

25. A 55-ft cable is to be cut into two pieces whose lengths have the ratio 4 to 7. Find the lengths of the two pieces. 20 ft, 35 ft

Cumulative Test for Chapters 4 to 6

This test is provided to help you in the process of review of the previous chapters. Answers are provided in the back of the book. If you missed any problems, be sure to go back and review the appropriate chapter sections.

Perform each of the indicated operations.

1. $(5x^2 - 9x + 3) + (3x^2 + 2x - 7)$
$8x^2 - 7x - 4$

2. $(8a^2 - 3a) - (9a^2 + 7)$
$-a^2 - 3a - 7$

3. Subtract $9w^2 + 5w$ from the sum of $8w^2 - 3w$ and $2w^2 - 4$. $w - 8w - 4$

4. $7xy(4x^2y - 2xy + 3xy^2)$
$28x^3y^2 - 14x^2y^2 + 21x^2y^3$

5. $(3s - 7)(5s + 4)$
$15s^2 - 23s - 28$

6. $(3a - b)(2a^2 + ab - b^2)$
$6a^3 + a^2b - 4ab^2 + b^3$

7. $\dfrac{5x^3y - 10x^2y^2 + 15xy^2}{-5xy}$
$-x^2 + 2xy - 3y$

8. $\dfrac{4x^2 + 6x - 4}{2x - 1}$
$2x + 4$

9. $\dfrac{3x^3 - 6x + 17}{3x + 6}$
$x^2 - 2x + 2 + \dfrac{5}{3x + 6}$

Solve the following equation for x.

10. $5 - 3(2x - 7) = 8 - 4x$ 9

Solve the following applications.

11. A bank teller has 83 \$5 and \$10 bills with a value of \$695. How many bills of each denomination does she have? 27 \$5 bills, 56 \$10 bills

12. A light plane makes a trip between two cities, against a steady headwind, in 7 h. Returning with the wind, the plane can travel 20 mi/h faster and makes the trip in 6 h. What is the plane's speed in each direction? 120 mi/h, 140 mi/h

Factor each of the following polynomials completely.

13. $24a^3 - 16a^2$
$8a^2(3a - 2)$

14. $7m^2n - 21mn - 49mn^2$
$7mn(m - 3 - 7n)$

Factor each of the following binomials completely.

15. $a^2 - 64b^2$ $(a + 8b)(a - 8b)$

16. $5p^3 - 80pq^2$ $5p(p + 4q)(p - 4q)$

Factor each of the following trinomials completely.

17. $a^2 - 14a + 48$
$(a - 6)(a - 8)$

18. $2w^3 - 8w^2 - 42w$
$2w(w - 7)(w + 3)$

19. $3r^2 + 5rs - 28s^2$
$(3r - 7s)(r + 4s)$

Solve each of the following equations.

20. $x^2 - 9x + 20 = 0$
4, 5

21. $2x^2 - 32 = 0$
$-4, 4$

22. $21x^2 - 28x = 28$
$-\frac{2}{3}, 2$

Solve the following applications.

23. Twice the square of a positive integer is 35 more than 9 times that integer. What is the integer? 7

24. The length of a rectangle is 2 in more than 3 times its width. If the area of the rectangle is 85 in^2, find the dimensions of the rectangle. 5 in by 17 in

Solve the following equation for the indicated variable.

25. $S_n = \frac{n}{2}(a_1 + a_n)$ for a_1 $\quad \frac{2S_n - na_n}{n}$

What value(s) of x, if any, must be excluded in the following algebraic fraction?

26. $\frac{3x - 1}{2x^2 - x - 3}$ $\quad -1, \frac{3}{2}$

Write each fraction in simplest form.

27. $\frac{m^2 - 4m}{3m - 12}$ $\quad \frac{m}{3}$

28. $\frac{a^2 - 49}{3a^2 + 22a + 7}$ $\quad \frac{a - 7}{3a + 1}$

Perform the indicated operations.

29. $\frac{3x^2 + 9x}{x^2 - 9} \cdot \frac{2x^2 - 9x + 9}{2x^3 - 3x^2}$ $\quad \frac{3}{x}$

30. $\frac{4w^2 - 25}{2w^2 - 5w} \div (6w + 15)$ $\quad \frac{1}{3w}$

31. $\frac{4}{3r} + \frac{1}{2r^2}$ $\quad \frac{8r + 3}{6r^2}$

32. $\frac{2}{x - 3} - \frac{5}{3x + 9}$

$\frac{x + 33}{3(x - 3)(x + 3)}$

33. $\frac{3y}{y^2 - 7y + 12} + \frac{9}{y - 3}$ $\quad \frac{12}{y - 4}$

Simplify the complex fractions.

34. $\frac{1 - \frac{1}{x}}{2 + \frac{1}{x}}$ $\quad \frac{x - 1}{2x + 1}$

35. $\frac{3 - \frac{m}{n}}{9 - \frac{m^2}{n^2}}$ $\quad \frac{n}{3n + m}$

Solve the following equations for x.

36. $\frac{5}{3x} + \frac{1}{x^2} = \frac{5}{2x}$ $\quad \frac{6}{5}$

37. $\frac{10}{x - 3} - 2 = \frac{5}{x + 3}$ $\quad -\frac{9}{2}, 7$

Solve the following applications.

38. If the reciprocal of 5 times a number is subtracted from the reciprocal of that number, the result is $\frac{2}{5}$. What is the number? 2

39. Jennifer drove 260 mi to attend a business conference. In returning from the conference along a different route, the trip was only 240 mi, but traffic slowed her speed by 4 mi/h. If her driving time was the same both ways, what was her speed each way?
52 mi/h, 48 mi/h

40. A letter-quality printer can print 80 form letters in 50 min. At that rate, how long will it take the printer to complete a job requiring 200 letters? 125 min

Chapter Seven

Graphing Linear Equations and Inequalities

7.1 Solutions of Equations in Two Variables

OBJECTIVES

1. To find solutions for an equation in two variables.
2. To use the ordered-pair notation to write solutions.

We discussed finding solutions for equations in Chapter 3. Recall that a solution is a value for the variable that "satisfies" the equation, or makes the equation a true statement. For instance, if

$$2x + 5 = 13$$

we can see that 4 is a solution for the equation because, replacing x with 4, we have

$$2 \cdot 4 + 5 = 13$$
$$8 + 5 = 13$$
$$13 = 13 \quad \text{(A true statement)}$$

We now want to consider *equations in two variables*. An example is

$$x + y = 5$$

What will the solution look like? It is not going to be a single number, because of the two variables. Here the solution will be a pair of numbers—one value for each of the letters, x and y. Suppose that x has the value 3. In the equation $x + y = 5$, you can substitute 3 for x.

$$3 + y = 5$$

Now solving for y gives

$$y = 2$$

An equation in two variables "pairs" two numbers, one for x and one for y.

So the pair of values $x = 3$ and $y = 2$ satisfies the equation because $3 + 2 = 5$. That pair of numbers is then a *solution* for the equation in two variables.

How many such pairs are there? Choose any value for x (or for y). You can always find the other *paired,* or *corresponding,* value in an equation of this form. We say that there are an *infinite* number of pairs that will satisfy the equation (or are solutions). To find some others for $x + y = 5$:

If $x = 5$,

$5 + y = 5$ or $y = 0$

If $y = 4$,

$x + 4 = 5$ or $x = 1$

CHECK YOURSELF 1

For the equation $x + y = 5$,

1. If $x = 4$, $y = ?$
2. If $y = 0$, $x = ?$

To simplify writing the pairs that satisfy an equation, we use the *ordered-pair* notation. The numbers are written in parentheses and are separated by a comma. For example, we know that the values $x = 3$ and $y = 2$ satisfy the equation $x + y = 5$. So we write the pair as

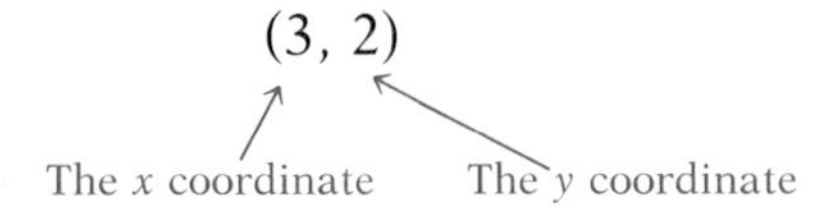

The first number of the pair is *always* the value for x and is called the *x coordinate*. The second number of the pair is *always* the value for y and is the *y coordinate*.

Be Careful!

(3, 2) means $x = 3$ and $y = 2$.
(2, 3) means $x = 2$ and $y = 3$.
(3, 2) and (2, 3) are entirely different. That's why we call them *ordered pairs*.

Using this ordered pair notation, we can say that (3, 2), (5, 0), and

(1, 4) are all *solutions* for the equation $x + y = 5$. Each pair gives values for x and y that will satisfy the equation.

Example 1

Which of the ordered pairs (2, 5), (5, −1), and (3, 4) are solutions for the following equation?

$$2x + y = 9$$

(*a*) To check whether (2, 5) is a solution, let $x = 2$ and $y = 5$ and see if the equation is satisfied.

$$2x + y = 9 \qquad \text{The original equation}$$

$$2 \cdot 2 + 5 = 9 \qquad \text{Substitute 2 for } x \text{ and 5 for } y.$$

$$4 + 5 = 9$$

$$9 = 9 \qquad \text{(True)}$$

(2, 5) is a solution because a *true statement* results.

(2, 5) is a solution for the equation.

(*b*) For (5, −1), let $x = 5$ and $y = -1$.

$$2 \cdot 5 - 1 = 9$$

$$10 - 1 = 9$$

$$9 = 9 \qquad \text{(True)}$$

So (5, −1) is a solution.

(*c*) For (3, 4), let $x = 3$ and $y = 4$. Then

$$2 \cdot 3 + 4 = 9$$

$$6 + 4 = 9$$

$$10 = 9 \qquad \text{(\textit{Not} a true statement)}$$

So (3, 4) is *not* a solution for the equation.

CHECK YOURSELF 2

Which of the ordered pairs (3, 4), (4, 3), (1, −2), and (0, −5) are solutions for the following equation?

$$3x - y = 5$$

Example 2

Complete the ordered pairs given below so that each is a solution for the equation

$x - 3y = 6$

(9,), (, −1), (0,), and (, 0)

(*a*) The first number, 9, appearing in (9,) represents the x value, so to complete the pair (9,), substitute 9 for x and then solve for y.

$$\begin{aligned} 9 - 3y &= 6 \\ -3y &= -3 \\ y &= 1 \end{aligned}$$

(9, 1) is a solution.

(*b*) To complete the pair (, −1), let y be −1 and solve for x.

$$\begin{aligned} x - 3(-1) &= 6 \\ x + 3 &= 6 \\ x &= 3 \end{aligned}$$

(3, −1) is a solution.

(*c*) To complete the pair (0,), let x be 0.

$$\begin{aligned} 0 - 3y &= 6 \\ -3y &= 6 \\ y &= -2 \end{aligned}$$

(0, −2) is a solution.

(*d*) To complete the pair (, 0), let y be 0.

$$\begin{aligned} x - 3 \cdot 0 &= 6 \\ x - 0 &= 6 \\ x &= 6 \end{aligned}$$

(6, 0) is a solution.

CHECK YOURSELF 3

Complete the ordered pairs below so that each is a solution for the equation $2x + 5y = 10$.

(10,), (, 4), (0,), and (, 0)

Example 3

Find four solutions for the equation

$2x + y = 8$

Generally you'll want to pick values for x (or for y) so that the resulting equation in one variable seems easy to solve.

In this case the values used to form the solutions are *up to you*. You can assign any value for x (or for y). We'll demonstrate with some possible choices.

Let $x = 2$:

$$\begin{aligned} 2 \cdot 2 + y &= 8 \\ 4 + y &= 8 \\ y &= 4 \end{aligned}$$

(2, 4) is a solution.

Let $y = 6$:

$$\begin{aligned} 2x + 6 &= 8 \\ 2x &= 2 \\ x &= 1 \end{aligned}$$

(1, 6) is a solution.

Let $x = 0$:

$$\begin{aligned} 2 \cdot 0 + y &= 8 \\ y &= 8 \end{aligned}$$

The solutions (0, 8) and (4, 0) will have special significance later in graphing. They are also easy to find!

(0, 8) is a solution.

Let $y = 0$:

$$\begin{aligned} 2x + 0 &= 8 \\ 2x &= 8 \\ x &= 4 \end{aligned}$$

(4, 0) is a solution.

CHECK YOURSELF 4

Find four solutions for $x - 3y = 12$.

CHECK YOURSELF ANSWERS

1. (1) $y = 1$; (2) $x = 5$.
2. (3, 4), (1, −2), and (0, −5) are solutions.
3. (10, −2), (−5, 4), (0, 2), and (5, 0)
4. (6, −2), (3, −3), (0, −4), and (12, 0) are four possibilities.

7.1 Exercises

Determine which of the ordered pairs are solutions for the given equation.

1. $x + y = 8$ $(5, 3), (-2, 6), (0, 8), (-3, 11)$ $(5, 3), (0, 8), (-3, 11)$

2. $x - y = 10$ $(9, 1), (9, -1), (10, 0), (5, 5)$ $(9, -1), (10, 0)$

3. $2x - y = 6$ $(4, 2), (3, 0), (0, 6), (5, 4)$ $(4, 2), (3, 0), (5, 4)$

4. $x + 5y = 15$ $(5, -2), (5, 2), (15, 0), (20, -1)$ $(5, 2), (15, 0), (20, -1)$

5. $3x + y = 9$ $(3, 0), (2, 2), (0, 3), (1, 6)$ $(3, 0), (1, 6)$

6. $x - 4y = 8$ $(8, 0), (0, 2), (4, -1), (12, -1)$ $(8, 0), (4, -1)$

7. $2x - 3y = 12$ $(0, 4), (6, 0), (9, 2), (3, -2)$ $(6, 0), (9, 2), (3, -2)$

8. $8x + 4y = 24$ $(3, 0), (6, -8), (0, 6), (5, -5)$ $(3, 0), (0, 6)$

9. $3x - 2y = 6$ $(2, 0), \left(\frac{1}{3}, -\frac{5}{2}\right), (0, 3), \left(1, \frac{3}{2}\right)$ $(2, 0), \left(\frac{1}{3}, -\frac{5}{2}\right)$

10. $4x + 3y = 12$ $(-3, 0), \left(\frac{5}{2}, \frac{2}{3}\right), (0, 4), \left(\frac{3}{2}, -2\right)$ $\left(\frac{5}{2}, \frac{2}{3}\right), (0, 4)$

11. $y = 3x$ $(0, 0), (1, 4), (2, 6), (3, 6)$ $(0, 0), (2, 6)$

12. $y = -2x + 1$ $(0, -2), (1, -1), \left(\frac{1}{2}, 0\right), (3, 5)$ $(1, -1), \left(\frac{1}{2}, 0\right)$

13. $x = 5$ $(5, 3), (0, 5), (5, 0), (5, 7)$ $(5, 3), (5, 0), (5, 7)$

14. $y = 7$ $(0, 7), (3, 7), (-2, -7), (5, 7)$ $(0, 7), (3, 7), (5, 7)$

Complete the ordered pairs so that each is a solution for the given equation.

15. $x + y = 12$ (5,), (, 3), (0,), (, 0) 7, 9, 12, 12

16. $x - y = 7$ (, 3), (9,), (0,), (, 0) 10, 2, −7, 7

17. $2x + y = 8$ (3,), (, 8), (, −2), (0,) 2, 0, 5, 8

18. $x + 4y = 12$ (0,), (, 2), (8,), (, 4) 3, 4, 1, −4

19. $5x - y = 10$ (, 5), (0,), (4,), (, −5) 3, −10, 10, 1

20. $x - 3y = 12$ (0,), (15,), (, 0), (, −2) −4, 1, 12, 6

21. $3x - 2y = 18$ (, 0), (, −9), (2,), (, 3) 6, 0, −6, 8

22. $2x + 5y = 10$ (0,), (10,), (, 0), (, 4) 2, −2, 5, −5

23. $x - 3y = 6$ $(0,\ \ )$, $\left(\ \ , \frac{4}{3}\right)$, $(\ \ , 0)$, $\left(\ \ , -\frac{2}{3}\right)$ −2, 10, 6, 4

24. $3x + 5y = 15$ $(0,\ \ )$, $\left(\ \ , \frac{9}{5}\right)$, $(\ \ , 0)$, $\left(\frac{10}{3},\ \ \right)$ 3, 2, 5, 1

25. $y = 3x - 2$ $(0,\ \ )$, $(\ \ , 7)$, $(\ \ , 0)$, $\left(\frac{5}{3},\ \ \right)$ $-2, 3, \frac{2}{3}, 3$

26. $y = -2x + 7$ $(0,\ \ )$, $(\ \ , -7)$, $\left(\frac{3}{2},\ \ \right)$, $(\ \ , 6)$ $7, 7, 4, \frac{1}{2}$

Find four solutions for each of the following equations. **Note:** Your answers may vary from those shown in the text.

27. $x - y = 9$
(0, −9), (2, −7), (4, −5), (6, −3)

28. $x + y = 15$
(0, 15), (5, 10), (10, 5), (15, 0)

29. $2x - y = 12$
(0, −12), (3, −6), (6, 0), (9, 6)

30. $3x - y = 18$
(0, −18), (3, −9), (6, 0), (9, 9)

31. $x + 4y = 16$
(0, 4), (4, 3), (8, 2), (12, 1)

32. $x + 3y = 18$
(0, 6), (3, 5), (6, 4), (9, 3)

33. $2x - 7y = 14$
(−7, −4), (0, −2), (7, 0), (14, 2)

34. $3x + 5y = 15$
(0, 3), (5, 0), (10, −3), (15, −6)

35. $y = 3x + 1$
(0, 1), (1, 4), (2, 7), (3, 10)

36. $y = 4x - 2$
(0, −2), (1, 2), (2, 6), (3, 10)

37. $x = -5$
(−5, 0), (−5, 1), (−5, 2), (−5, 3)

38. $y = 3$
(0, 3), (1, 3), (2, 3), (3, 3)

Skillscan (Section 2.1)

Plot points with the following coordinates on the number line shown below.

a. −4 **b.** 8 **c.** 0 **d.** −9

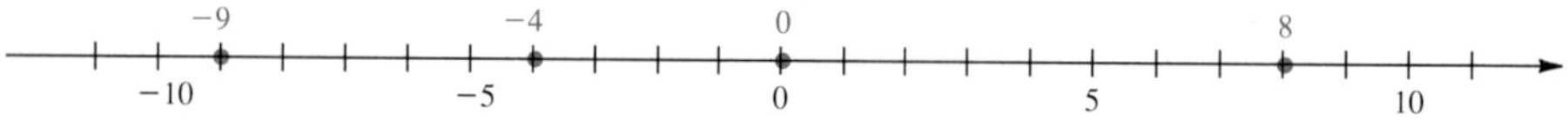

Give the coordinates of each of the following points.

e. *A* −8 **f.** *B* −2 **g.** *C* 3 **h.** *D* 9

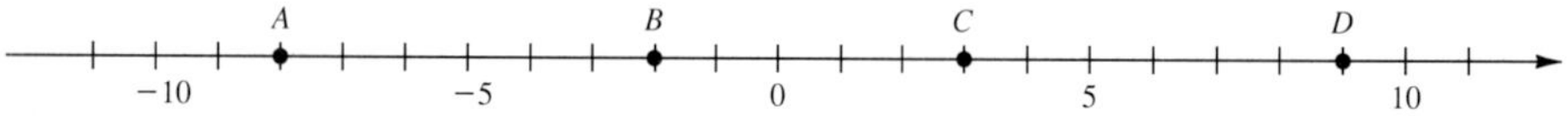

Answers

1. (5, 3), (0, 8), (−3, 11) **3.** (4, 2), (3, 0), (5, 4) **5.** (3, 0), (1, 6) **7.** (6, 0), (9, 2), (3, −2) **9.** (2, 0), $\left(\frac{1}{3}, -\frac{5}{2}\right)$ **11.** (0, 0), (2, 6) **13.** (5, 3), (5, 0), (5, 7) **15.** (5, 7), (9, 3), (0, 12), (12, 0) **17.** (3, 2), (0, 8), (5, −2), (0, 8) **19.** (3, 5), (0, −10), (4, 10), (1, −5) **21.** (6, 0), (0, −9), (2, −6), (8, 3) **23.** (0, −2), $\left(10, \frac{4}{3}\right)$, (6, 0), $\left(4, -\frac{2}{3}\right)$ **25.** (0, −2), (3, 7), $\left(\frac{2}{3}, 0\right)$, $\left(\frac{5}{3}, 3\right)$ **27.** (0, −9), (2, −7), (4, −5), (6, −3) **29.** (0, −12), (3, −6), (6, 0), (9, 6) **31.** (0, 4), (4, 3), (8, 2), (12, 1) **33.** (−7, −4), (0, −2), (7, 0), (14, 2) **35.** (0, 1), (1, 4), (2, 7), (3, 10) **37.** (−5, 0), (−5, 1), (−5, 2), (−5, 3)

a–d. 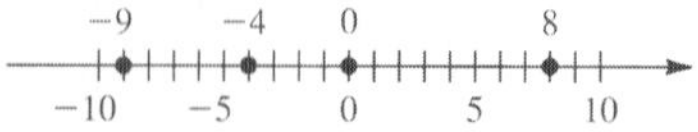

e. −8 **f.** −2 **g.** 3 **h.** 9

7.2 The Rectangular Coordinate System

OBJECTIVES

1. To graph or plot points corresponding to ordered pairs.
2. To give the coordinates of a point in a plane.

In Section 7.1 we saw that ordered pairs could be used to write the solutions of equations in two variables. The next step is to graph those ordered pairs as points in a plane.

Since there are two numbers (one for x and one for y), we will need two number lines. One line is drawn horizontally, and the other is drawn vertically; their point of intersection (at their respective zero points) is called the *origin*. The horizontal line is called the *x axis*, while the vertical line is called the *y axis*. Together the lines form the *rectangular coordinate system*.

This system is also called the *cartesian* coordinate system, named in honor of its inventor, René Descartes (1596–1650) a French mathematician and philosopher.

The axes divide the plane into four regions called *quadrants*, which are numbered (usually by Roman numerals) counterclockwise from the upper right.

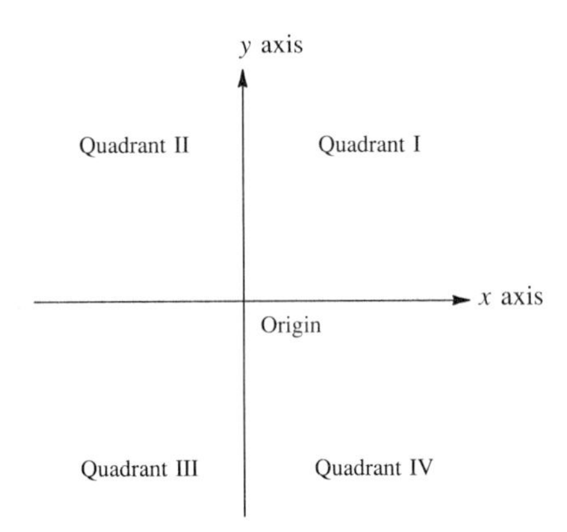

We now want to establish a correspondence between ordered pairs of numbers (x, y) and points in the plane.

First, for any ordered pair

$$(x, y)$$

x coordinate ↗ ↖ y coordinate

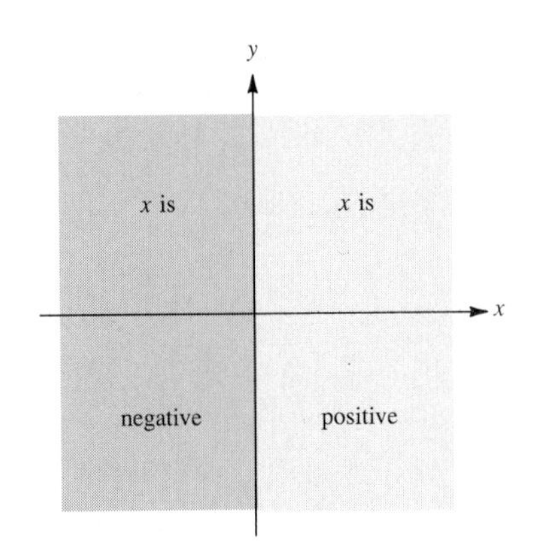

1. If the x coordinate is
   Positive, the point corresponding to that pair is located x units to the *right* of the y axis.

   Negative, the point is x units to the *left* of the y axis.

   Zero, the point is on the y axis.

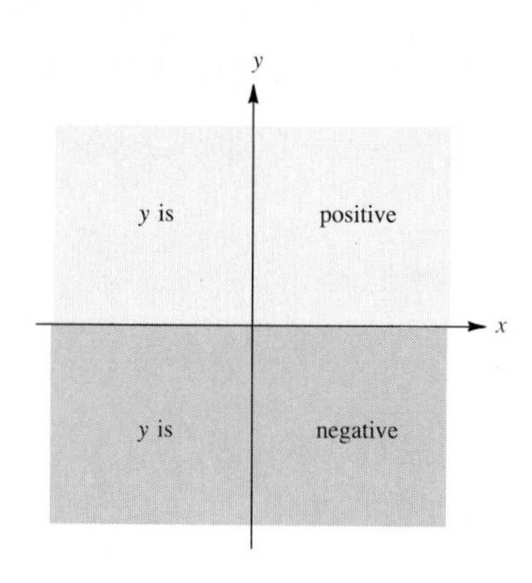

2. If the y coordinate is
Positive, the point is y units *above* the x axis.

Negative, the point is y units *below* the x axis.

Zero, the point is on the x axis.

The next example illustrates how to use these guidelines to give coordinates to points in the plane.

Example 1

Remember: The x coordinate gives the horizontal distance from the y axis. The y coordinate gives the vertical distance from the x axis.

(*a*)

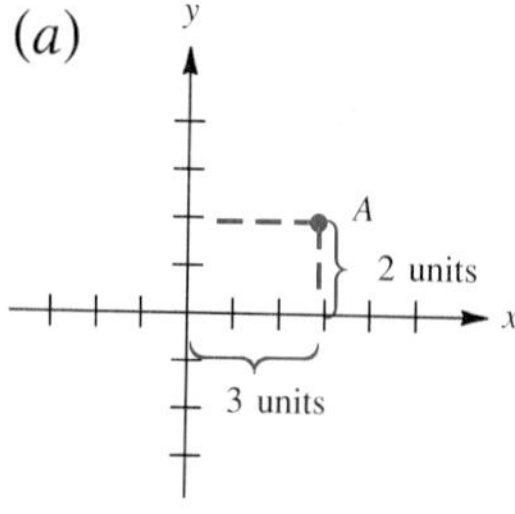

Point A is 3 units to the *right* of the y axis and 2 units *above* the x axis. Point A has coordinates (3, 2).

(*b*)

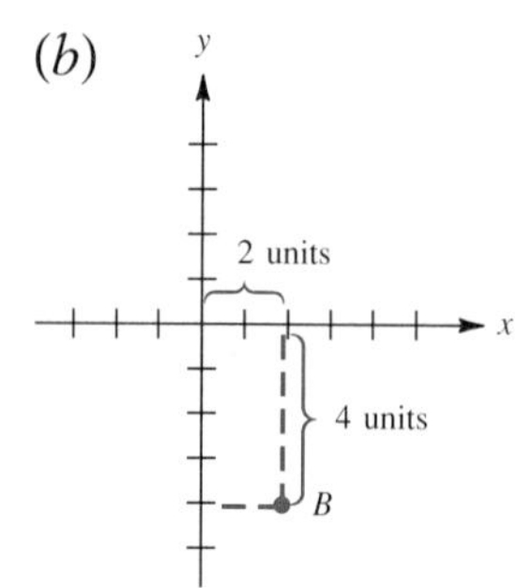

Point B is 2 units to the *right* of the y axis and 4 units *below* the x axis. Point B has coordinates (2, −4).

(*c*)

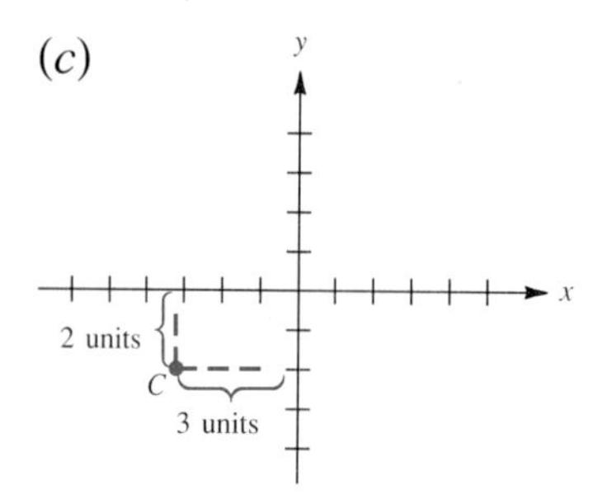

Point C is 3 units to the *left* of the y axis and 2 units *below* the x axis. C has coordinates (−3, −2).

(*d*)

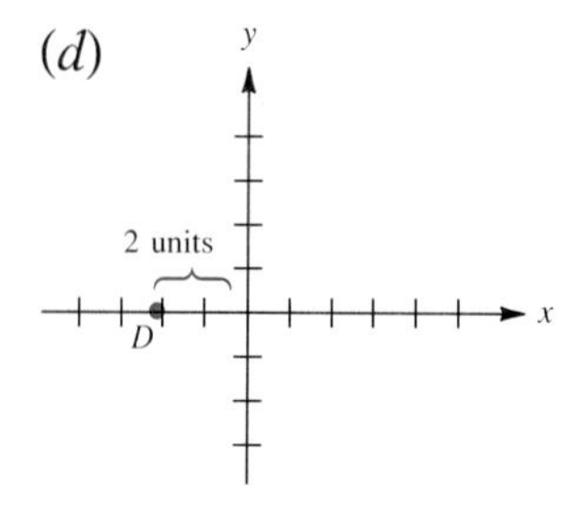

Point D is 2 units to the *left* of the y axis and *on* the x axis. D has coordinates (−2, 0).

CHECK YOURSELF 1

Give the coordinates of points P, Q, R, and S.

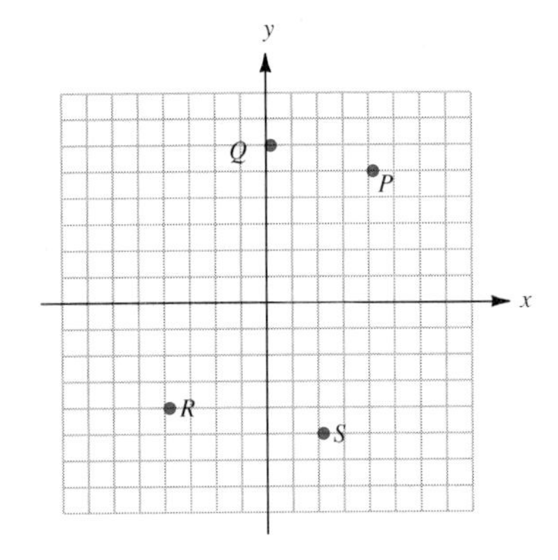

P ________

Q ________

R ________

S ________

Reversing the process above will allow us to graph (or plot) a point in the plane given the coordinates of the point. You can use the following steps.

TO GRAPH A POINT IN THE PLANE

STEP 1 Start at origin.

STEP 2 Move right or left according to value of x coordinate.

STEP 3 Move up or down according to value of y coordinate.

Example 2

(*a*) Graph the point corresponding to the ordered pair (4, 3).

Since the x coordinate is 4, move 4 units to the right on the x axis. Then, because the y coordinate is 3, move 3 units up from the point you stopped at on the x axis. This locates the point corresponding to (4, 3).

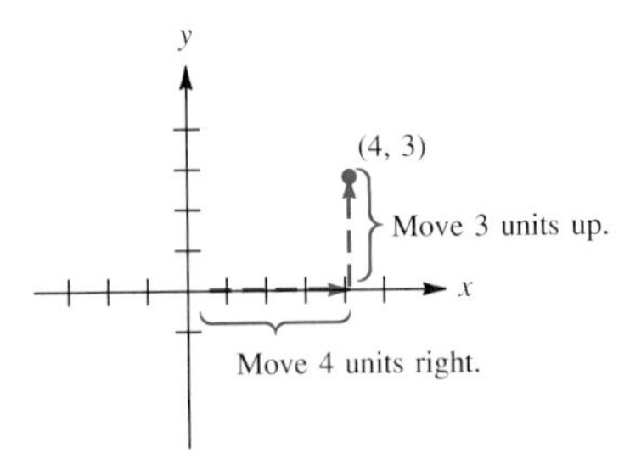

(*b*) Graph the point corresponding to the ordered pair $(-5, 2)$.

In this case move 5 units *left* (because the x coordinate is negative) and then 2 units *up*.

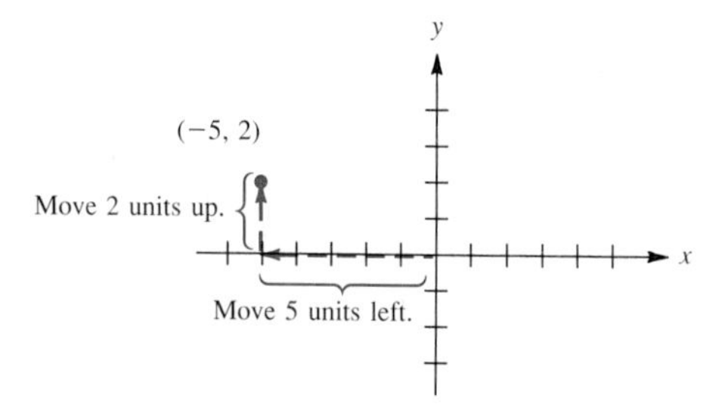

(*c*) Graph the point corresponding to $(-4, -2)$.

Here move 4 units *left* and then 2 units *down* (the y coordinate is negative.)

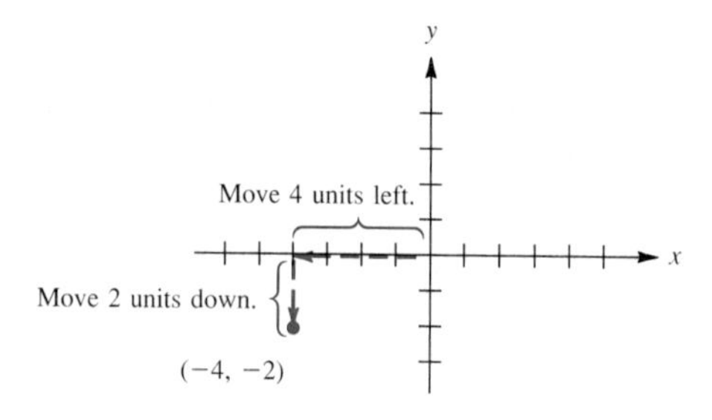

(*d*) Graph the point corresponding to $(0, -3)$.

There is *no* horizontal movement because the x coordinate is 0. Move 3 units *down*.

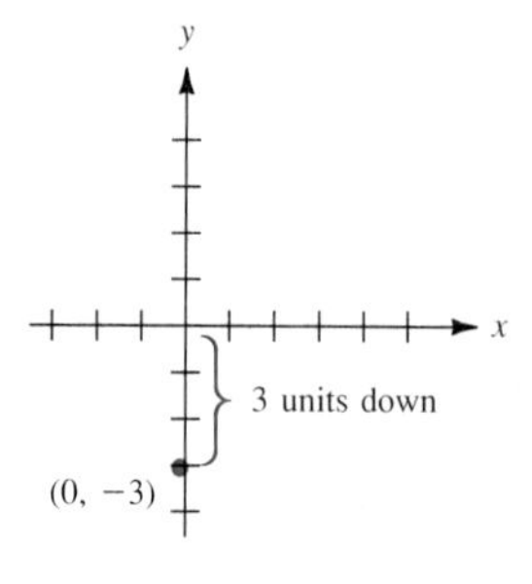

(*e*) Graph the point corresponding to $(5, 0)$. Move 5 units *right*. The desired point is on the x axis because the y coordinate is 0.

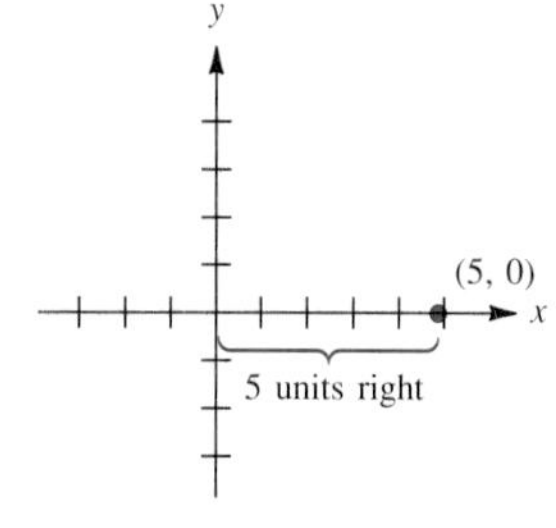

CHECK YOURSELF 2

Graph the points corresponding to $M(5, 2)$, $N(-4, 6)$, $P(-7, -5)$, and $Q(0, -6)$.

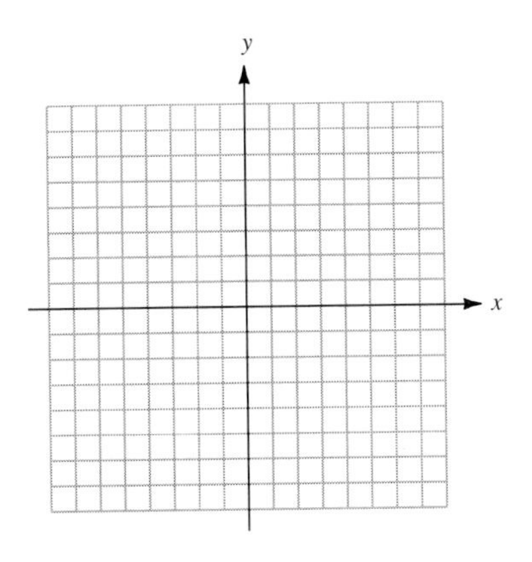

CHECK YOURSELF ANSWERS

1. $P(4, 5)$, $Q(0, 6)$, $R(-4, -4)$, and $S(2, -5)$
2.

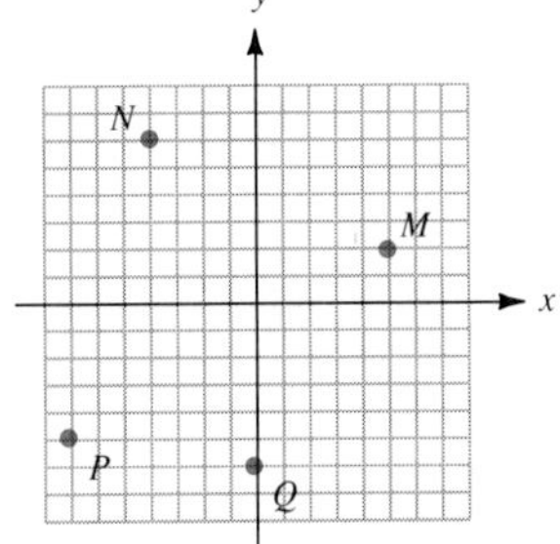

7.2 Exercises

1–5. Give the coordinates of the points graphed below.

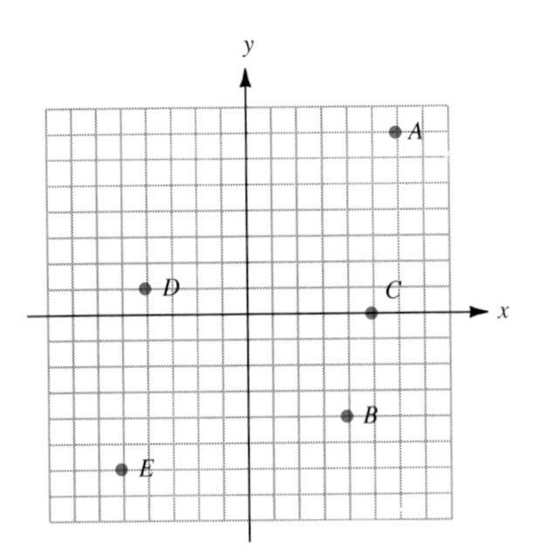

1. A (6, 7)

2. B (4, −4)

3. C (5, 0)

4. D (−4, 1)

5. E (−5, −6)

6–10. Give the coordinates of the points graphed below.

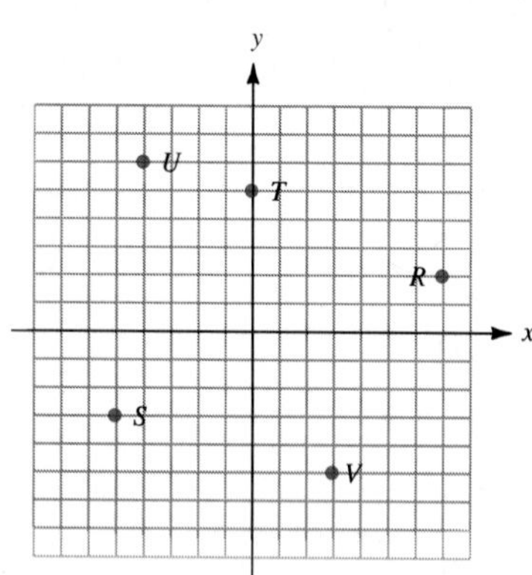

6. R (7, 2) **7.** S (−5, −3)

8. T (0, 5) **9.** U (−4, 6)

10. V (3, −5)

11–16. Plot points with the following coordinates on the graph below.

11. $M(6, 2)$ **12.** $N(0, -2)$

13. $P(-3, 5)$ **14.** $Q(6, 0)$

15. $R(-5, -7)$ **16.** $S(-4, -3)$

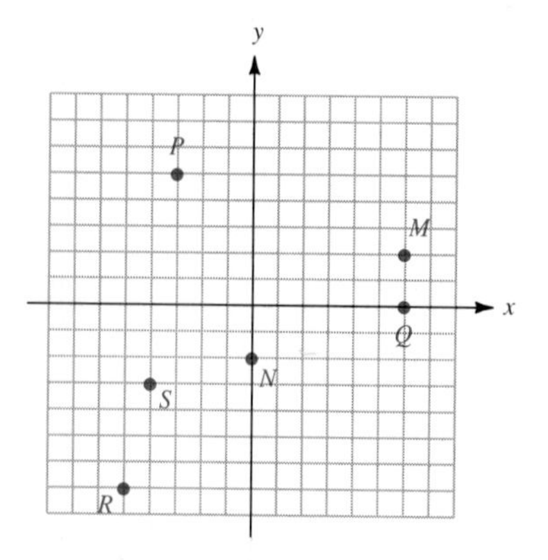

17–22. Plot points with the following coordinates on the graph below.

17. $F(-2, -1)$ **18.** $G(5, 3)$

19. $H(4, -2)$ **20.** $I(-4, 0)$

21. $J(-6, 4)$ **22.** $K(0, 7)$

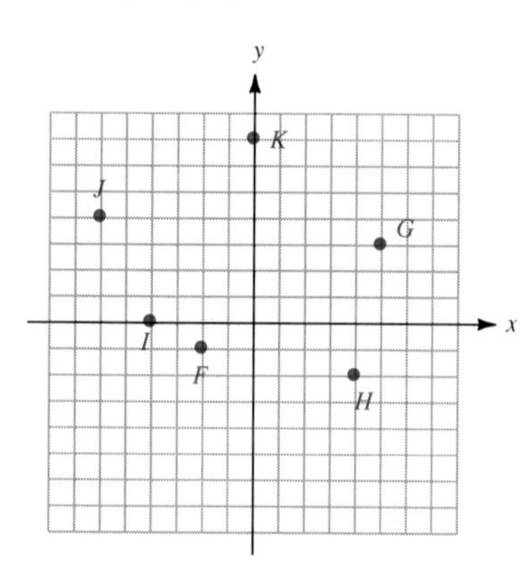

Going Beyond

23. Graph points with coordinates (1, 2), (2, 3), and (3, 4) below. What do you observe? Can you give the coordinates of another point with the same property?
(4, 5)

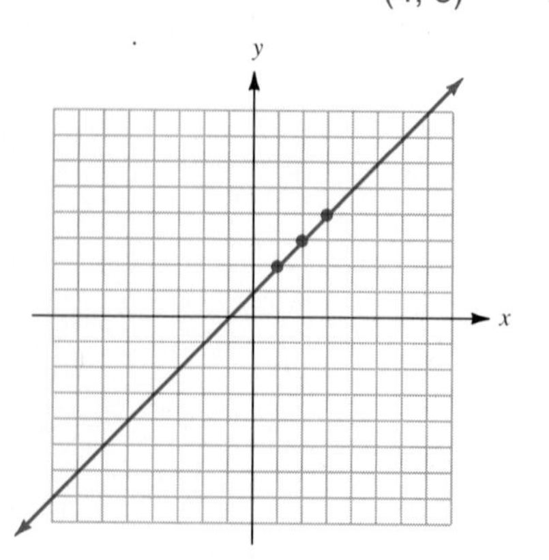

24. Graph points with coordinates (1, 2), (2, 4), and (3, 6) below. What do you observe? Can you give the coordinates of another point with the same property? (4, 8)

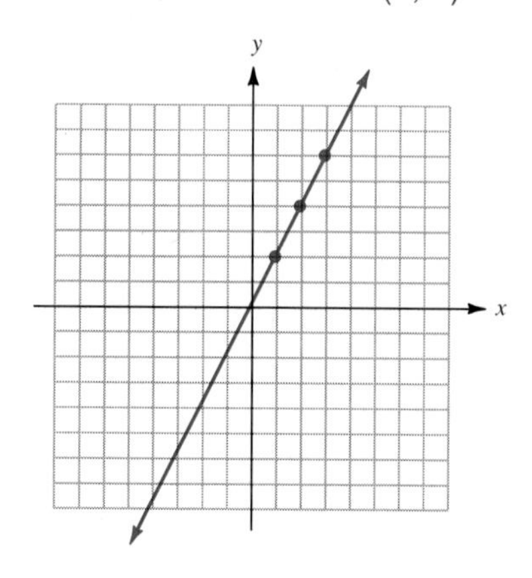

25. Graph points with coordinates $(-1, -3)$, $(0, 0)$, and $(1, 3)$ below. What do you observe? Can you give the coordinates of another point with the same property? (2, 6)

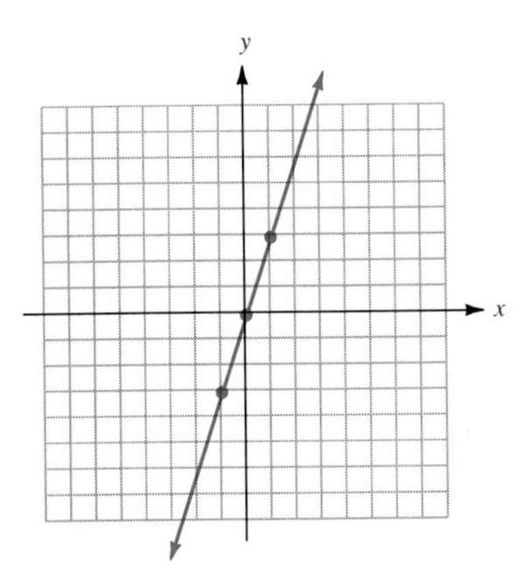

26. Graph points with coordinates $(-1, 3)$, $(1, 5)$, and $(3, 7)$ below. What do you observe? Can you give the coordinates of another point with the same property? $(-3, 1)$

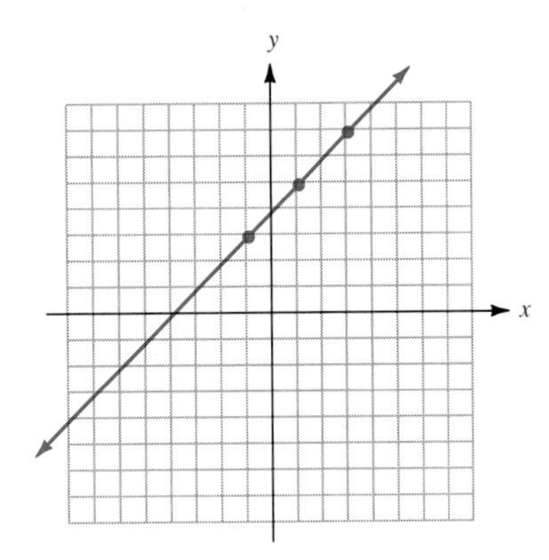

Skillscan (Section 3.4)

Solve each of the following equations.

a. $2x - 4 = 6$ 5

b. $3 - 5y = 18$ -3

c. $7x + 12 = -2$ -2

d. $-2 + 5y = 4$ $\frac{6}{5}$

e. $2 - 3y = 6$ $-\frac{4}{3}$

f. $-3x + 8 = 6$ $\frac{2}{3}$

Answers

1. (6, 7) **3.** (5, 0) **5.** $(-5, -6)$ **7.** $(-5, -3)$ **9.** $(-4, 6)$

11–21.

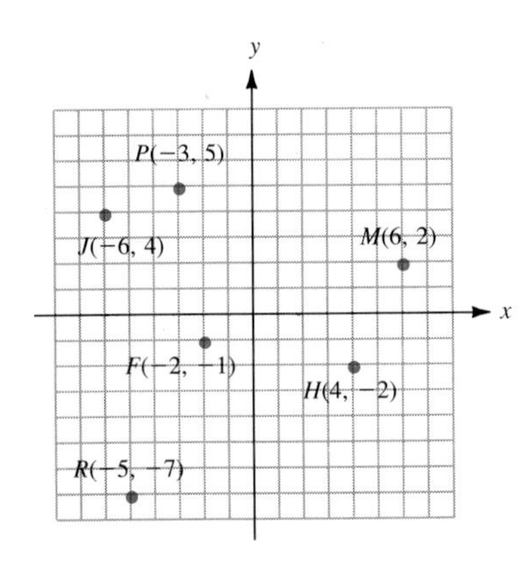

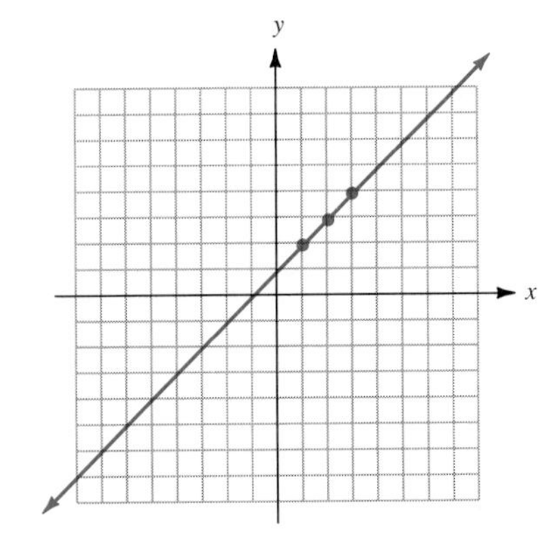

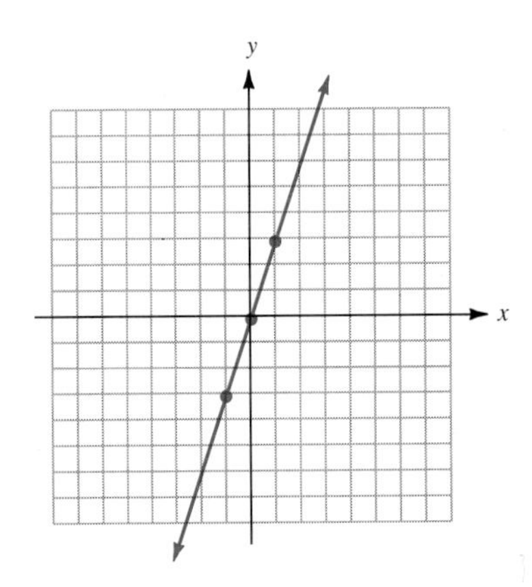

23. The points lie on a line; (4, 5) **25.** The points lie on a line; (2, 6)

a. 5 **b.** -3 **c.** -2 **d.** $\frac{6}{5}$ **e.** $-\frac{4}{3}$ **f.** $-\frac{2}{3}$

7.3 Graphing Linear Equations

OBJECTIVES

1. To graph linear equations.
2. To graph by the intercept method.
3. To graph by solving the equation for *y*.

We are now ready to combine our work of the last two sections. In Section 7.1 you learned to write the solutions of equations in two variables as ordered pairs. Then, in Section 7.2, these ordered pairs were graphed in the plane. Putting these ideas together will let us graph certain equations. The following example illustrates.

Example 1

Graph $x + 2y = 4$.

We are going to find *three* solutions for the equation. We'll point out the reason why shortly.

Step 1 Find solutions for $x + 2y = 4$.

To find solutions, we choose any convenient values for x, say $x = 0$, $x = 2$, and $x = 4$. Given these values for x, we can substitute and then solve for the corresponding value for y. So

If $x = 0$, $y = 2$, so (0, 2) is a solution.
If $x = 2$, $y = 1$, so (2, 1) is a solution.
If $x = 4$, $y = 0$, so (4, 0) is a solution.

A handy way to show this information is in a table such as this:

The table is just a convenient way to display the information. It is the same as writing (0, 2), (2, 1), and (4, 0).

x	y
0	2
2	1
4	0

CHECK YOURSELF 1

Use a table to show three solutions for the equations $2x - y = 6$.

Step 2 We now graph the solutions found in step 1.

$x + 2y = 4$

x	y
0	2
2	1
4	0

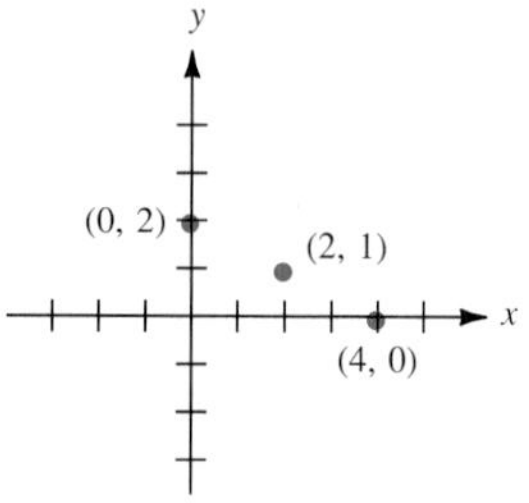

y
x
(0, 2)
(2, 1)
(4, 0)

What pattern do you see? It appears that the three points lie on a straight line, and that is in fact the case.

CHECK YOURSELF 2

Graph the three solutions you found in Check Yourself 1.

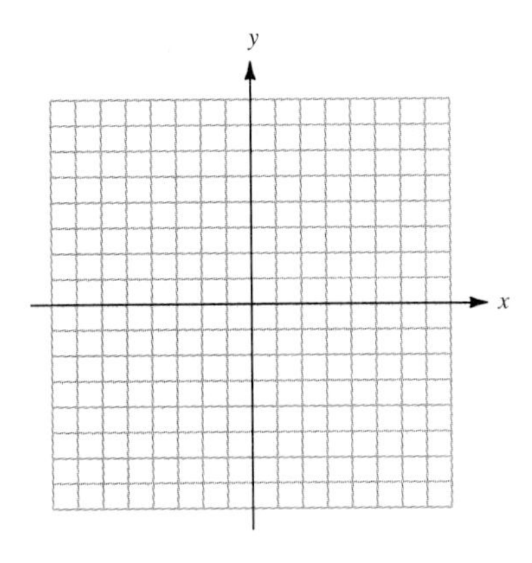

Step 3 Draw a straight line through the three points determined in step 2.

The arrows on the end of the line mean that the line extends indefinitely in either direction.

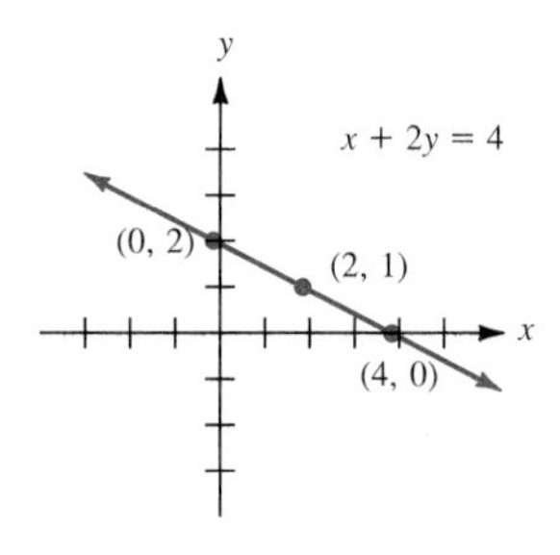

The graph is a "picture" of the solutions for the given equation.

The line shown is the *graph* of the equation $x + 2y = 4$. It represents *all* the ordered pairs that are solutions (an infinite number) for that equation.

> Every ordered pair which is a solution will have its graph on this line. Any point on the line will have coordinates that are a solution for the equation.

Note: Why did we suggest finding *three* solutions in step 1? Two points determine a line, so technically you need only two. The third point that we find is a check to catch any possible errors.

CHECK YOURSELF 3

Complete the graph of $2x - y = 6$ by drawing a line through the three points graphed in Check Yourkself 2. (Use a straightedge.)

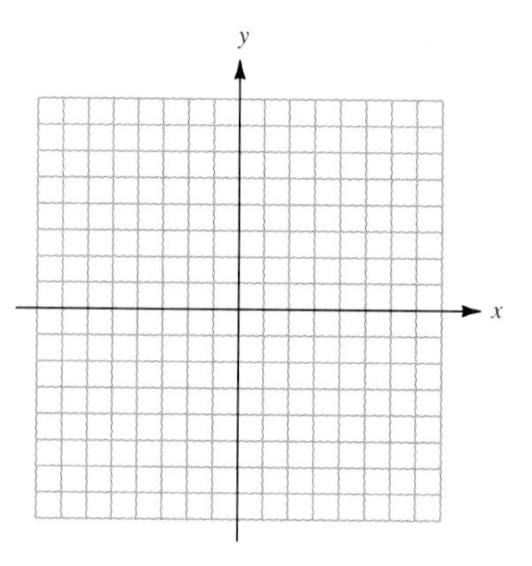

Let's summarize. An equation that can be written in the form

$$Ax + By = C$$

where A, B, and C are real numbers and A and B cannot both be 0 is called a *linear equation in two variables*. The graph of this equation is a *straight line*.

The steps of graphing:

> **TO GRAPH A LINEAR EQUATION**
>
> STEP 1 Find at least three solutions for the equation, and put your results in tabular form.
>
> STEP 2 Graph the solutions found in step 1.
>
> STEP 3 Draw a straight line through the points determined in step 2 to form the graph of the equation.

Example 2

Graph $y = 3x$.

Step 1 Some solutions are

x	y
0	0
1	3
2	6

Let $x = 0$, 1, and 2, and substitute to determine the corresponding y values. Again the choices for x are simply convenient. Other values for x would serve the same purpose.

Step 2 Graph the points.

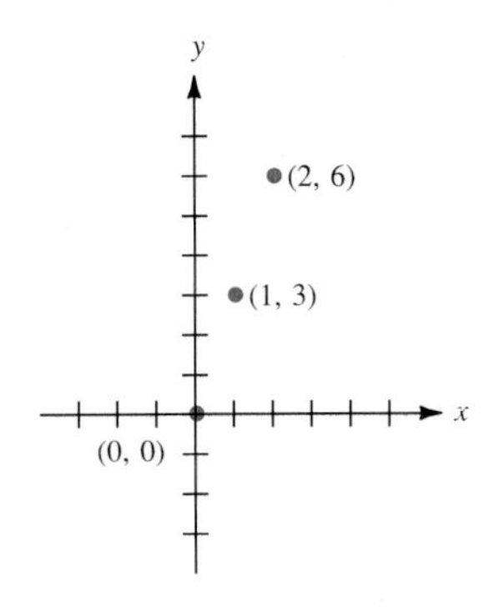

Step 3 Draw a line through the points.

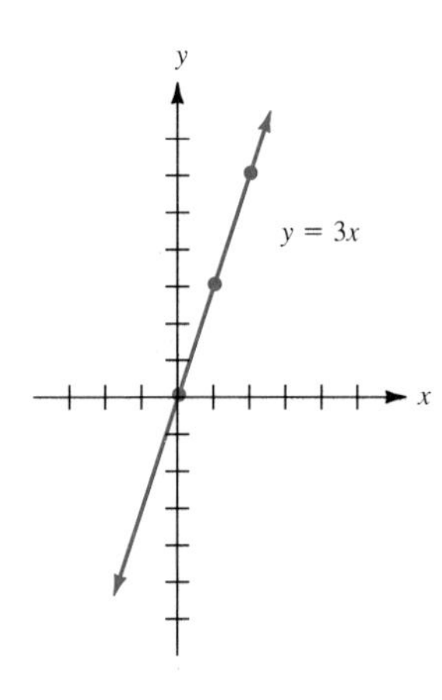

CHECK YOURSELF 4

Graph the equation $y = -2x$ after completing the table of values.

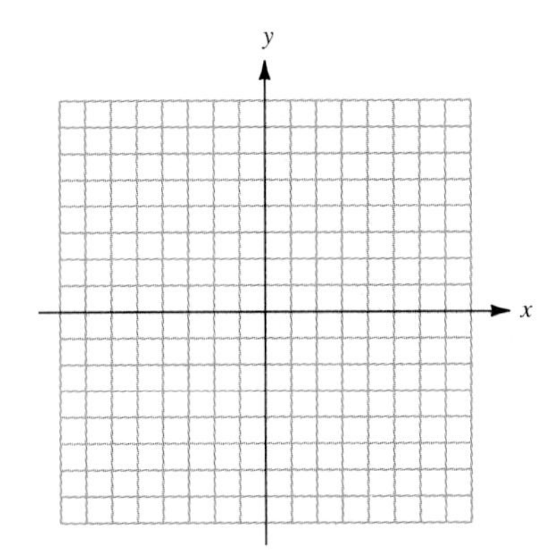

x	y
0	
1	
2	

Let's work through another example of graphing a line from its equation.

Example 3

Graph $y = 2x + 3$.

Step 1 Some solutions are

x	y
0	3
1	5
2	7

Step 2 Graph the points corresponding to these values.

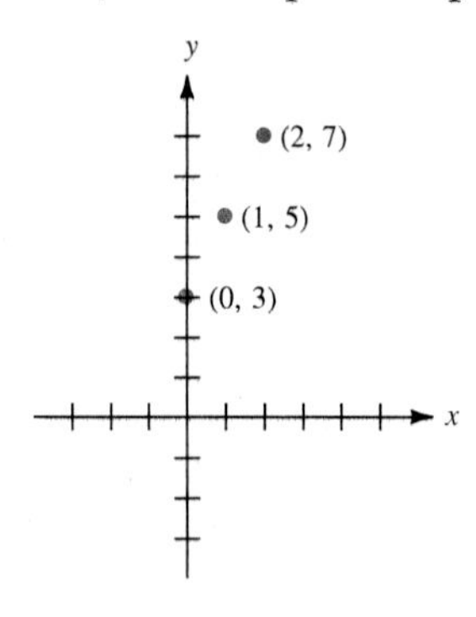

Step 3 Draw a line through the points.

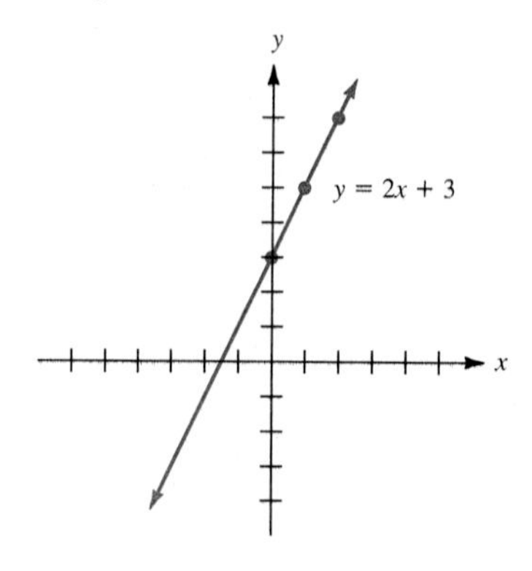

CHECK YOURSELF 5

Graph the equation $y = 3x - 2$ after completing the table of values.

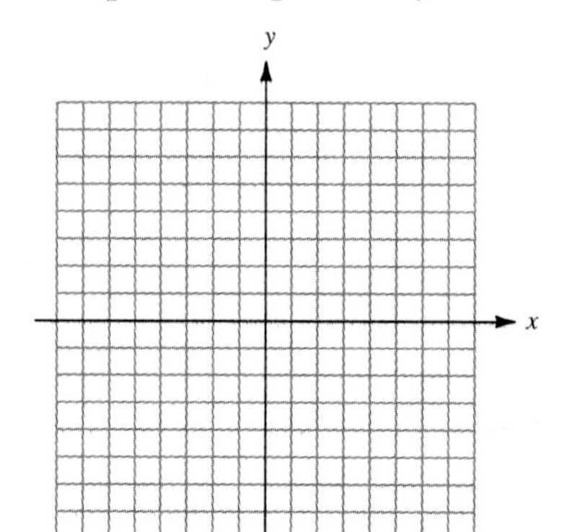

x	y
0	
1	
2	

In graphing equations, particularly when fractions are involved, a careful choice of values for x can simplify the process. Consider the following example.

Example 4

Graph

$$y = \frac{3}{2}x - 2$$

As before, we want to find solutions for the given equation by picking convenient values for x. Note that in this case, choosing *multiples of 2* will avoid fractional values for y and make the plotting of those solutions much easier. For instance, here we might choose values of -2, 0, and 2 for x.

Step 1
If $x = -2$,

$$y = \frac{3}{2}(-2) - 2$$

$$= -3 - 2 = -5$$

If $x = 0$,

$$y = \frac{3}{2}(0) - 2$$

$$= 0 - 2 = -2$$

Suppose we do *not* choose a multiple of 2, say $x = 3$. Then

$$y = \frac{3}{2}(3) - 2$$

$$= \frac{9}{2} - 2$$

$$= \frac{5}{2}$$

$\left(3, \frac{5}{2}\right)$ is still a valid solution but we must graph a point with fractional coordinates.

If $x = 2$,

$$y = \frac{3}{2}(2) - 2$$

$$= 3 - 2 = 1$$

In tabular form, the solutions are

x	y
-2	-5
0	-2
2	1

Step 2 Graph the points determined above.

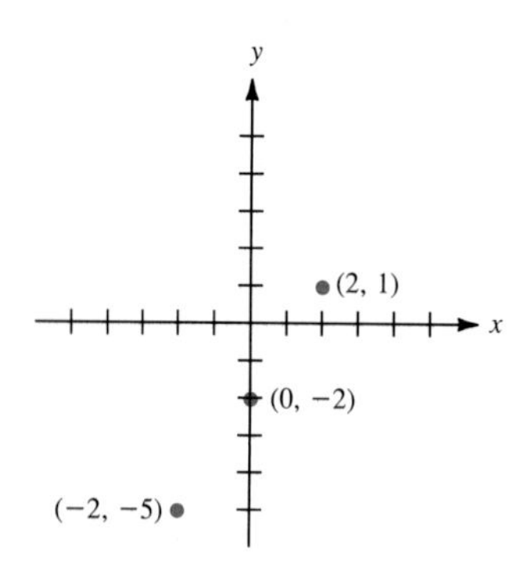

Step 3 Draw a line through the points.

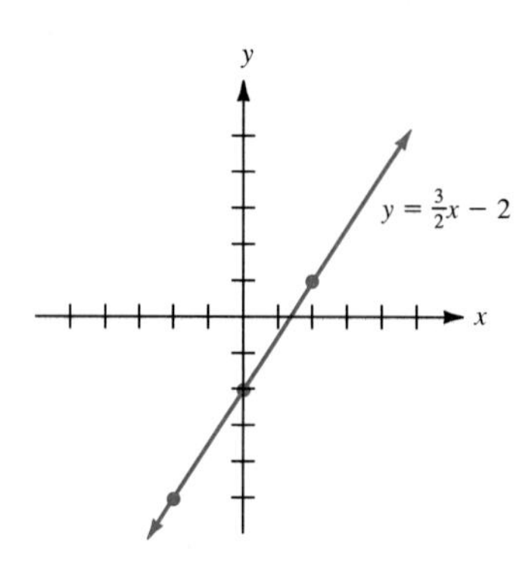

CHECK YOURSELF 6

Graph the equation $y = -\dfrac{1}{3}x + 3$ after completing the table of values.

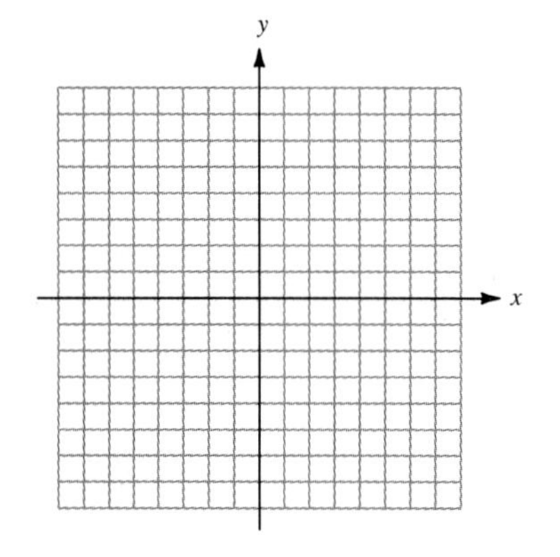

x	y
−3	
0	
3	

Some special cases of linear equations are illustrated in Examples 5 and 6.

Example 5

Graph $x = 3$.

The equation $x = 3$ is equivalent to $x + 0 \cdot y = 3$. Let's look at some solutions.

If $y = 1$,

$$\begin{aligned} x + 0 \cdot 1 &= 3 \\ x &= 3 \end{aligned}$$

If $y = 4$,

$$\begin{aligned} x + 0 \cdot 4 &= 3 \\ x &= 3 \end{aligned}$$

If $y = -2$,

$$\begin{aligned} x + 0(-2) &= 3 \\ x &= 3 \end{aligned}$$

In tabular form,

x	y
3	1
3	4
3	−2

What do you observe? The variable x has the value 3, regardless of the value of y. Look at the graph.

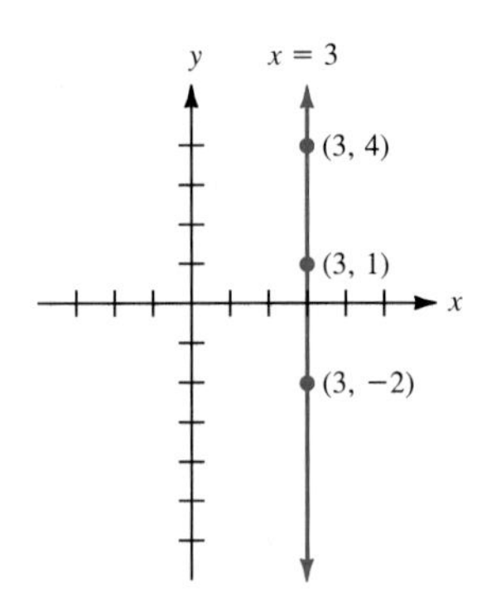

The graph of $x = 3$ is a vertical line crossing the x axis at (3, 0).

Note that plotting points in this case is not really necessary. Simply recognize that the graph of $x = 3$ *must* be a vertical line (parallel to the y axis) which intercepts the x axis at 3.

CHECK YOURSELF 7

Graph the equation $x = -2$.

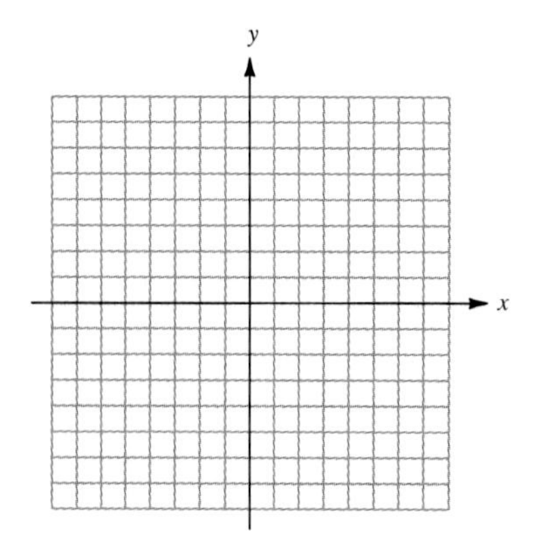

Here is a related example.

Example 6

Graph $y = 4$.

Since $y = 4$ is equivalent to $0 \cdot x + y = 4$, any value for x paired with 4 for y will form a solution. A table of values might be

x	y
-2	4
0	4
2	4

Here is the graph.

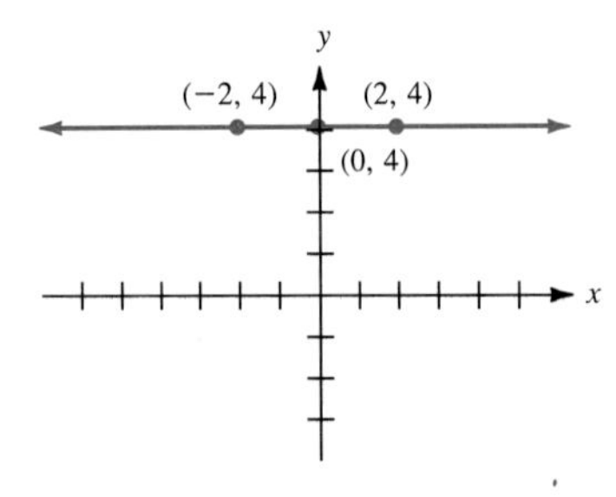

This time the graph is a horizontal line that crosses the y axis at $(0, 4)$. Again point plotting is not required. The graph of $y = 4$ *must* be horizontal (parallel to the x axis) which intercepts the y axis at 4.

CHECK YOURSELF 8

Graph the equation $y = -3$.

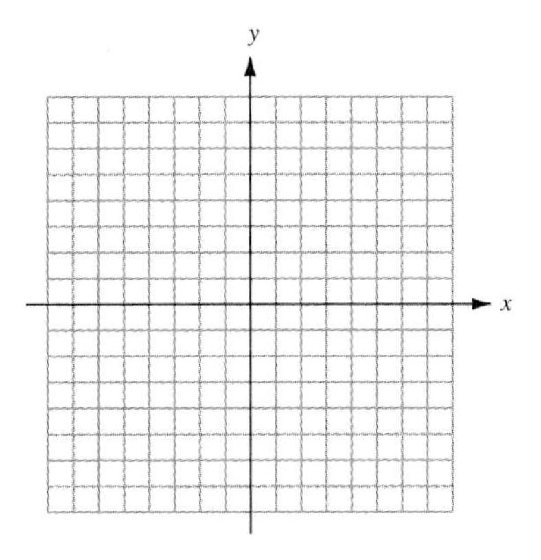

To summarize our work in the previous two examples:

VERTICAL AND HORIZONTAL LINES

1. The graph of $x = a$ is a *vertical line* crossing the x axis at $(a, 0)$.
2. The graph of $y = b$ is a *horizontal line* crossing the y axis at $(0, b)$.

To simplify the graphing of certain linear equations, some students prefer the *intercept* method of graphing. This method makes use of the fact that the solutions that are easiest to find are those with an x coordinate or a y coordinate of 0. For instance, let's graph the equation

$$4x + 3y = 12$$

With practice all of this can be done mentally, which is the big advantage of this method.

First, let $x = 0$ and solve for y.

$$\begin{aligned} 4 \cdot 0 + 3y &= 12 \\ 3y &= 12 \\ y &= 4 \end{aligned}$$

So $(0, 4)$ is one solution. Now let $y = 0$ and solve for x.

$$\begin{aligned} 4x + 3 \cdot 0 &= 12 \\ 4x &= 12 \\ x &= 3 \end{aligned}$$

A second solution is $(3, 0)$.

The two points corresponding to these solutions can now be used to graph the equation.

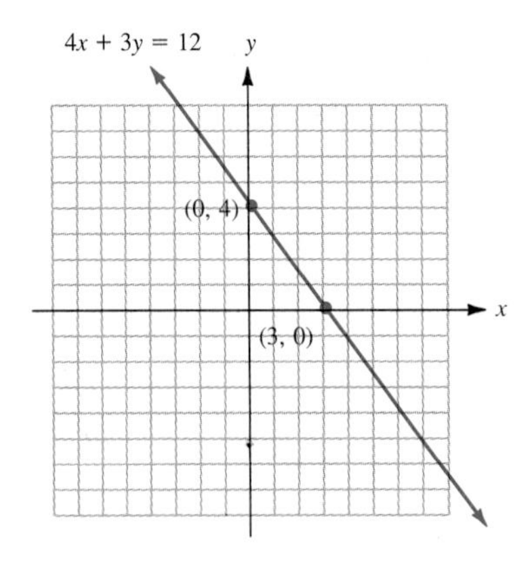

The intercepts are the points where the line cuts the x and y axes.

The number 3 is called the x *intercept,* and the number 4 is the y *intercept* of the graph. Using these points to draw the graph gives the name to this method. Let's look at a second example of graphing by the intercept method.

Example 7

Graph $3x - 5y = 15$, using the intercept method.

To find the x intercept, let $y = 0$.

$$3x - 5 \cdot 0 = 15$$
$$x = 5$$

The x intercept

To find the y intercept, let $x = 0$.

$$3 \cdot 0 - 5y = 15$$
$$y = -3$$

The y intercept

So (5, 0) and (0, −3) are solutions for the equation, and we can use the corresponding points to graph the equation.

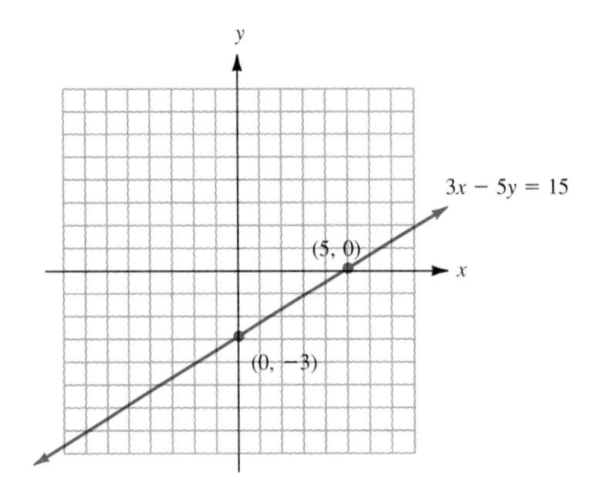

CHECK YOURSELF 9

Graph $4x + 5y = 20$, using the intercept method.

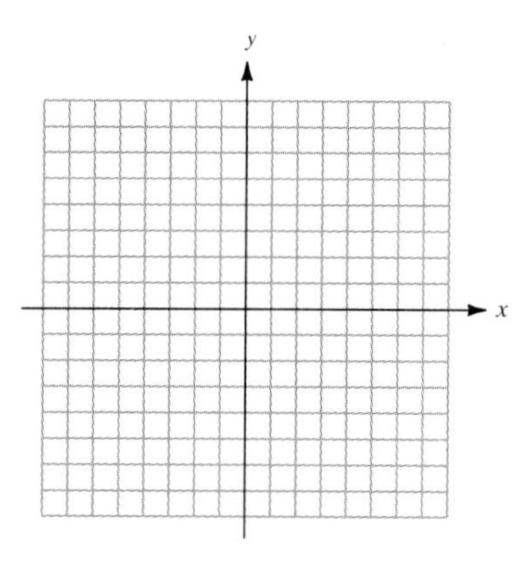

This all looks quite easy, and for many equations it is. What are the drawbacks?

Finding the third "checkpoint" is always a good idea.

For one, you don't have a third checkpoint, and it is possible for errors to occur. You can, of course, still find a third point (other than the two intercepts) to be sure your graph is correct. A second difficulty occurs when the x and y intercepts are very close to each other (or are actually the same point—the origin). For instance, if we have the equation

$$3x + 2y = 1$$

the intercepts are $\left(\frac{1}{3}, 0\right)$ and $\left(0, \frac{1}{2}\right)$. It is hard to draw a line accurately through these intercepts, so choose other solutions farther away from the origin for your points.

Let's summarize the steps of graphing by the intercept method for appropriate equations.

GRAPHING A LINE BY THE INTERCEPT METHOD

STEP 1 To find the x intercept: Let $y = 0$, then solve for x.

STEP 2 To find the y intercept: Let $x = 0$, then solve for y.

STEP 3 Graph the x and y intercepts.

STEP 4 Draw a straight line through the intercepts.

A third method of graphing linear equations involves "solving the equation for y." The reason we use this extra step is that it often will make finding solutions for the equation much easier. Let's look at an example.

Example 8

Graph $2x + 3y = 6$.

Remember that solving for y means that we want to leave y isolated on the left.

Rather than finding solutions for the equation in this form, we solve for y.

$$2x + 3y = 6$$

$$3y = 6 - 2x \quad \text{Subtract } 2x.$$

$$y = \frac{6 - 2x}{3} \quad \text{Divide by 3.}$$

or $$y = 2 - \frac{2}{3}x$$

Again, to pick convenient values for x, we suggest you look at the equation carefully. Here, for instance, picking multiples of 3 for x will make the work much easier.

Now find your solutions by picking convenient values for x. If $x = -3$,

$$y = 2 - \frac{2}{3}(-3)$$

$$= 2 + 2 = 4$$

So $(-3, 4)$ is a solution.

If $x = 0$,

$$y = 2 - \frac{2}{3} \cdot 0$$

$$= 2$$

So $(0, 2)$ is a solution.

If $x = 3$,

$$y = 2 - \frac{2}{3} \cdot 3$$

$$= 2 - 2 = 0$$

So $(3, 0)$ is a solution.

We can now plot the points that correspond to these solutions and form the graph of the equation as before.

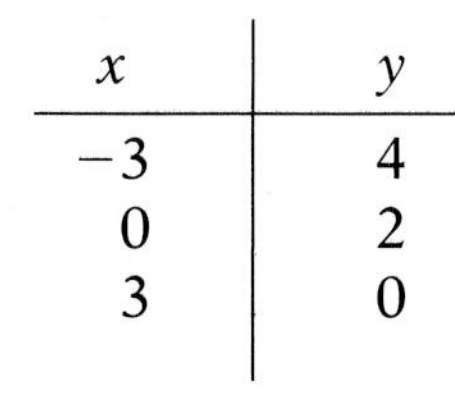

x	y
−3	4
0	2
3	0

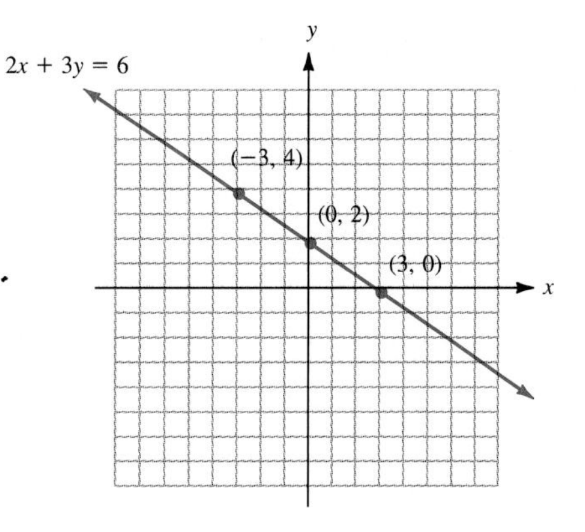

CHECK YOURSELF 10

Graph the equation $5x + 2y = 10$. Solve for y to determine solutions.

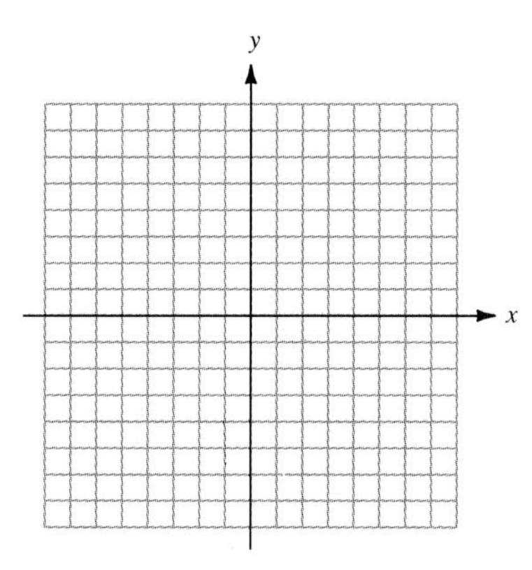

x	y

CHECK YOURSELF ANSWERS

1.

x	y
1	−4
2	−2
3	0

2.

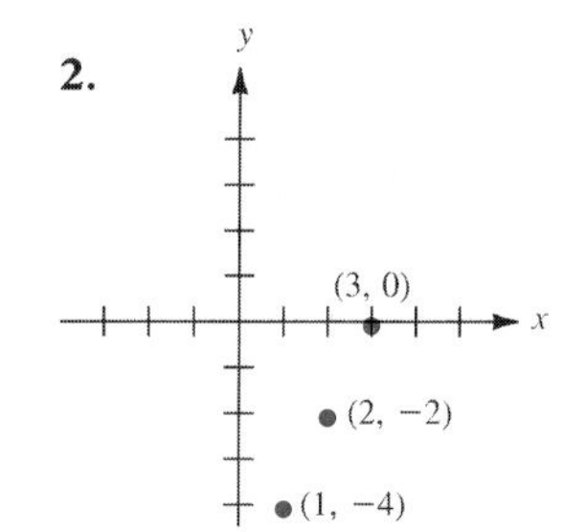

3.

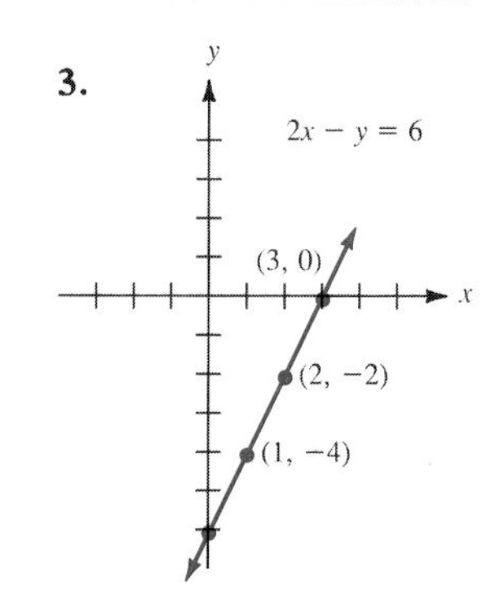

4.

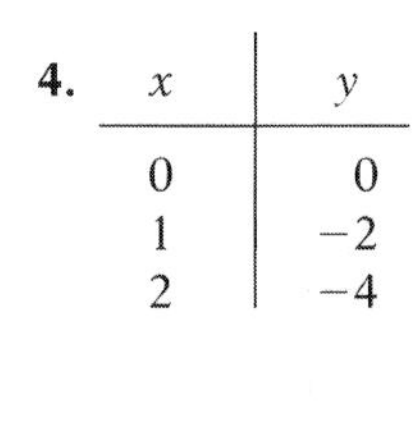

x	y
0	0
1	−2
2	−4

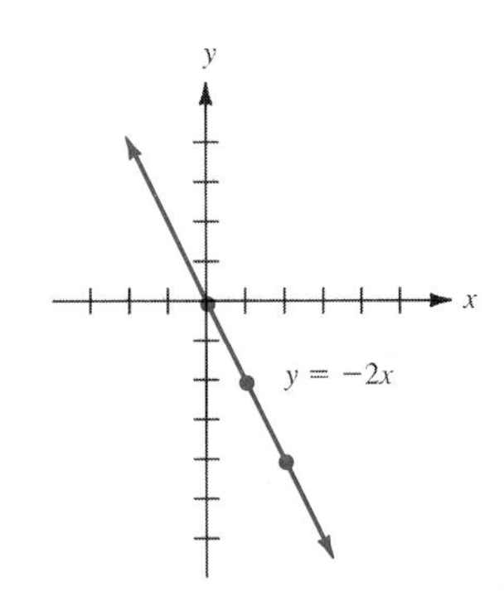

5.

x	y
0	−2
1	1
2	4

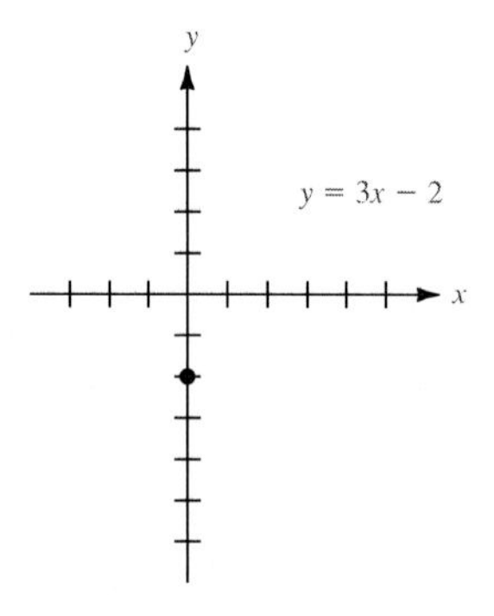

6.

x	y
−3	4
0	3
3	2

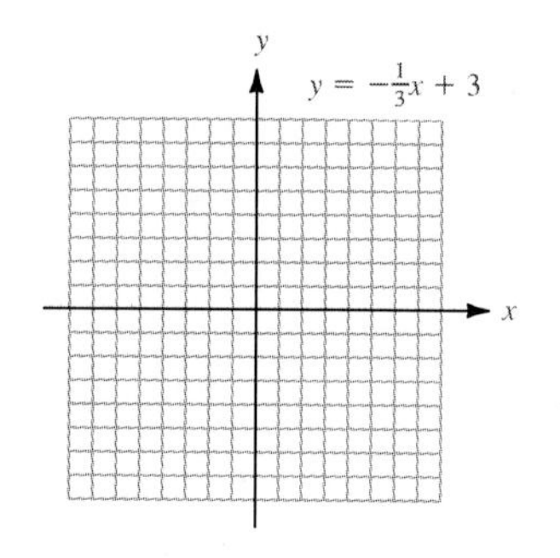

7.

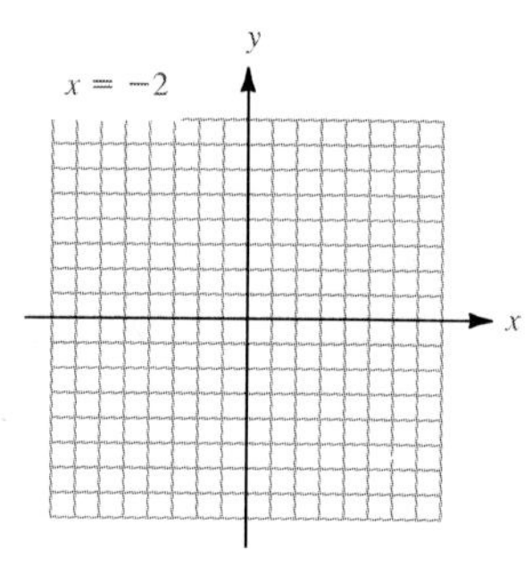

8.

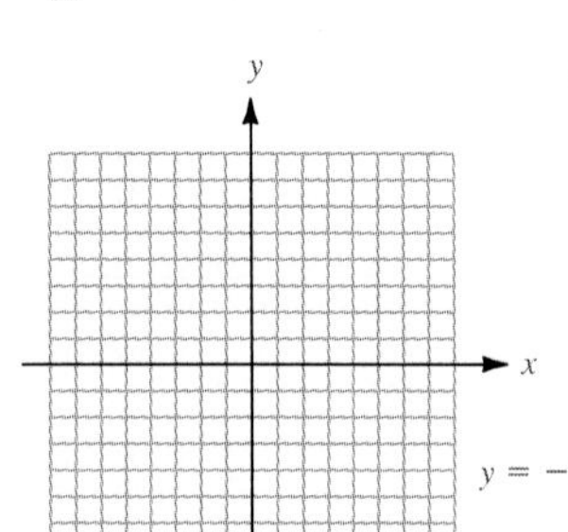

9.

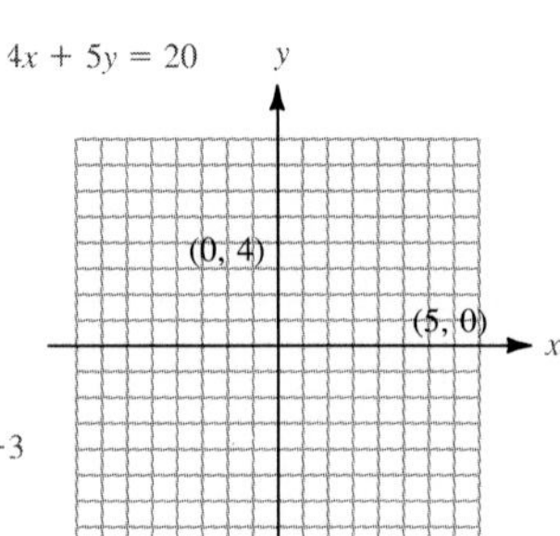

10.

$y = -\frac{5}{2}x + 5$ y

x

7.3 Exercises

Graph each of the following equations.

1. $x + y = 5$

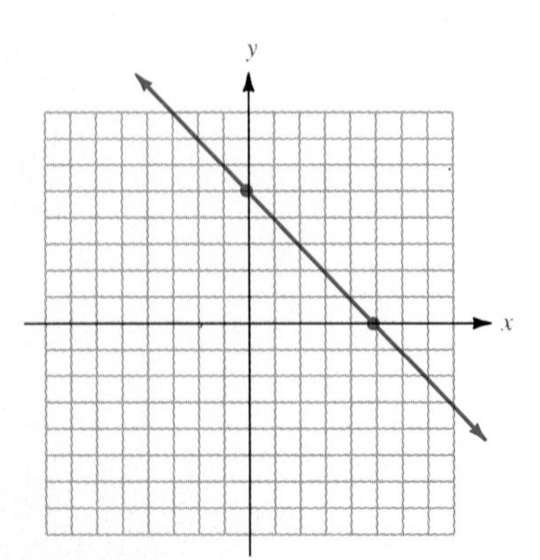

2. $x - y = 6$

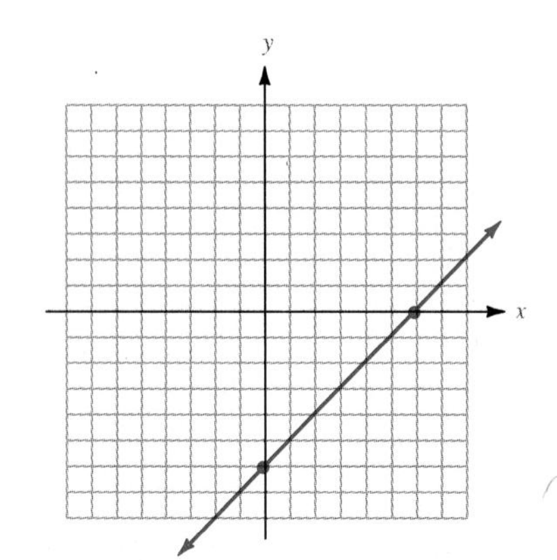

3. $x - y = -3$

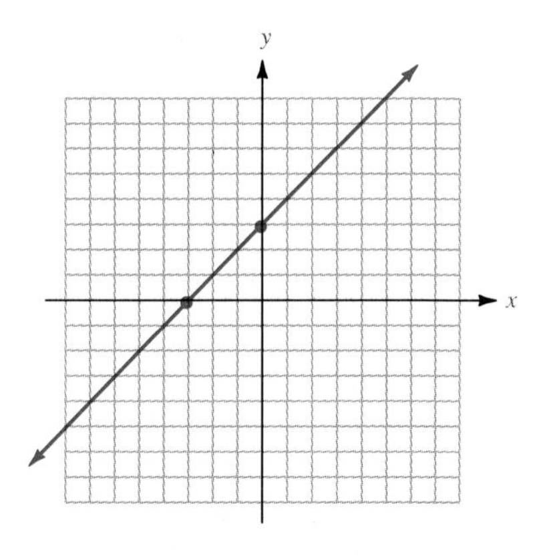

4. $x + y = -2$

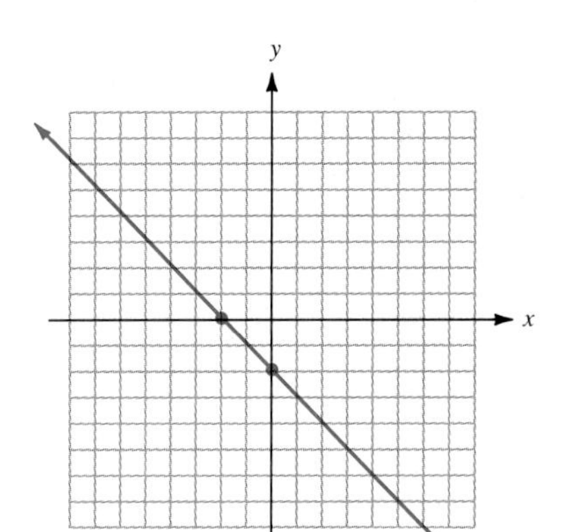

5. $3x + y = 3$

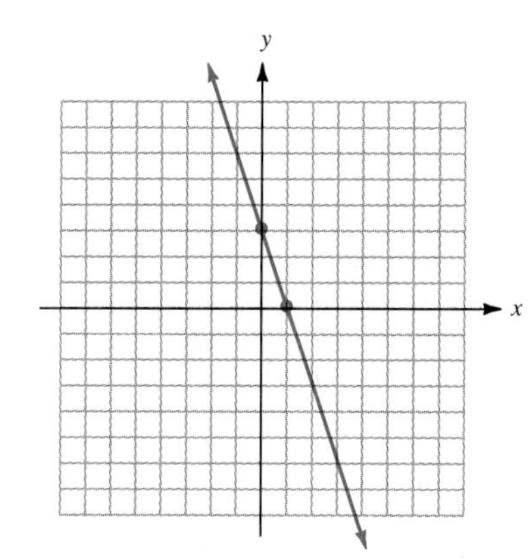

6. $x - 2y = 4$

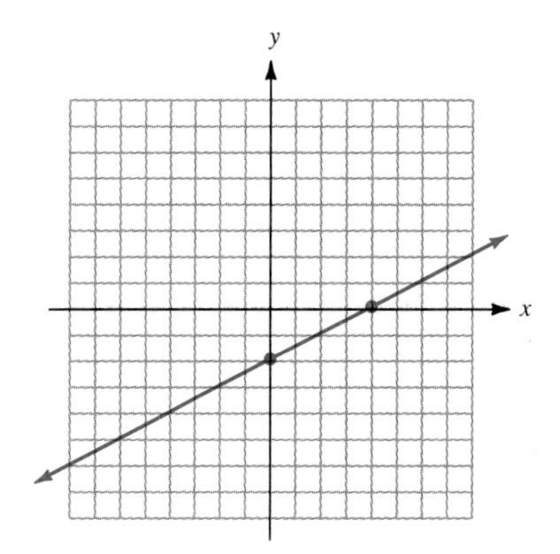

7. $2x + y = 0$

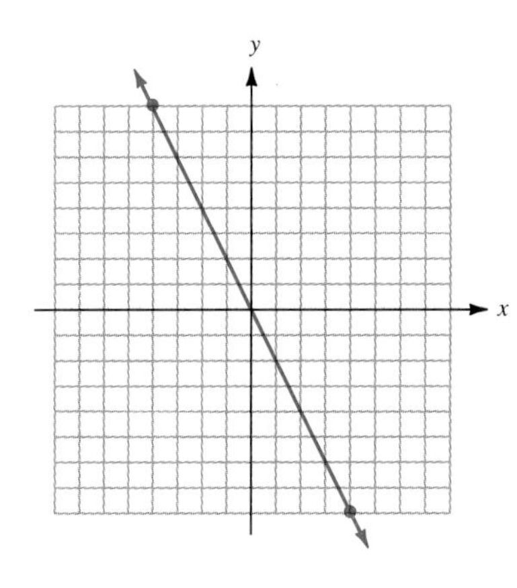

8. $4x - y = 8$

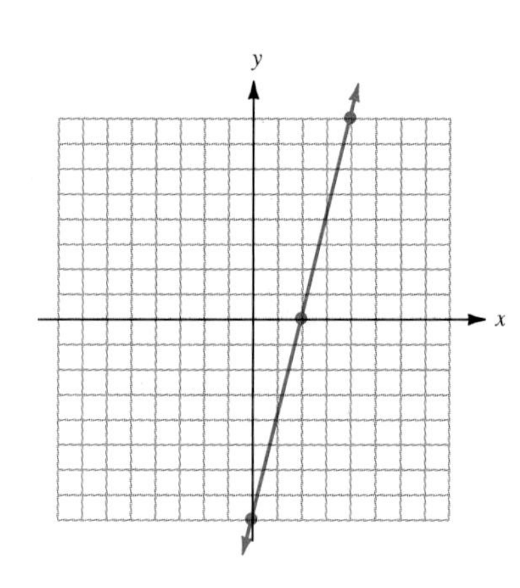

9. $x + 5y = 10$

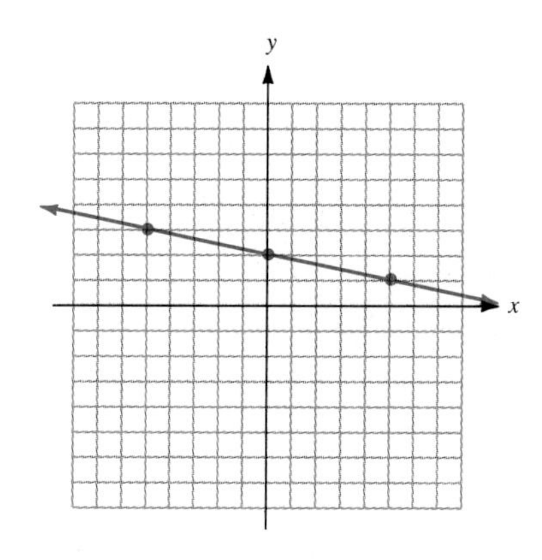

10. $2x - 5y = 10$

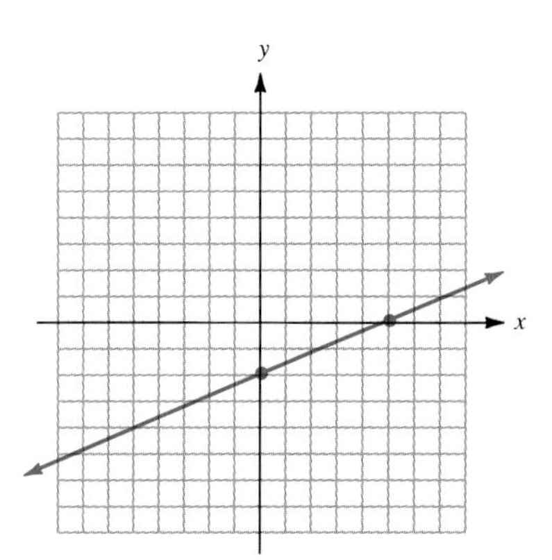

11. $y = 4x$

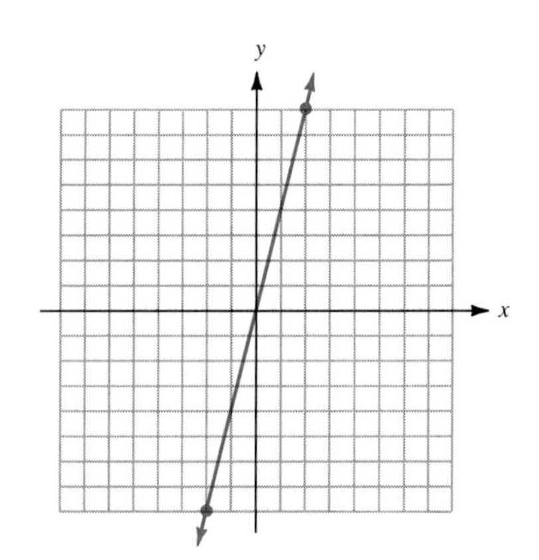

12. $y = -3x$

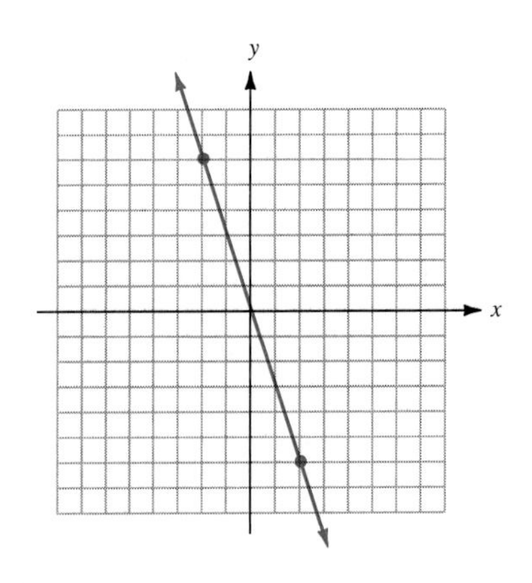

13. $y = 2x + 1$

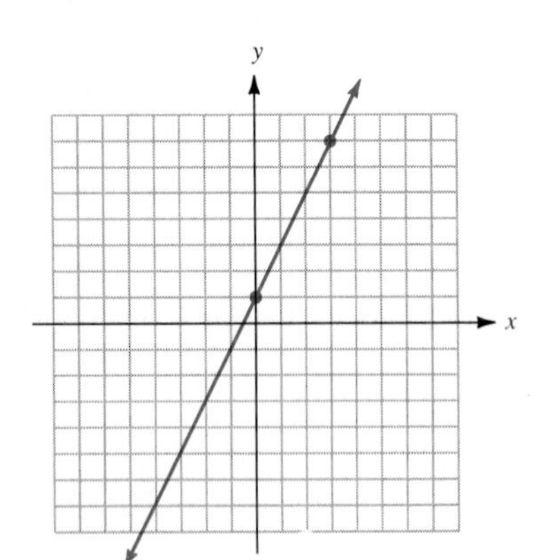

14. $y = 4x - 3$

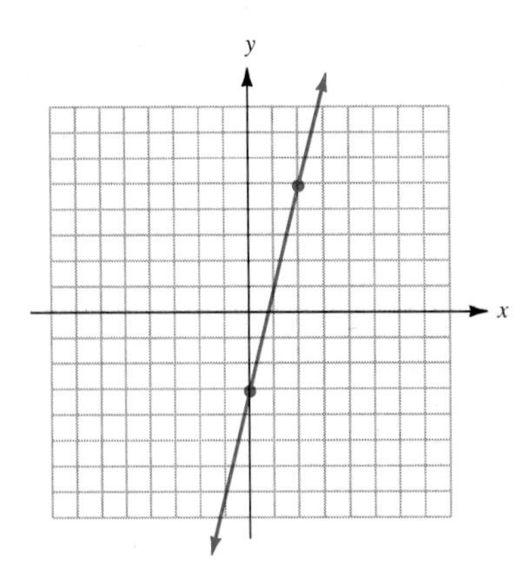

15. $y = -2x + 1$

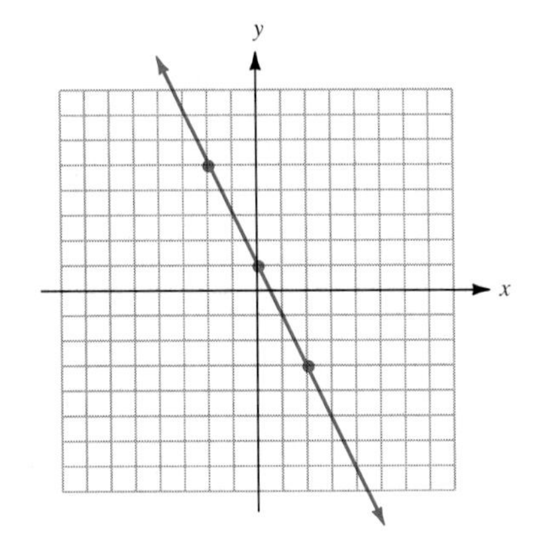

16. $y = -4x - 3$

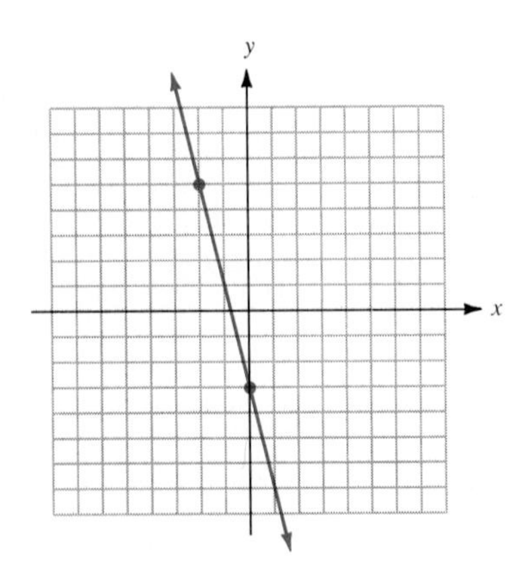

17. $y = \frac{1}{4}x$

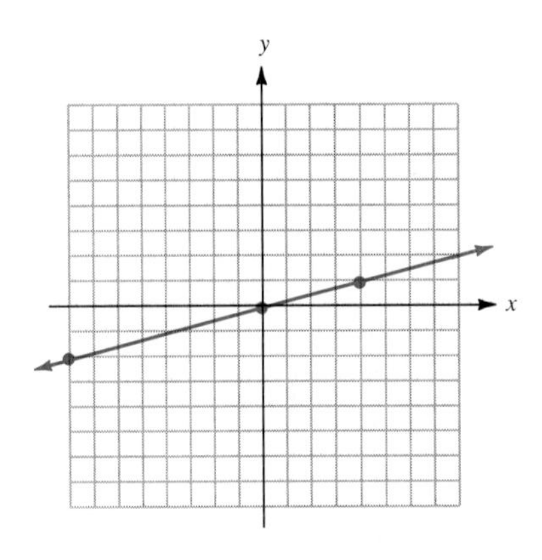

18. $y = -\frac{1}{3}x$

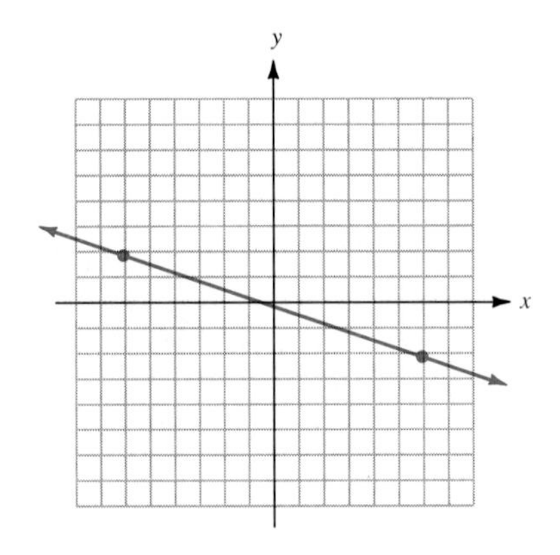

19. $y = \frac{2}{3}x + 3$

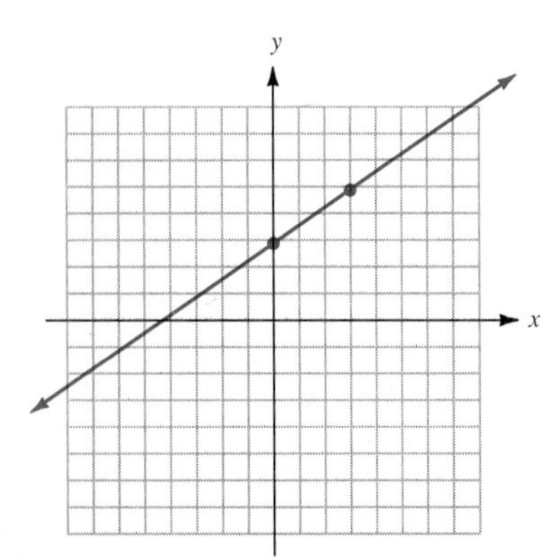

20. $y = \frac{3}{4}x - 2$

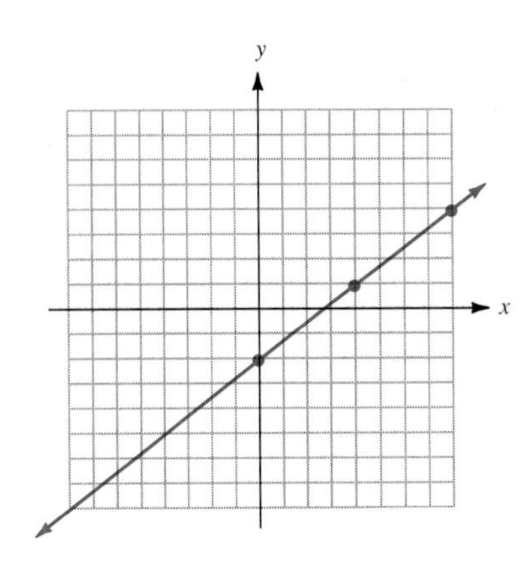

21. $x = 2$

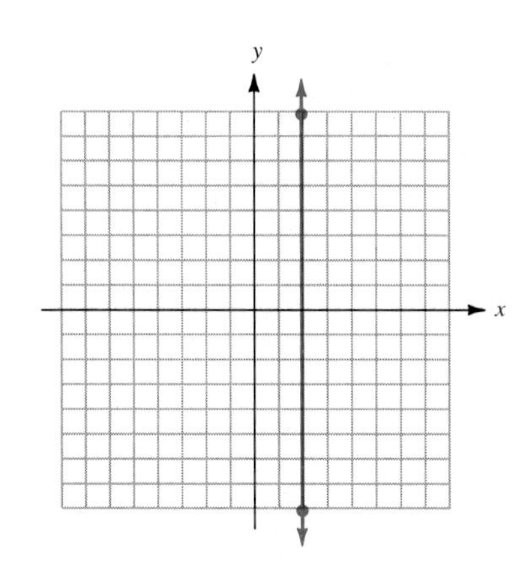

22. $y = -1$

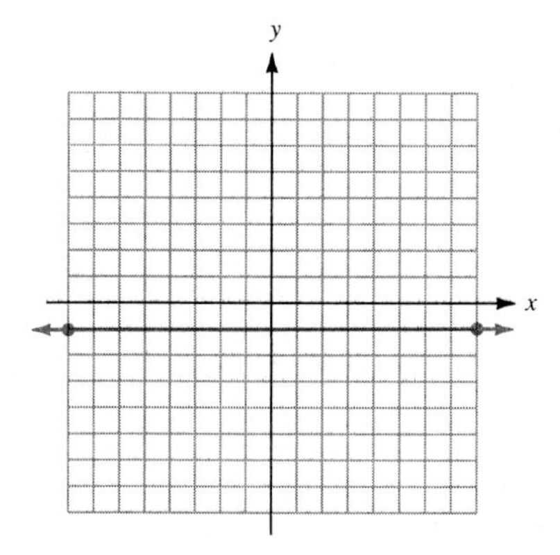

23. $y = 4$

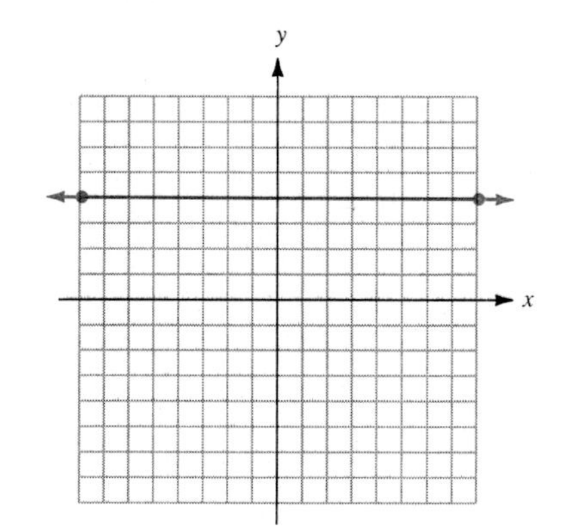

24. $x = -3$

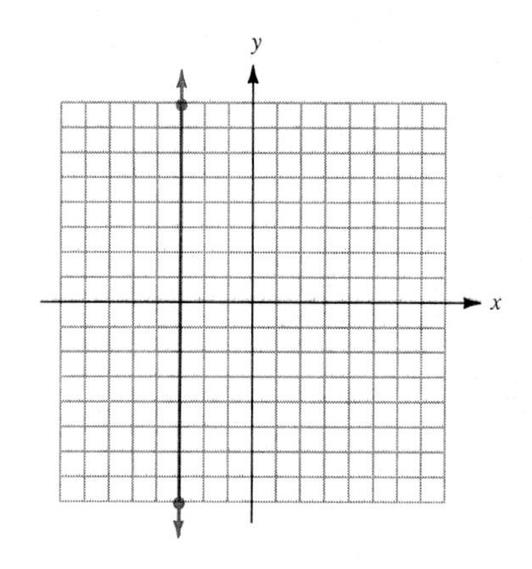

Graph each of the following equations, using the intercept method.

25. $x - 2y = 6$

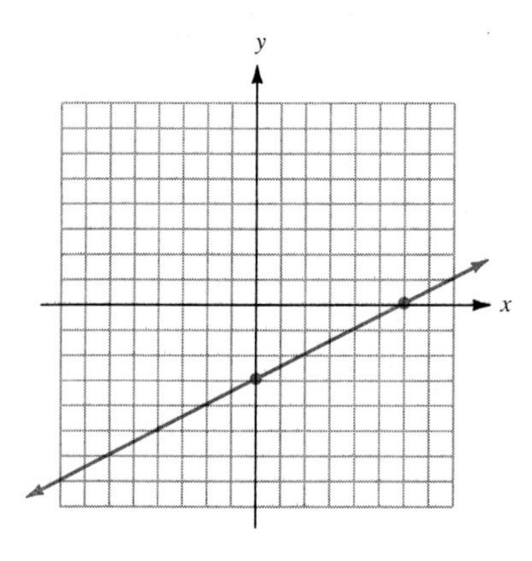

26. $5x + y = 5$

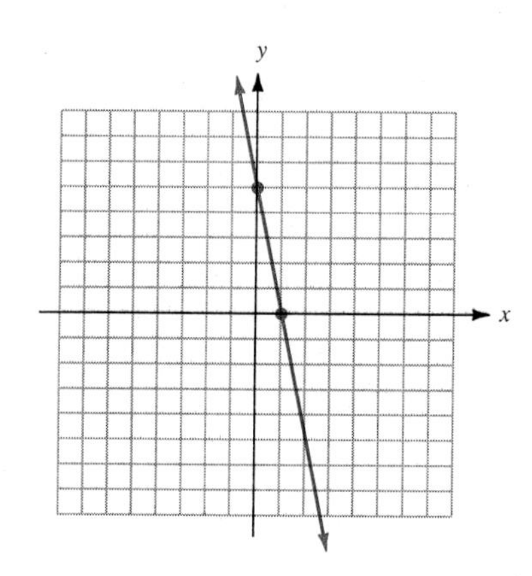

27. $4x + y = 8$

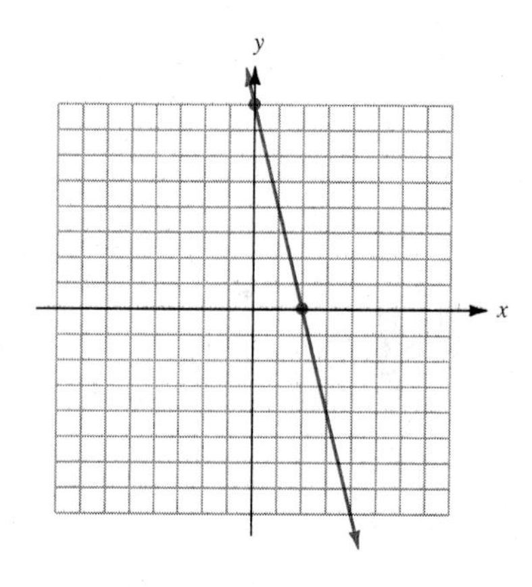

28. $2x + 5y = 10$

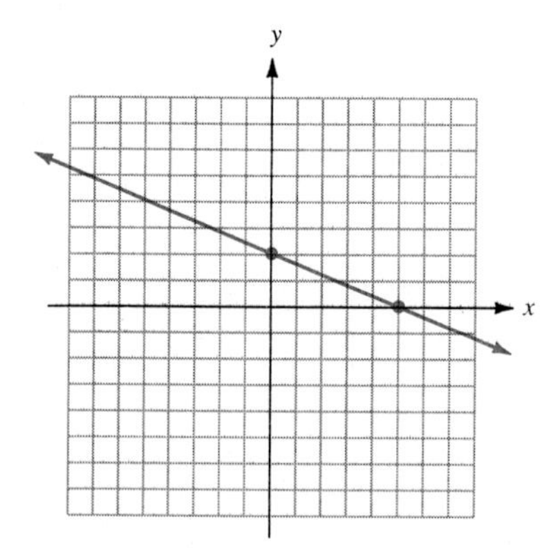

29. $3x + 4y = 12$

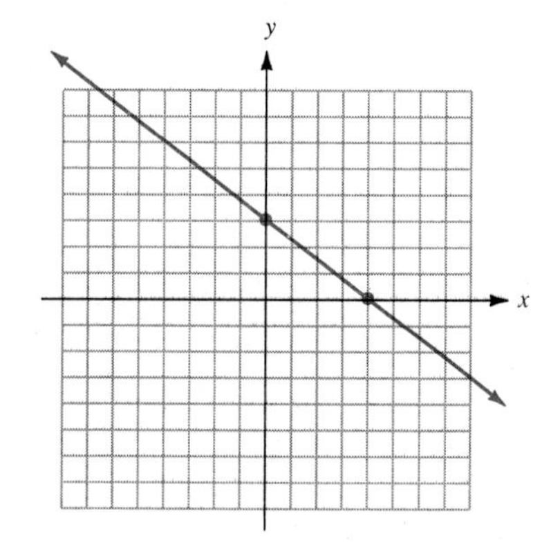

30. $5x - 3y = 15$

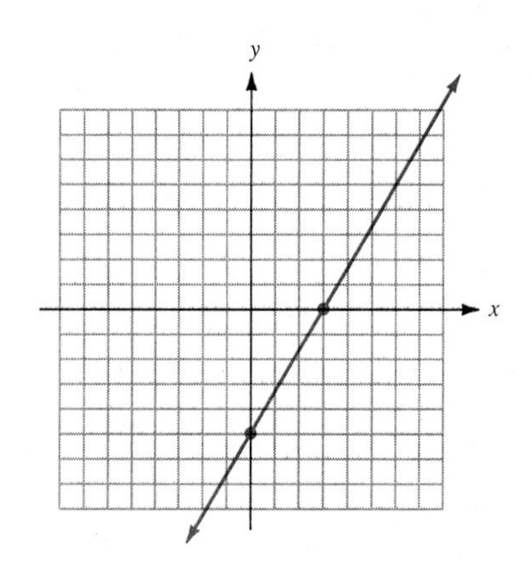

Graph each of the following equations by first solving for y.

31. $x + 2y = 8$ $\quad y = 4 - \frac{x}{2}$

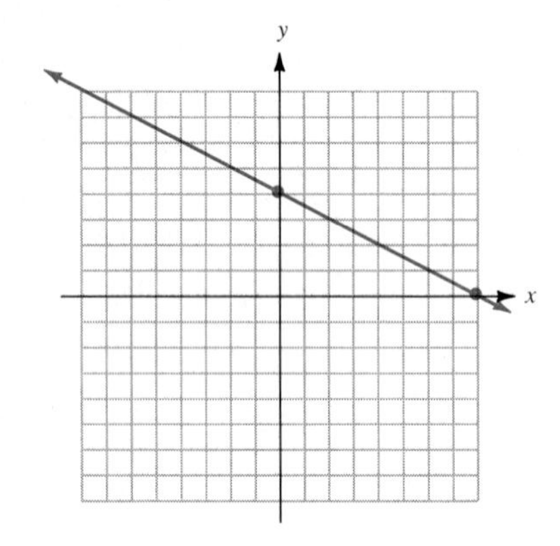

32. $x - 3y = 6$ $\quad y = -2 + \frac{x}{3}$

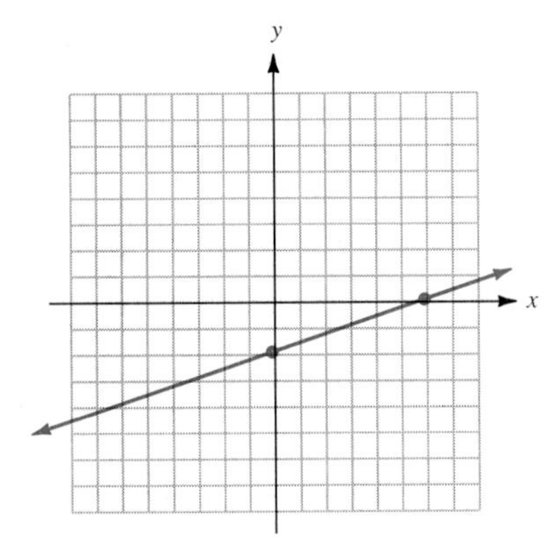

33. $2x + 3y = 12$ $\quad y = 4 - \frac{2}{3}x$

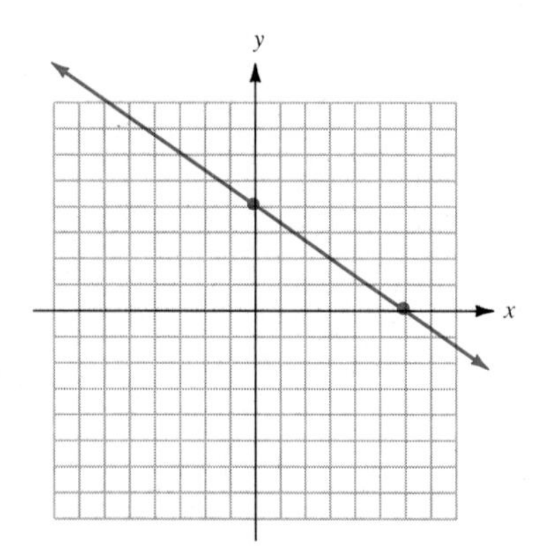

34. $3x - 4y = 12$ $\quad y = -3 + \frac{3}{4}x$

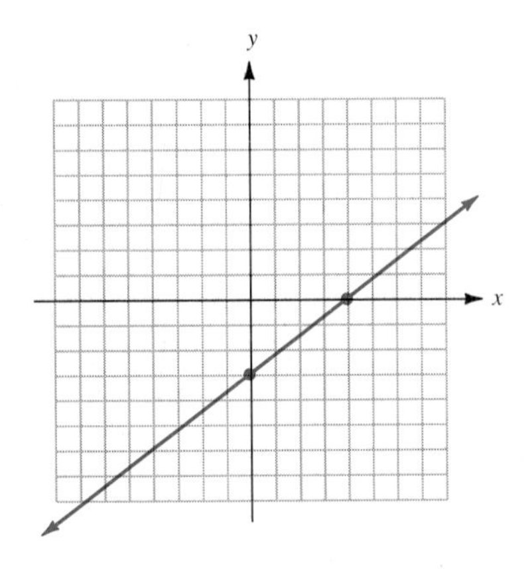

35. $7x - 3y = 21$ $\quad y = -7 + \frac{7}{3}x$

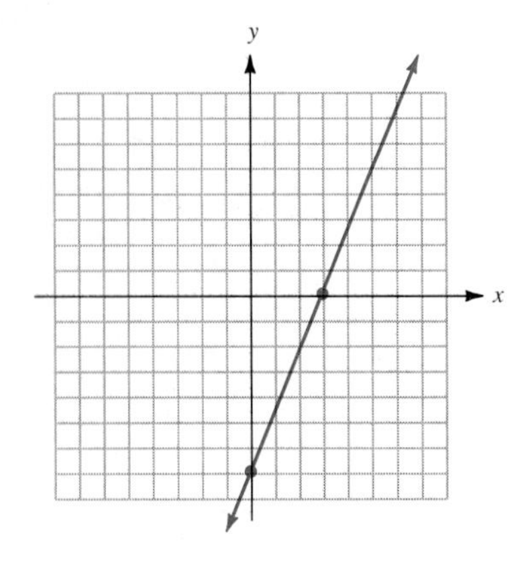

36. $5x + 4y = 20$ $\quad y = 5 - \frac{5}{4}x$

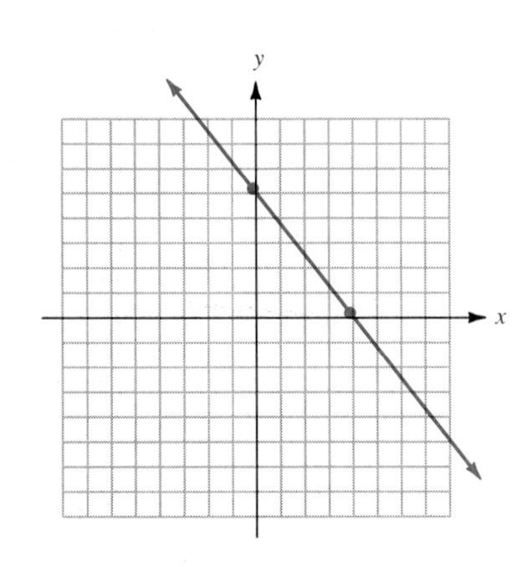

Going Beyond

Write an equation that describes the following relationships between x and y. Then graph each relationship.

37. y is 3 times x. $y = 3x$

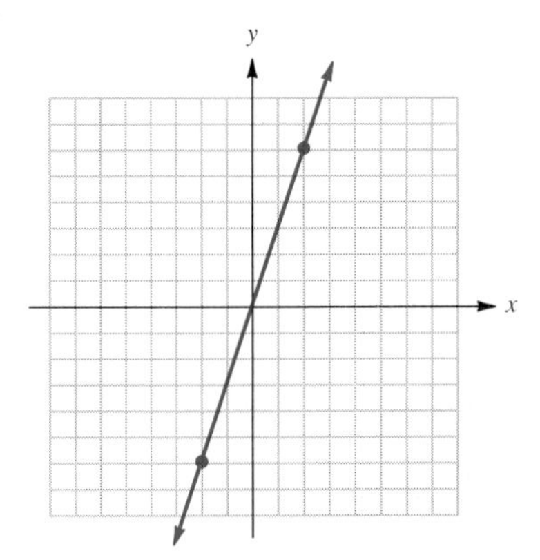

38. y is twice x. $y = 2x$

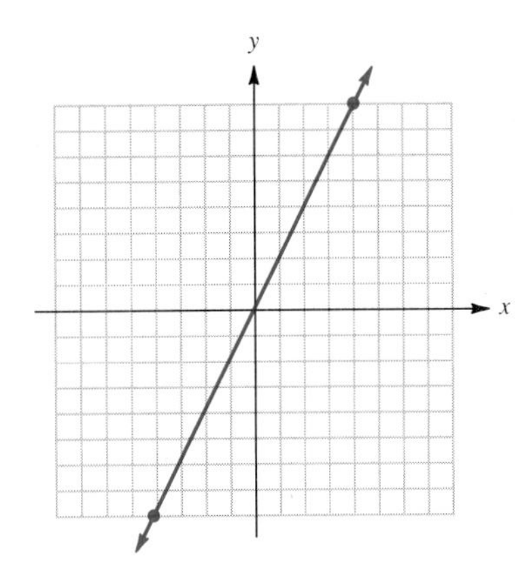

39. y is 2 more than x. $y = x + 2$

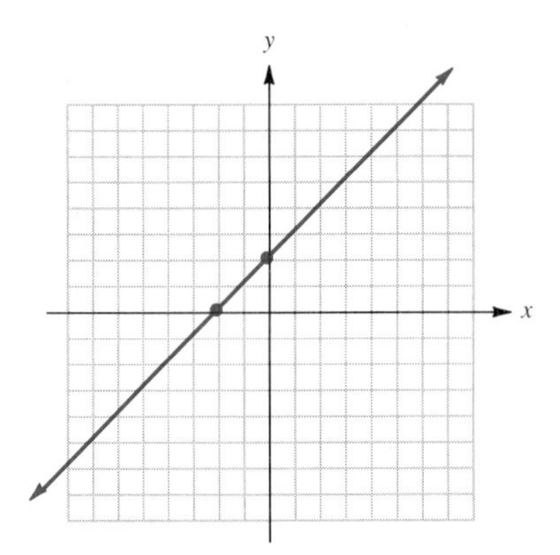

40. y is 3 less than x. $y = x - 3$

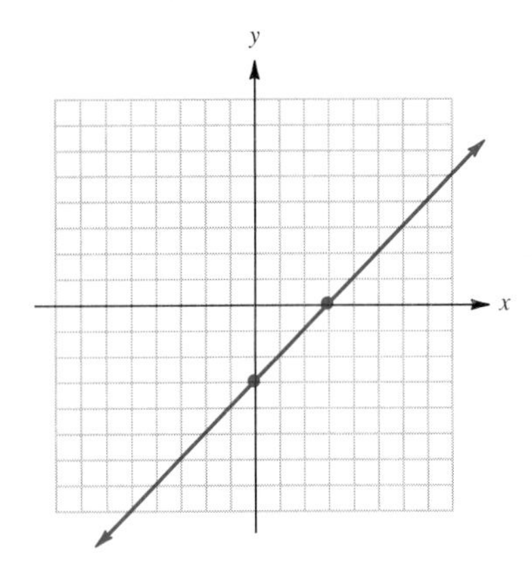

41. y is 2 less than 3 times x.
$y = 3x - 2$

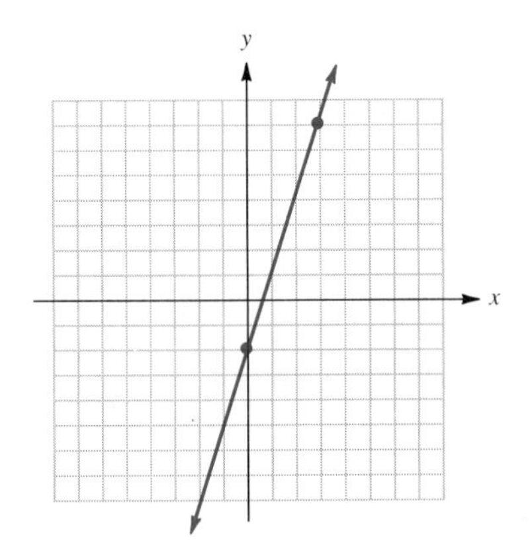

42. y is 3 more than twice x.
$y = 2x + 3$

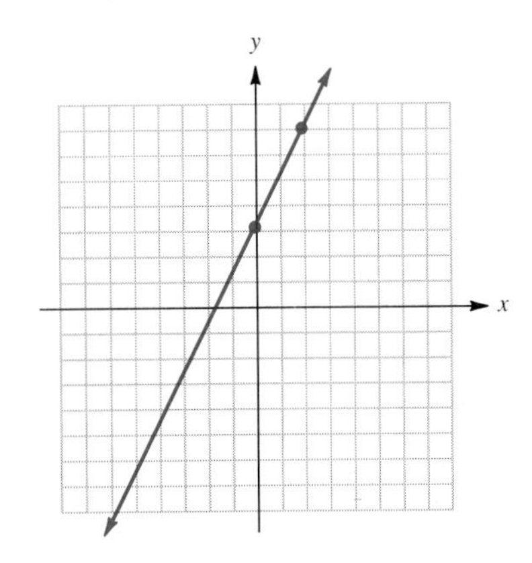

43. The sum of x and y is 6.
$x + y = 6$

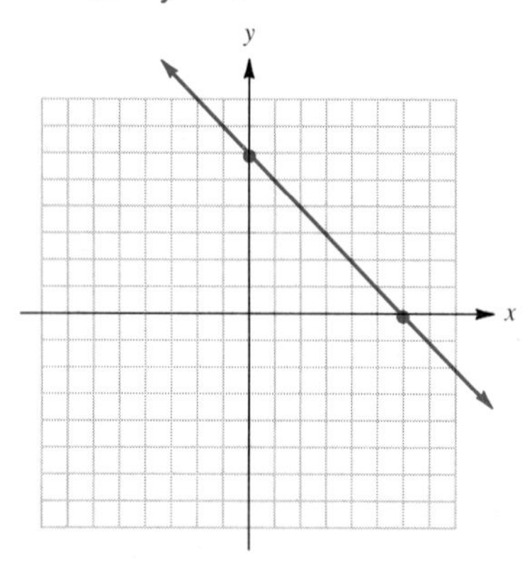

44. The sum of twice x and y is 8.
$2x + y = 8$

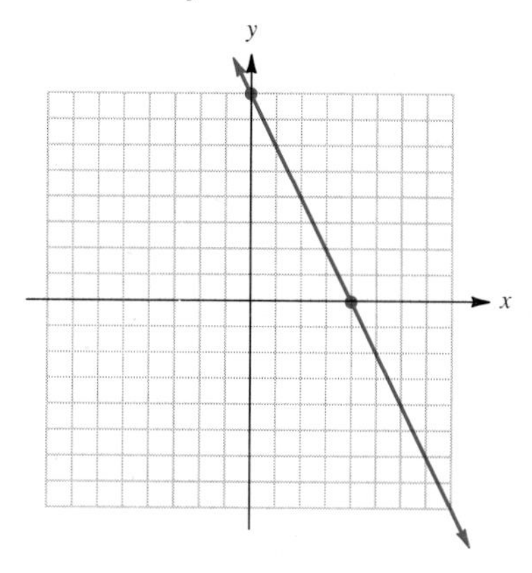

45. The difference of x and the product of 4 and y is 8.
$x - 4y = 8$

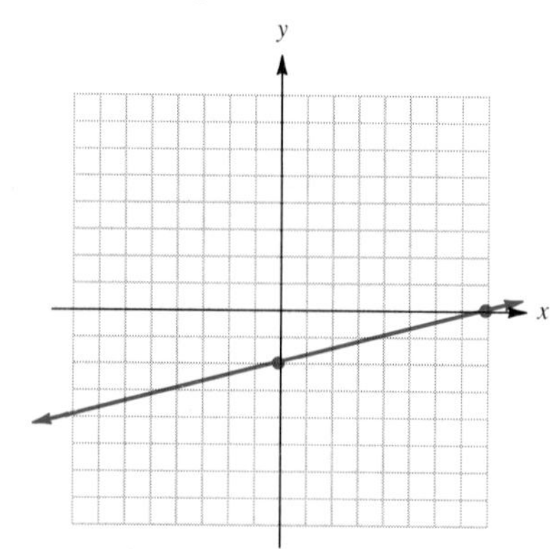

46. The difference of twice x and y is 4.
$2x - y = 4$

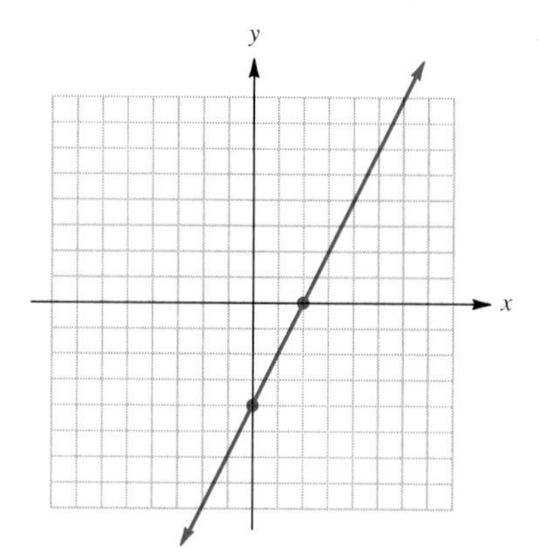

Graph each pair of equations on the same axes. Give the coordinates of the point where the lines intersect.

47. $x + y = 6$
$x - y = 2$ (4, 2)

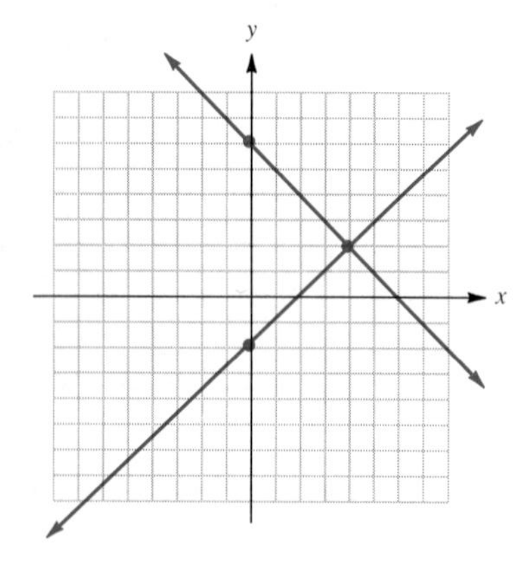

48. $x - y = 5$
$x + y = 7$ (6, 1)

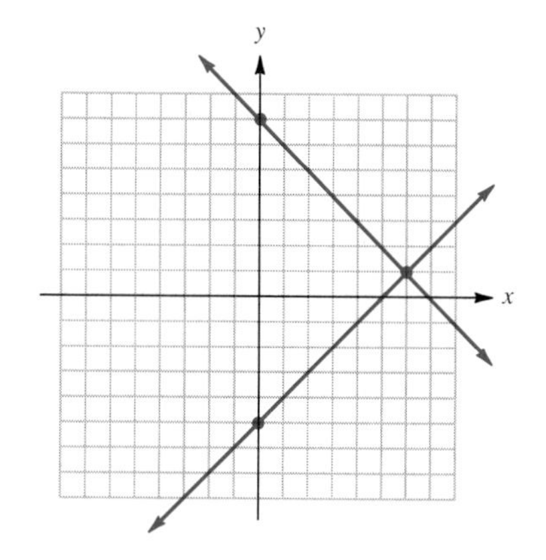

Graph each set of equations on the same coordinate system. Do the lines intersect? What are the y intercepts?

49. $y = 2x$
$y = 2x + 3$
$y = 2x - 4$
The lines do not intersect. The y intercepts are 0, 3, and −4.

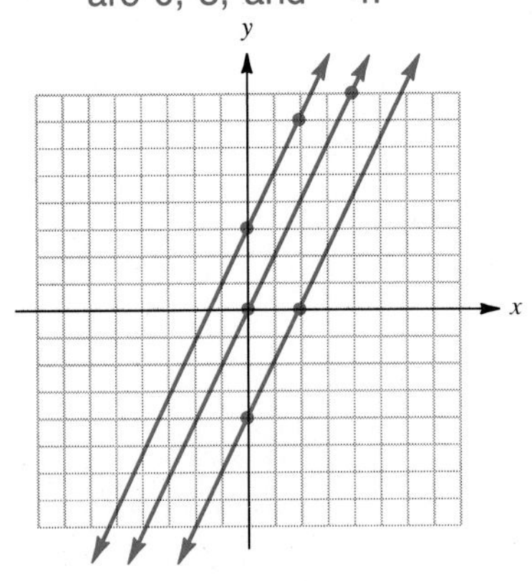

50. $y = -3x$
$y = -3x + 2$
$y = -3x - 3$
The lines do not intersect. The y intercepts are 0, 2, −3.

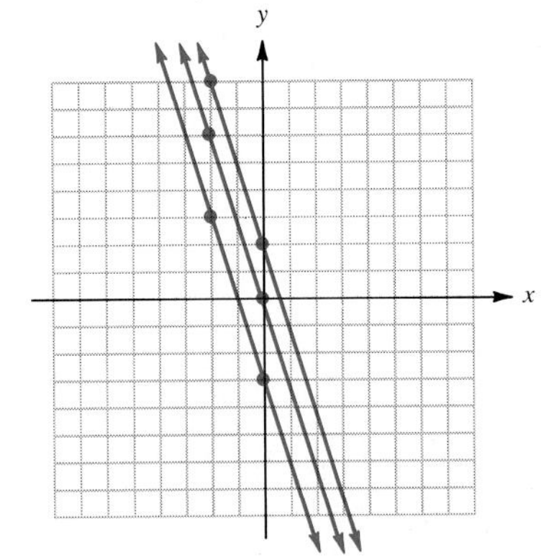

Skillscan (Section 2.5)

Evaluate the following expressions.

a. $\dfrac{6-3}{7-4}$ 1

b. $\dfrac{8-2}{6-3}$ 2

c. $\dfrac{-8-4}{-3-3}$ 2

d. $\dfrac{-14-2}{2-6}$ 4

e. $\dfrac{5-(-2)}{4-2}$ $\dfrac{7}{2}$

f. $\dfrac{-3-(-4)}{2-(-2)}$ $\dfrac{1}{4}$

g. $\dfrac{-5-(-5)}{7-2}$ 0

h. $\dfrac{-8+(-8)}{-4+4}$ Undefined

Answers

1. $x + y = 5$

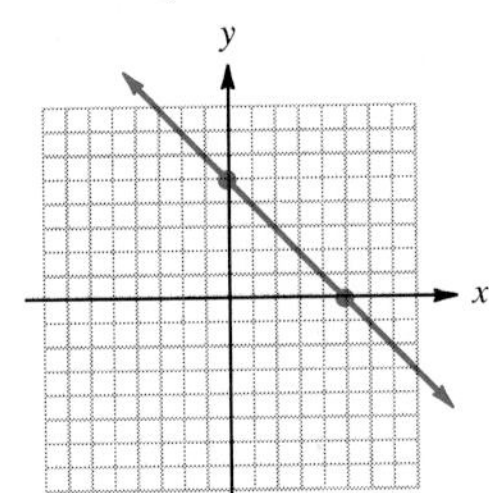

3. $x - y = -3$

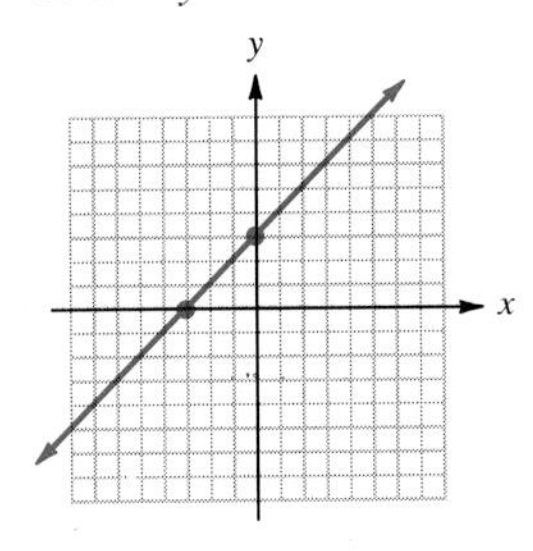

5. $3x + y = 3$

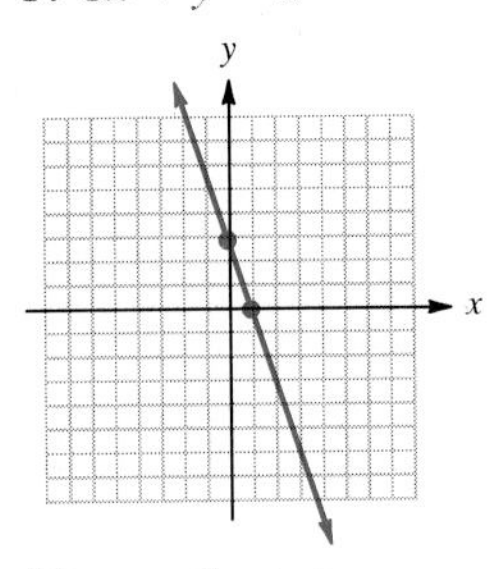

7. $2x + y = 0$

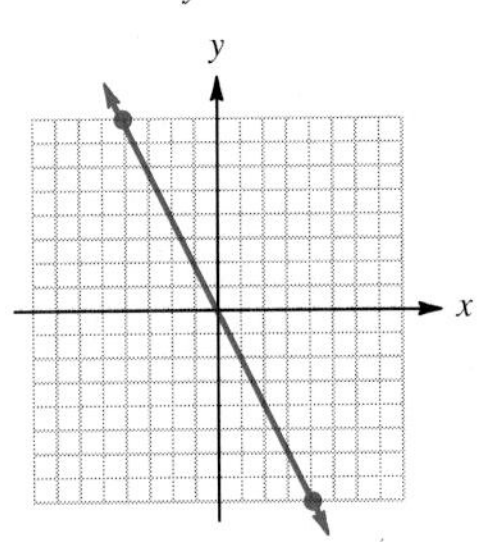

9. $x + 5y = 10$

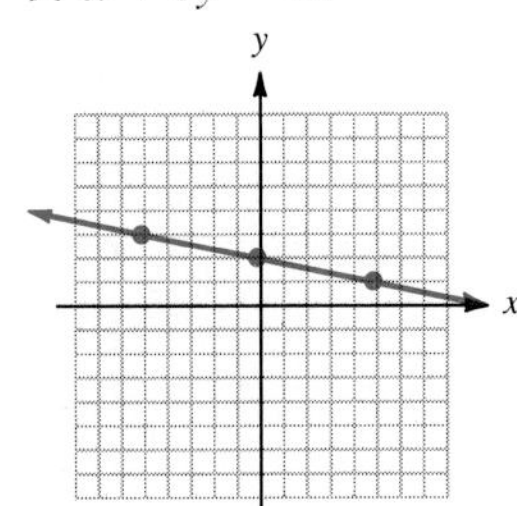

11. $y = 4x$

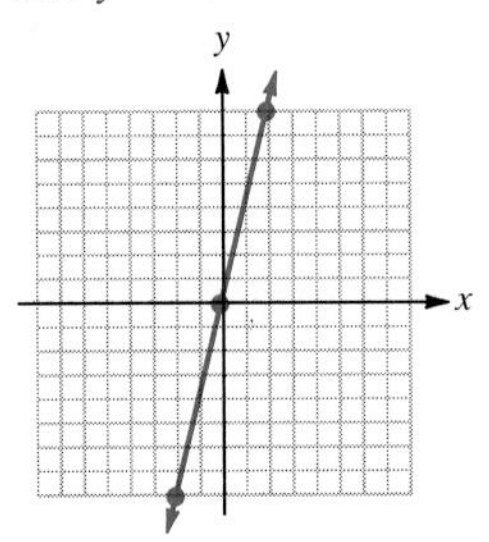

13. $y = 2x + 1$

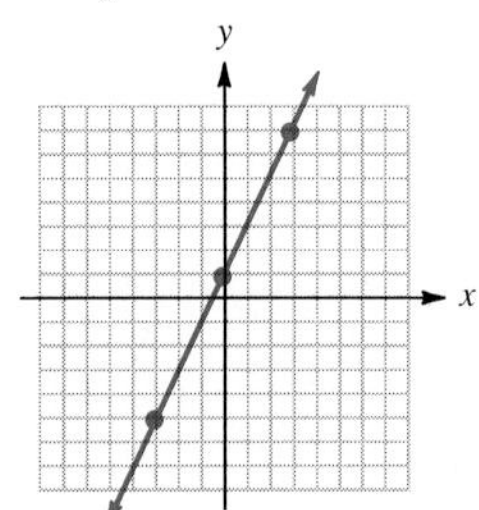

15. $y = -2x + 1$

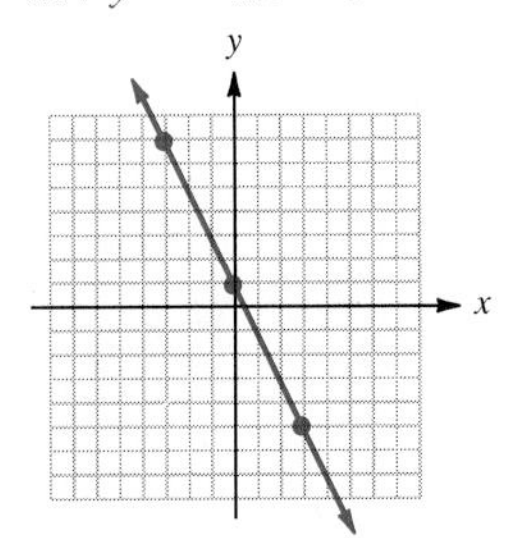

17. $y = \frac{1}{4}x$

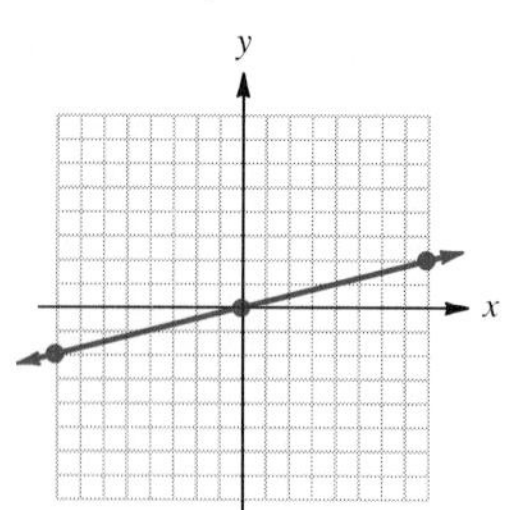

19. $y = \frac{2}{3}x + 3$

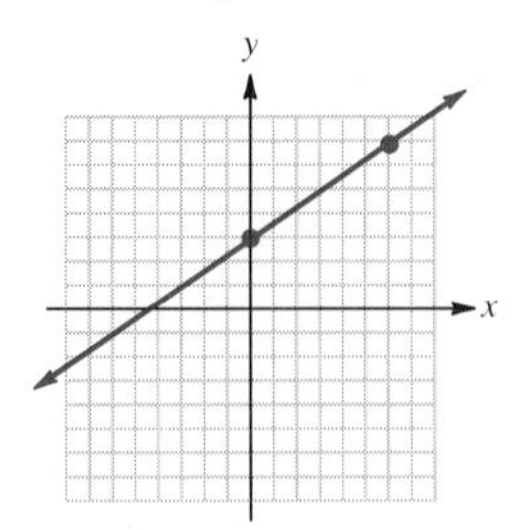

21. $x = 2$

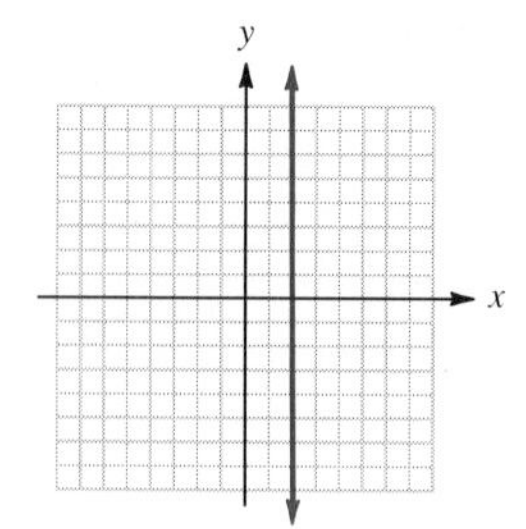

23. $y = 4$

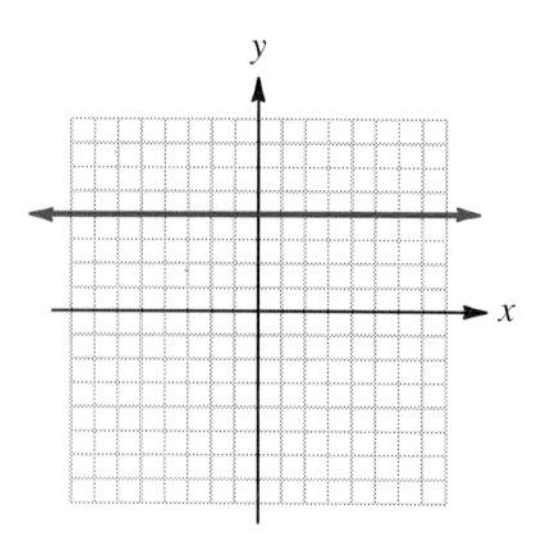

25. $x - 2y = 6$

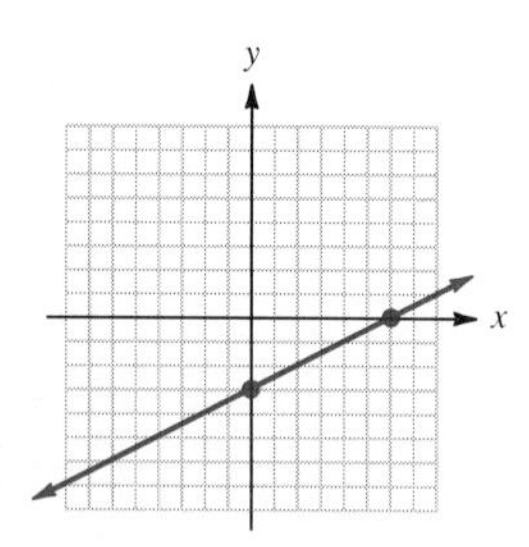

27. $4x + y = 8$

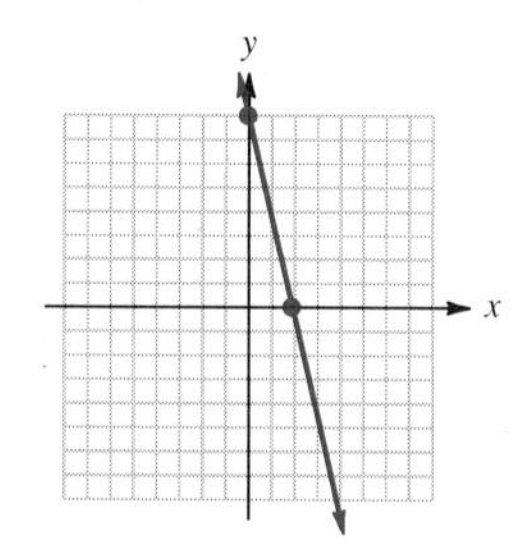

29. $3x + 4y = 12$

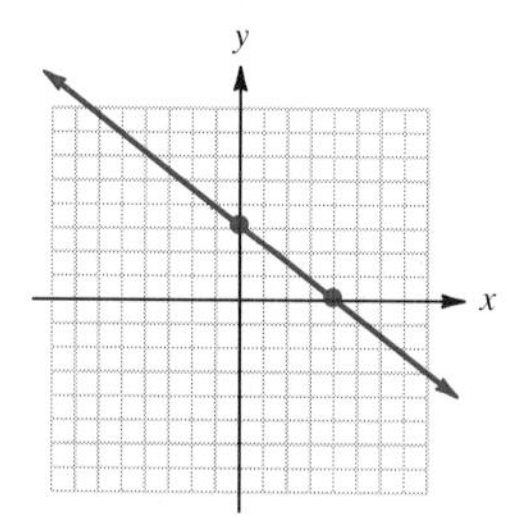

31. $y = 4 - \frac{x}{2}$

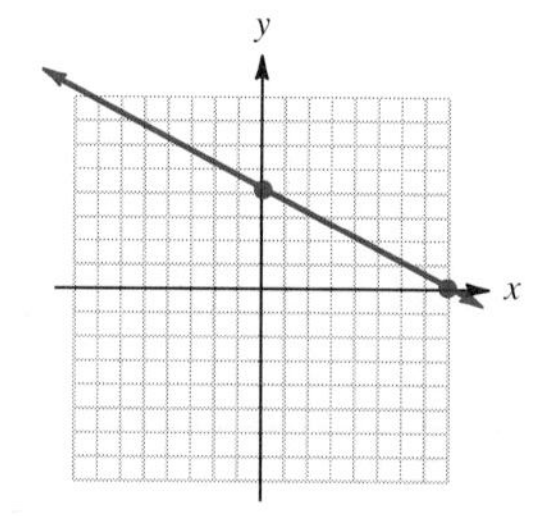

33. $y = 4 - \frac{2}{3}x$

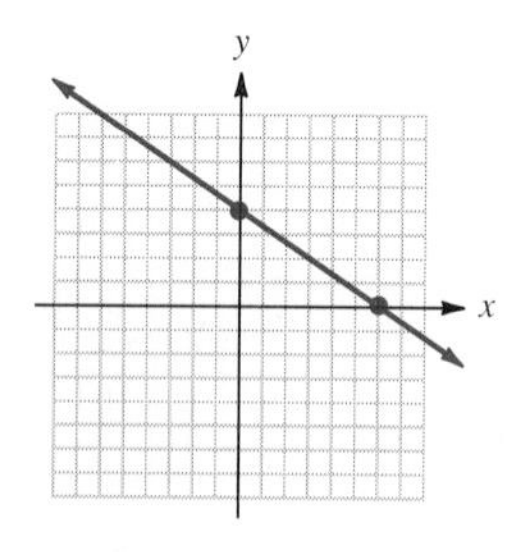

35. $y = -7 + \frac{7}{3}x$

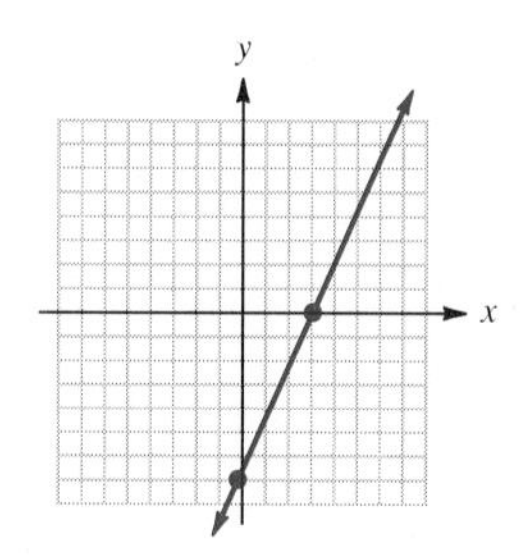

37. $y = 3x$

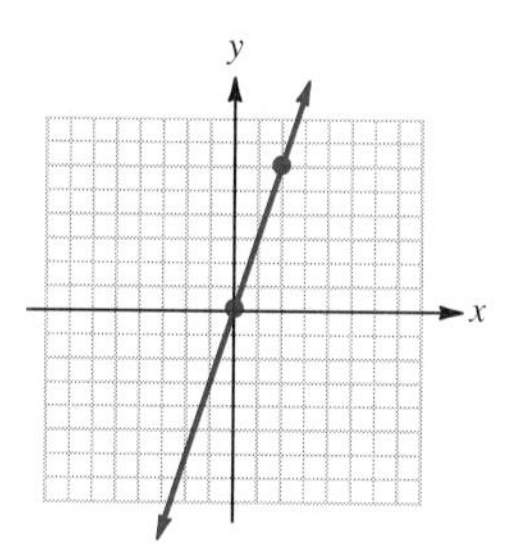

39. $y = x + 2$

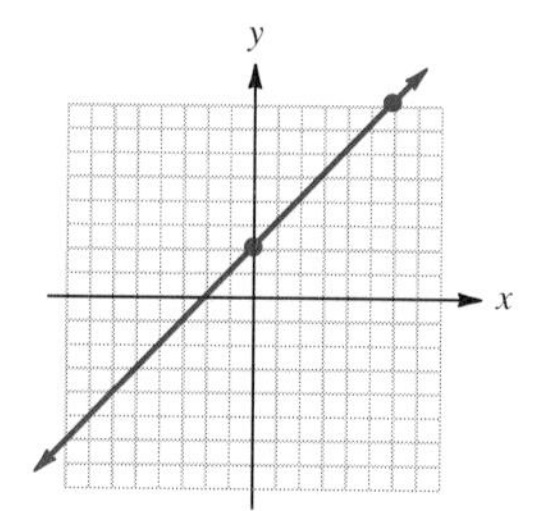

41. $y = 3x - 2$

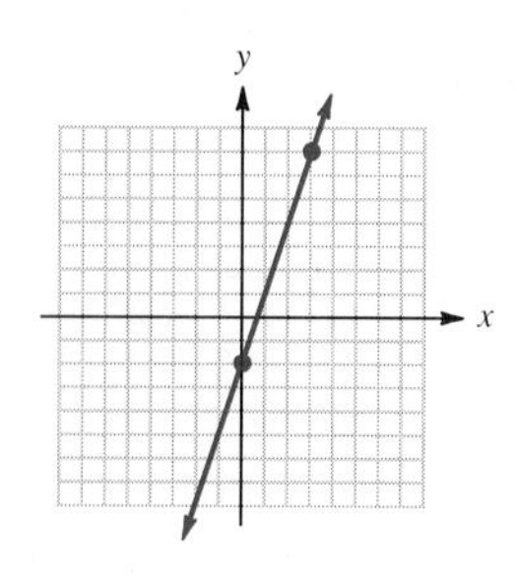

43. $x + y = 6$

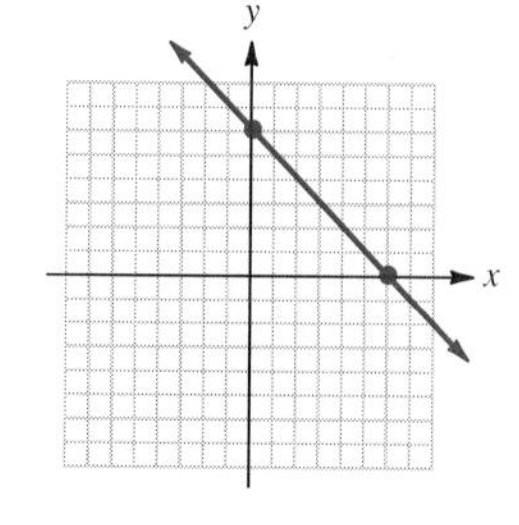

45. $x - 4y = 8$

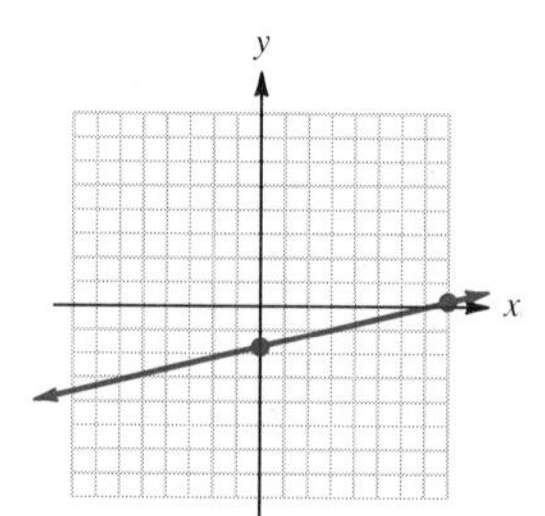

47. $\left.\begin{array}{r} x + y = 6 \\ x - y = 2 \end{array}\right\} (4, 2)$

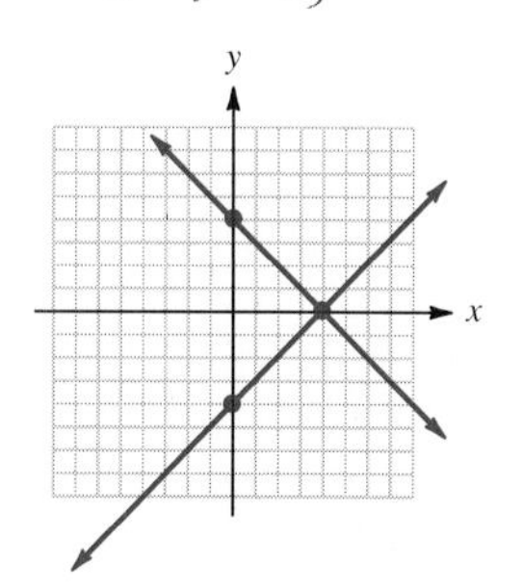

49.

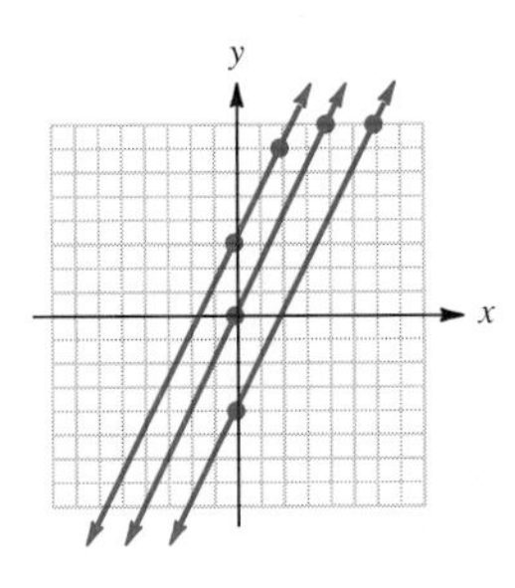

The lines do not intersect. The y intercepts are 0, 3, and −4.

a. 1 **b.** 2 **c.** 2 **d.** 4 **e.** $\frac{7}{2}$ **f.** $\frac{1}{4}$ **g.** 0 **h.** Undefined

7.4 The Slope of a Line

OBJECTIVES

1. To find the slope of a line through two given points
2. To use the slope-intercept form for a line.

We know from our work in Section 7.3 that the graph of an equation like

$$y = 2x + 3$$

is a straight line. In this section we want to develop an important idea related to the equation of a line and its graph, called the *slope* of a line. Finding the slope of a line gives us a numerical measure of the "steepness" or inclination of that line. It will also allow us to write the equation of a line, given certain properties of that line.

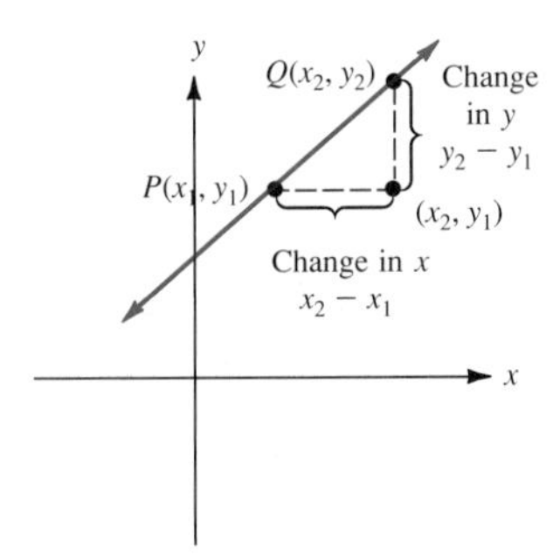

Recall that an equation like $y = 2x + 3$ is a *linear equation in two variables*. Its graph is always a straight line.

Note: x_1 is read "x sub one," x_2 is read "x sub two," and so on. The 1 in x_1 and the 2 in x_2 are called *subscripts*.

To find the slope of a line, we first let $P(x_1, y_1)$ and $Q(x_2, y_2)$ be any two distinct points on that line. The *horizontal change* (or the change in x) between the points is $x_2 - x_1$. The *vertical change* (or the change in y) between the points is $y_2 - y_1$.

We call the ratio of the vertical change, $y_2 - y_1$, to the horizontal change, $x_2 - x_1$, the *slope* of the line as we move along the line from P to Q. That ratio is usually denoted by the letter m, and so we have the following formula:

The difference $x_2 - x_1$ is sometimes called the *run* between points P and Q. The difference $y_2 - y_1$ is called the *rise*. So the slope may be thought of as "rise over run."

THE SLOPE OF A LINE

If $P(x_1, y_1)$ and $Q(x_2, y_2)$ are any two points on a line; then m, the slope of the line, is given by

$$m = \frac{\text{vertical change}}{\text{horizontal change}} = \frac{y_2 - y_1}{x_2 - x_1} \quad \text{where } x_2 \neq x_1$$

As will be clear from our following examples, this definition provides exactly the numerical measure of "steepness" that we want. If a line "rises" as we move from left to right, the slope will be positive—the steeper the line, the larger the numerical value of the slope. If the line "falls" from left to right, the slope will be negative.

Let's proceed to some examples.

Example 1

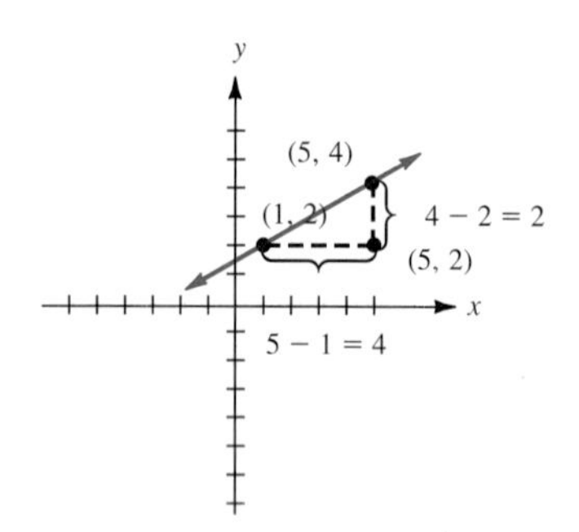

Find the slope of the line containing points with coordinates (1, 2) and (5, 4).

Let $P(x_1, y_1) = (1, 2)$ and $Q(x_2, y_2) = (5, 4)$. By the definition above, we have

$$m = \frac{y_2 - y_1}{x_2 - x_1} = \frac{4 - 2}{5 - 1} = \frac{2}{4} = \frac{1}{2}$$

Note: We would have found the same slope if we had reversed P and Q and subtracted in the other order. In that case,

$$m = \frac{2 - 4}{1 - 5} = \frac{-2}{-4} = \frac{1}{2}$$

It makes no difference which point is labeled (x_1, y_1) and which is (x_2, y_2); the resulting slope will be the same. You must simply stay with your choice once it is made and *not* reverse the order of the subtraction in your calculations.

CHECK YOURSELF 1

Find the slope of the line containing points with coordinates (2, 3) and (5, 5).

Example 2

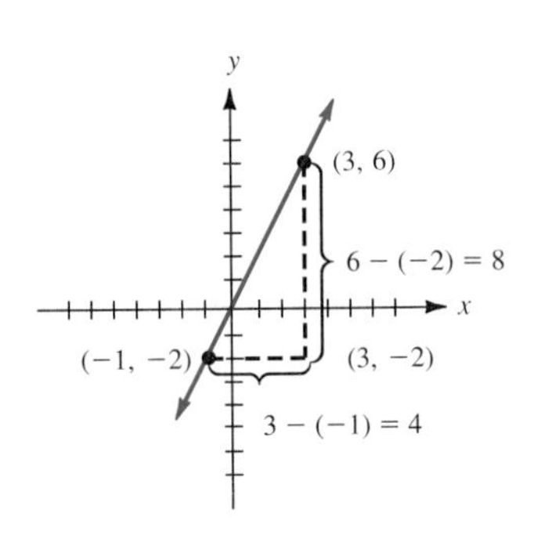

Find the slope of the line containing points with the coordinates (−1, −2) and (3, 6).

Again, applying the definition, we have

$$m = \frac{6 - (-2)}{3 - (-1)} = \frac{6 + 2}{3 + 1} = \frac{8}{4} = 2$$

The figure below compares the slopes found in the two previous examples. Line l_1, from Example 1, had slope $\frac{1}{2}$. Line l_2, from Example 2, had slope 2. Do you see the idea of slope measuring steepness? The greater the slope, the more steeply the line is inclined upward.

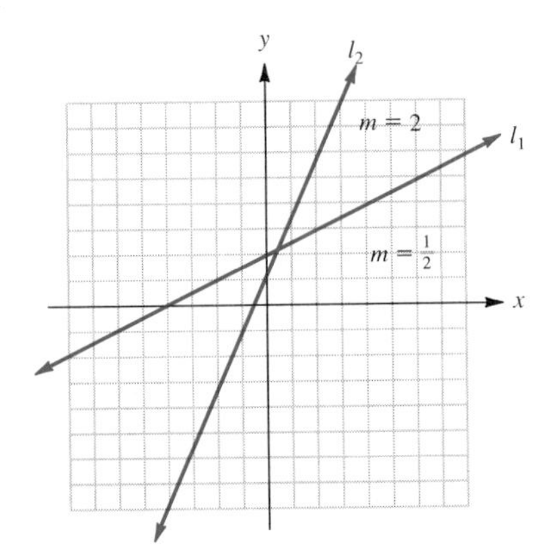

CHECK YOURSELF 2

Find the slope of the line containing points with coordinates (−1, 2) and (2, 7). Draw a sketch of this line and the line of Check Yourself 1. Compare the lines and the two slopes.

Let's look at another fact about slope in our next example.

Example 3

Find the slope of the line containing points with coordinates (−2, 3) and (1, −3).

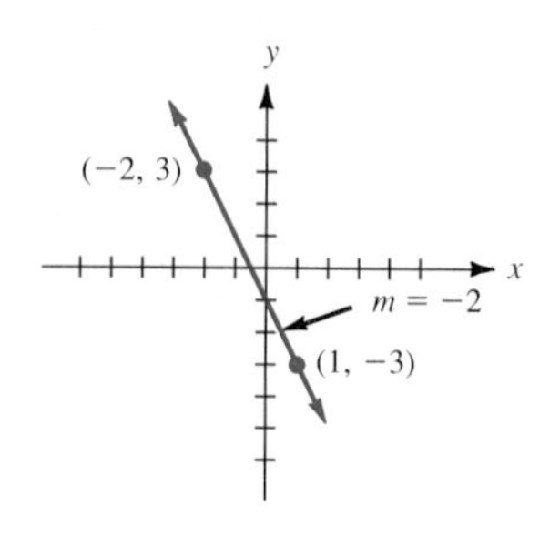

By the definition,

$$m = \frac{-3 - 3}{1 - (-2)} = \frac{-6}{3} = -2$$

Do you see the difference between this and our previous examples? In Examples 1 and 2, the lines were rising from left to right and the slope was positive. Here the line has a *negative* slope. The line *falls* as we move from left to right.

CHECK YOURSELF 3

Find the slope of the line containing points with coordinates (−1, 3) and (1, −3).

Example 4

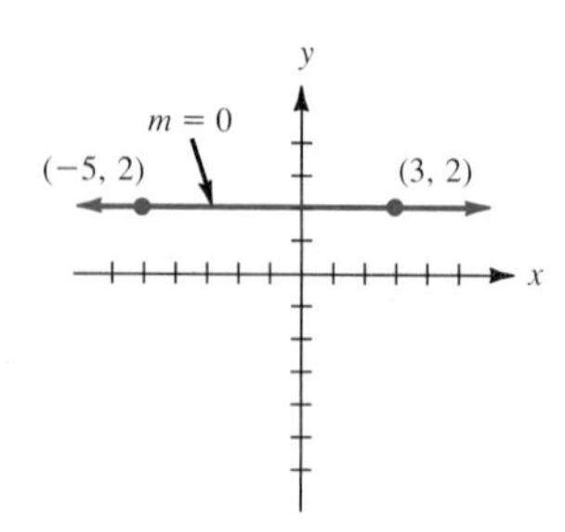

Find the slope of the line containing points with coordinates (−5, 2) and (3, 2).

By the definition,

$$m = \frac{2 - 2}{3 - (-5)} = \frac{0}{8} = 0$$

The slope of the line is 0. In fact, that will be the case for any horizontal line. Since any two points on the line have the same y coordinate, the vertical change $y_2 - y_1$ must always be 0, and so the resulting slope is 0.

CHECK YOURSELF 4

Find the slope of the line containing points with coordinates (−2, −4) and (3, −4).

Example 5

Find the slope of the line containing points with coordinates (2, −5) and (2, 5).

By the definition,

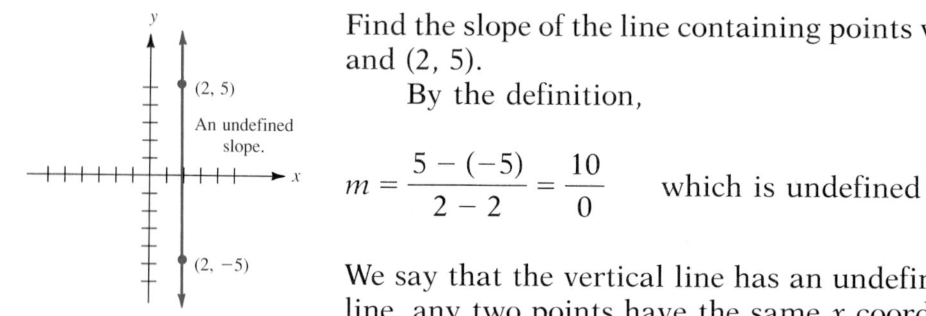

$$m = \frac{5 - (-5)}{2 - 2} = \frac{10}{0} \qquad \text{which is undefined}$$

We say that the vertical line has an undefined slope. On a vertical line, any two points have the same x coordinate. This means that the horizontal change $x_2 - x_1$ must always be 0 and since division by 0 is undefined, m will be undefined.

CHECK YOURSELF 5

Find the slope of the line containing points with the coordinates $(-3, -5)$ and $(-3, 2)$.

The following sketch will summarize the results of the previous examples.

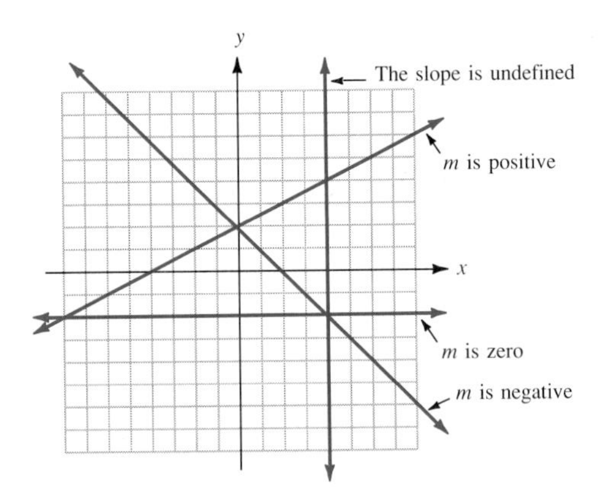

Four lines are illustrated in the figure. Note that

1. The slope of a line that rises from left to right is positive.
2. The slope of a line that falls from left to right is negative.
3. The slope of a horizontal line is 0.
4. A vertical line has an undefined slope.

We now want to consider finding the equation of a line when its slope and y intercept are known. Suppose that the y intercept of a line is b. Then the point at which the line crosses the y axis must have coordinates $(0, b)$. Look at the sketch below.

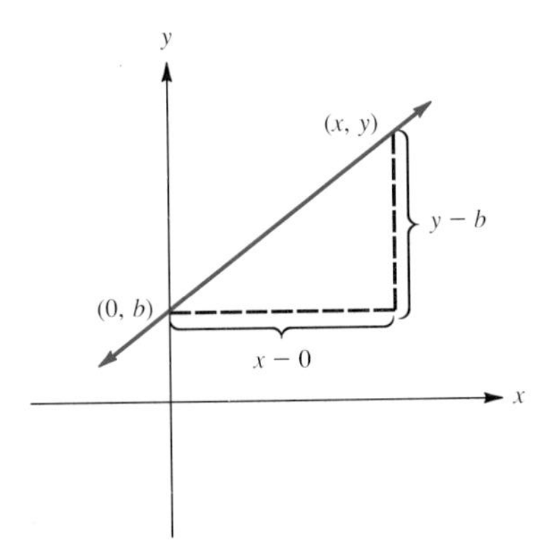

Now, using any other point (x, y) on the line and using our definition of slope, we can write

Change in y

$$m = \frac{y - b}{x - 0} \qquad (1)$$

Change in x

or

$$m = \frac{y - b}{x} \tag{2}$$

Multiplying both sides of Equation (2) by x, we have

$$mx = y - b \tag{3}$$

Finally, adding b to both sides of Equation (3) gives

$$mx + b = y$$

or

$$y = mx + b \tag{4}$$

We can summarize the above discussion as follows:

THE SLOPE-INTERCEPT FORM FOR A LINE

An equation of the line with slope m and y intercept b is

$$y = mx + b$$

Note: In this form, the equation is *solved for y*. The coefficient of x will give you the slope of the line, and the constant term gives the y intercept. This is applied in the following example.

Example 6

(*a*) Find the slope and y intercept for the graph of the equation

$$y = \underset{\substack{\uparrow \\ m}}{3}x + \underset{\substack{\uparrow \\ b}}{4}$$

The graph has slope 3 and y intercept 4.

(*b*) Find the slope and y intercept for the graph of the equation

$$y = \underset{\substack{\uparrow \\ m}}{-\frac{2}{3}}x - \underset{\substack{\uparrow \\ b}}{5}$$

The slope of the line is $-\frac{2}{3}$; the y intercept is -5.

CHECK YOURSELF 6

Find the slope and y intercept for the graph of each of the following equations.

1. $y = -3x - 7$ **2.** $y = \frac{3}{4}x + 5$

As the following example illustrates, we may have to solve for y as the first step in determining the slope and y intercept for the graph of an equation.

Example 7

Find the slope and y intercept for the graph of the equation

$3x + 2y = 6$

First, we must solve the equation for y.

$3x + 2y = 6$

$2y = -3x + 6$ Subtract $3x$ from both sides.

$y = -\frac{3}{2}x + 3$ Divide each term by 2.

The equation is now in slope-intercept form. The slope is $-\frac{3}{2}$, and the y intercept is 3.

CHECK YOURSELF 7

Find the slope and y intercept for the graph of the equation

$2x - 5y = 10$

As we mentioned earlier, knowing certain properties of a line (namely, its slope and y intercept) will also allow us to write the equation of the line by using the slope-intercept form. The following example illustrates.

Example 8

(a) Write the equation of a line with slope 3 and y intercept 5.

We know that $m = 3$ and $b = 5$. Using the slope-intercept form, we have

$$y = \underset{\substack{\uparrow \\ m}}{3}x + \underset{\substack{\uparrow \\ b}}{5}$$

which is the desired equation.

(*b*) Write the equation of a line with slope $-\frac{3}{4}$ and y intercept -3.

We know that $m = -\frac{3}{4}$ and $b = -3$. In this case,

$$y = \overset{\substack{m \\ \swarrow}}{-\frac{3}{4}}x + \overset{\substack{b \\ \downarrow}}{(-3)}$$

or

$$y = -\frac{3}{4}x - 3$$

which is the desired equation.

CHECK YOURSELF 8

Write the equation of a line

1. With slope -2 and y intercept 7

2. With slope $\frac{2}{3}$ and y intercept -3

We can also use the slope and y intercept of a line in drawing its graph. Consider the following example.

Example 9

Graph the line with slope $\frac{2}{3}$ and y intercept 2.

Since the y intercept is 2, we begin by plotting the point (0, 2). Now since the horizontal change (or run) is 3, we move 3 units to the right *from that y intercept.* Then since the vertical change (or rise) is 2, we move 2 units up to locate another point on the desired graph. Note that we will have located that second point at (3, 4). The final step is to simply draw a line through that point and the y intercept.

Note:
$m = \frac{2}{3} = \frac{\text{rise}}{\text{run}}$

The line rises from left to right because the slope is positive.

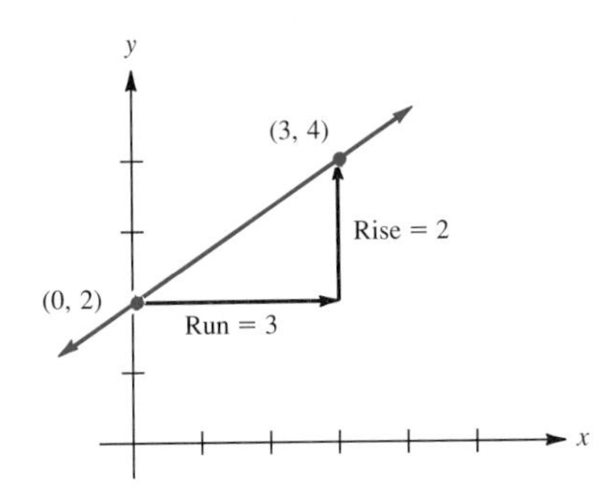

CHECK YOURSELF 9

Graph the equation of a line with slope $\frac{3}{5}$ and y intercept -2.

Let's work through one final example, using the slope and y intercept of a line.

Example 10

Write the equation of a line with slope $-\frac{3}{2}$ and y intercept 4. Then graph the line.

Using our slope-intercept form, we know that the equation of the line must be

$$y = -\frac{3}{2}x + 4$$

Now to graph the line, we again start at the y intercept, here (0, 4). We move 2 units to the right (the run is 2) and then 3 units *down* (the vertical change, or rise, is -3) to locate a second point on the line at (2, 1). We then draw a line through that point and the y intercept to complete the graph.

Think of m as $\frac{-3}{2}$. The line falls from left to right because the slope is negative.

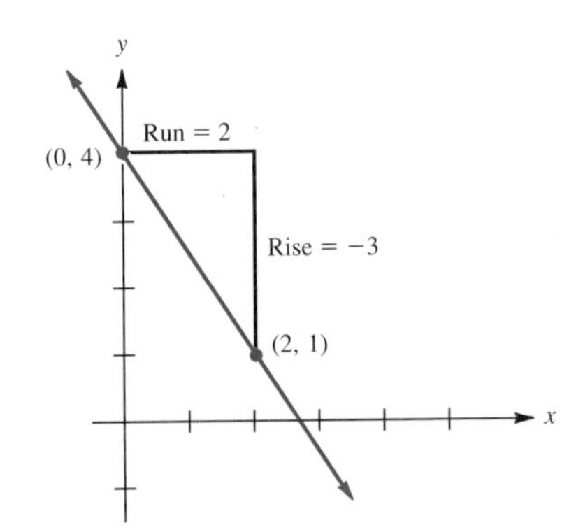

CHECK YOURSELF 10

Write the equation of a line with slope -3 and y intercept 3. Then graph the equation.

Hint: Think of the slope m as $\frac{-3}{1}$.

The following algorithm will summarize the use of graphing through the use of the slope-intercept form.

GRAPHING BY USING THE SLOPE-INTERCEPT FORM

1. Write the original equation of the line in slope-intercept form.
2. Determine the slope m and the y intercept b.
3. Plot the y intercept at $(0, b)$.
4. Use m (the change in y over the change in x) to determine a second point on the desired line.
5. Draw a line through the two points determined above to complete the graph.

CHECK YOURSELF ANSWERS

1. $m = \frac{2}{3}$ **2.** $m = \frac{5}{3}$ **3.** $m = -3$ **4.** $m = 0$ **5.** m is undefined.

6. $m = -3, b = -7$; (2) $m = \frac{3}{4}, b = 5$.

7. $y = \frac{2}{5}x - 2$; the slope is $\frac{2}{5}$; the y intercept is -2. **8.** (1) $y = -2x + 7$; (2) $y = \frac{2}{3}x - 3$.

9.

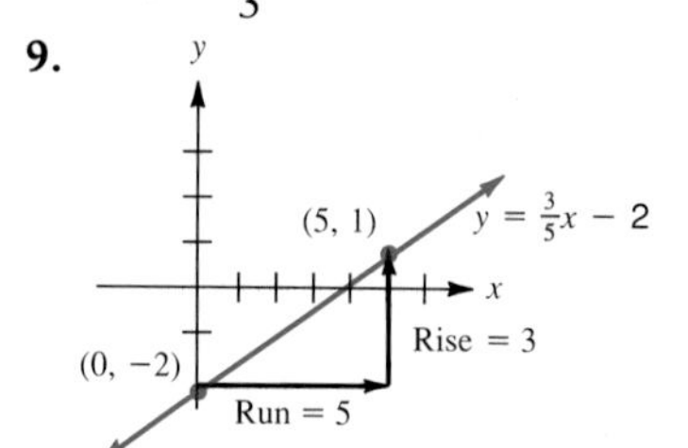

10.

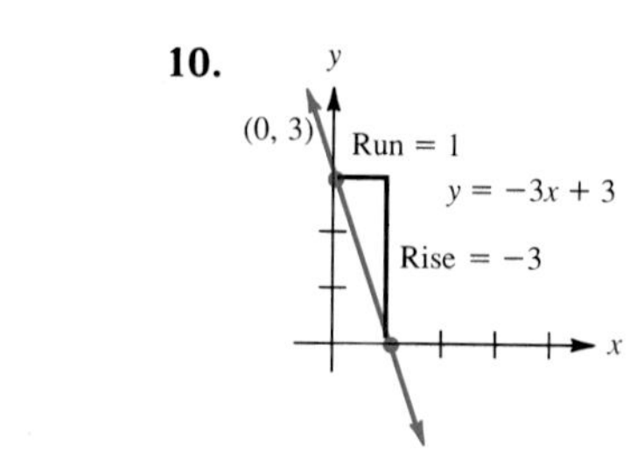

7.4 Exercises

Find the slope of the line through the following pairs of points.

1. (2, 3) and (5, 6)
1

2. (3, 1) and (7, 9)
2

3. (−1, −2) and (1, 8)
5

4. $(-2, 3)$ and $(0, 11)$

4

5. $(-2, 1)$ and $(3, 7)$

$\frac{6}{5}$

6. $(-2, -3)$ and $(4, -1)$

$\frac{1}{3}$

7. $(-3, 2)$ and $(4, -5)$

-1

8. $(-4, 5)$ and $(2, -7)$

-2

9. $(2, 3)$ and $(4, 0)$

$-\frac{3}{2}$

10. $(-2, 3)$ and $(3, 1)$

$-\frac{2}{5}$

11. $(-4, 5)$ and $(2, 5)$

0

12. $(4, -5)$ and $(4, 2)$

undefined

13. $(-3, -2)$ and $(3, 3)$

$\frac{5}{6}$

14. $(-3, -5)$ and $(2, -5)$

0

15. $(-4, -3)$ and $(-4, 2)$

undefined

16. $(-5, 7)$ and $(3, -2)$

$-\frac{9}{8}$

17. $(-3, 7)$ and $(2, 3)$

$-\frac{4}{5}$

18. $(-4, -2)$ and $(4, 4)$

$\frac{3}{4}$

Find the slope and y intercept of the line represented by each of the following equations.

19. $y = 3x + 2$ 3, 2

20. $y = -5x + 3$ -5, 3

21. $y = -7x - 4$ -7, -4

22. $y = 6x - 8$ 6, -8

23. $y = \frac{2}{3}x + 1$ $\frac{2}{3}$, 1

24. $y = -5x$ -5, 0

25. $y = \frac{3}{4}x$ $\frac{3}{4}$, 0

26. $y = -\frac{2}{5}x - 3$ $-\frac{2}{5}$, -3

27. $3x + 4y = 12$ $-\frac{3}{4}$, 3

28. $5x + 2y = 10$ $-\frac{5}{2}$, 5

29. $y = 8$ 0, 8

30. $3x - 2y = 6$ $\frac{3}{2}$, -3

31. $2x - 3y = 9$ $\frac{2}{3}$, -3

32. $x = 2$ Undefined, no y intercept

Write the equation of the line with given slope and y intercept. Then graph each line, *using* the slope and y intercept.

33. $m = 4$, $b = 2$ $y = 4x + 2$

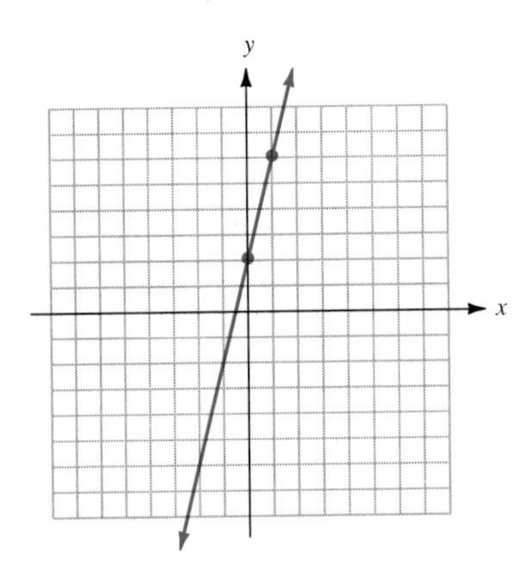

34. $m = -3$, $b = 5$ $y = -3x + 5$

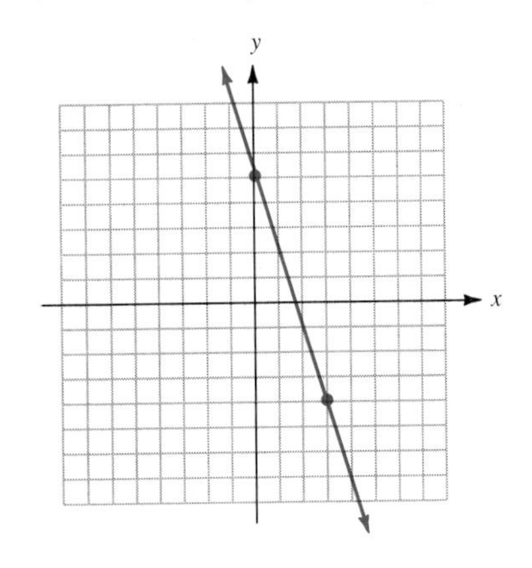

35. $m = -2, b = 5$ $\quad y = -2x + 5$

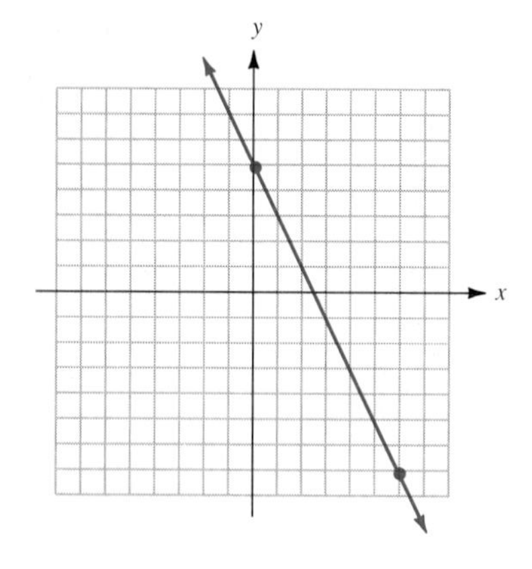

36. $m = 5, b = -3$ $\quad y = 5x - 3$

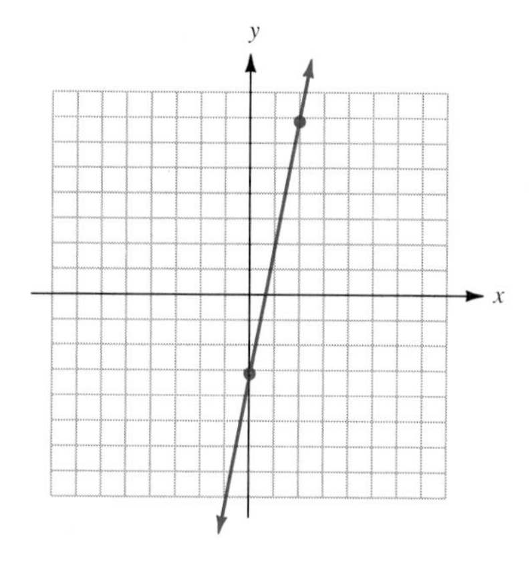

37. $m = \frac{1}{3}, b = -2$ $\quad y = \frac{1}{3}x - 2$

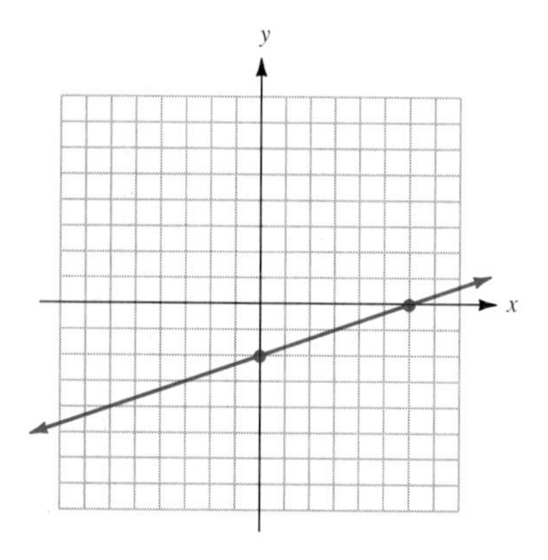

38. $m = -\frac{4}{5}, b = 6$ $\quad y = -\frac{4}{5}x + 6$

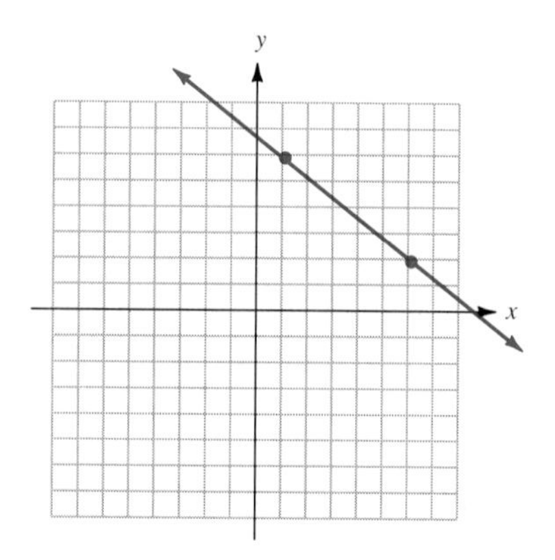

39. $m = -\frac{3}{4}, b = 0$ $\quad y = -\frac{3}{4}x$

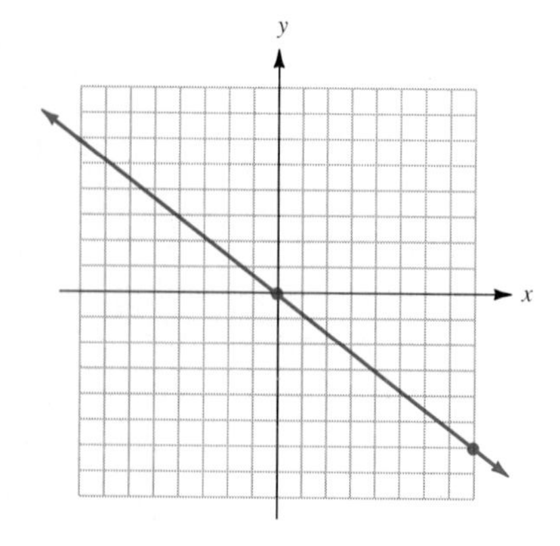

40. $m = \frac{3}{4}, b = -2$ $\quad y = \frac{3}{4}x - 2$

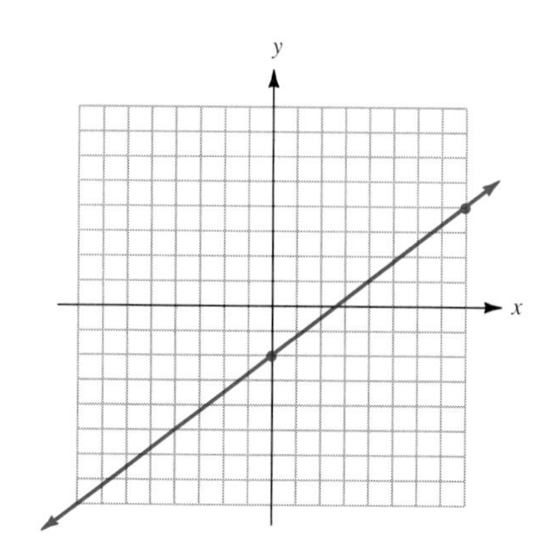

41. $m = \frac{3}{2}$, $b = 3$ $\quad y = \frac{3}{2}x + 3$

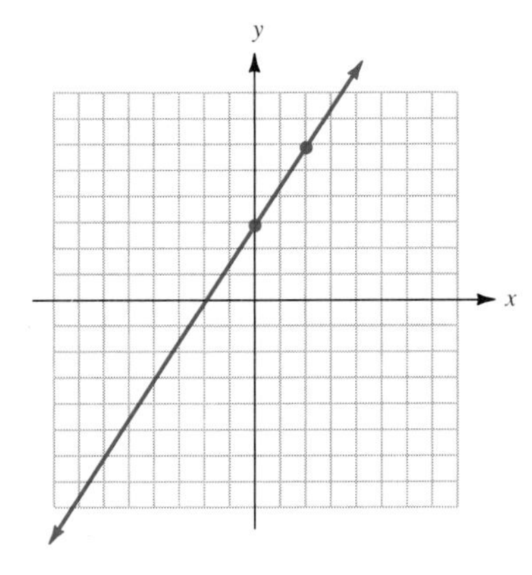

42. $m = -2$, $b = 0$ $\quad y = -2x$

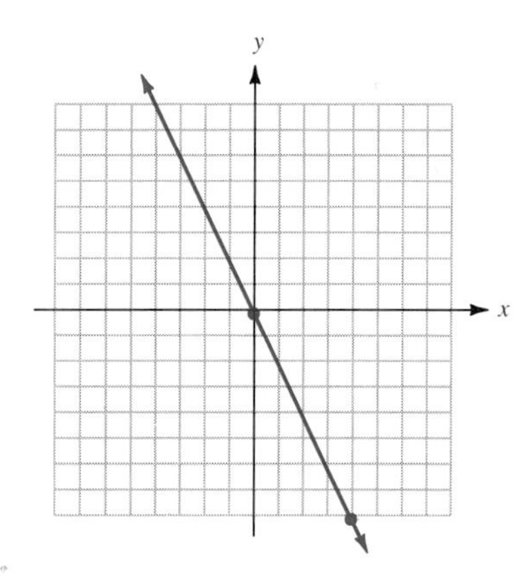

In Exercises 43 to 50, match the graph with one of the equations on the right.

43. e

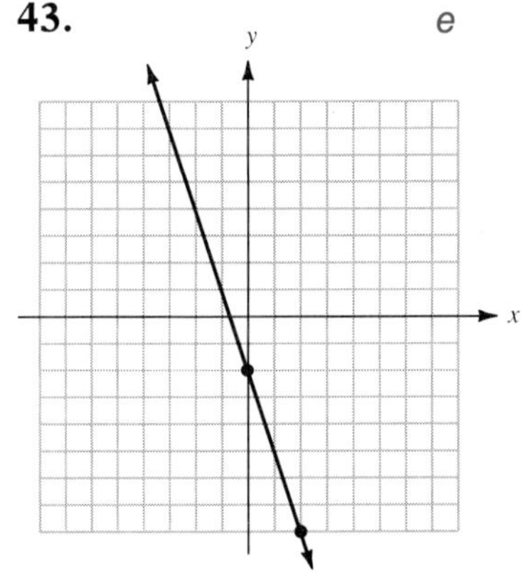

44. g

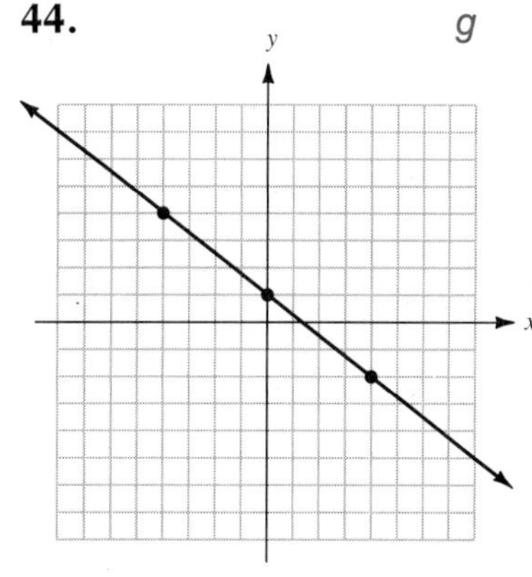

(*a*) $y = 2x$

(*b*) $y = x + 1$

(*c*) $y = -x + 3$

(*d*) $y = 2x + 1$

(*e*) $y = -3x - 2$

(*f*) $y = \frac{2}{3}x + 1$

(*g*) $y = -\frac{3}{4}x + 1$

(*h*) $y = -4x$

45. a

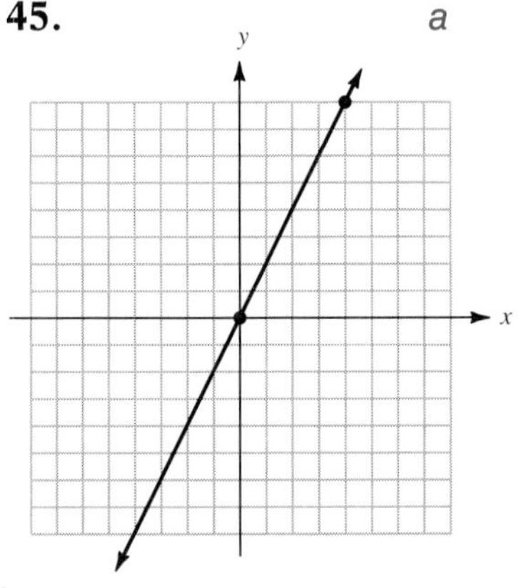

46. d

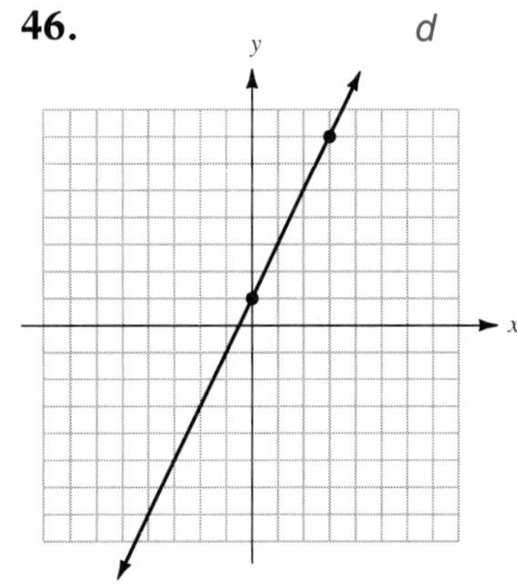

47. *b*

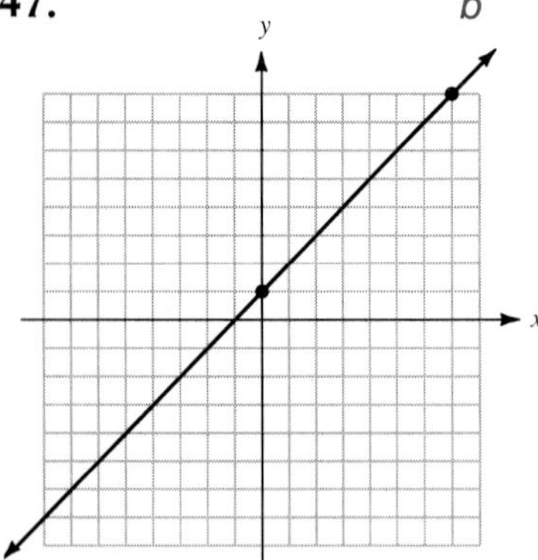

48. *c*

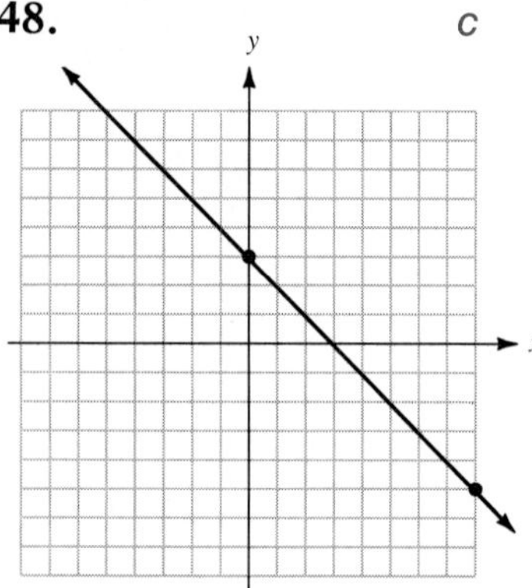

49. *h*

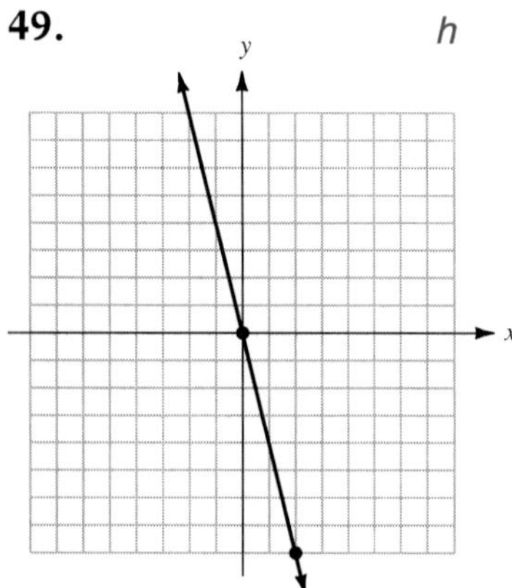

50. *f*

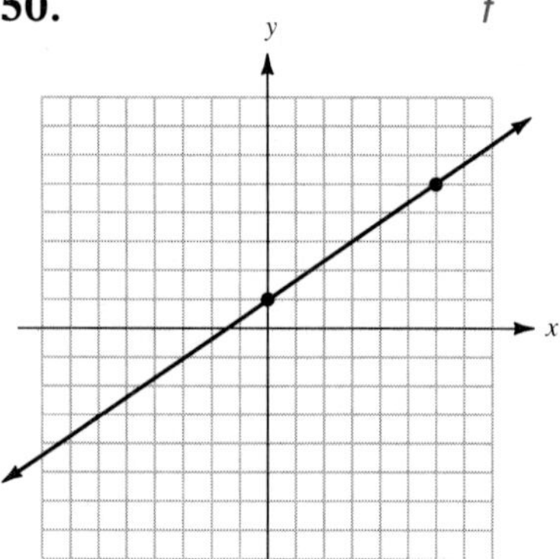

Skillscan (Section 3.7)

Graph each of the following inequalities.

a. $x < 3$

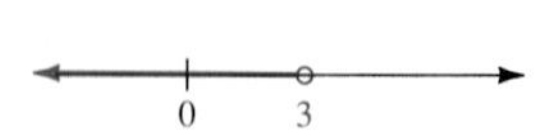

b. $x \geq -2$

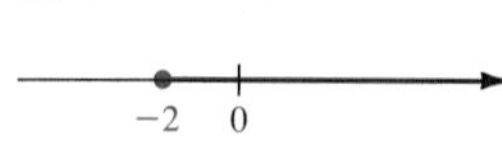

c. $2x \leq 8$

d. $3x > -9$

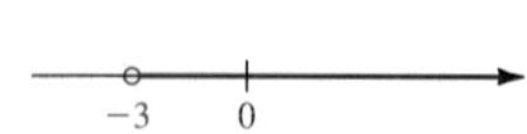

e. $-3x < 12$

f. $-2x \leq 10$

g. $\frac{2}{3}x \leq 4$

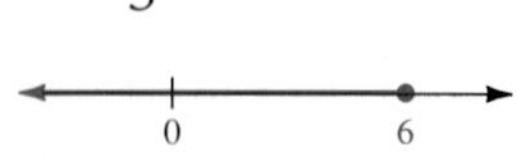

h. $-\frac{3}{4}x \geq 6$

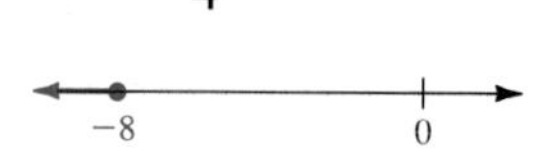

Answers

1. 1 **3.** 5 **5.** $\frac{6}{5}$ **7.** -1 **9.** $-\frac{3}{2}$ **11.** 0 **13.** $\frac{5}{6}$ **15.** Undefined **17.** $-\frac{4}{5}$

19. Slope 3, y intercept 2 **21.** Slope -7, y intercept -4 **23.** Slope $\frac{2}{3}$, y intercept 1

25. Slope $\frac{3}{4}$, y intercept 0 **27.** Slope $-\frac{3}{4}$, y intercept 3 **29.** Slope 0, y intercept 8

31. Slope $\frac{2}{3}$, y intercept -3

33. $y = 4x + 2$

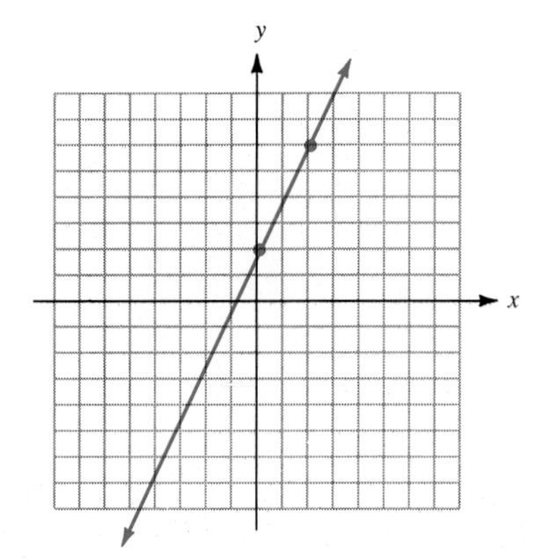

35. $y = -2x + 5$

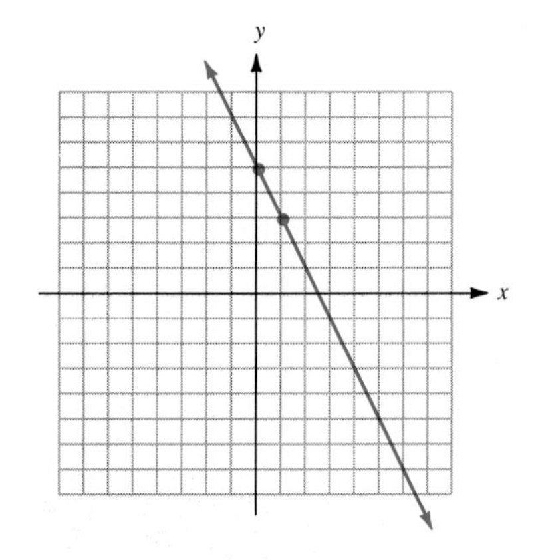

37. $y = \frac{1}{3}x - 2$

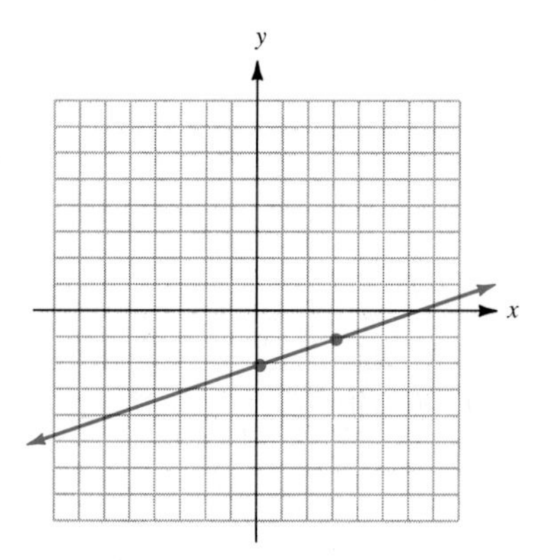

39. $y = -\frac{3}{4}x$

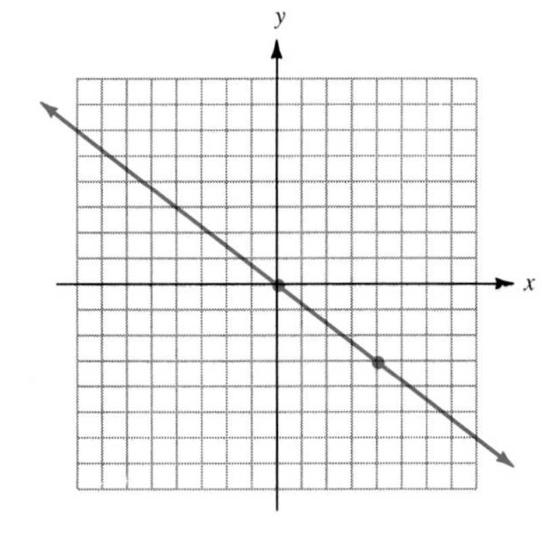

41. $y = \frac{3}{2}x + 3$

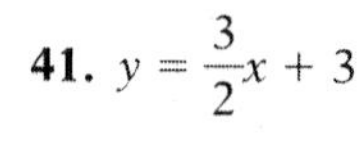

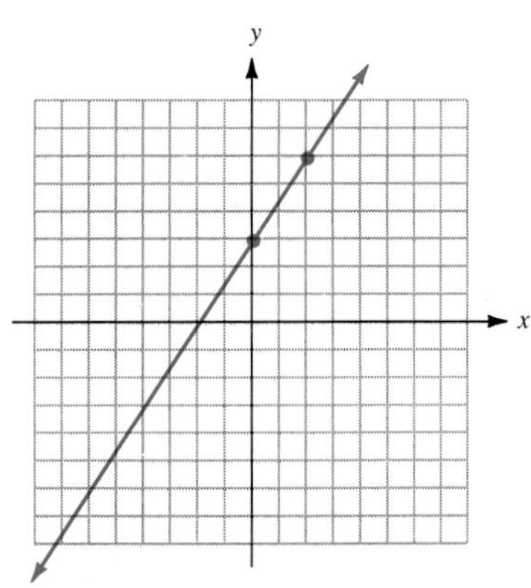

43. *e* **45.** *a* **47.** *b* **49.** *h*

a. $x < 3$

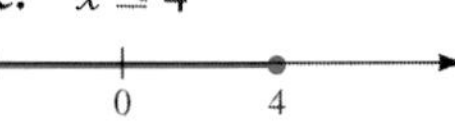

b. $x \geq -2$

−2 0

c. $x \leq 4$

0 4

d. $x > -3$

−3 0

e. $x > -4$

−4 0

f. $x \geq -5$

−5 0

g. $x \leq 6$

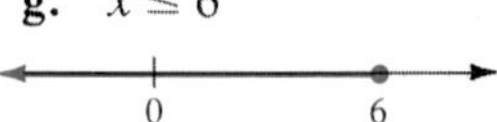

h. $x \leq -8$

−8 0

7.5 Graphing Linear Inequalities

OBJECTIVE
To graph a linear inequality in two variables.

In Section 3.6 you learned to graph inequalities in one variable on a number line. We now want to extend our work with graphing to include linear inequalities in two variables.

First, consider a definition.

The inequality symbols $\leq$, $>$, and $\geq$ can also be used.

> An inequality that can be written in the form
>
> $Ax + By < C$
>
> where A and B are not both 0, is called a *linear inequality in two variables.*

Some examples of linear inequalities in two variables are

$x + 3y > 6 \qquad y \leq 3x + 1 \qquad 2x - y \geq 3.$

The *graph* of a linear inequality is always a region (actually a half plane) of the plane whose boundary is a straight line. Let's look at an example of graphing such an inequality.

Example 1

Graph $2x + y < 4$.

First, replace the inequality symbol ($<$) with an equals sign. We then have $2x + y = 4$. This equation forms the *boundary line* of the graph of the original inequality. You can graph the line by any of the methods discussed earlier in this chapter.

> **Note:** When equality is *not included* ($<$ or $>$), use a *dotted line* for the graph of the boundary line. This means that the line is not included in the graph of the linear inequality.

The boundary line for our inequality is shown below.

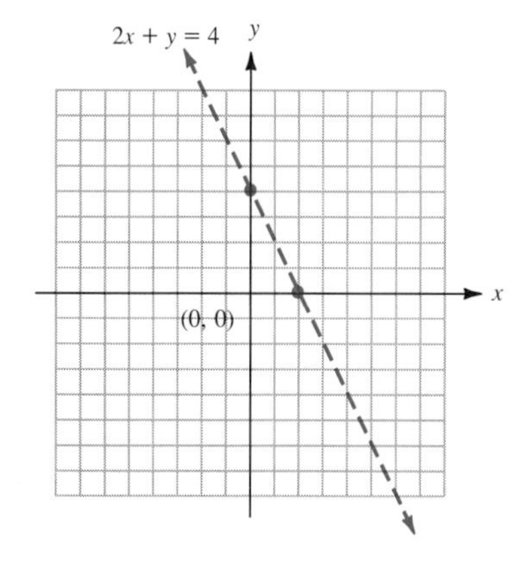

You can always use the origin for a test point unless the boundary line passes through the origin.

We see that the boundary line separates the plane into two regions, each of which is called a *half plane*.

We now need to choose the correct half plane. Choose any convenient test point not on the boundary line. The origin (0, 0) is a good choice because it makes for easy calculation.

Substitute $x = 0$ and $y = 0$ into the inequality.

$$2 \cdot 0 + 0 < 4$$
$$0 + 0 < 4$$
$$0 < 4 \quad \text{A true statement}$$

Since the inequality is *true* for the test point, we shade the half plane containing that test point (here the origin).

The origin and all other points *below* the boundary line represent solutions for our original inequality.

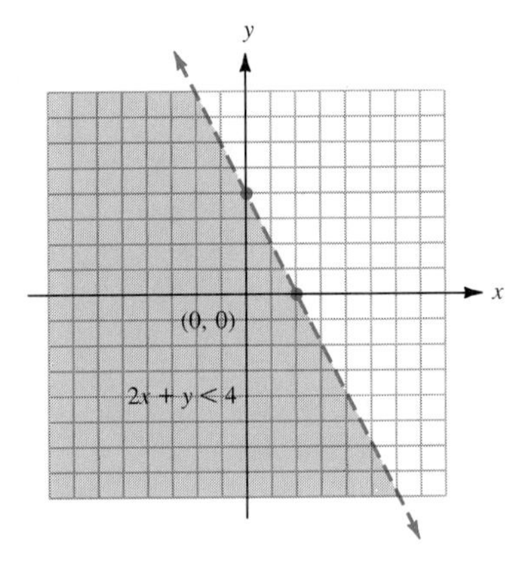

CHECK YOURSELF 1

Graph the inequality $x + 3y < 3$.

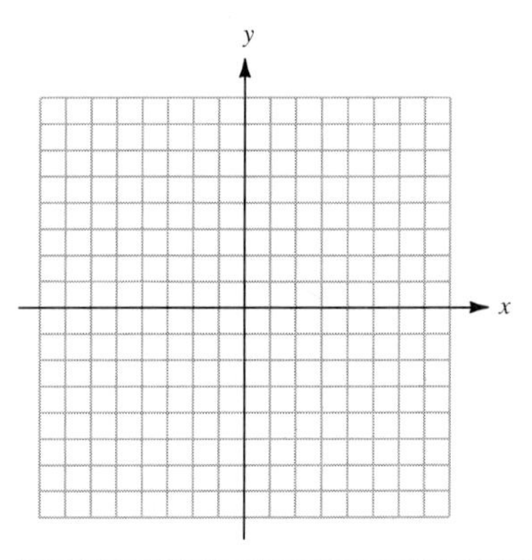

Let's consider a second example.

Example 2

Graph $4x - 3y \geq 12$.

Again we replace the inequality symbol ($\geq$) with an equals sign to write the equation for our boundary line.

First, graph the boundary line, $4x - 3y = 12$.

Note: When equality *is included* ($\leq$ or $\geq$), use a *solid line* for the graph of the boundary line. This means the line is included in the graph of the linear inequality.

The graph of our boundary line (a solid line here) is shown below.

Although any of our graphing methods can be used here, the intercept method is probably the most efficient.

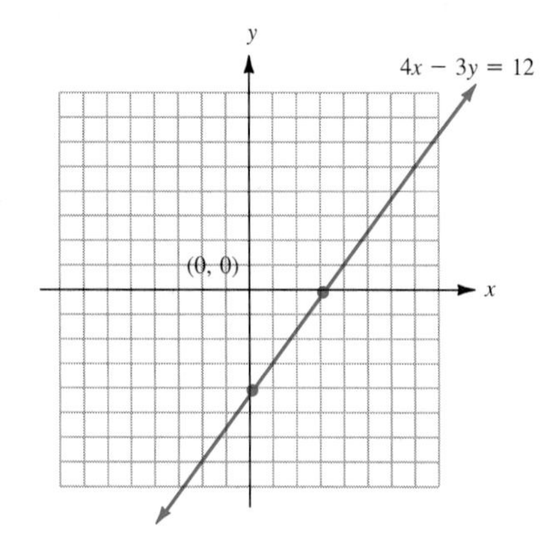

Again, we use (0, 0) as a convenient test point. Substituting 0 for x and for y in the original inequality, we have

$$4 \cdot 0 - 3 \cdot 0 \geq 12$$

$$0 \geq 12 \quad \Big\} \text{ A false statement}$$

Since the inequality is *false* for the test point, we shade the half plane that does *not* contain that test point, here (0, 0).

All points *on and below* the boundary line represent solutions for our original inequality.

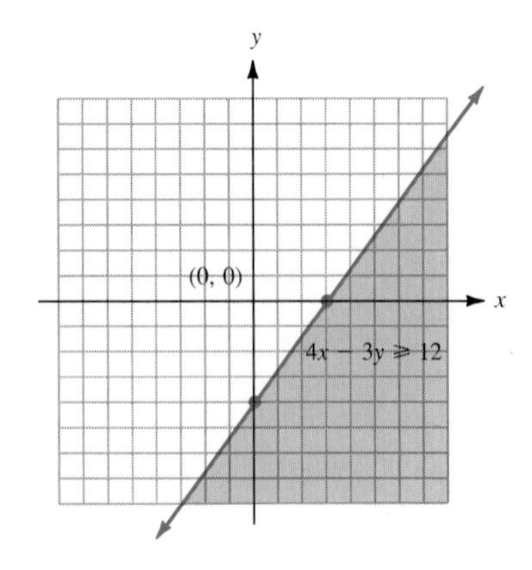

CHECK YOURSELF 2

Graph the inequality $3x + 2y \geq 6$.

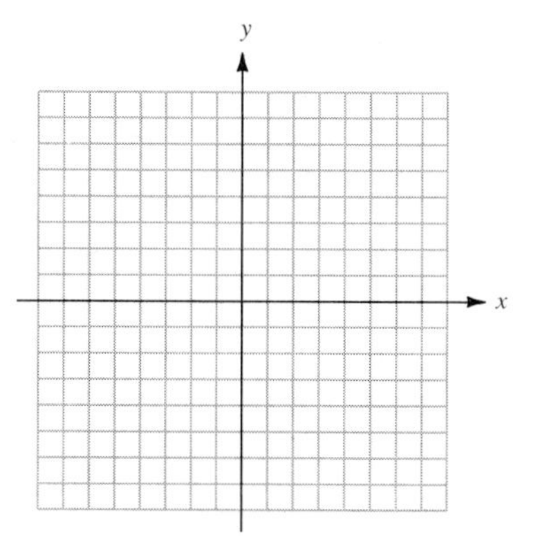

Example 3

Graph $x \leq 5$.

The boundary line is $x = 5$. Its graph is a solid line because equality is included. Using (0, 0) as a test point, we substitute 0 for x with the result

$0 \leq 5$ } A true statement

Since the inequality is *true* for the test point, we shade the half plane containing the origin.

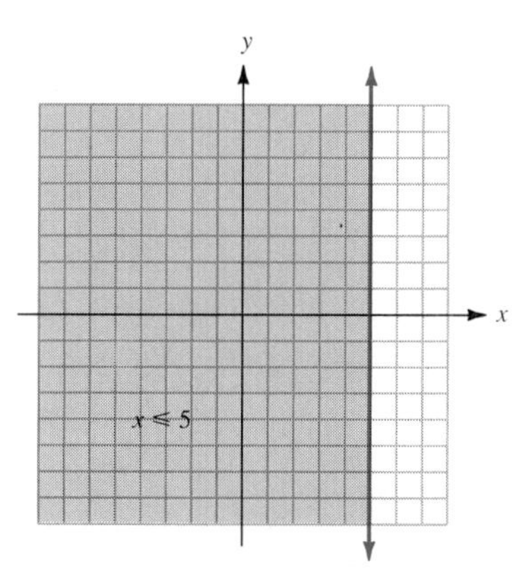

Note: If the correct half plane is obvious, you may not need to use a test point. Did you know without testing which half plane to shade in this example?

CHECK YOURSELF 3

Graph the inequality $y < 2$.

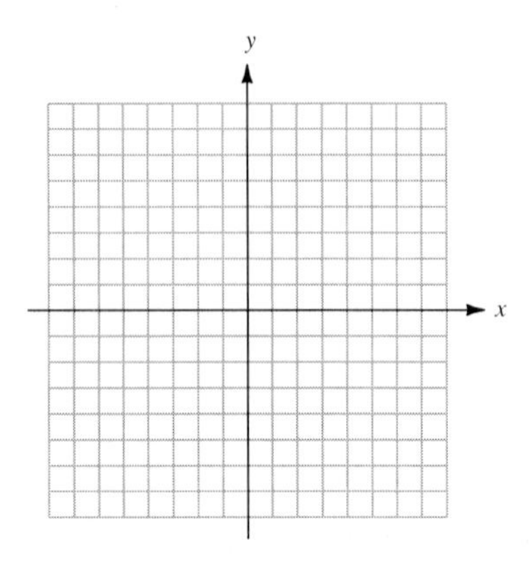

As we mentioned earlier, we may have to use a point other than the origin as our test point. The following example illustrates.

Example 4

Graph $2x + 3y < 0$.

The boundary line is $2x + 3y = 0$. Its graph is shown below.

We use a dotted line for our boundary line since equality is not included.

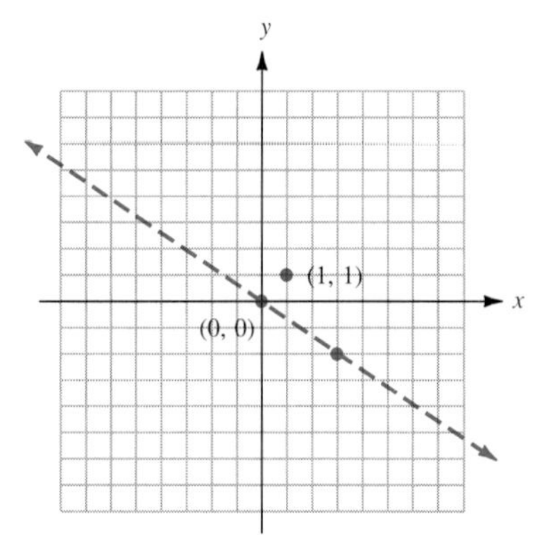

We can't use (0, 0) as our test point in this case. Do you see why?

Choose any other point *not* on the line. For instance, we have picked (1, 1) as a test point. Substituting 1 for x and 1 for y gives

$$\begin{aligned} 2 \cdot 1 + 3 \cdot 1 &< 0 \\ 2 + 3 &< 0 \\ 5 &< 0 \end{aligned} \quad \text{A false statement}$$

Since the inequality is *false* at our test point, we shade the half plane *not* containing (1, 1). This is shown in the following.

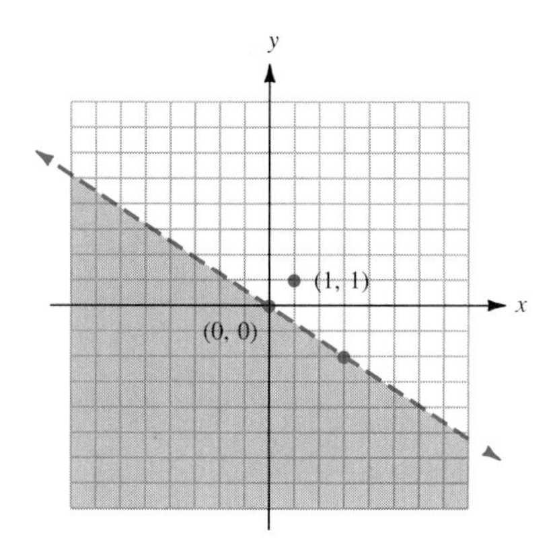

CHECK YOURSELF 4

Graph the inequality $x - 2y < 0$.

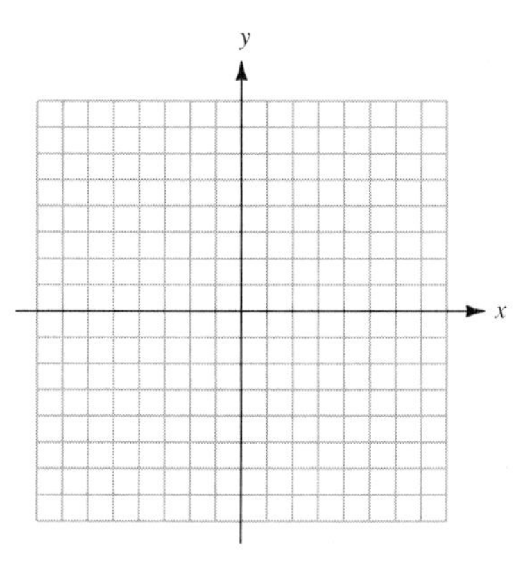

The following steps summarize our work in graphing linear inequalities in two variables.

TO GRAPH A LINEAR INEQUALITY

STEP 1 Replace the inequality symbol with an equals sign to form the equation of the boundary line of the graph.

STEP 2 Graph the boundary line. Use a dotted line if equality is not included ($<$ or $>$). Use a solid line if equality is included ($\le$ or $\ge$).

STEP 3 Choose any convenient test point *not* on the line.

STEP 4 If the inequality is *true* at the checkpoint, shade the half plane including the test point. If the inequality is *false* at the checkpoint, shade the half plane not including the test point.

CHECK YOURSELF ANSWERS

1.

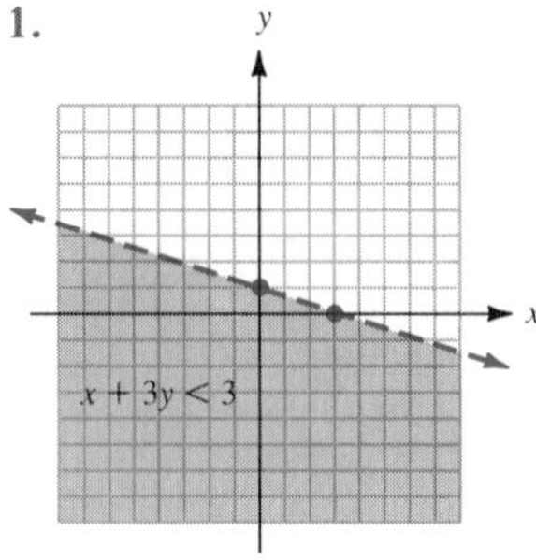

2.

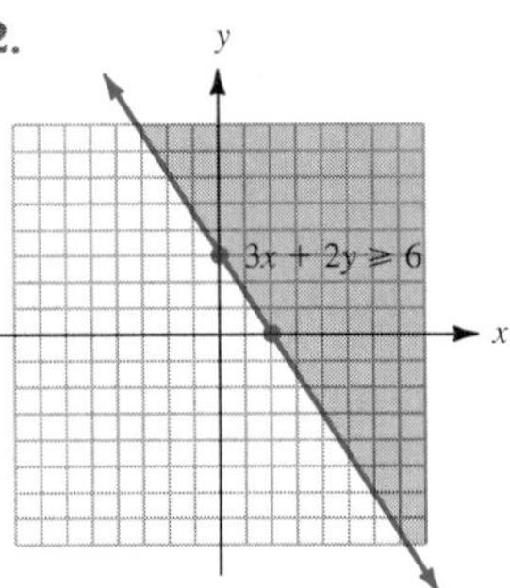

3.

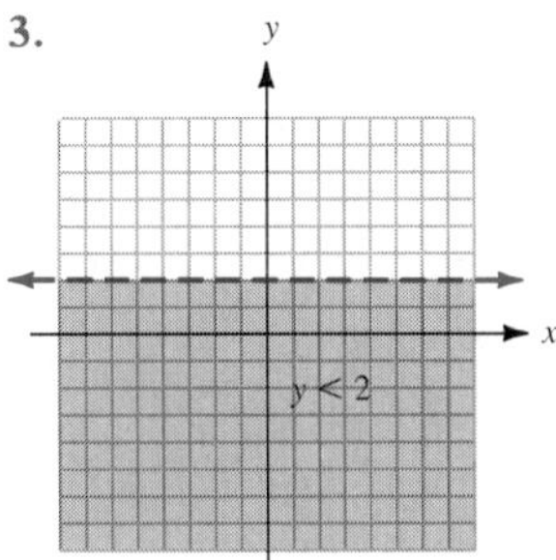

4.

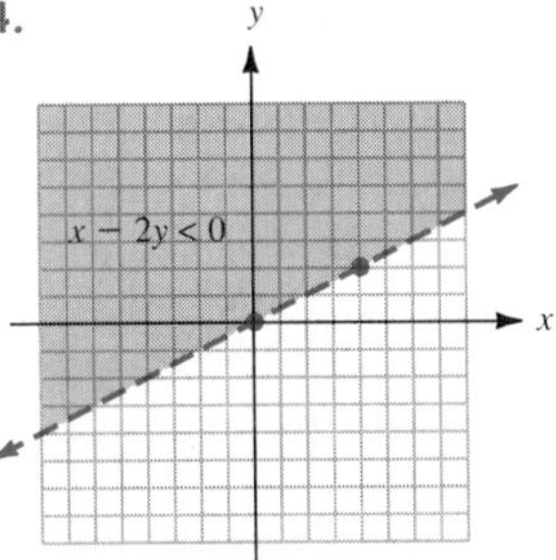

7.5 Exercises

In Exercises 1 to 8, we have graphed the boundary line for the linear inequality. Determine the correct half plane in each case, and complete the graph.

1. $x + y < 6$

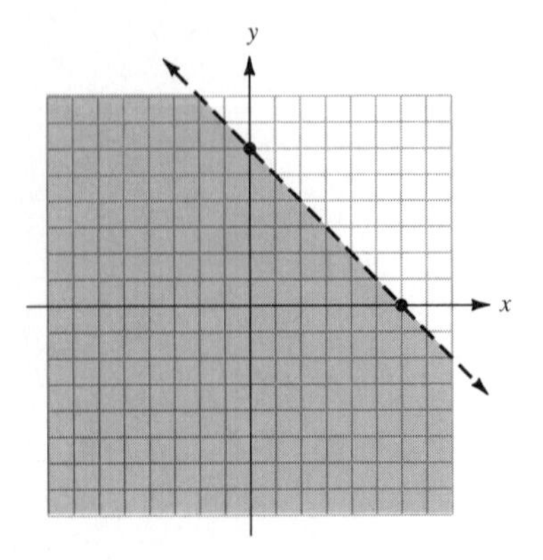

2. $x - y \geq 3$

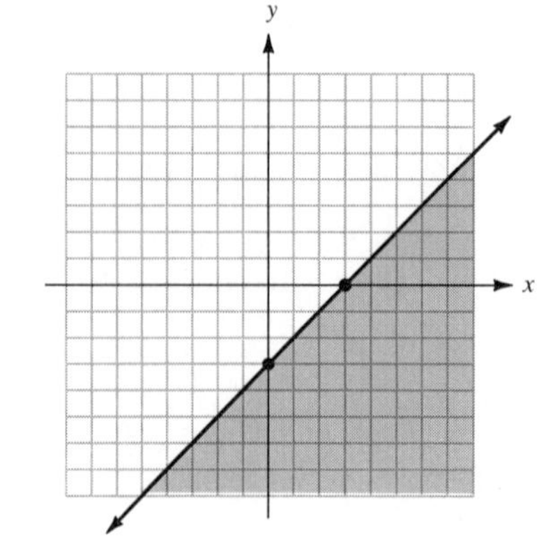

3. $x - 2y \geq 6$

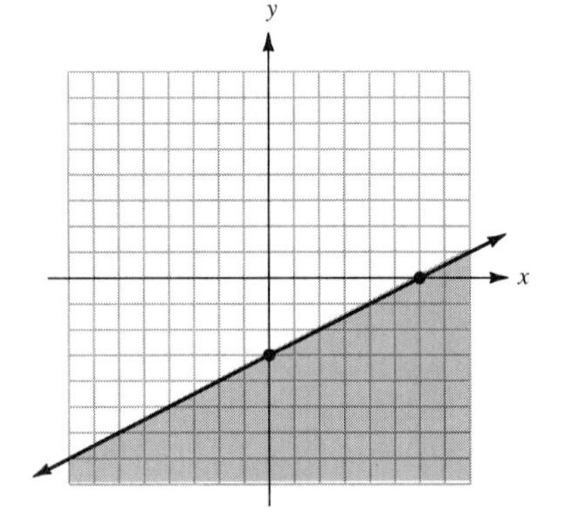

4. $2x + y < 4$

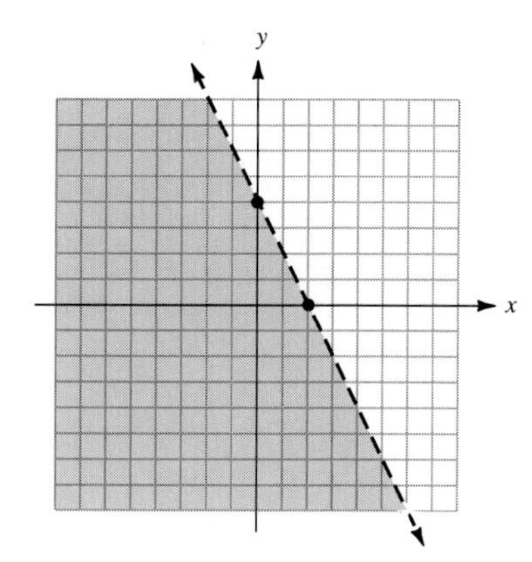

5. $x \leq -2$

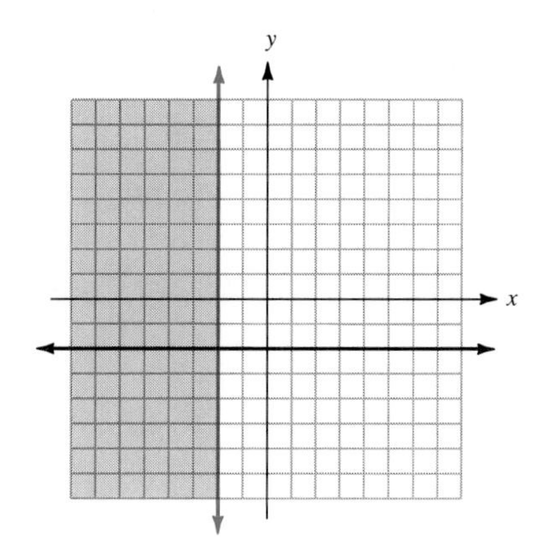

6. $y \geq 3x$

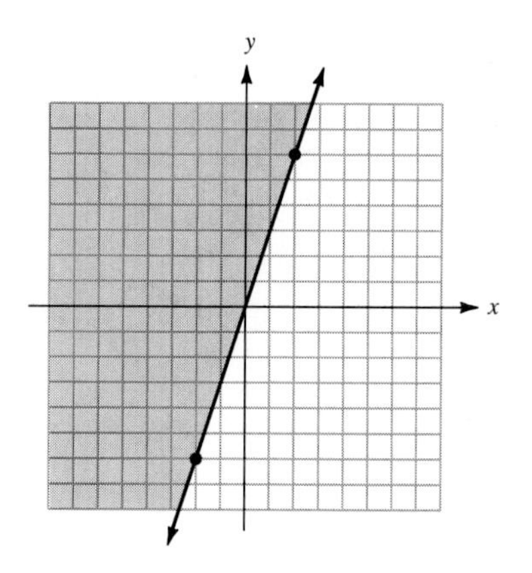

7. $y < 2x - 8$

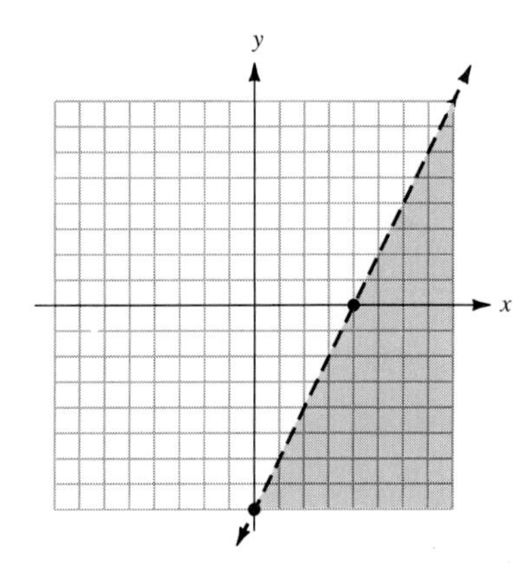

8. $y > 4$

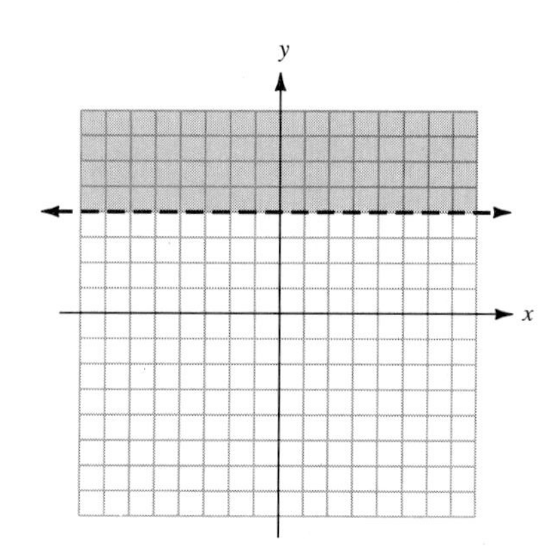

Graph each of the following inequalities.

9. $x + y < 4$

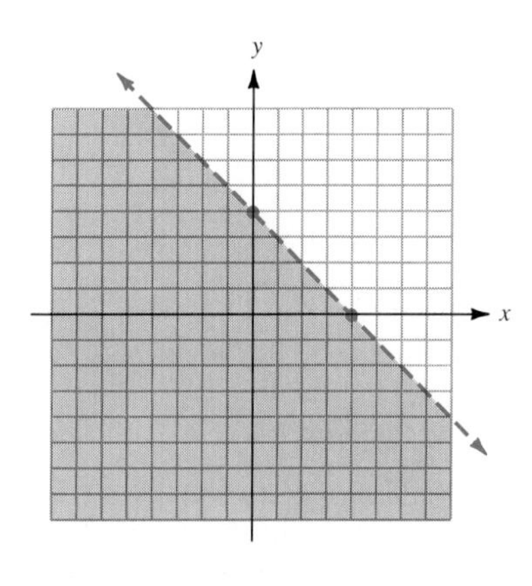

10. $x - y \geq 5$

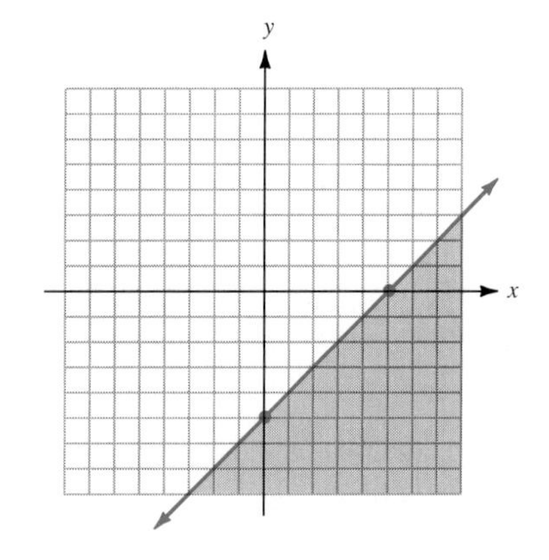

11. $x - y \leq 2$

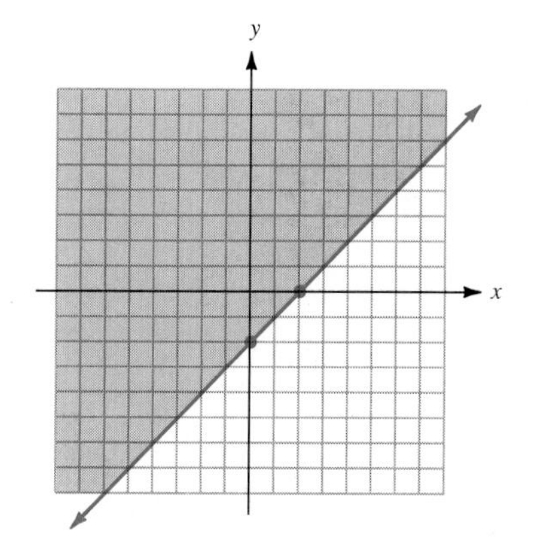

12. $x + y > 6$

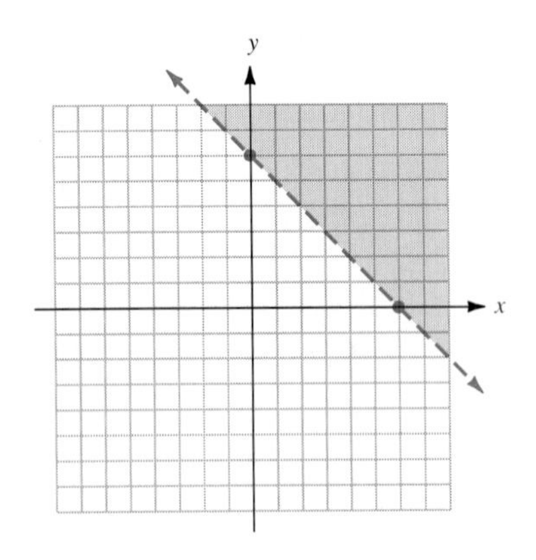

13. $2x + y \geq 4$

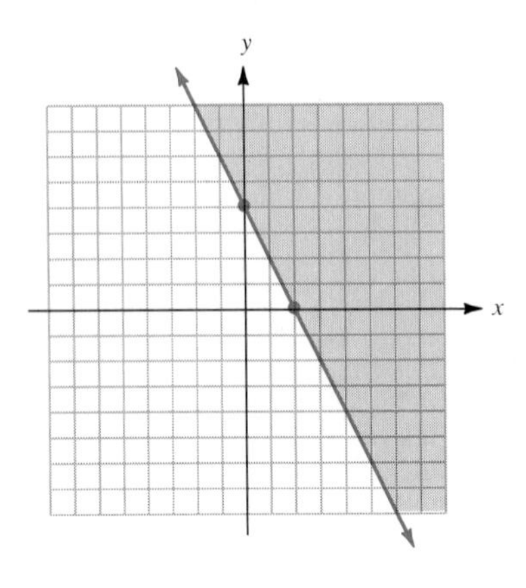

14. $3x - y < 6$

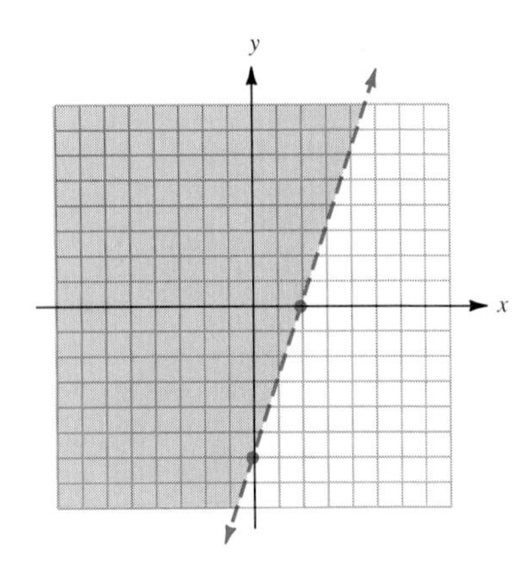

15. $x \leq 2$

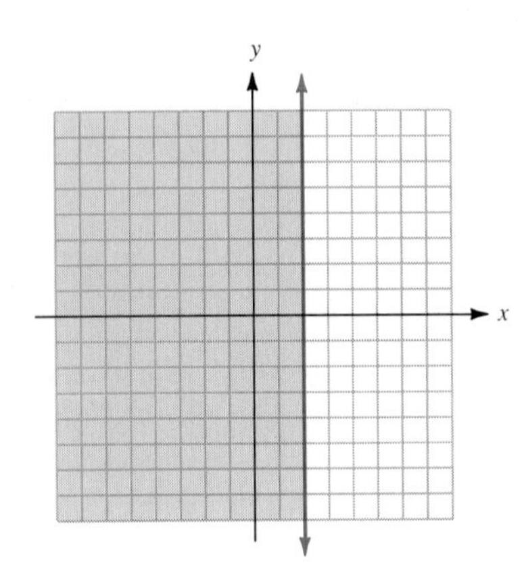

16. $5x + y \geq 5$

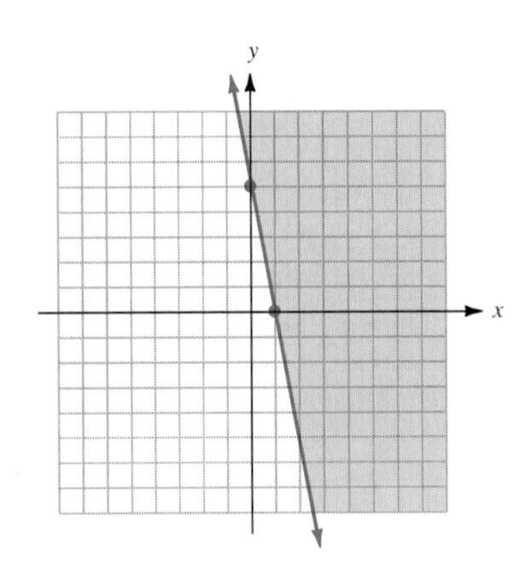

17. $x - 4y < 4$

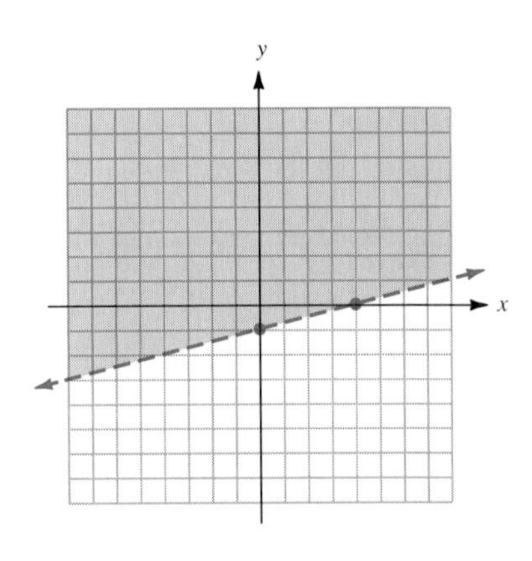

18. $y > 4$

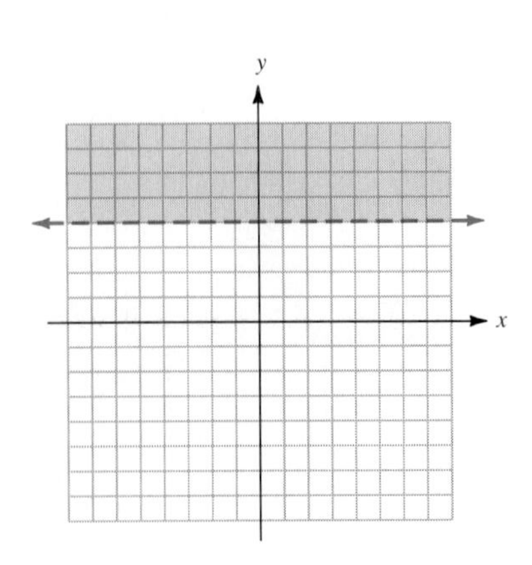

19. $y < -3$

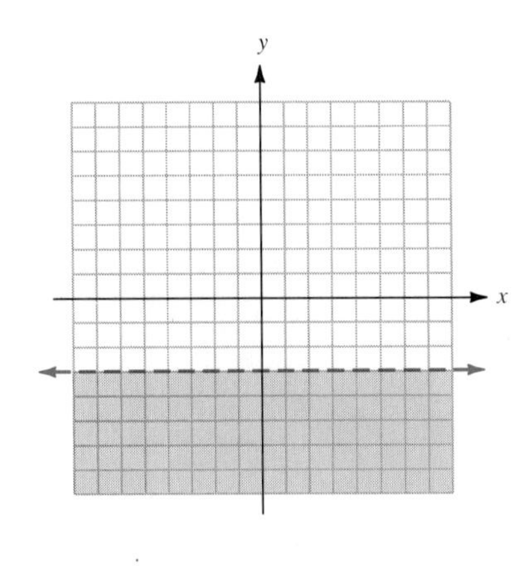

20. $3x + 4y > 12$

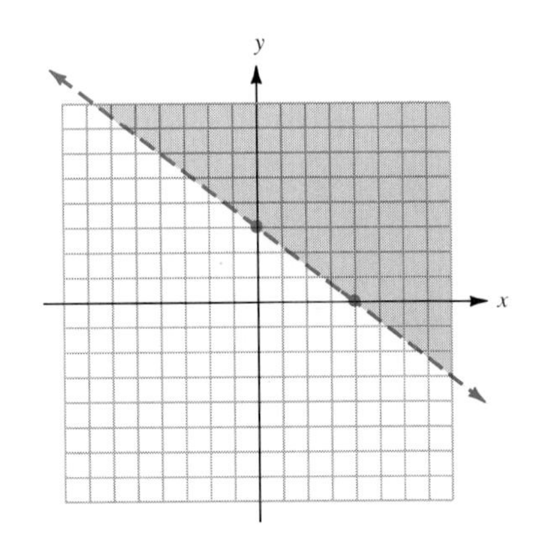

21. $3x - 2y \geq 6$

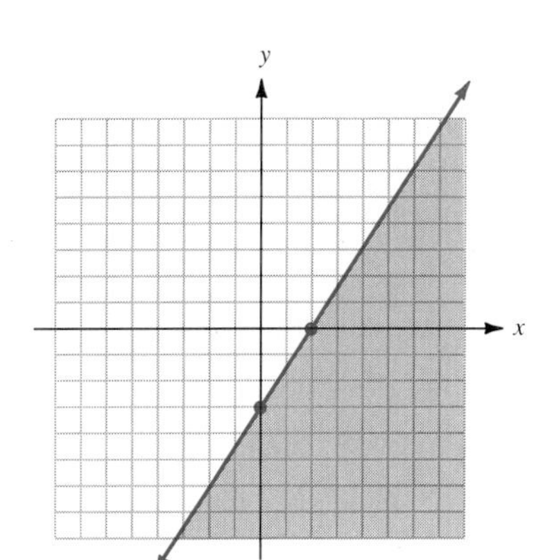

22. $x \geq -3$

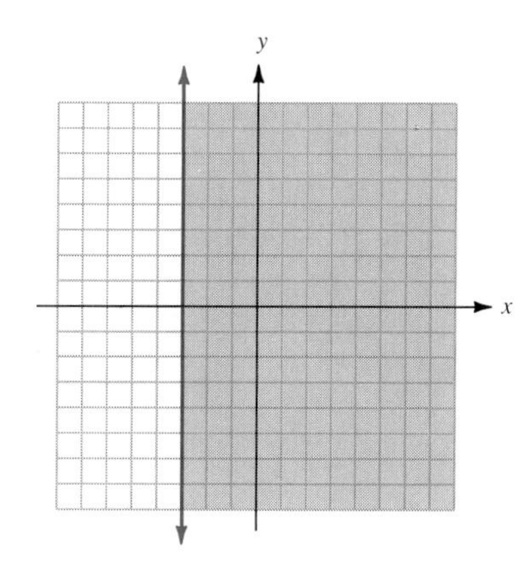

23. $2x + 3y \geq 0$

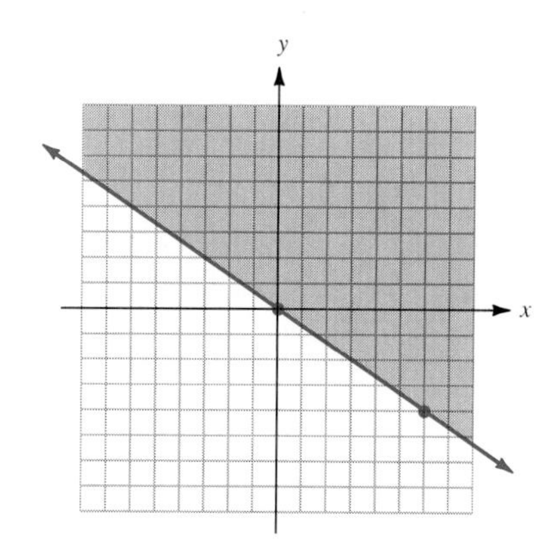

24. $2x + 5y < 10$

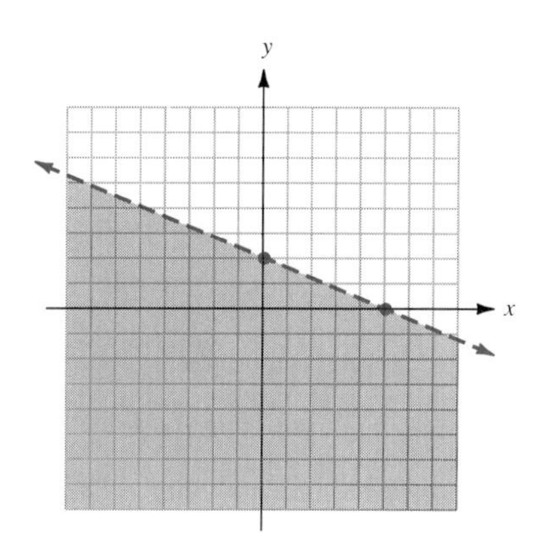

25. $5x + 3y > 15$

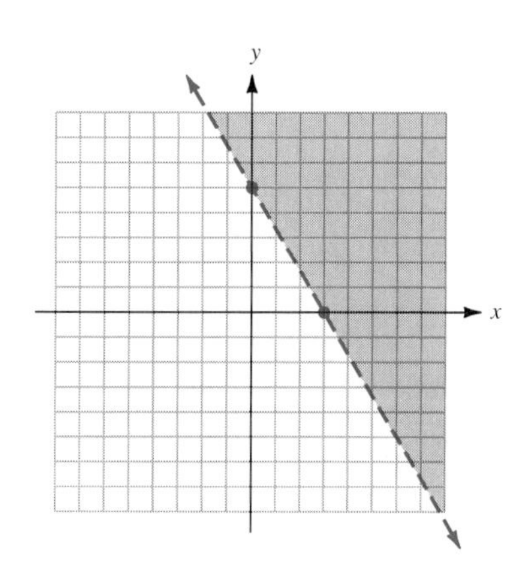

26. $x - 4y \geq 0$

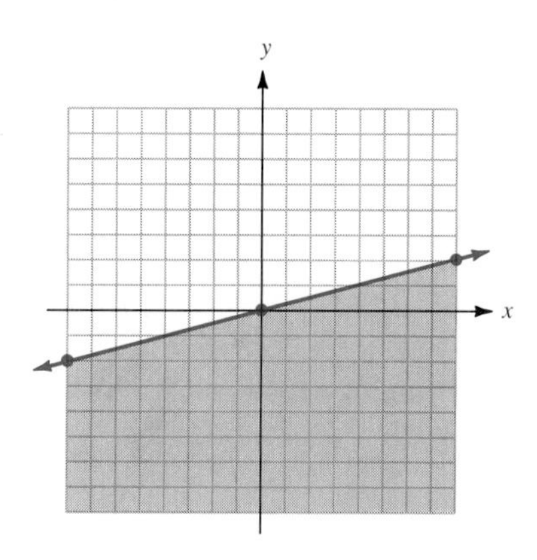

27. $y \leq 3x$

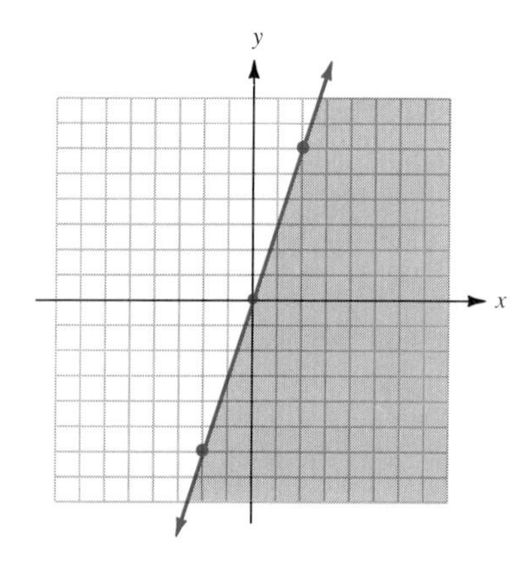

28. $4x - 3y < 12$

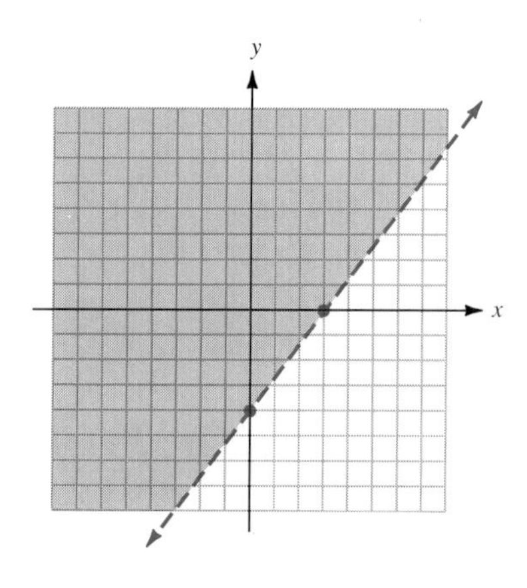

29. $y > 3x - 2$

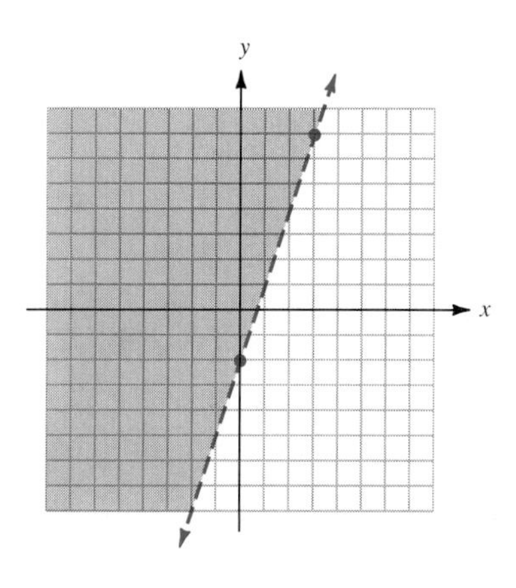

30. $y \geq -2x$

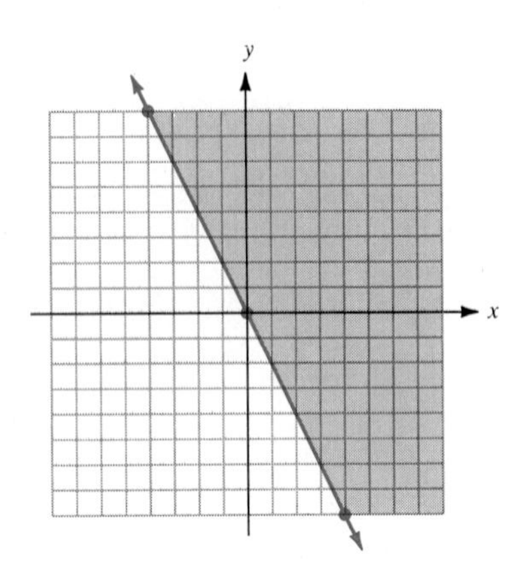

31. $y < -2x - 3$

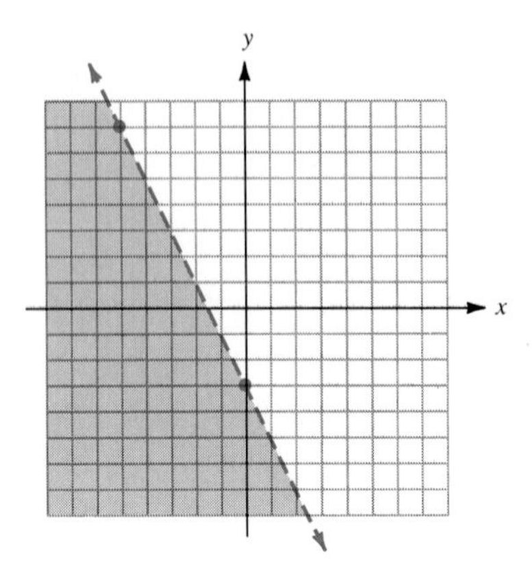

32. $y \leq 3x + 4$

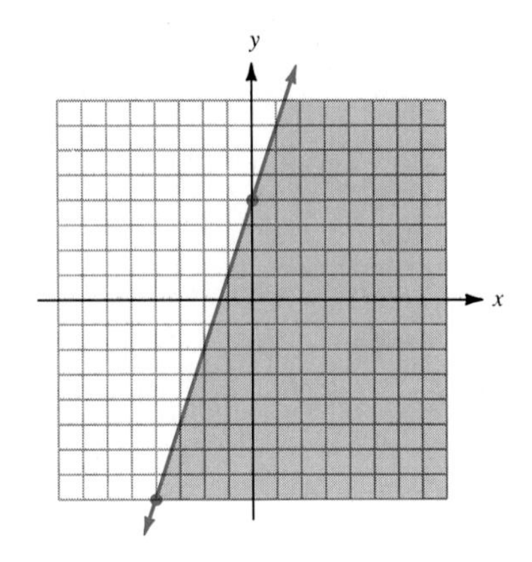

Answers

1. $x + y < 6$

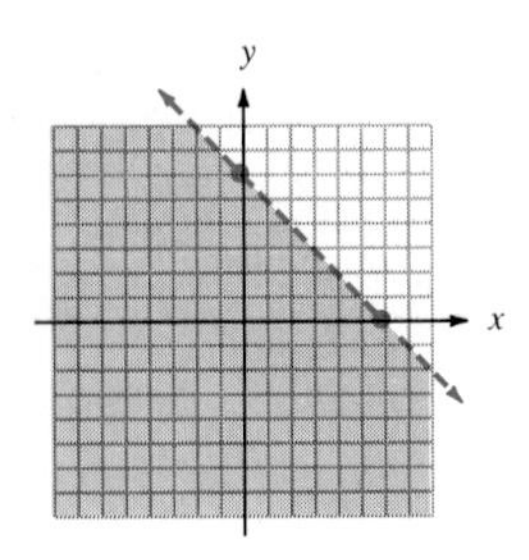

3. $x - 2y \geq 6$

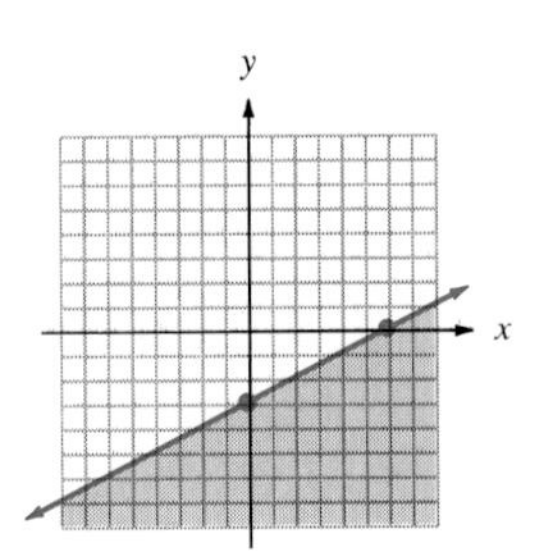

5. $x \leq -2$

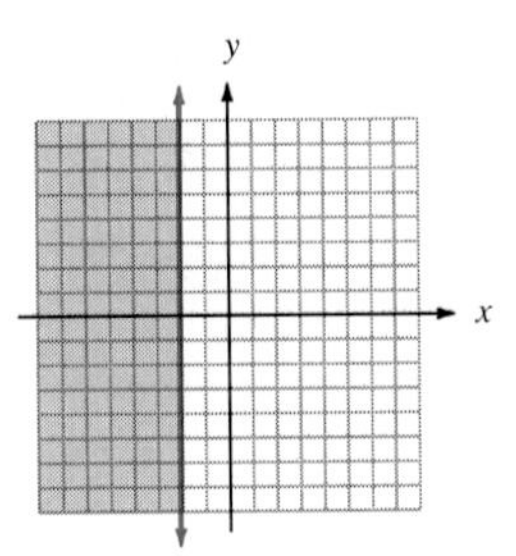

7. $y < 2x - 8$

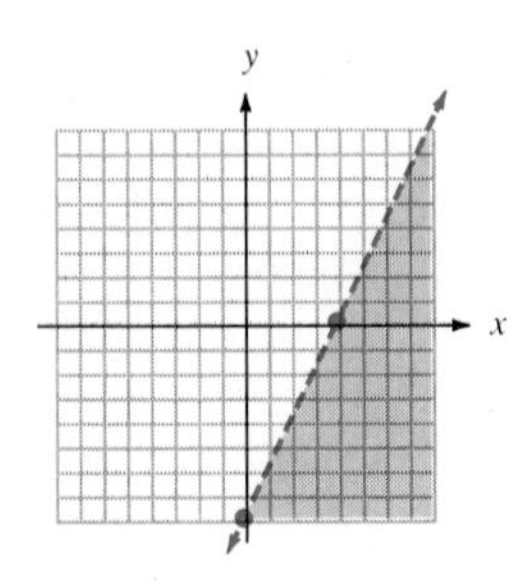

9. $x + y < 4$

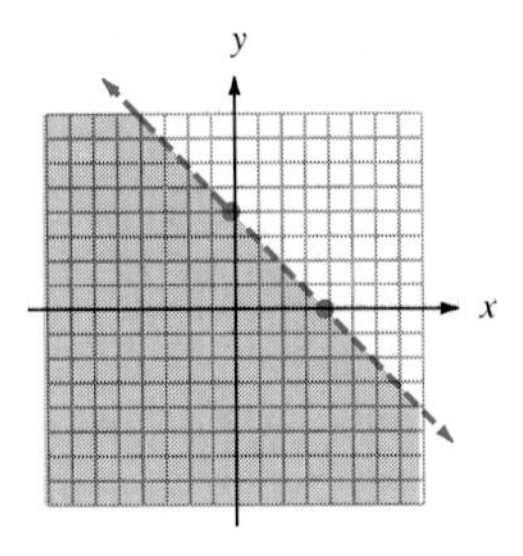

11. $x - y \leq 2$

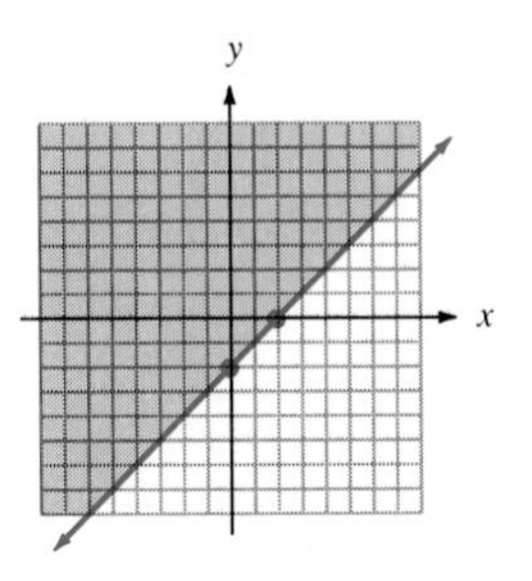

13. $2x + y \geq 4$

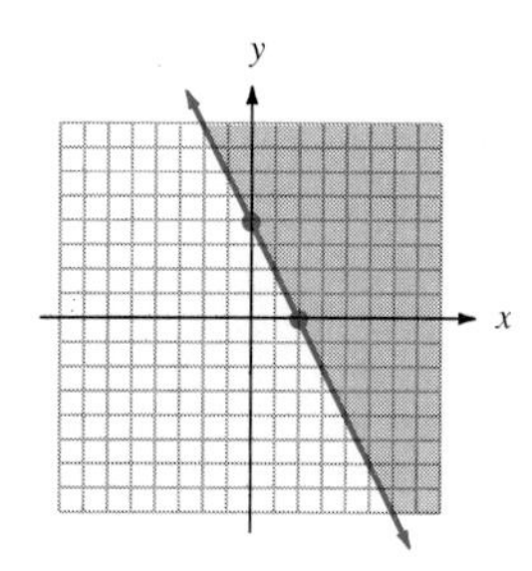

15. $x \leq 2$

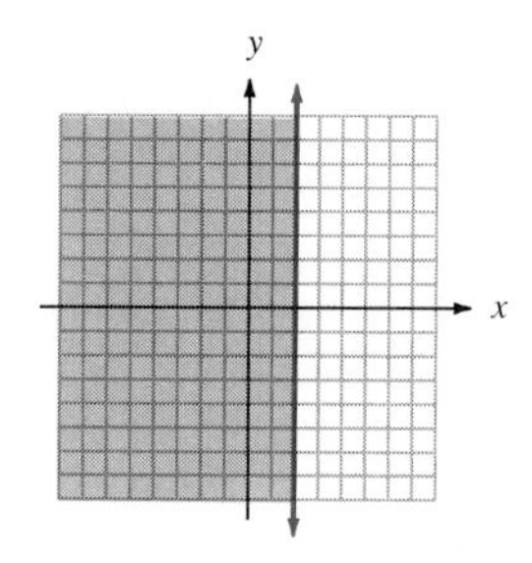

17. $x - 4y < 4$

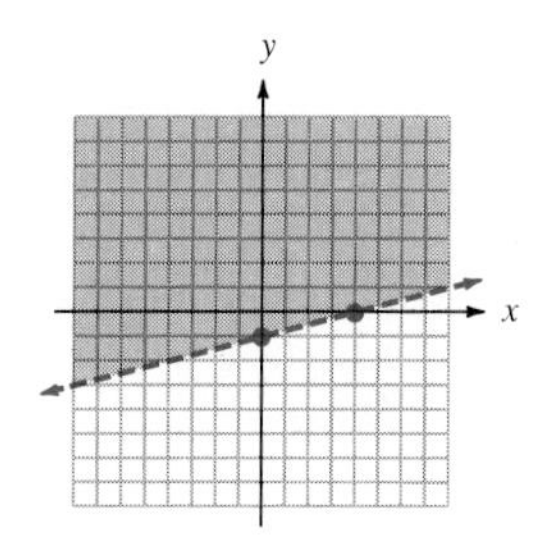

19. $y < -3$

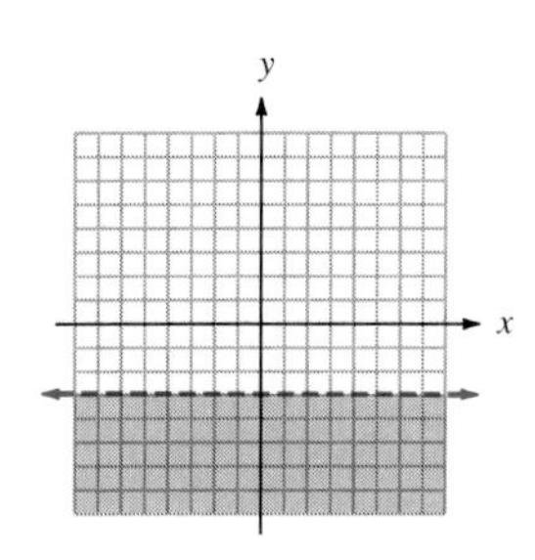

21. $3x - 2y \geq 6$

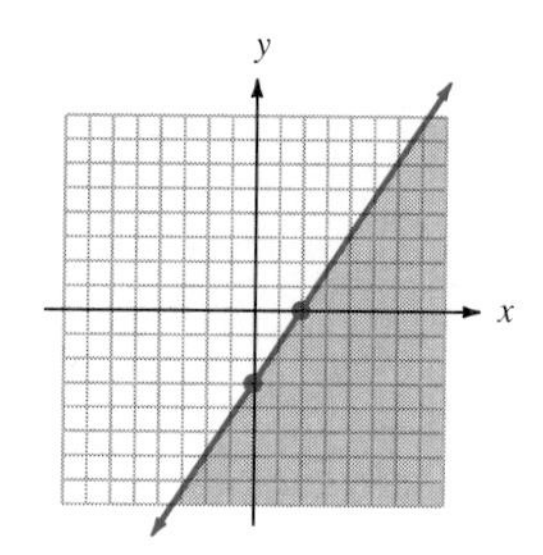

23. $2x + 3y \geq 0$

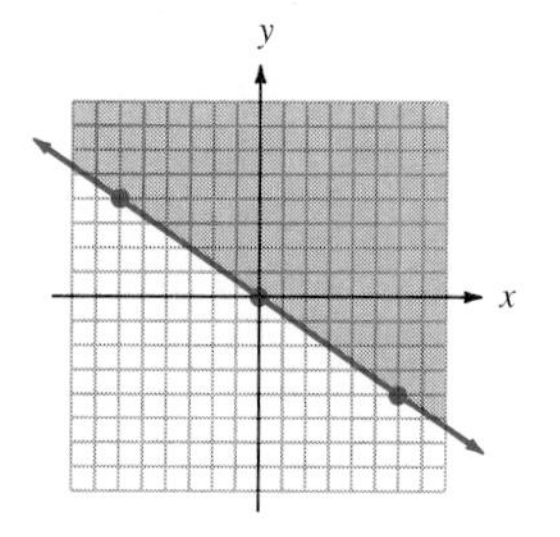

25. $5x + 3y > 15$

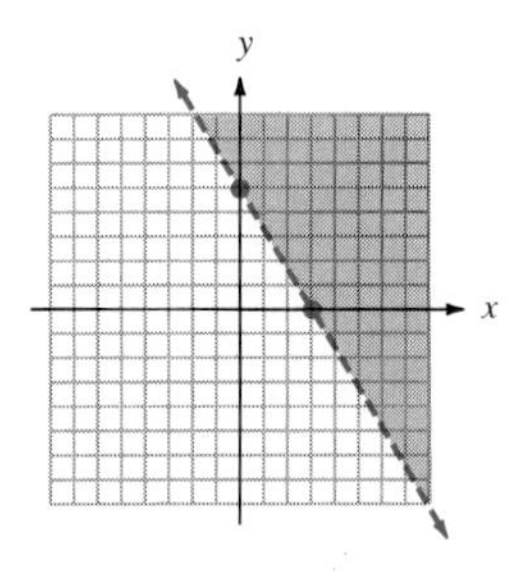

27. $y \leq 3x$

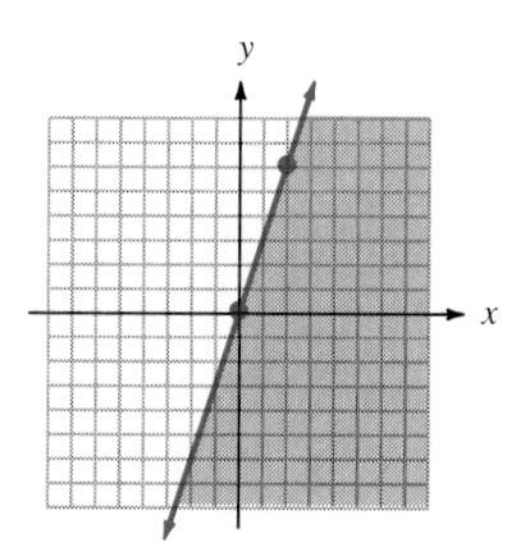

29. $y > 3x - 2$

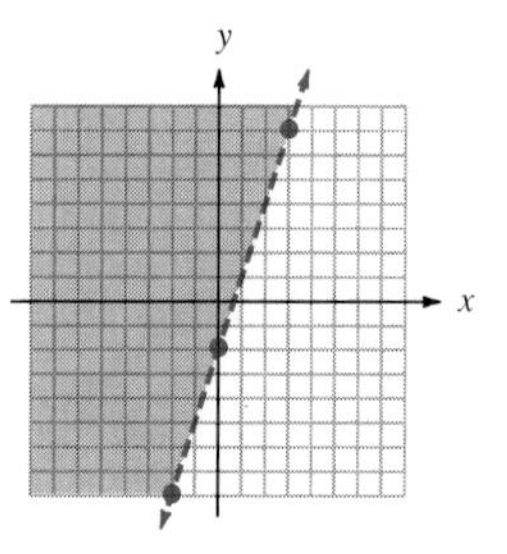

31. $y < -2x - 3$

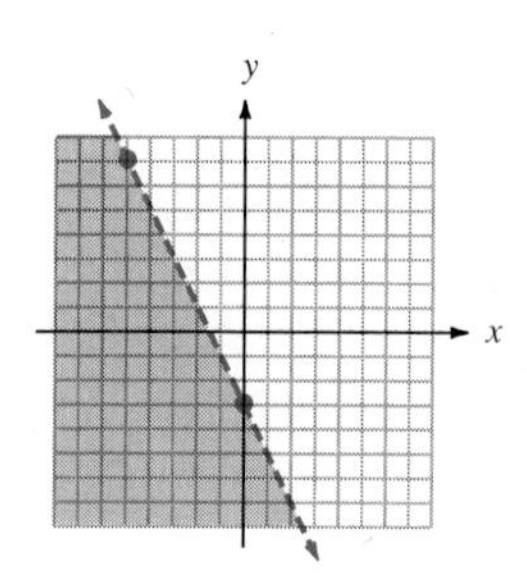

Summary

The Rectangular Coordinate System [7.1] to [7.2]

Solutions of Linear Equations A pair of values that satisfy the equation. Solutions for linear equations in two variables are written as *ordered pairs*. An ordered pair has the form

$$(x, y)$$

x coordinate ↗ ↖ y coordinate

If $2x - y = 10$,
(6, 2) is a solution for the equation, because substituting 6 for x and 2 for y gives a true statement.

The Rectangular Coordinate System A system formed by two perpendicular axes which intersect at a point called the *origin*. The horizontal line is called the *x axis*. The vertical line is called the *y axis*.

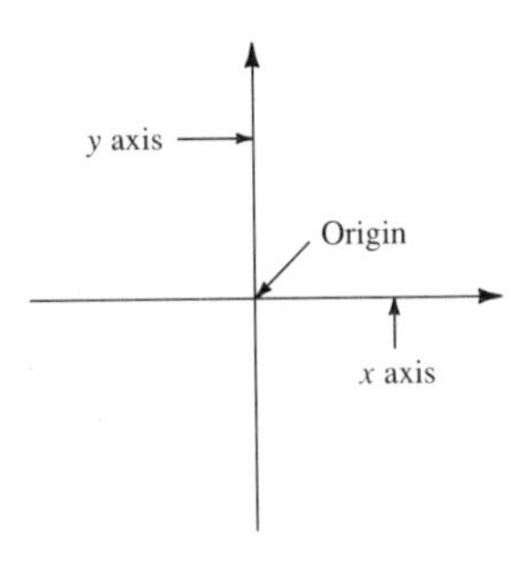

Graphing Points from Ordered Pairs The coordinates of an ordered pair allow you to associate a point in the plane with every ordered pair.

To graph a point in the plane,

1. Start at the origin.
2. Move right or left according to the value of the x coordinate, to the right if x is positive or to the left if x is negative.
3. Then move up or down according to the value of the y coordinate, up if y is positive and down if y is negative.

To graph the point corresponding to (2, 3):

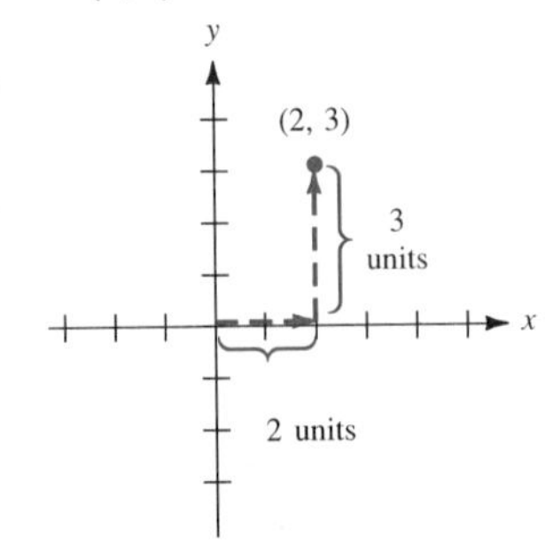

Graphing Linear Equations [7.3]

Linear Equation An equation that can be written in the form

$$Ax + By = C$$

where A and B are not both 0.

$2x + 3y = 4$ is a linear equation.

Graphing Linear Equations

1. Find at least three solutions for the equation, and put your results in tabular form.
2. Graph the solutions found in step 1.
3. Draw a straight line through the points determined in step 2 to form the graph of the equation.

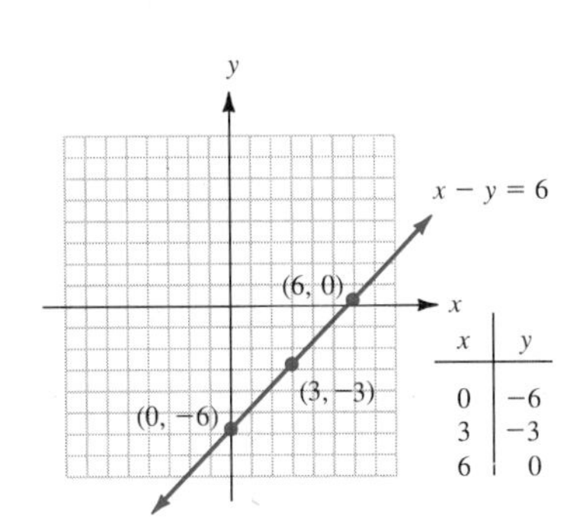

x and y Intercepts of a Line The *x intercept* is the x coordinate of the point where the line intersects the x axis. The *y intercept* is the y coordinate of the point where the line intersects the y axis.

Graphing by the Intercept Method

1. Find the x intercept by letting $y = 0$. Then solve for x.
2. Find the y intercept by letting $x = 0$. Then solve for y.
3. Graph the x and y intercepts.
4. Draw a straight line through the intercepts.

If $2x + 4y = 8$, $(4, 0)$ gives the x intercept and $(0, 2)$ gives the y intercept.

To graph $2x + 4y = 8$:

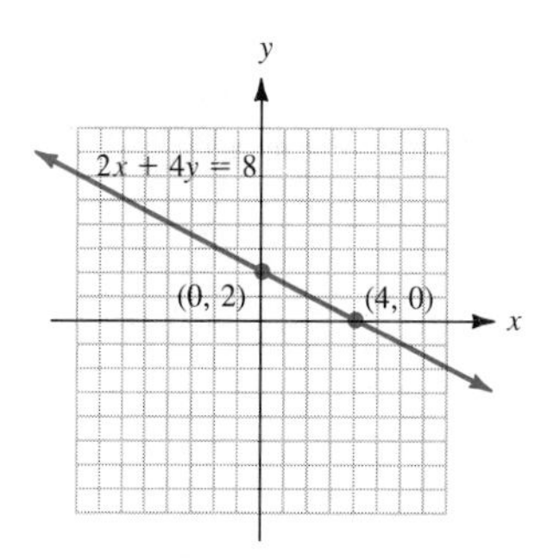

Graphing by Solving for y

1. Solve the given equation for y.
2. Use the equivalent equation (solved for y) to determine solutions.
3. Graph as before.

To graph $x + 2y = 6$, we can solve for y.

$$y = -\frac{1}{2}x + 3$$

Some solutions are

x	y
−2	4
0	3
2	2

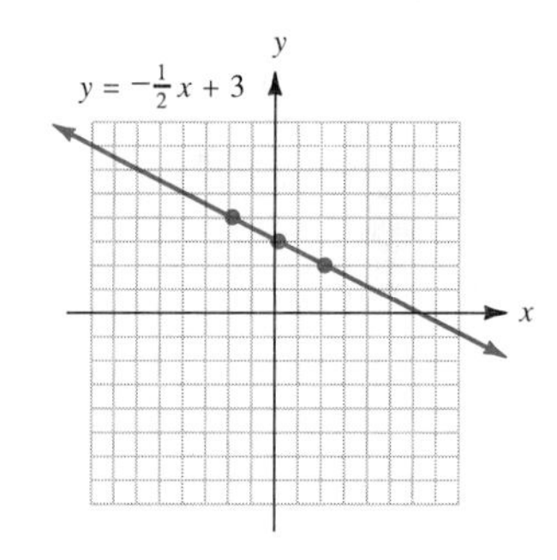

The Slope of a Line [7.4]

Slope The slope of a line gives a numerical measure of the steepness of the line. The slope m of a line containing the distinct points in the plane $P(x_1, y_1)$ and $Q(x_2, y_2)$ is given by

$$m = \frac{y_2 - y_1}{x_2 - x_1} \quad \text{where } x_2 \neq x_1$$

To find the slope of the line through $(-2, -3)$ and $(4, 6)$,

$$m = \frac{6 - (-3)}{4 - (-2)} = \frac{6 + 3}{4 + 2} = \frac{9}{6} = \frac{3}{2}$$

Slope-Intercept Form The slope-intercept form for the equation of a line is

$$y = mx + b$$

where the line has slope m and y intercept b.

For the equation

$$y = \frac{2}{3}x - 3$$

the slope m is $\frac{2}{3}$ and b, the y intercept, is -3.

Graphing Linear Inequalities [7.5]

The Graphing Steps

1. Replace the inequality symbol with an equals sign to form the equation of the boundary line of the graph.
2. Graph the boundary line. Use a dotted line if equality is not included ($<$ or $>$). Use a solid line if equality is included ($\leq$ or $\geq$).
3. Choose any convenient test point not on the line.
4. If the inequality is *true* at the checkpoint, shade the half plane including the test point. If the inequality is *false* at the checkpoint, shade the half plane that does not include the checkpoint.

To graph $x - 2y < 4$: $x - 2y = 4$ is the boundary line. Using $(0, 0)$ as the checkpoint, we have

$$0 - 2 \cdot 0 < 4$$
$$0 < 4 \quad \text{(true)}$$

Shade the half plane that includes $(0, 0)$.

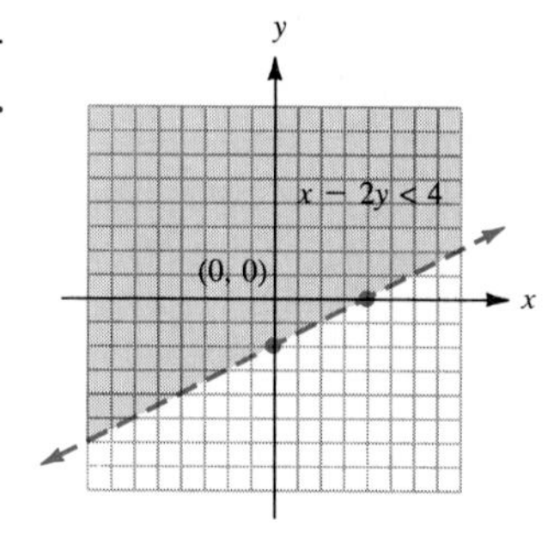

Summary Exercises Chapter 7

This summary exercise set is provided to give you practice with each of the objectives of the chapter. Each exercise is keyed to the appropriate chapter section. The answers are provided in the instructor's manual. Your instructor will give you guidelines on how to best use these exercises in your instructional setting.

[7.1] Determine which of the ordered pairs are solutions for the given equations.

1. $x - y = 8$ $(8, 0), (4, 4), (4, -4), (0, -8)$ (8, 0), (4, −4), (0, −8)

2. $3x + y = 6$ $(2, 0), (3, 2), (1, 3), (3, 1)$ (2, 0), (1, 3)

3. $2x + 5y = 10$ $(5, 0), (10, 2), (-5, 4), (0, -2)$ (5, 0), (−5, 4)

4. $2x - 3y = 6$ $(3, 0), \left(\frac{3}{2}, -1\right), \left(2, \frac{2}{3}\right), (0, -2)$ $(3, 0), \left(\frac{3}{2}, -1\right), (0, -2)$

[7.1] Complete the ordered pairs so that each is a solution for the given equation.

5. $x + y = 10$ $(5,\ ), (\ , 10), (10,\ ), (8,\ )$ 5, 0, 0, 2

6. $x - 2y = 8$ $(0,\ ), (10,\ ), (\ , -2), (6,\ )$ −4, 1, 4, −1

7. $2x - 3y = 12$ $(6,\ ), (9,\ ), (\ , -4), (\ , -6)$ 0, 2, 0, −3

8. $y = 3x - 4$ $(2,\ ), (\ , 5), \left(\frac{1}{3},\ \right), \left(\frac{4}{3},\ \right)$ 2, 3, −3, 0

[7.1] Find four solutions for each of the following equations.

9. $x + y = 8$ (0, 8), (2, 6), (4, 4), (6, 2)

10. $2x + y = 4$ (0, 4), (2, 0), (4, −4), (6, −8)

11. $3x - 2y = 6$ (0, −3), (2, 0), (4, 3), (6, 6)

12. $y = -\frac{2}{3}x + 3$ (0, 3), (3, 1), (6, −1), (9, −3)

[7.2] Give the coordinates of the points graphed below.

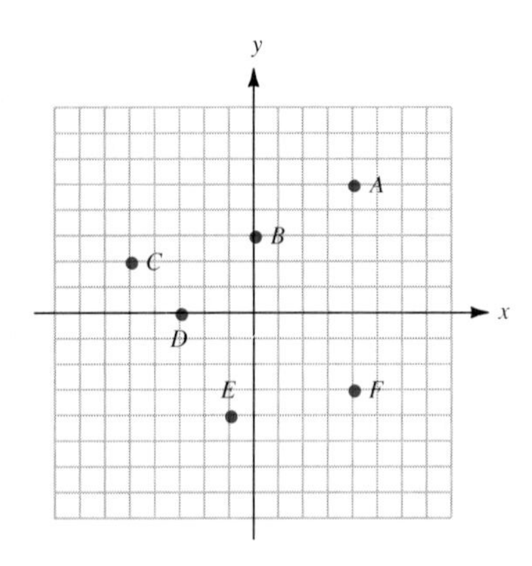

13. A (4, 5)

14. B (0, 3)

15. C (−5, 2)

16. D (−3, 0)

17. E (−1, −4)

18. F (4, −3)

[7.2] Plot points with the coordinates shown.

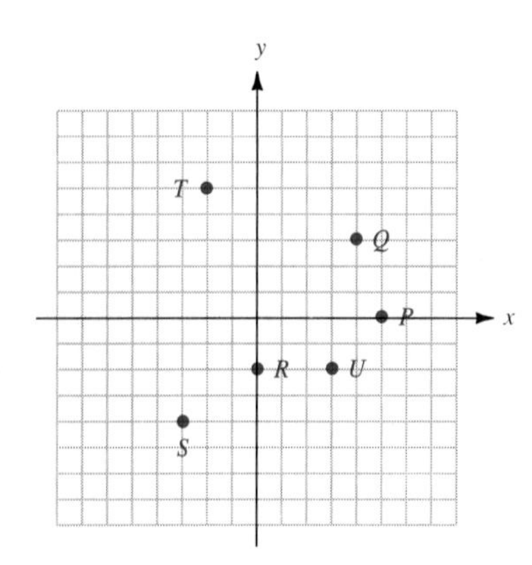

19. P (5, 0)

20. Q (4, 3)

21. R (0, −2)

22. S (−3, −4)

23. T (−2, 5)

24. U (3, −2)

[7.3] Graph each of the following equations.

25. $x + y = 4$

26. $x - y = 5$

27. $y = 3x$

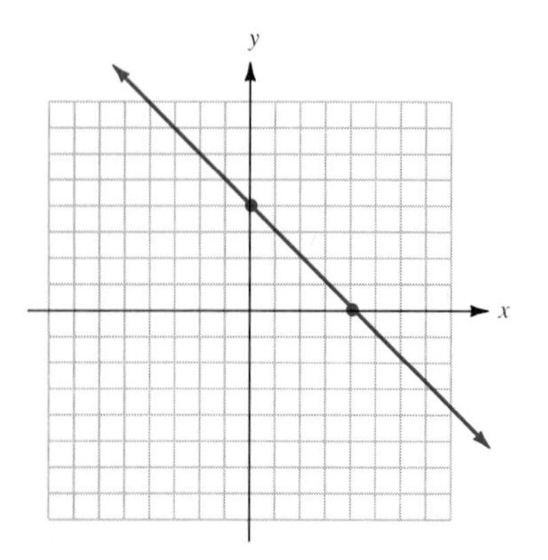

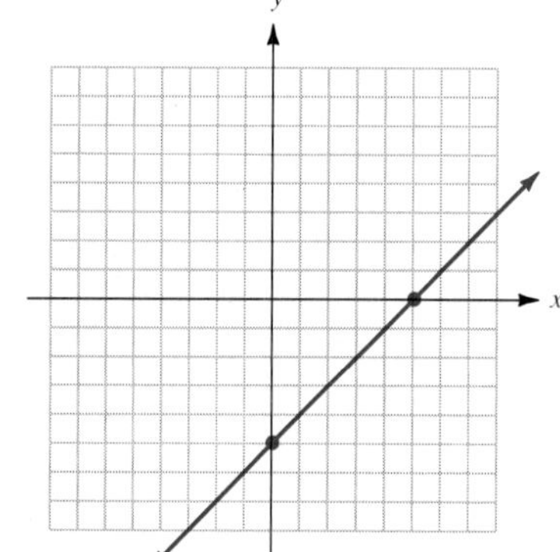

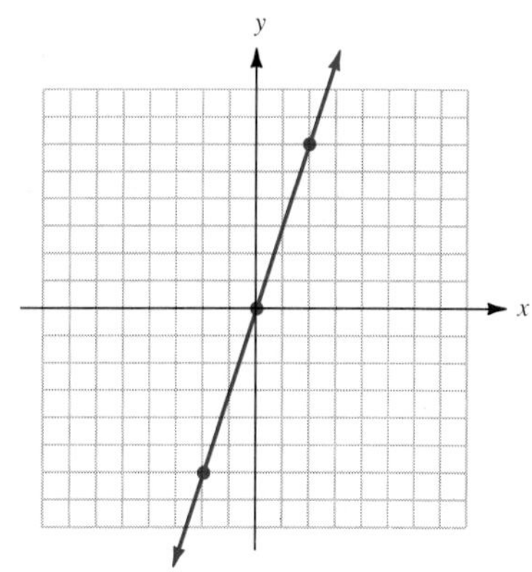

28. $y = -2x$

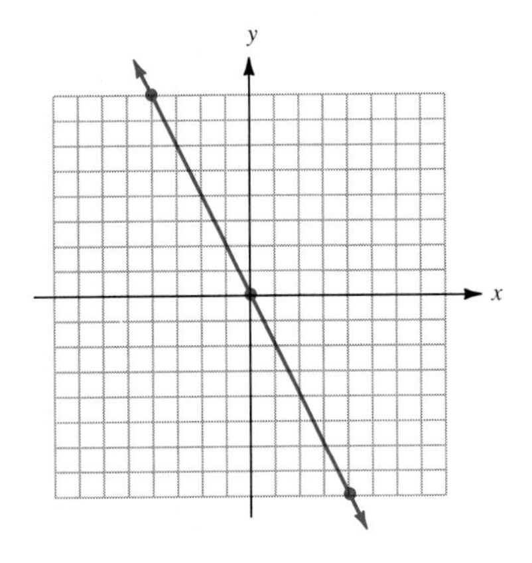

29. $y = \frac{2}{3}x$

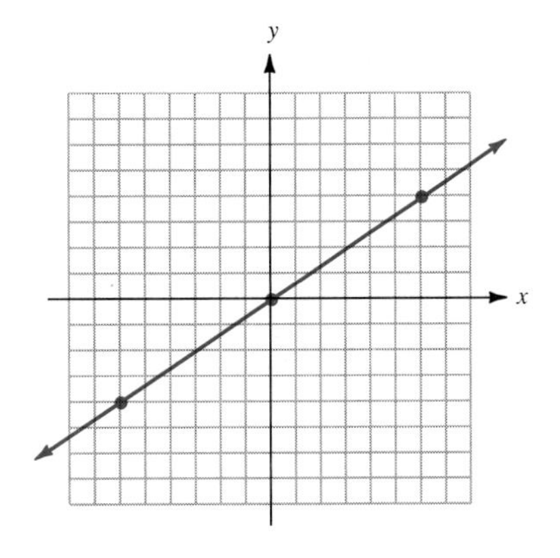

30. $y = 2x + 3$

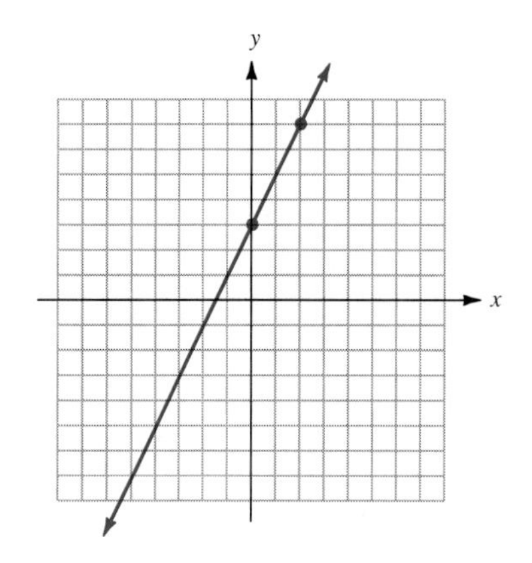

31. $y = 3x - 2$

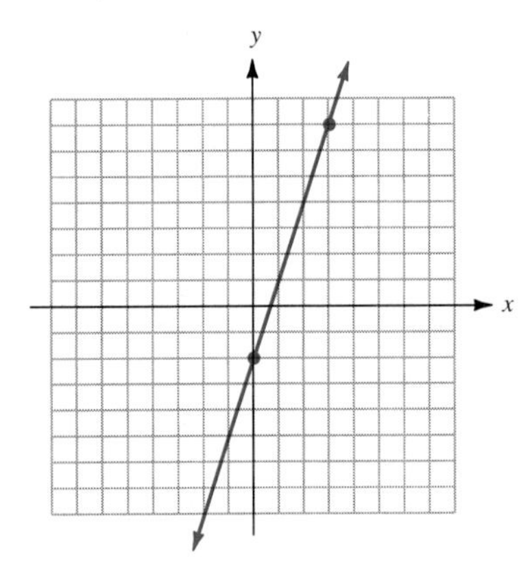

32. $y = -4x + 3$

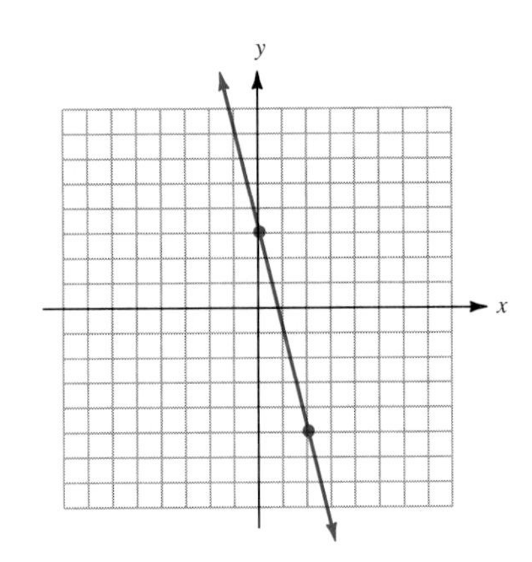

33. $y = \frac{2}{3}x - 2$

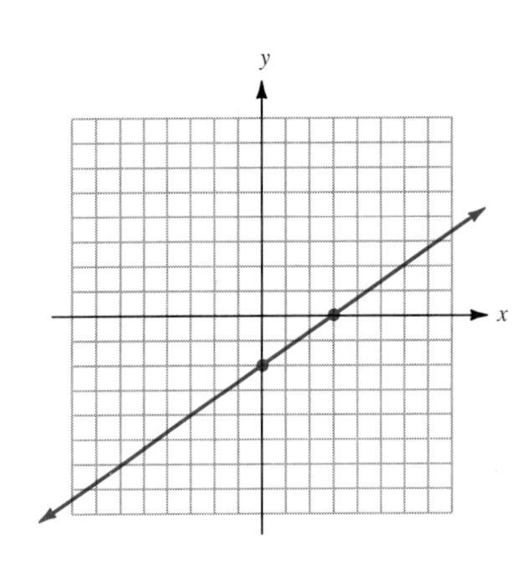

34. $2x - y = 2$

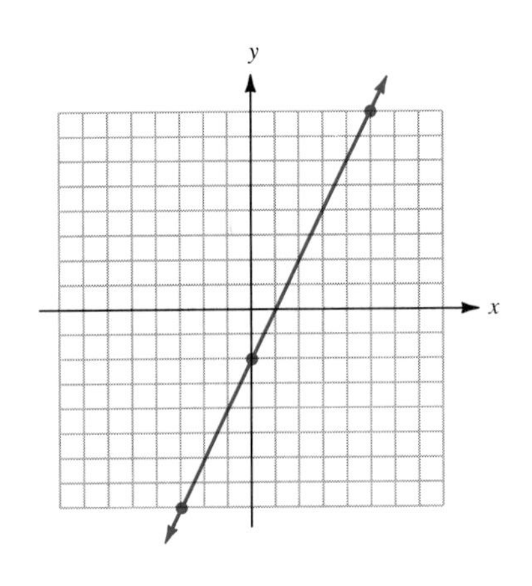

35. $3x + y = 6$

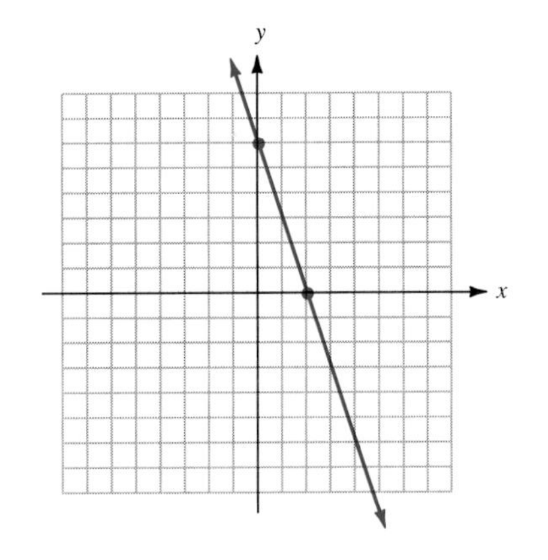

36. $2x + 3y = 12$

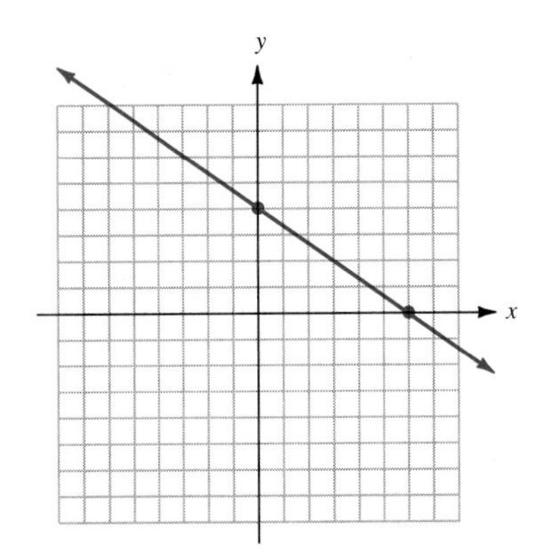

37. $3x - 5y = 15$

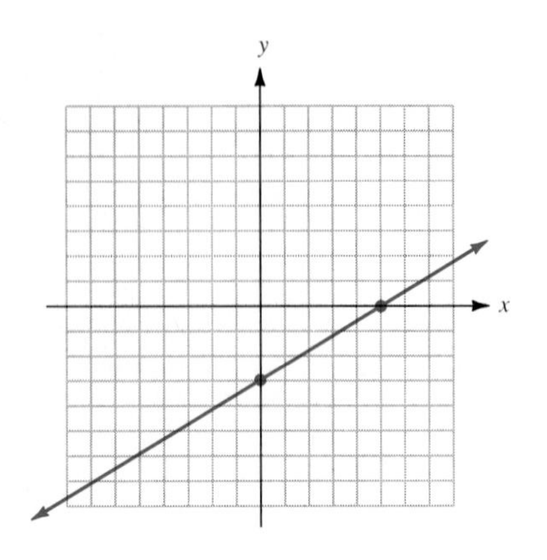

38. $x = 2$

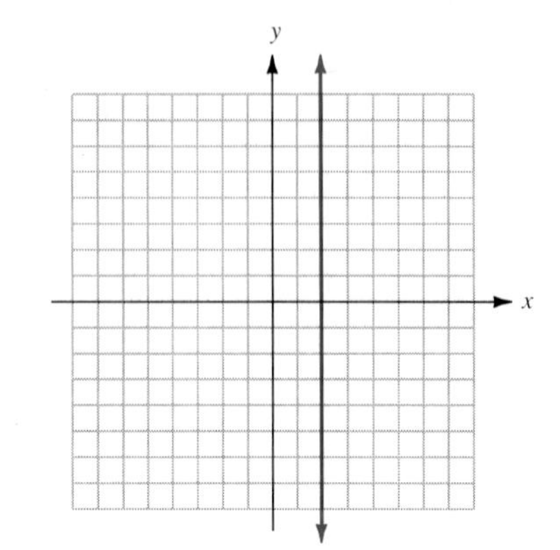

39. $y = -3$

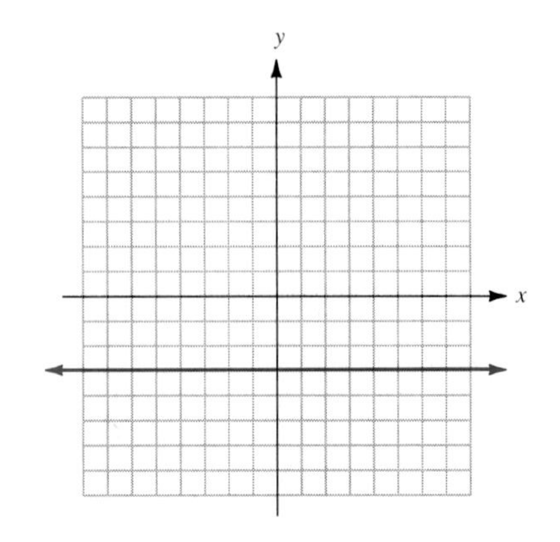

[7.3] Graph each of the following equations.

40. $3x - 4y = 12$

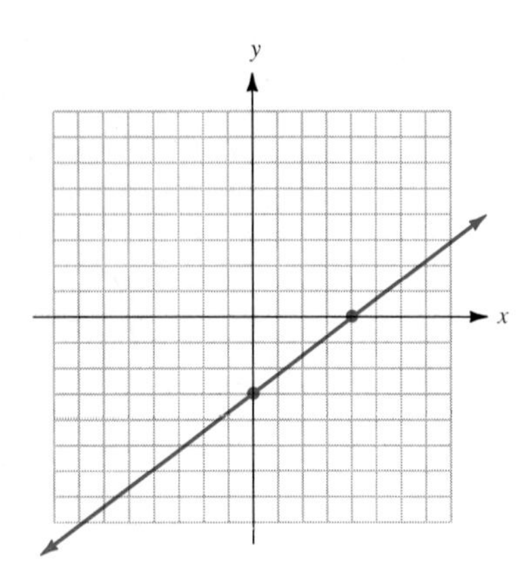

41. $3x + 5y = 15$

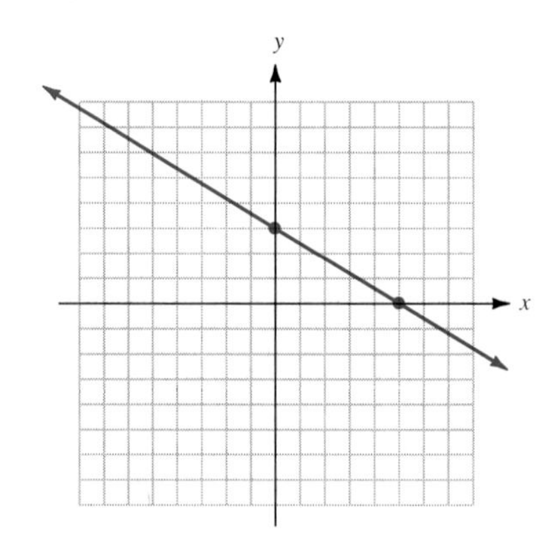

[7.3] Graph each equation by first solving for y.

42. $3x + y = 6$ $\quad y = -3x + 6$

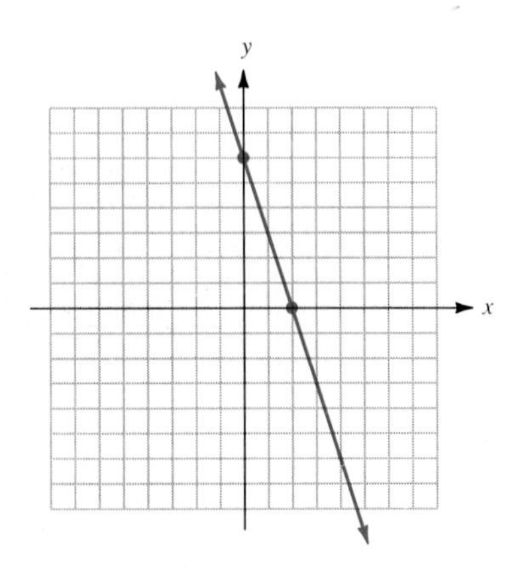

43. $2x + 3y = 6$

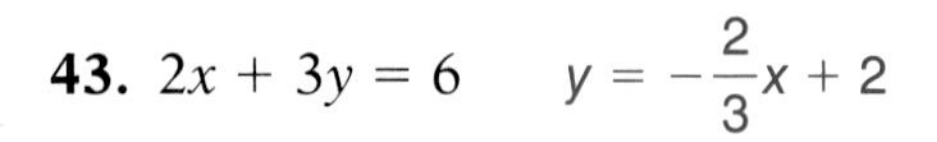

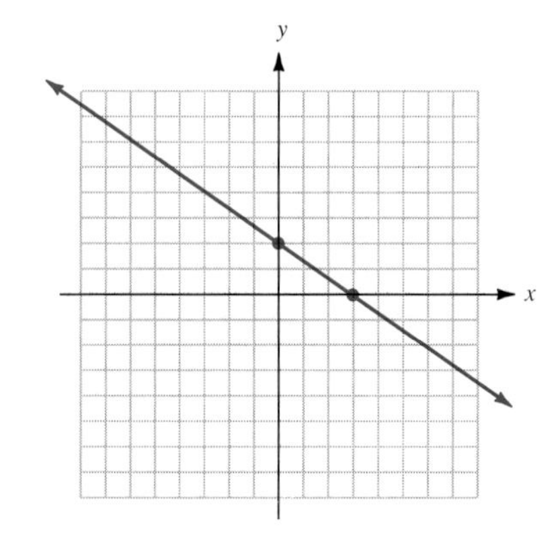

[7.4] Find the slope of the lines through the following pairs of points.

44. (3, 2) and (5, 8)

3

45. (−2, 3) and (1, −3)

−2

46. (−1, 5) and (3, 3)

$-\frac{1}{2}$

47. (−2, −5) and (2, 1)

$\frac{3}{2}$

48. (−2, 5) and (5, 5)

0

49. (−3, 4) and (−1, −3)

$-\frac{7}{2}$

50. (−3, −4) and (5, −2)

$\frac{1}{4}$

51. (−4, −2) and (−4, 3)

Undefined

[7.4] Find the slope and *y* intercept of the line represented by each of the following equations.

52. $y = 3x + 5$

3, 5

53. $y = -3x - 4$

−3, −4

54. $y = -\frac{2}{3}x$

$-\frac{2}{3}$, 0

55. $y = \frac{3}{4}x + 3$

$\frac{3}{4}$, 3

56. $2x + 5y = 10$

$-\frac{2}{5}$, 2

57. $3x - 2y = 6$

$\frac{3}{2}$, −3

58. $y = -2$

0, −2

59. $x = 3$

undefined slope, no *y* intercept

[7.4] Write the equation of the line with the given slope and *y* intercept. Then graph each line, *using* the slope and *y* intercept.

60. $m = 3$, $b = 2$

$y = 3x + 2$

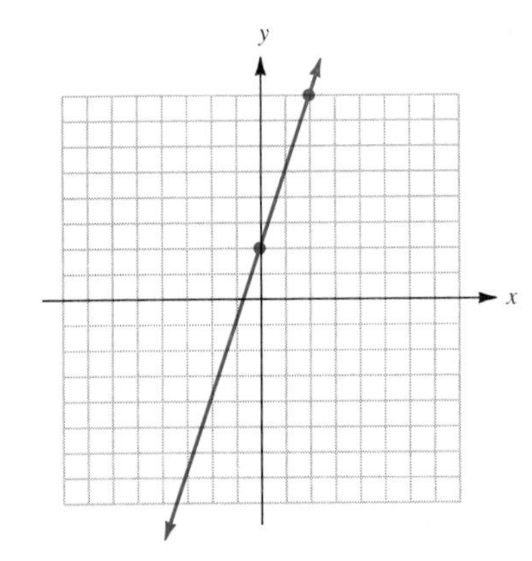

61. $m = \frac{2}{3}$, $b = -3$

$y = \frac{2}{3}x - 3$

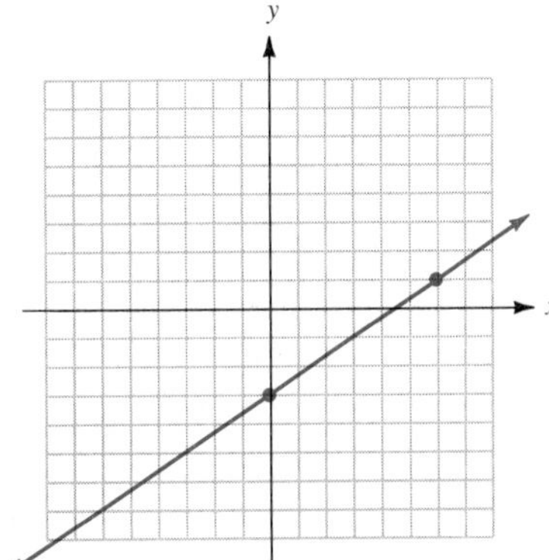

62. $m = -\frac{5}{2}$, $b = 3$

$y = -\frac{5}{2}x + 3$

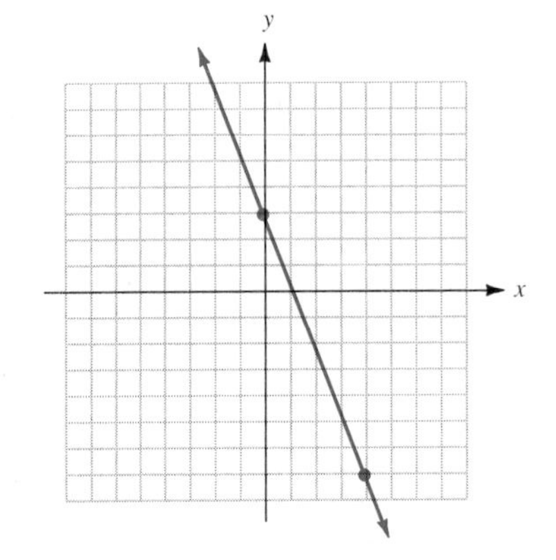

[7.5] Graph each of the following inequalities.

63. $x + y \leq 5$

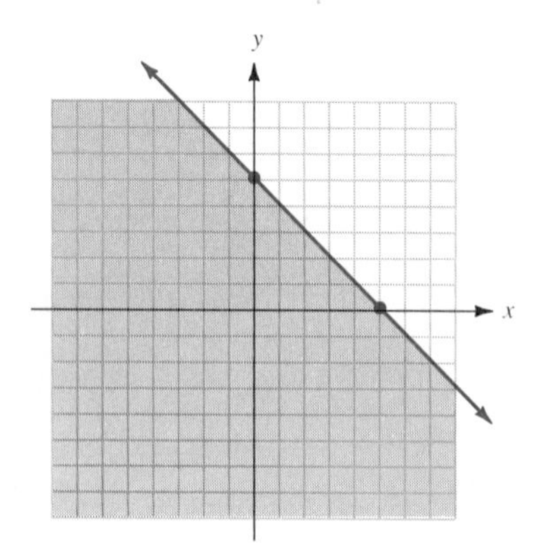

64. $x - y > 6$

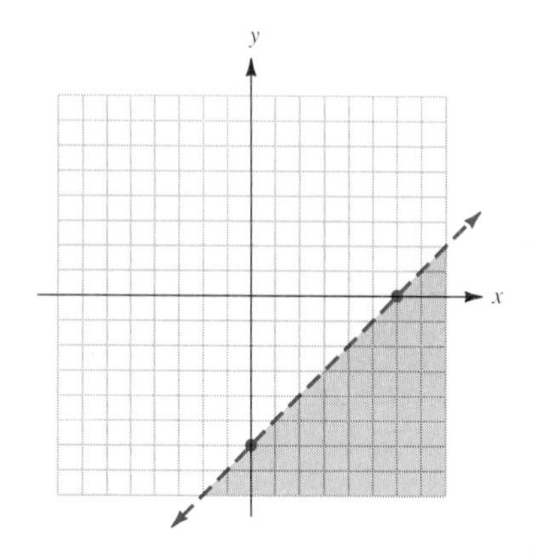

65. $2x + y < 4$

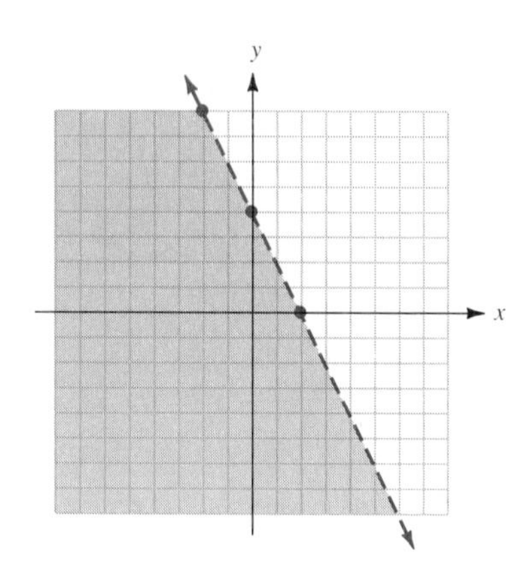

66. $3x - y \geq 6$

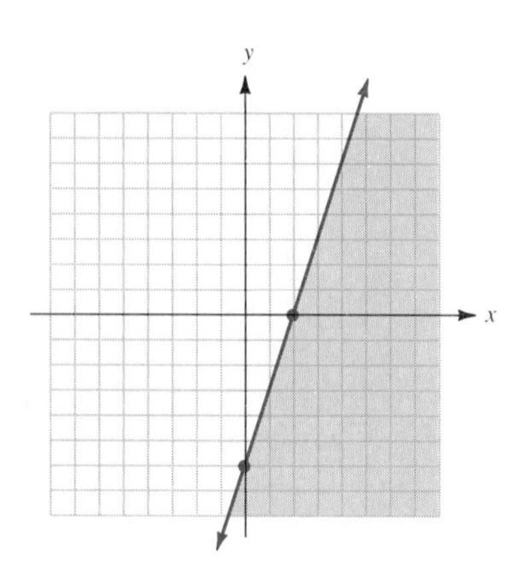

67. $x - 2y > 8$

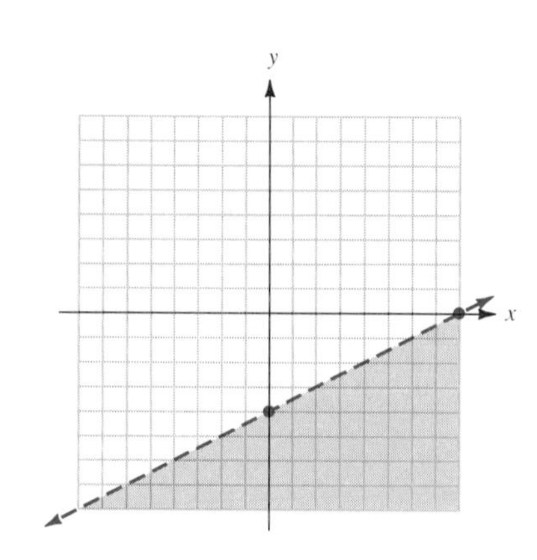

68. $2x - y \leq 0$

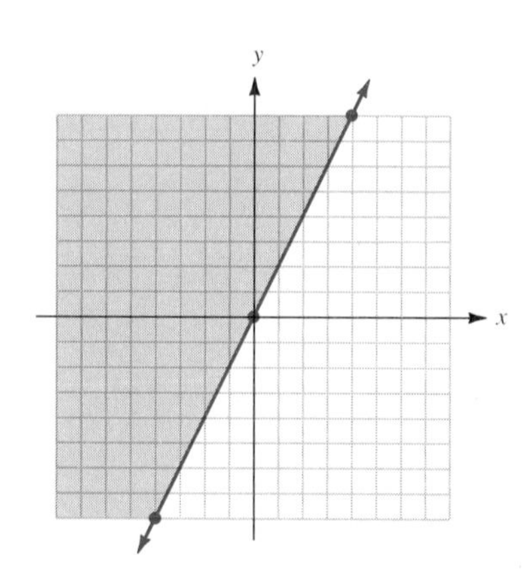

69. $x > 2$

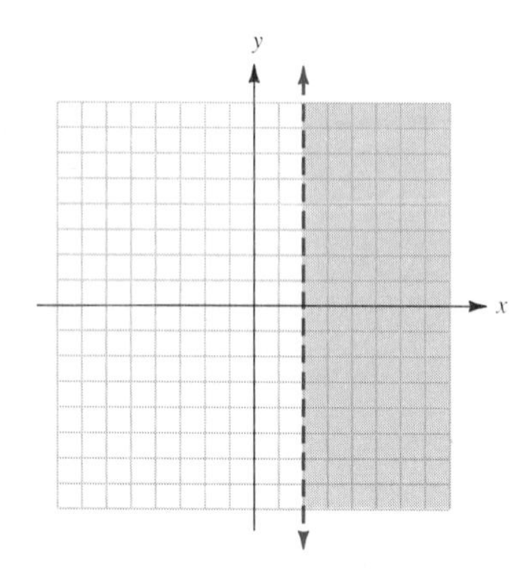

70. $y \leq 3$

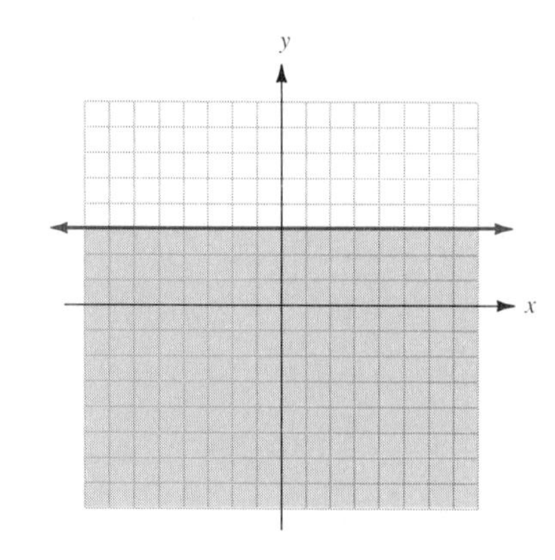

Self-Test for Chapter Seven

The purpose of this self-test is to help you check your progress and to review for a chapter test in class. Allow yourself about an hour to take the test. When you are done, check your answers in the back of the book. If you missed any problems, be sure to go back and review the appropriate sections in the chapter and the exercises that are provided.

Determine which of the ordered pairs are solutions for the given equations.

1. $x + y = 6$ $(3, 3), (2, 6), (6, 0), (4, 1)$
(3, 3), (6, 0)

2. $2x - y = 8$ $(4, 0), (2, -2), (3, -2), (0, 8)$
(4, 0), (3, −2)

Complete the ordered pairs so that each is a solution for the given equation:

3. $x + 2y = 6$ $(4, \ ), \ (\ , 3), \ (6, \ ), \ (\ , 2)$
1, 0, 0, 2

4. $3x + 2y = 12$ $(4, \ ), (\ , 3), (\ , 6), \left(\frac{2}{3}, \ \right)$
0, 2, 0, 5

Find four solutions for each of the following equations.

5. $x - y = 8$ (0, −8), (2, −6), (4, −4), (6, −2)

6. $3x - 2y = 6$ (0, −3), (2, 0), (4, 3), (6, 6)

Give the coordinates of the points graphed below.

7. A (4, 5)

8. B (−7, 1)

9. C (0, −4)

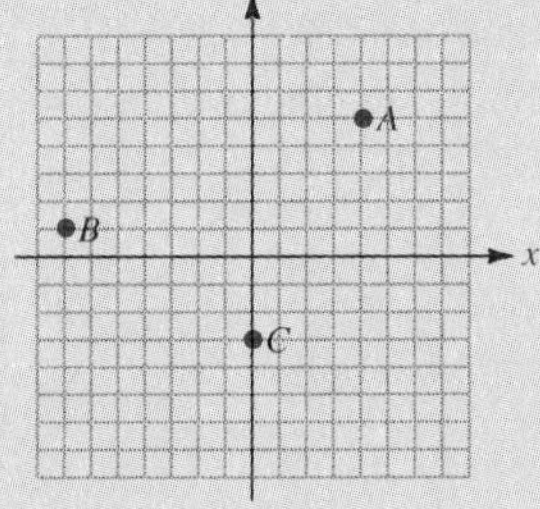

Plot points with the coordinates shown.

10. $S\ (3, -4)$

11. $T\ (0, 5)$

12. $U\ (-4, -5)$

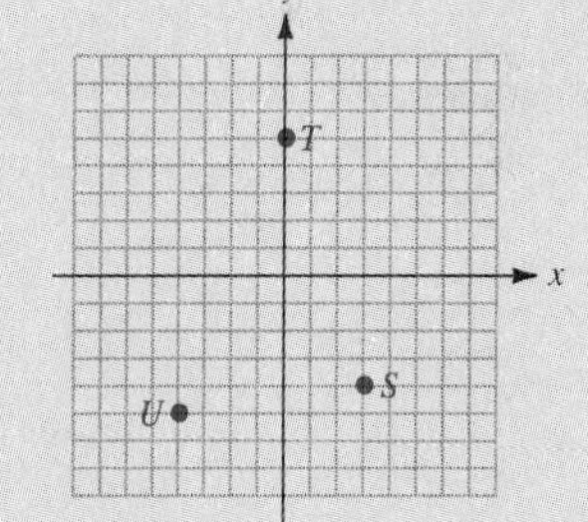

Graph each of the following equations.

13. $x + y = 6$

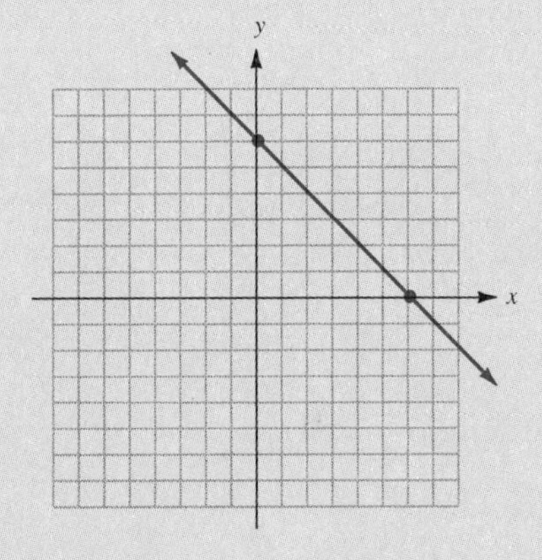

14. $y = 4x$

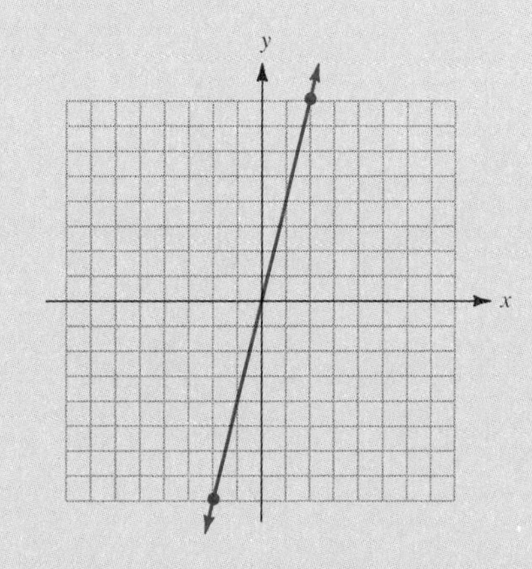

15. $y = \frac{2}{3}x - 3$

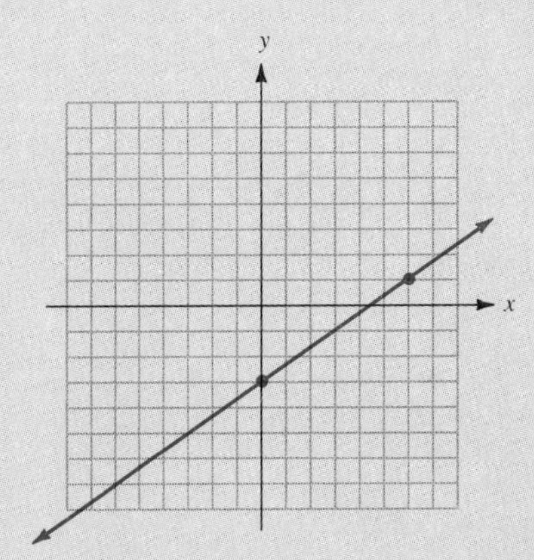

16. $x + 2y = 4$

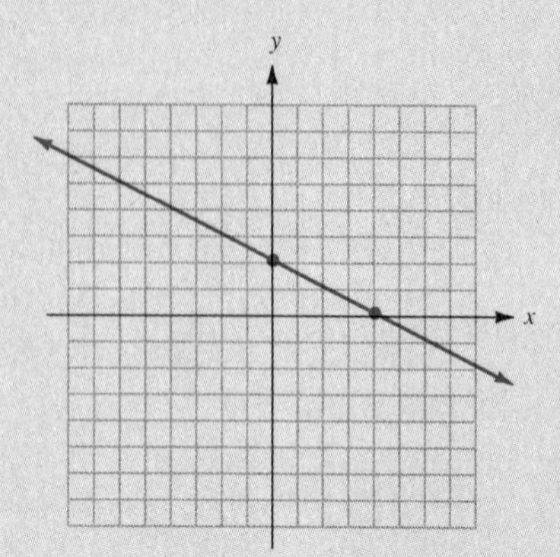

17. $3x + 5y = 15$

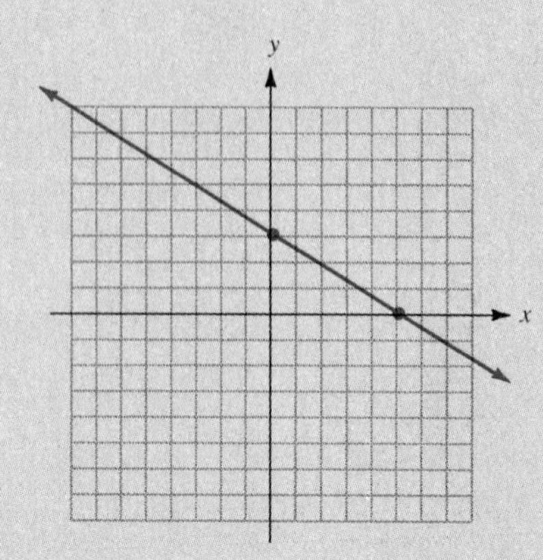

18. $y = -2$

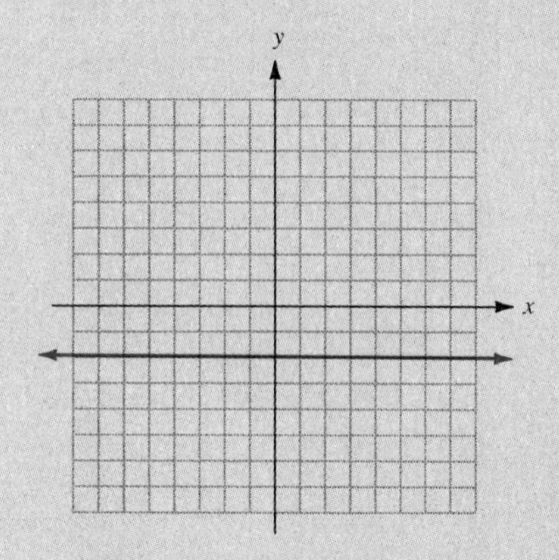

Find the slope of the line through the following pairs of points.

19. $(-2, 3)$ and $(3, 8)$ $\quad 1$

20. $(-3, 7)$ and $(3, 4)$ $\quad -\frac{1}{2}$

Write the equation of the line with the given slope and y intercept. Then graph each line, *using* the slope and y intercept.

21. $m = -2, b = 5$ $\quad y = -2x + 5$

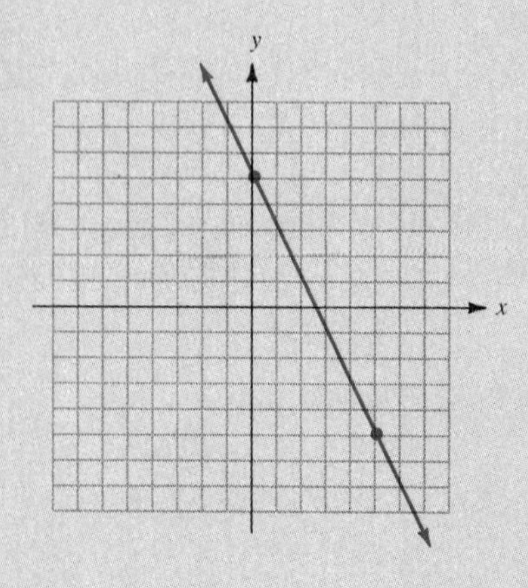

22. $m = \frac{3}{4}, b = -2$ $\quad y = \frac{3}{4}x - 2$

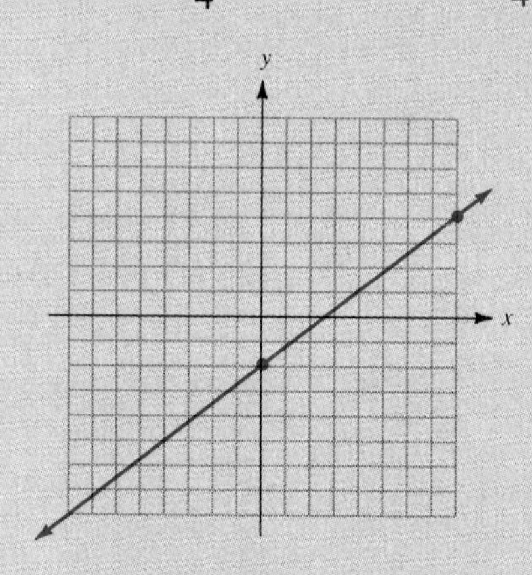

Graph each of the following inequalities.

23. $x + y < 6$

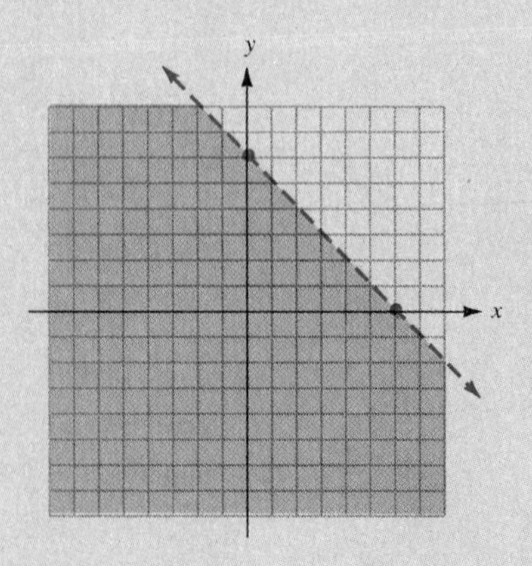

24. $2x + y \geq 8$

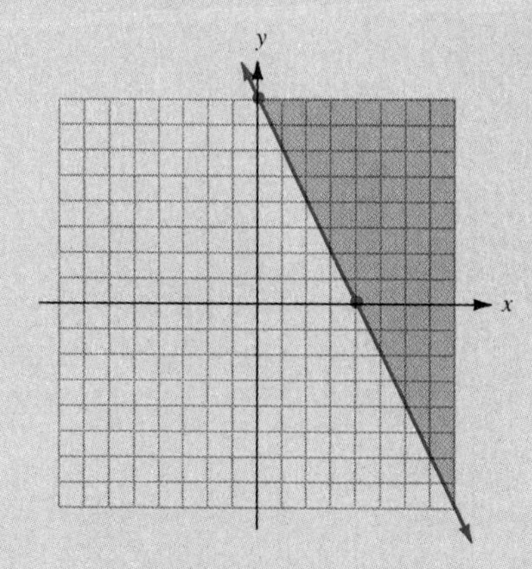

25. $x \leq 4$

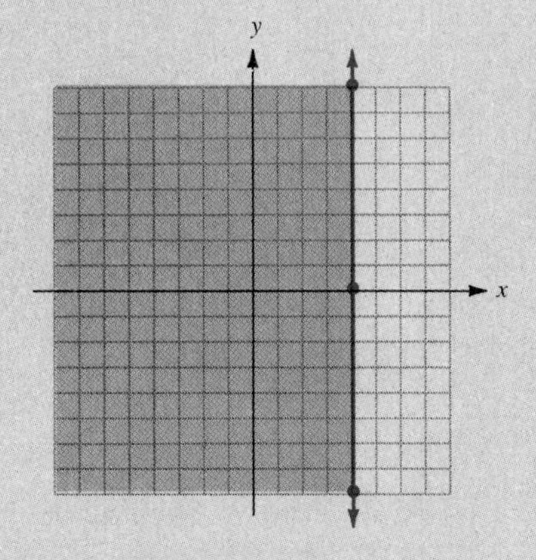

Chapter Eight

Systems of Linear Equations

8.1 Systems of Linear Equations: Solving by Graphing

OBJECTIVE
To find the solution of a system of equations by graphing.

From our work in Section 7.3, we know that an equation of the form $x + y = 3$ is a linear equation. Remember that its graph is a straight line. Often we will want to consider two equations together. They then form a *system of linear equations.* An example of such a system is

$$x + y = 3$$
$$3x - y = 5$$

You know that a solution for a linear equation in two variables is any ordered pair (x, y) that satisfies the equation. Often you will want to find a single ordered pair that satisfies both of the equations of the system. It is called the *solution for the system.* For instance, the solution for the system above is (2, 1) because, replacing x with 2 and y with 1, we have

$x + y = 3$	$3x - y = 5$
$2 + 1 = 3$	$3 \cdot 2 - 1 = 5$
$3 = 3$	$6 - 1 = 5$
	$5 = 5$

There is no other ordered pair that satisfies both equations.

Since both statements are true, the ordered pair (2, 1) satisfies both equations.

One approach to finding the solution for a system of linear equations is called the *graphical method.* Using this approach, we graph the two equations on the same coordinate system. The coordinates of the point where the lines intersect will be the solution for the given system.

Example 1

Solve the system by graphing.

$$x + y = 6$$
$$x - y = 4$$

First, we determine solutions for the equations of our system. For $x + y = 6$, two solutions are (6, 0) and (0, 6). For $x - y = 4$, two solutions are (4, 0) and (0, −4). Using these intercepts, we graph the two equations. The lines intersect at the point (5, 1).

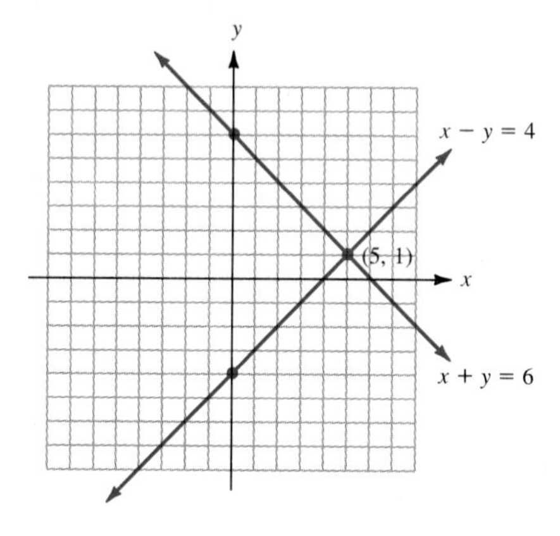

(5, 1) is the solution of the system. It is the only point whose graph lies on both lines.

Note: By substituting 5 for x and 1 for y into the two original equations, we can check that (5, 1) is indeed the solution for our system.

$$\begin{array}{r} x + y = 6 \\ \hline 5 + 1 = 6 \\ 6 = 6 \end{array} \qquad \begin{array}{r} x - y = 4 \\ \hline 5 - 1 = 4 \\ 4 = 4 \end{array}$$

Because both statements are true, (5, 1) is the solution for the system.

CHECK YOURSELF 1

Solve by graphing.

$$2x - y = 4$$
$$x + y = 5$$

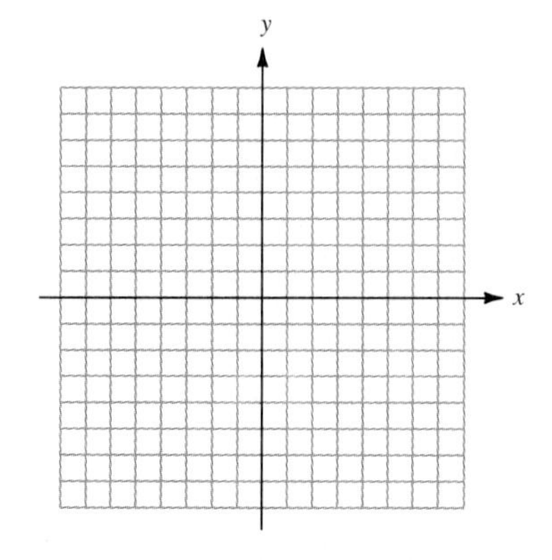

A system such as the one of Example 1 has exactly one solution and is called a *consistent* system. It is possible that a system of equations will have no solution. Look at the following example.

Example 2

Solve by graphing.

$$2x + y = 2$$
$$2x + y = 4$$

We can graph the two lines as before.

Note: In slope-intercept form, our equations are

$y = -2x + 2$

and

$y = -2x + 4$

Both lines have slope -2.

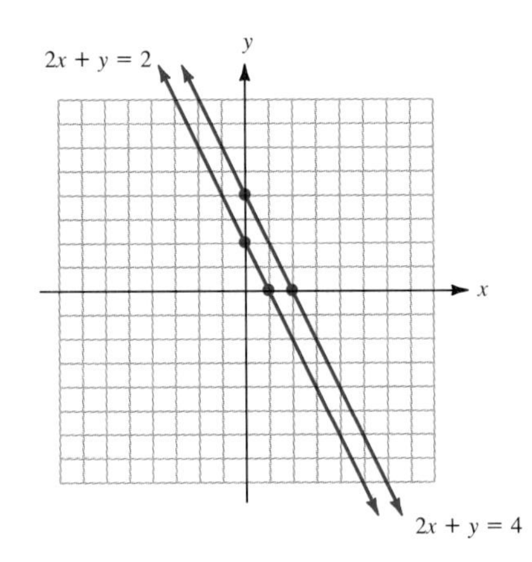

Notice that the slope for each of these lines is -2, but that they have different y intercepts. This means that the lines are parallel (they will never intersect). Since the lines have no points in common, there is no ordered pair that will satisfy both equations. The system is called *inconsistent* and has no solution.

CHECK YOURSELF 2

Solve by graphing (if possible).

$$x - 3y = 3$$
$$x - 3y = 6$$

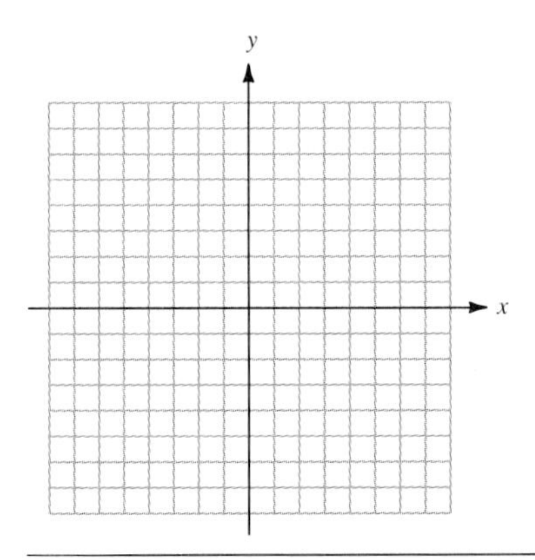

There is one more possibility for linear systems, as the following example illustrates.

Example 3

Solve by graphing.

Note that multiplying the first equation by 2 results in the second equation.

$$x - 2y = 4$$
$$2x - 4y = 8$$

Graphing as before, we find

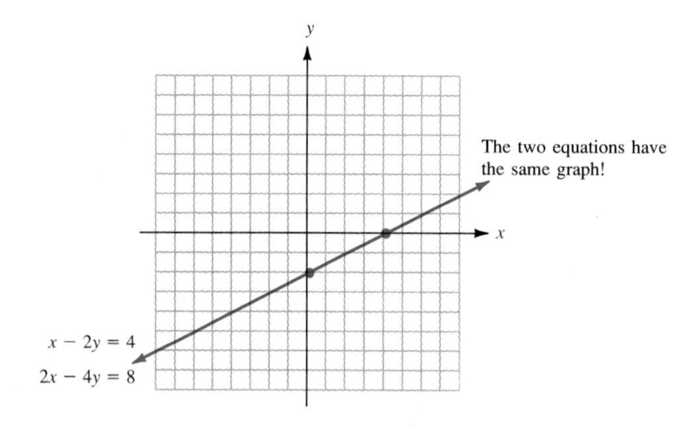

Since the graphs coincide, there are *infinitely many* solutions for this system. Every point on the graph of $x - 2y = 4$ is also on the graph of $2x - 4y = 8$, so any ordered pair satisfying $x - 2y = 4$ also satisfies $2x - 4y = 8$. This is called a *dependent* system, and any point on the line is a solution.

CHECK YOURSELF 3

Solve by graphing.

$$x + y = 4$$
$$2x + 2y = 8$$

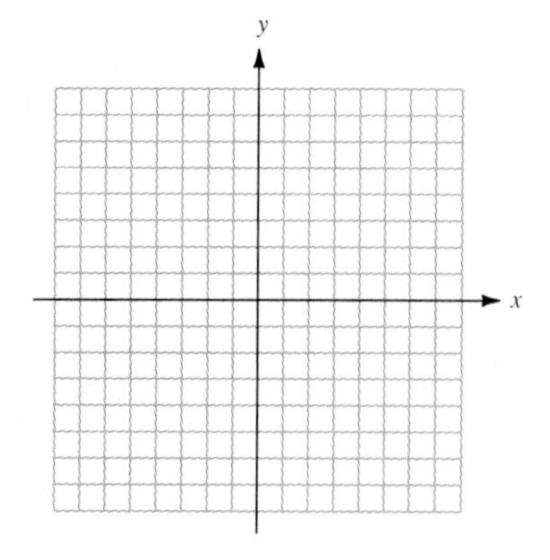

The following will summarize our work in this section.

SOLVING A SYSTEM OF EQUATIONS BY GRAPHING

1. Graph both equations on the same coordinate system.
2. The system may have
   a. *One solution.* The lines intersect at one point. The solution is the ordered pair corresponding to that point.
   b. *No solution.* The lines are parallel.
   c. *Infinitely many solutions.* The two equations have the same graph. Any ordered pair corresponding to a point on the line is a solution.

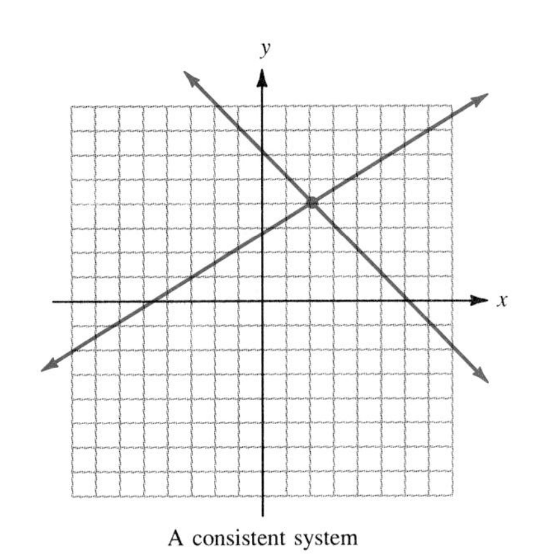

A consistent system

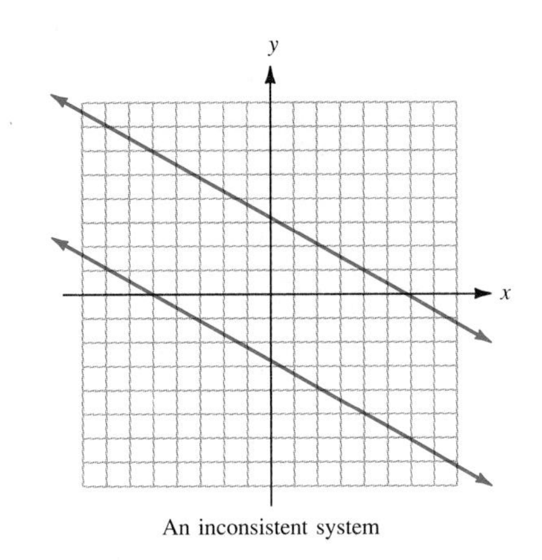

An inconsistent system

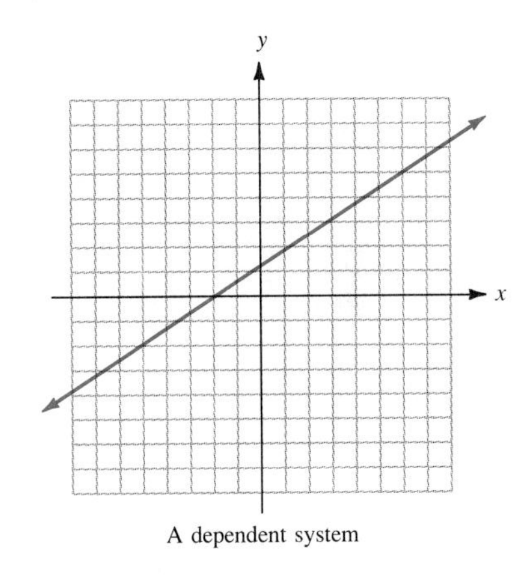

A dependent system

CHECK YOURSELF ANSWERS

1. $x + y = 5$

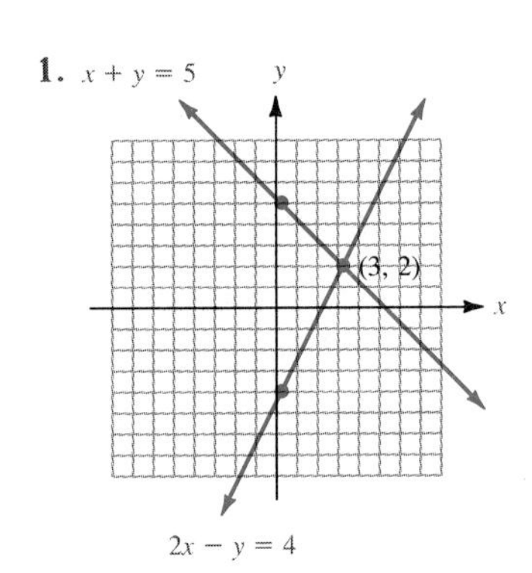

2. There is no solution. The lines are parallel, so the system is inconsistent.

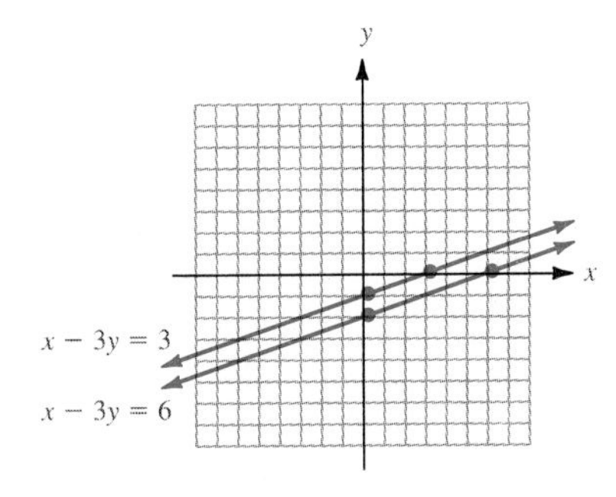

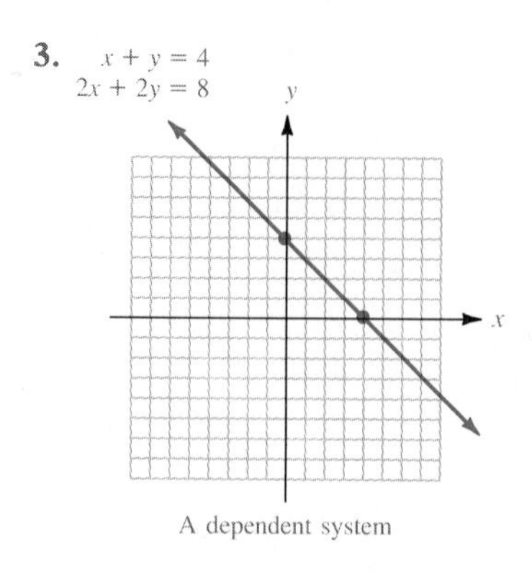

A dependent system

8.1 Exercises

Solve each of the following systems by graphing.

1. $x + y = 6$
$x - y = 4$

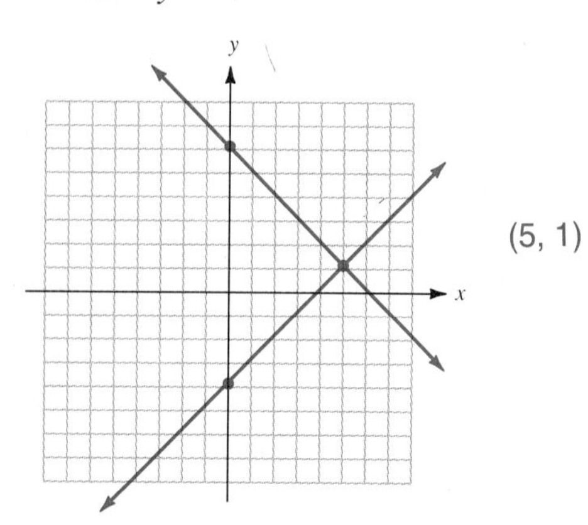

(5, 1)

2. $x - y = 8$
$x + y = 2$

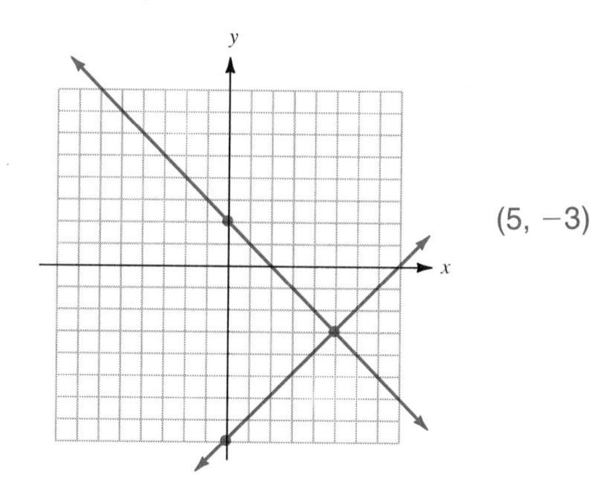

(5, −3)

3. $-x + y = 3$
$x + y = 5$

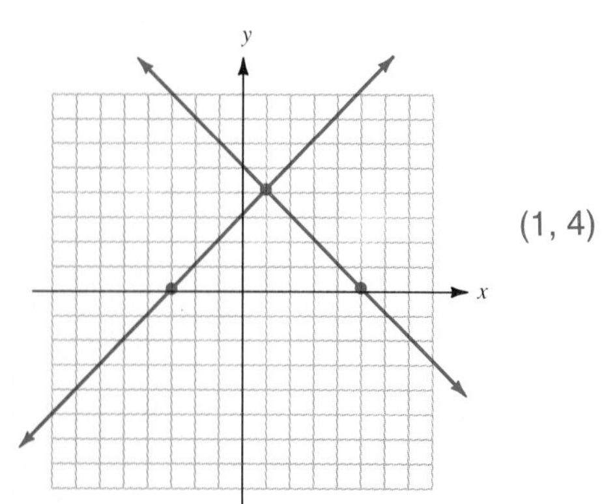

(1, 4)

4. $x + y = 7$
$-x + y = 3$

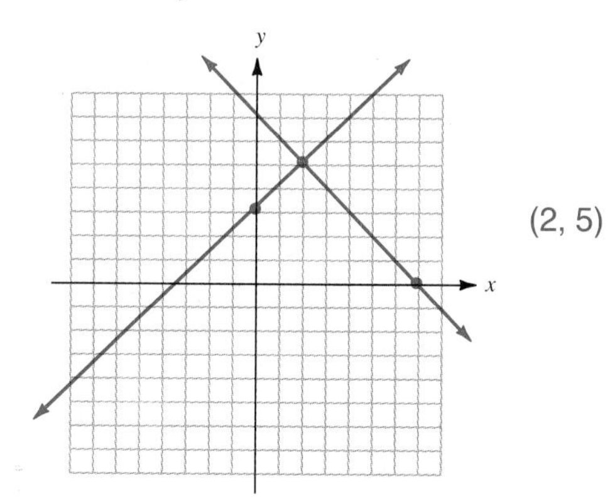

(2, 5)

5. $x + 2y = 4$
$x - y = 1$

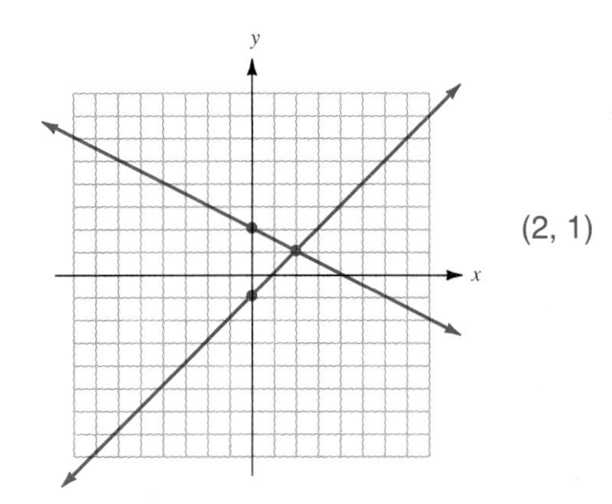

(2, 1)

6. $3x + y = 6$
$x + y = 4$

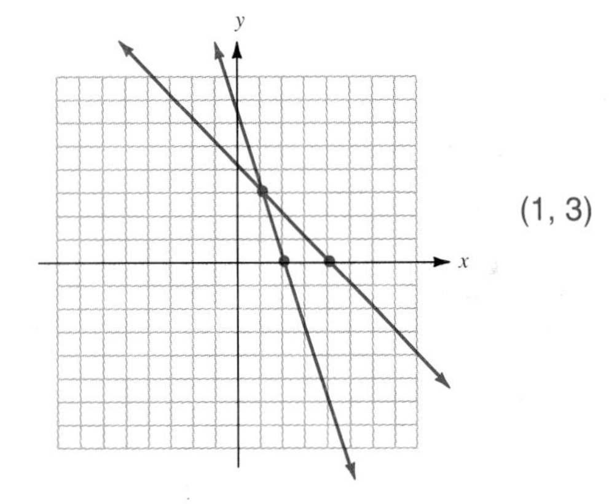

(1, 3)

7. $2x + y = 8$
$2x - y = 0$

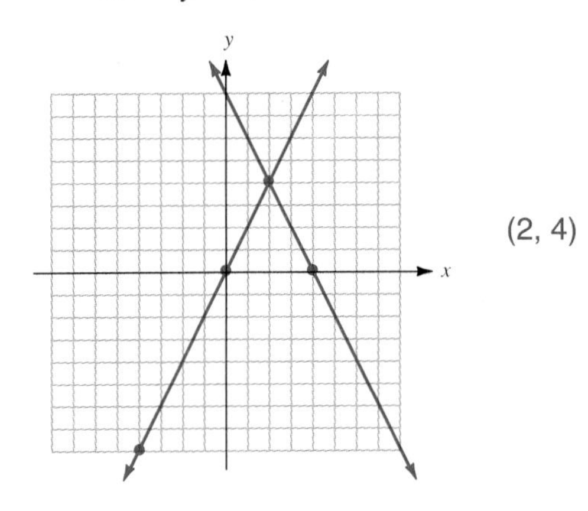

(2, 4)

8. $x - 2y = -2$
$x + 2y = 6$

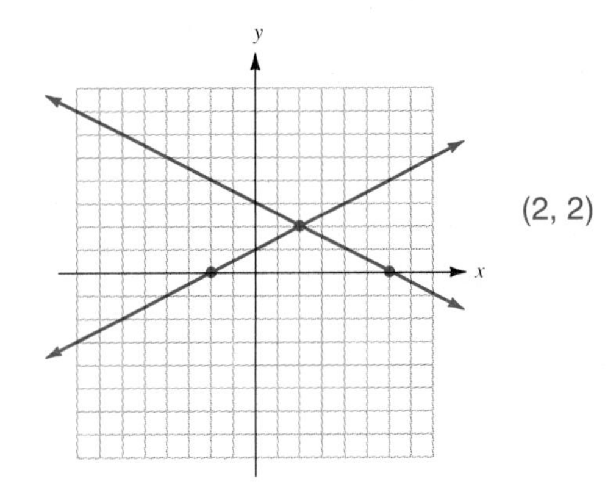

(2, 2)

9. $x + 3y = 12$
$2x - 3y = 6$

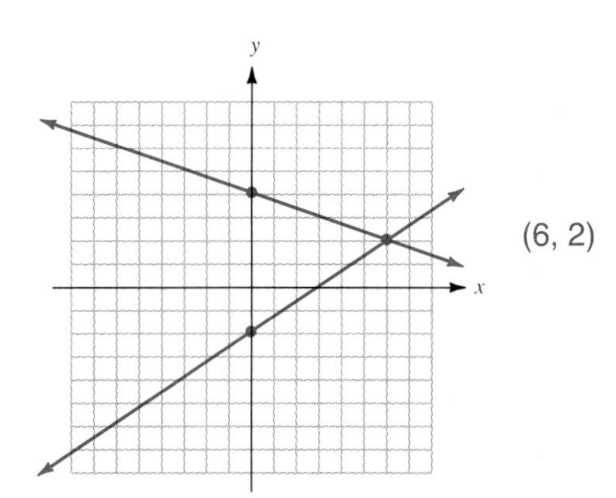

(6, 2)

10. $2x - y = 4$
$2x - y = 6$

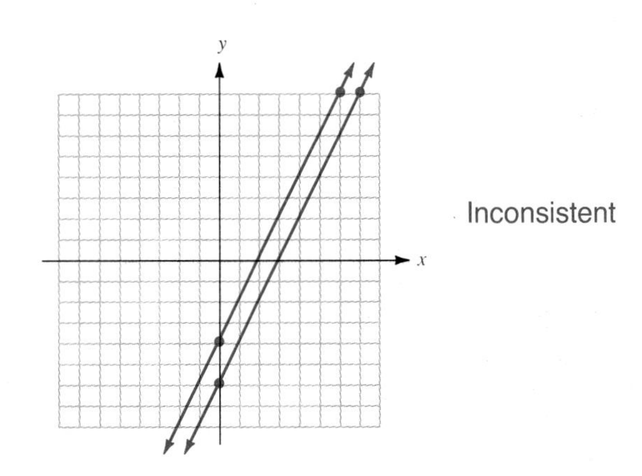

Inconsistent

11. $3x + 2y = 12$
$y = 3$

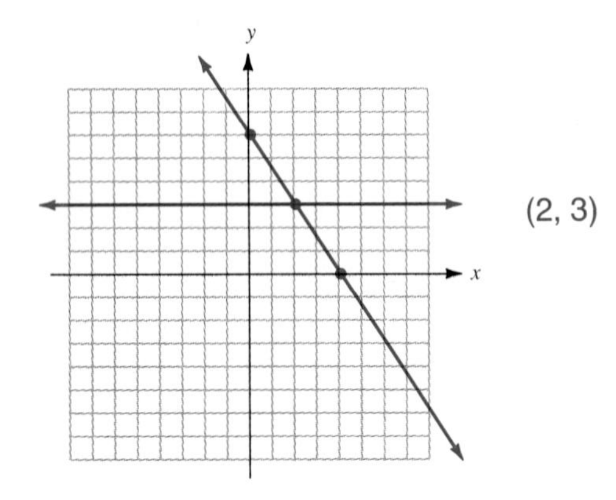

(2, 3)

12. $x - 2y = 8$
$3x - 2y = 12$

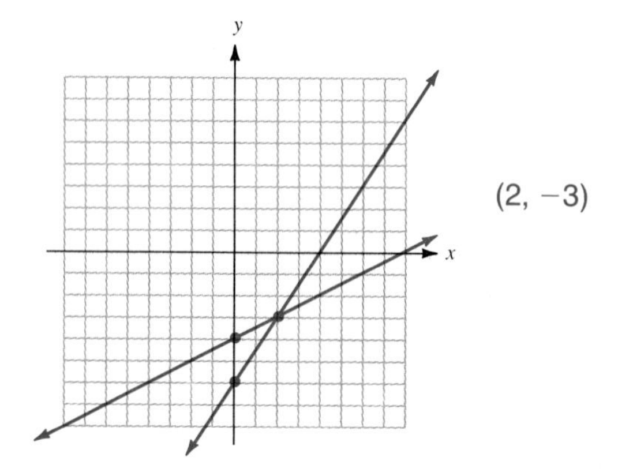

(2, −3)

13. $x - y = 4$
$2x - 2y = 8$

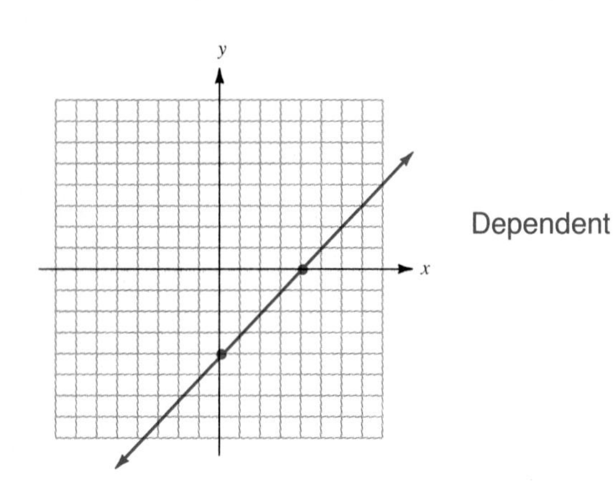

Dependent

14. $2x - y = 8$
$x = 2$

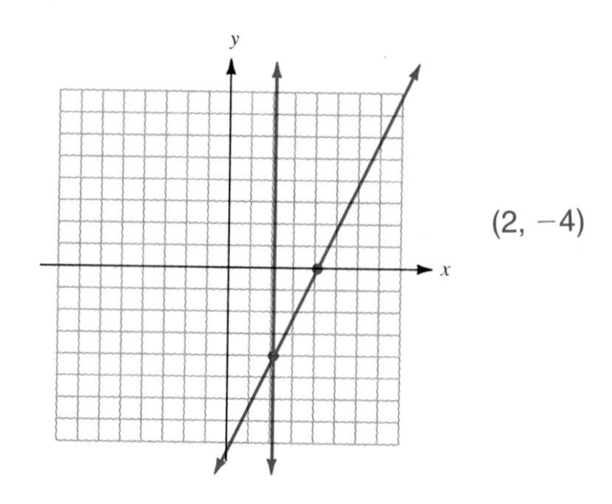

(2, −4)

15. $x - 4y = -4$
$x + 2y = 8$

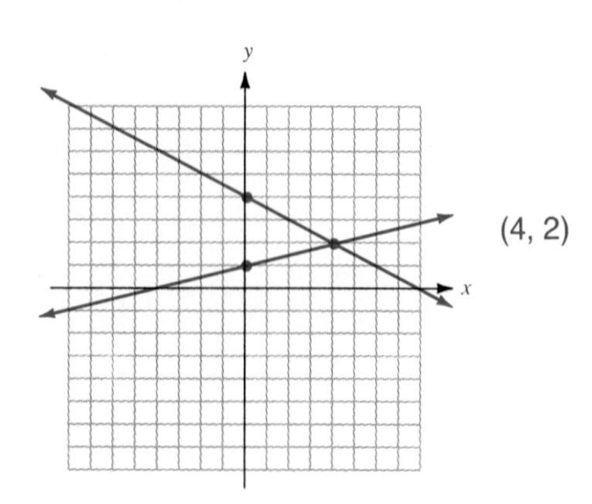

(4, 2)

16. $x - 6y = 6$
$-x + y = 4$

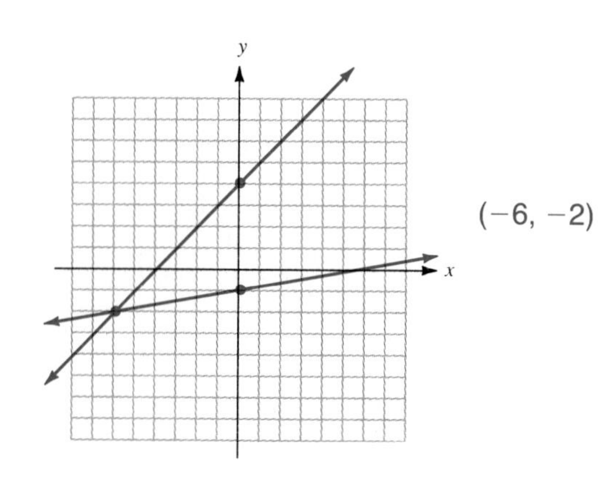

(−6, −2)

17. $3x - 2y = 6$
$2x - y = 5$

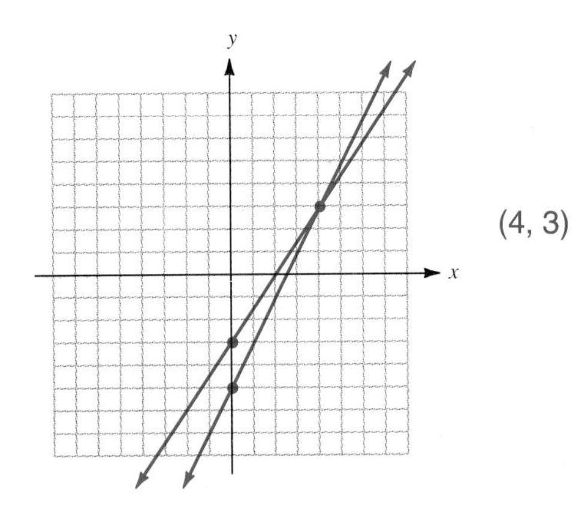

(4, 3)

18. $4x + 3y = 12$
$x + y = 2$

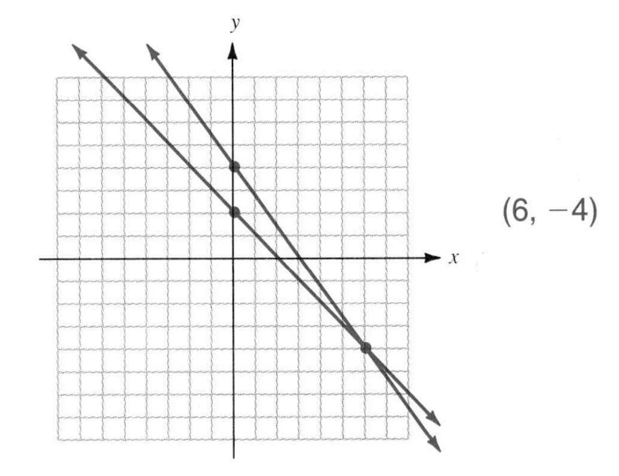

(6, −4)

19. $3x - y = 3$
$3x - y = 6$

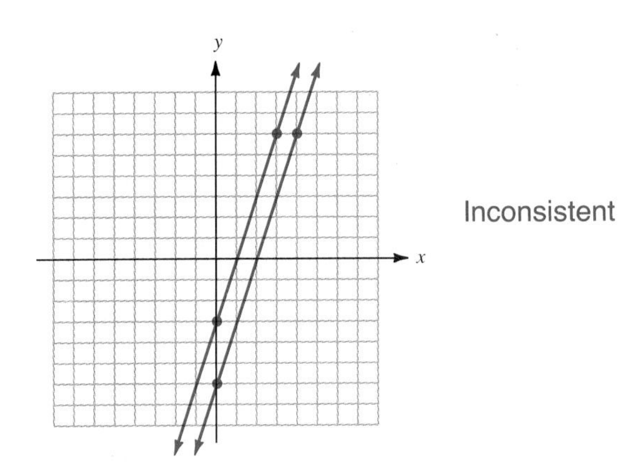

Inconsistent

20. $3x - 6y = 9$
$x - 2y = 3$

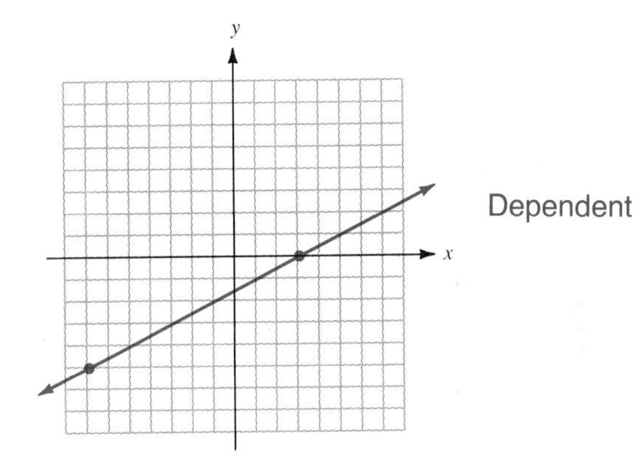

Dependent

21. $2y = 3$
$x - 2y = -3$

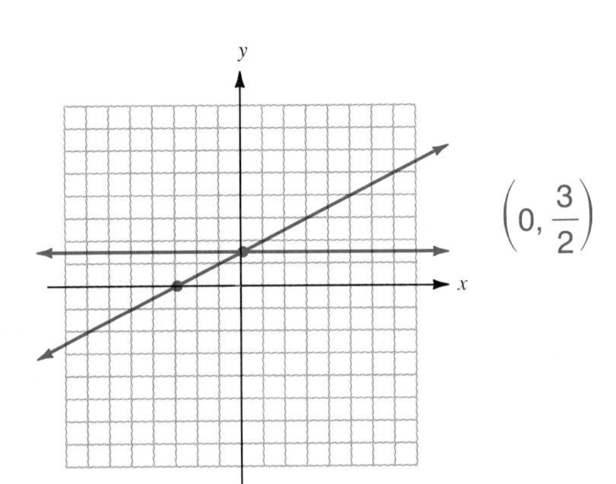

$\left(0, \frac{3}{2}\right)$

22. $x + y = -6$
$-x + 2y = 6$

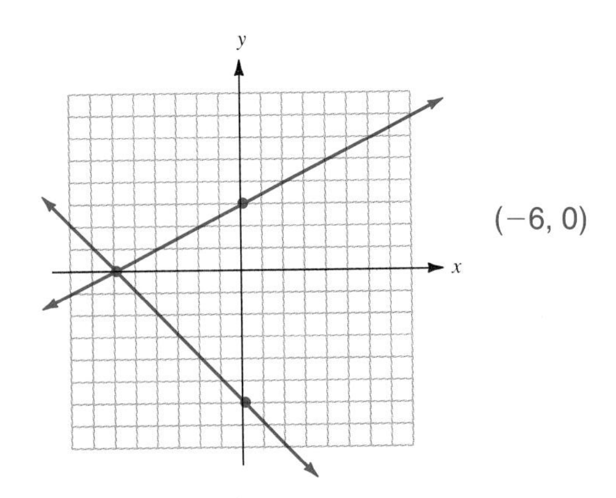

(−6, 0)

Skillscan (Section 1.4)
Simplify each of the following expressions.

a. $(2x + y) + (x - y)$ $3x$

b. $(x + y) + (-x + y)$ $2y$

c. $(3x + 2y) + (-3x - 3y)$ $-y$

d. $(x - 5y) + (2x + 5y)$ $3x$

e. $2(x + y) + (3x - 2y)$ $5x$

f. $2(2x - y) + (-4x - 3y)$ $-5y$

g. $3(2x + y) + 2(-3x + y)$ $5y$

h. $3(2x - 4y) + 4(x + 3y)$ $10x$

Answers

1. $\left.\begin{aligned} x + y &= 6 \\ x - y &= 4 \end{aligned}\right\}$ (5, 1)

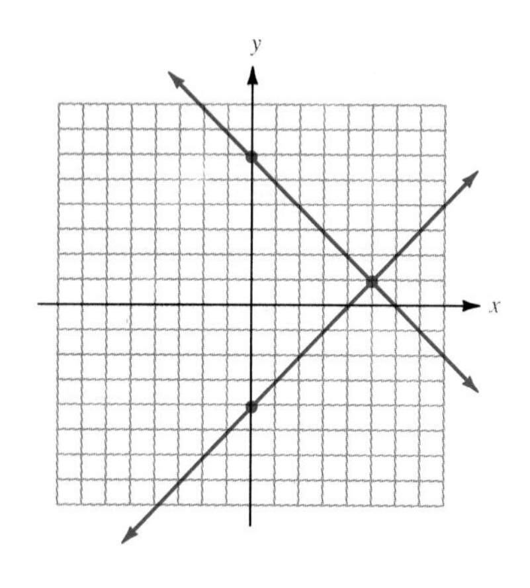

3. $\left.\begin{aligned} -x + y &= 3 \\ x + y &= 5 \end{aligned}\right\}$ (1, 4)

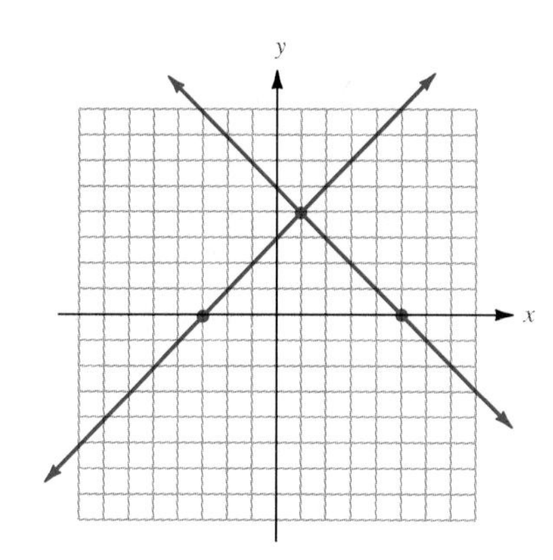

5. $\left.\begin{aligned} x + 2y &= 4 \\ x - y &= 1 \end{aligned}\right\}$ (2, 1)

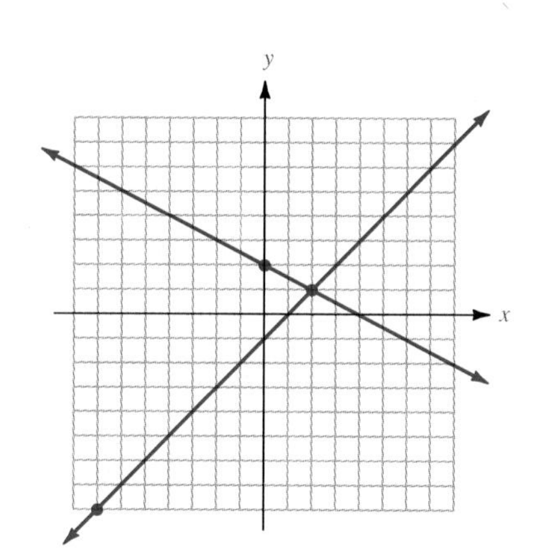

7. $\left.\begin{aligned} 2x + y &= 8 \\ 2x - y &= 0 \end{aligned}\right\}$ (2, 4)

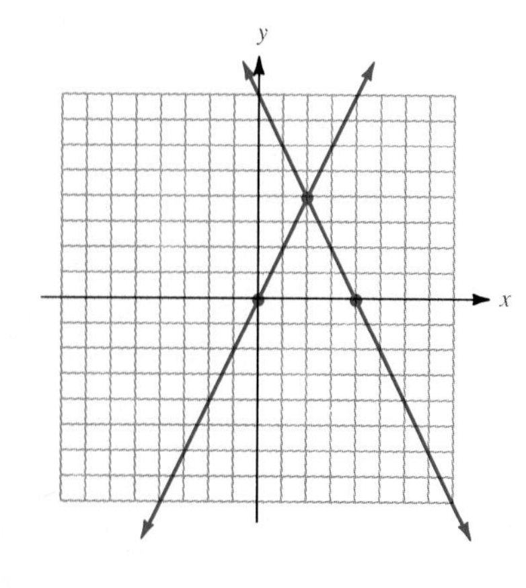

9. $\left.\begin{aligned} x + 3y &= 12 \\ 2x - 3y &= 6 \end{aligned}\right\}$ (6, 2)

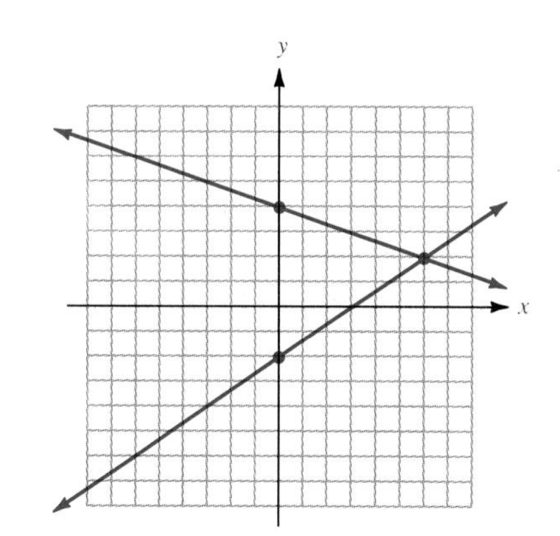

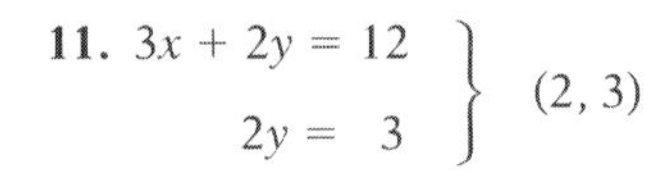

11. $\left.\begin{aligned} 3x + 2y &= 12 \\ 2y &= 3 \end{aligned}\right\}$ (2, 3)

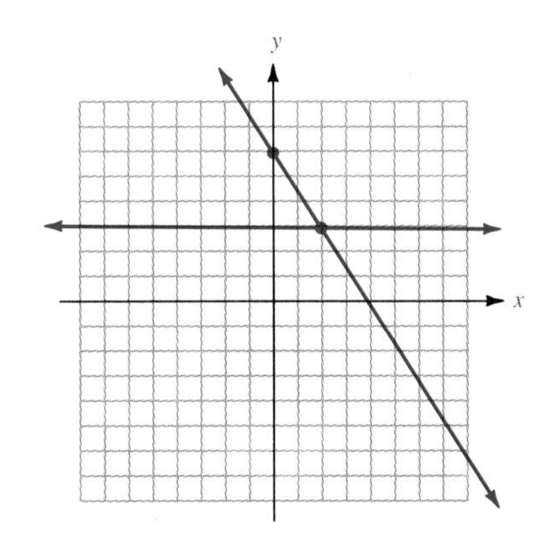

13. $\left.\begin{aligned} x - y &= 4 \\ 2x - 2y &= 8 \end{aligned}\right\}$ Dependent

15. $\left.\begin{aligned} x - 4y &= -4 \\ x + 2y &= 8 \end{aligned}\right\}$ $(4, 2)$

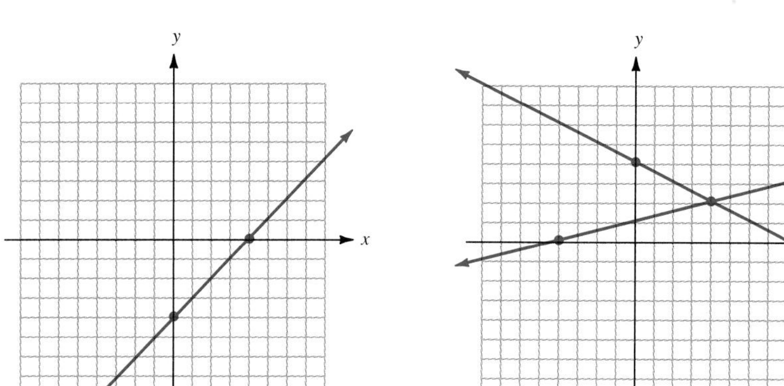

17. $\left.\begin{aligned} 3x - 2y &= 6 \\ 2x - y &= 5 \end{aligned}\right\}$ $(4, 3)$

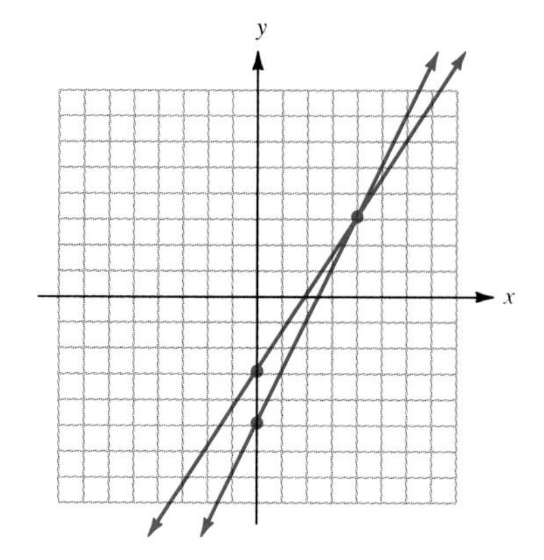

19. $\left.\begin{aligned} 3x - y &= 3 \\ 3x - y &= 6 \end{aligned}\right\}$ Inconsistent

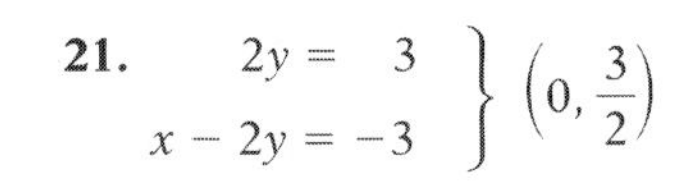

21. $\left.\begin{aligned} 2y &= 3 \\ x - 2y &= -3 \end{aligned}\right\}$ $\left(0, \frac{3}{2}\right)$

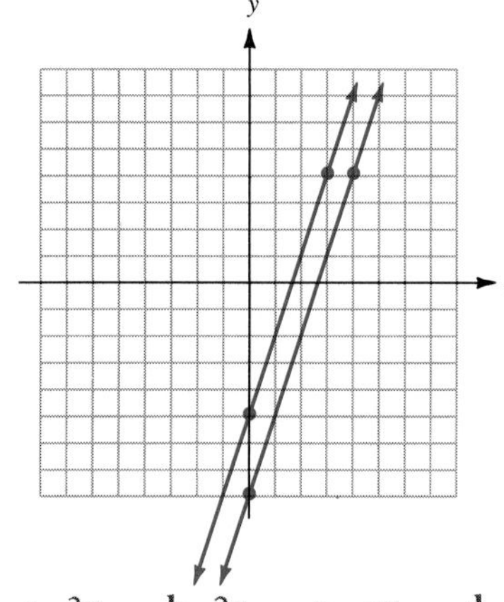

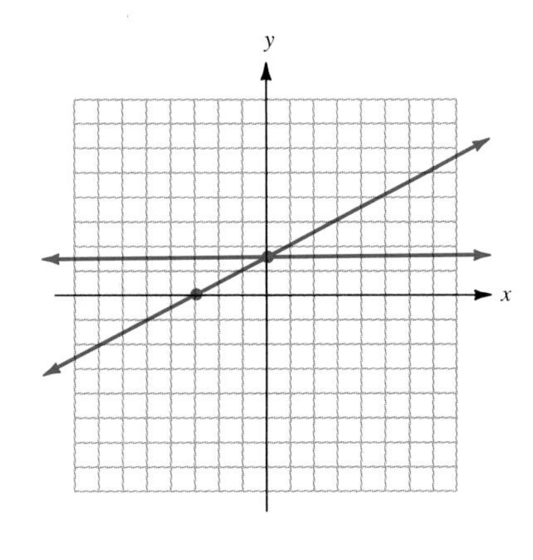

a. $3x$ **b.** $2y$ **c.** $-y$ **d.** $3x$ **e.** $5x$ **f.** $-5y$ **g.** $5y$ **h.** $10x$

8.2 Systems of Linear Equations: Solving by Adding

OBJECTIVE
To find the solution of a system of linear equations by adding.

The graphical method of solving equations, shown in Section 8.1, has two definite disadvantages. First, it is time-consuming to graph each system that you want to solve. Second (and a bigger problem), the graphical method is not precise. For instance, look at the graph of the system

$$\begin{aligned} x - 2y &= 4 \\ 3x + 2y &= 6 \end{aligned}$$

which is shown below.

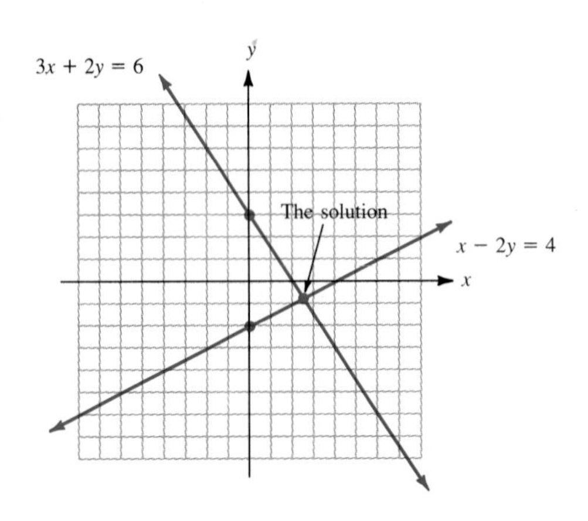

The exact solution for the system happens to be $\left(\frac{5}{2}, -\frac{3}{4}\right)$, but that would be difficult to read from the graph. Fortunately, there are algebraic methods that do not have this disadvantage and will allow you to find exact solutions for a system of equations.

Let's illustrate an algebraic method of finding a solution. It is called the *addition method.*

Example 1

This method uses the fact that if

$a = b$ and $c = d$

then

$a + c = b + d$

This is the *additive property* of equality. Note that by the additive property, if equals are added to equals, the resulting sums are equal.

This is also called *solution by elimination* for this reason.

Solve the system.

$$x + y = 8$$
$$x - y = 2$$

Note that the coefficients of the y terms are the opposites of each other (1 and -1) and that adding the two equations will "eliminate" the variable y. That addition step is shown below.

$$\begin{aligned} x + y &= 8 \\ x - y &= 2 \\ \hline 2x \quad &= 10 \\ x &= 5 \end{aligned}$$

By adding, we eliminate the variable y. The resulting equation contains *only* the variable x.

We now know that 5 is the x coordinate of our solution. Substitute 5 for x into *either* of the original equations.

$$x + y = 8$$
$$5 + y = 8$$
$$y = 3$$

So (5, 3) is the solution.

To check, replace x and y with these values in *both* of the original equations.

$$\begin{array}{ll} \underline{x + y = 8} & \underline{x - y = 2} \\ 5 + 3 = 8 & 5 - 3 = 2 \\ 8 = 8 \quad \text{(True)} & 2 = 2 \quad \text{(True)} \end{array}$$

Since (5, 3) satisfies both equations, it is the solution.

CHECK YOURSELF 1

Solve the system by adding.

$$x - y = -2$$
$$x + y = 6$$

Example 2

Solve the system.

$$-3x + 2y = 12$$
$$3x - y = -9$$

In this case, adding will eliminate the x terms.

Note that we don't care which variable is eliminated. Choose the one that requires the least work.

$$\begin{array}{r} -3x + 2y = 12 \\ 3x - y = -9 \\ \hline y = 3 \end{array}$$

Now substitute 3 for y in either equation. From the first equation

$$-3x + 2 \cdot 3 = 12$$
$$-3x = 6$$
$$x = -2$$

and $(-2, 3)$ is the solution.

Show that you get the same x coordinate by substituting 3 for y in the second equation rather than in the first. Then check the solution.

CHECK YOURSELF 2

Solve the system by adding.

$$5x - 2y = 9$$
$$-5x + 3y = -11$$

Note that in both of the above examples we found an equation in a single variable by adding. We could do this because the coefficients of one of the variables were opposites. This gave 0 as a coefficient for one of the variables after we added the two equations. In some systems, you will not be able to directly eliminate either variable by adding. However, an equivalent system can always be written by multiplying one or both of the equations by a nonzero constant so that the coefficients of x (or of y) are opposites.

The following example illustrates.

Example 3

Solve the system.

$$2x + y = 13 \qquad (1)$$
$$3x + y = 18 \qquad (2)$$

Remember that multiplying both sides of an equation by some nonzero number does not change the solutions. So even though we have "altered" the equations, they are equivalent and will have the same solutions.

Note that adding the equations in this form won't help. You will still have terms in x and in y. However, look at what happens if we multiply both sides of equation (2) by -1 as the first step.

$$2x + y = 13 \longrightarrow \quad 2x + y = 13$$
$$3x + y = 18 \xrightarrow[\text{by } -1]{\text{Multiply}} -3x - y = -18$$

Now we can add.

$$\begin{array}{rcr} 2x + y & = & 13 \\ -3x - y & = & -18 \\ \hline -x & = & -5 \\ x & = & 5 \end{array}$$

Substitute 5 for x in equation (1).

$$2 \cdot 5 + y = 13$$
$$y = 3$$

(5, 3) is the solution. We will leave it to the reader to check this solution.

CHECK YOURSELF 3

Solve the system by adding.

$$x - 2y = 9$$
$$x + 3y = -1$$

To summarize, multiplying both sides of one of the equations by a nonzero constant can yield an equivalent system in which the coefficients of the x terms or the y terms are opposites. This means that a variable can be eliminated by adding. Let's look at another example.

Example 4

Solve the system.

$$\begin{aligned} x + 4y &= 2 \quad (1) \\ 3x - 2y &= -22 \quad (2) \end{aligned}$$

One approach is to multiply both sides of equation (2) by 2. Do you see that the coefficients of the y terms will then be opposites?

$$\begin{aligned} x + 4y &= 2 & \longrightarrow & \quad x + 4y = 2 \\ 3x - 2y &= -22 & \xrightarrow[\text{by 2}]{\text{Multiply}} & \quad 6x - 4y = -44 \end{aligned}$$

If we add the resulting equations, the variable y will be eliminated and we can solve for x.

$$\begin{aligned} x + 4y &= 2 \\ 6x - 4y &= -44 \\ \hline 7x \quad &= -42 \\ x &= -6 \end{aligned}$$

Now substitute -6 for x in Equation (1) to find y.

$$\begin{aligned} -6 + 4y &= 2 \\ 4y &= 8 \\ y &= 2 \end{aligned}$$

So $(-6, 2)$ is the solution.

Again you should check this result. As is often the case, there are several ways to solve the system. For example, what if we multiply both sides of equation (1) by -3? The coefficients of the x terms will then be opposites, and adding will eliminate the variable x so that we can solve for y. Try that for yourself in the following Check Yourself exercise.

CHECK YOURSELF 4

Solve the system by eliminating x.

$$\begin{aligned} x + 4y &= 2 \\ 3x - 2y &= -22 \end{aligned}$$

It may be necessary to multiply in each equation separately so that one of the variables will be eliminated when the equations are added. The following example illustrates.

Example 5

Solve the system.

$$4x + 3y = 11 \quad (1)$$
$$3x - 2y = 4 \quad (2)$$

Do you see that multiplying in one equation will not help in this case? We will have to multiply in both equations.

To eliminate x in this case, we can multiply both sides of equation (1) by 3 and both sides of equation (2) by -4. The coefficients of the x terms will then be opposites.

$$4x + 3y = 11 \xrightarrow[\text{by } 3]{\text{Multiply}} 12x + 9y = 33$$

$$3x - 2y = 4 \xrightarrow[\text{by } -4]{\text{Multiply}} -12x + 8y = -16$$

Adding the resulting equations gives

$$17y = 17$$
$$y = 1$$

Now substituting 1 for y in equation (1), we have

$$4x + 3 \cdot 1 = 11$$
$$4x = 8$$
$$x = 2$$

and (2,1) is the solution.

Again there are different ways to approach the solution. You could choose to eliminate y. Try that in the exercise that follows.

CHECK YOURSELF 5

Solve the system by eliminating y.

$$4x + 3y = 11$$
$$3x - 2y = 4$$

Let's summarize the solution steps that we have illustrated.

TO SOLVE A SYSTEM OF LINEAR EQUATIONS BY ADDING

STEP 1 Multiply both sides of one or both equations by nonzero numbers to form an equivalent system in which the coefficients of one of the variables are opposites, if necessary.

STEP 2 Add the equations of the new system.

STEP 3 Solve the resulting equation for the remaining variable.

STEP 4 Substitute the value found in step 3 into either of the original equations to find the value of the second variable.

STEP 5 Check your solution in both of the original equations.

In Section 8.1 we saw that certain systems had either no solution or infinitely many solutions. Let's see how this situation is indicated when we are using the addition method of solving equations.

Example 6

Solve.

$$x + 3y = -2 \qquad (1)$$
$$3x + 9y = -6 \qquad (2)$$

We will multiply both sides of equation (1) by -3.

$$x + 3y = -2 \xrightarrow[\text{by } -3]{\text{Multiply}} -3x - 9y = 6$$
$$3x + 9y = -6 \longrightarrow \underline{3x + 9y = -6}$$
$$0 = 0$$

Adding, we see that both variables have been eliminated, and we have the true statement $0 = 0$.

Look at the graph of the system.

The lines coincide. That will be the case whenever *adding eliminates both variables* and a true statement results.

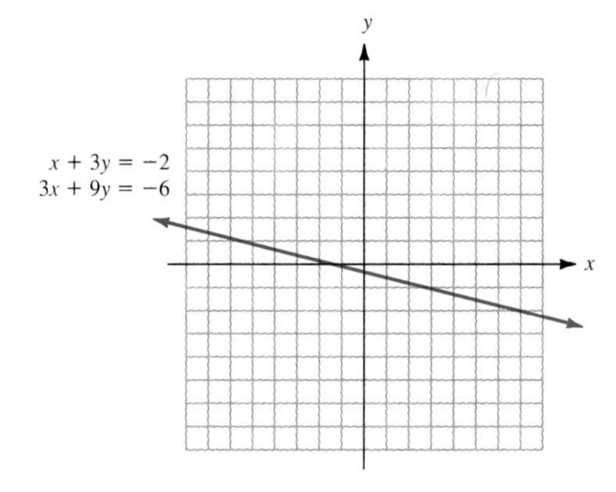

As we see, the two equations have the *same* graph. This means that the system is *dependent*, and there are *infinitely many solutions*. Any (x, y) that satisfies $x + 3y = -2$ will also satisfy $3x + 9y = -6$.

CHECK YOURSELF 6

Solve the system by adding.

$$\begin{aligned} x - 2y &= 3 \\ -2x + 4y &= -6 \end{aligned}$$

In the previous section we also encountered systems that had *no* solutions. The final example illustrates what happens when we try to solve such a system with the algebraic methods of this section.

Example 7

Solve the system.

$$3x - y = 4 \qquad (1)$$
$$-6x + 2y = -5 \qquad (2)$$

We will multiply both sides of equation (1) by 2.

$$\begin{aligned} 3x - y &= 4 \xrightarrow[\text{by 2}]{\text{Multiply}} & 6x - 2y &= 8 \\ -6x + 2y &= -5 \longrightarrow & -6x + 2y &= -5 \\ & \text{Adding, we have} & 0 &= 3 \end{aligned}$$

Again both variables have been eliminated by addition. But this time we have the *false* statement $0 = 3$. This is because we tried to solve a system whose graph consists of two parallel lines, as we see in the graph below. There is *no* solution for the system. It is *inconsistent*.

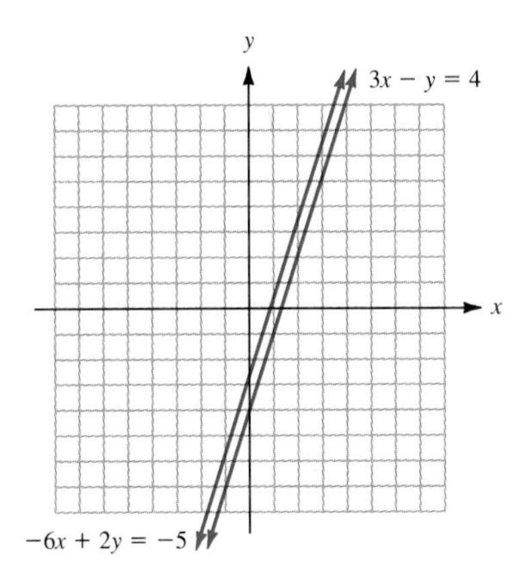

CHECK YOURSELF 7

Solve the system by adding.

$5x + 15y = 20$
$x + 3y = 3$

CHECK YOURSELF ANSWERS

1. (2, 4). **2.** (1, −2). **3.** (5, −2). **4.** (−6, 2). **5.** (2, 1).
6. A dependent system. **7.** There is no solution. The system is inconsistent.

8.2 Exercises

Solve each of the following systems by adding.

1. $x + y = 6$
$x - y = 4$
(5, 1)

2. $x - y = 8$
$x + y = 2$
(5, −3)

3. $-x + y = 3$
$x + y = 5$
(1, 4)

4. $x + y = 7$
$-x + y = 3$
(2, 5)

5. $2x - y = 1$
$-2x + 3y = 5$
(2, 3)

6. $x - 2y = 2$
$x + 2y = -14$
(−6, −4)

7. $x + 3y = 12$
$2x - 3y = 6$
(6, 2)

8. $-3x + y = 8$
$3x - 2y = -10$
(−2, 2)

9. $x + 2y = -2$
$3x + 2y = -12$
$\left(-5, \frac{3}{2}\right)$

10. $4x - 3y = 22$
$4x + 5y = 6$
(4, −2)

11. $4x - 3y = 6$
$4x + 5y = 22$
(3, 2)

12. $2x + 3y = 1$
$5x + 3y = 16$
(5, −3)

13. $2x + y = 8$
$2x + y = 2$
Inconsistent

14. $5x + 4y = 7$
$5x - 2y = 19$
(3, −2)

15. $3x - 5y = 2$
$2x - 5y = -2$
(4, 2)

16. $2x - y = 4$
$2x - y = 6$
Inconsistent

17. $x + y = 3$
$3x - 2y = 4$
(2, 1)

18. $x - y = -2$
$2x + 3y = 21$
(3, 5)

19. $-5x + 2y = -3$
$x - 3y = -15$
(3, 6)

20. $x + 5y = 10$
$-2x - 10y = -20$
Dependent

21. $7x + y = 10$
$2x + 3y = -8$
(2, −4)

22. $3x - 4y = 2$
$4x - y = 20$
(6, 4)

23. $5x + 2y = 28$
$x - 4y = -23$
$\left(3, \frac{13}{2}\right)$

24. $7x + 2y = 17$
$x - 5y = 13$
(3, −2)

25. $3x - 4y = 2$
$-6x + 8y = -4$
Dependent

26. $-x + 5y = 19$
$4x + 3y = -7$
$(-4, 3)$

27. $5x - 2y = 31$
$4x + 3y = 11$
$(5, -3)$

28. $7x + 3y = -13$
$5x + 2y = -8$
$(2, -9)$

29. $3x - 2y = 12$
$5x - 3y = 21$
$(6, 3)$

30. $-4x + 5y = -6$
$5x - 2y = 16$
$(4, 2)$

31. $-2x + 7y = 2$
$3x - 5y = -14$
$(-8, -2)$

32. $3x + 4y = 0$
$5x - 3y = -29$
$(-4, 3)$

33. $7x + 4y = 20$
$5x + 6y = 19$
$\left(2, \frac{3}{2}\right)$

34. $5x + 4y = 5$
$7x - 6y = 36$
$\left(3, -\frac{5}{2}\right)$

35. $2x - 7y = 6$
$-4x + 3y = -12$
$(3, 0)$

36. $3x + 2y = -18$
$7x - 6y = -42$
$(-6, 0)$

37. $5x - y = 20$
$4x + 3y = 16$
$(4, 0)$

38. $3x + y = -5$
$5x - 4y = 20$
$(0, -5)$

39. $3x + y = 1$
$5x + y = 2$
$\left(\frac{1}{2}, -\frac{1}{2}\right)$

40. $2x - y = 2$
$2x + 5y = -1$
$\left(\frac{3}{4}, -\frac{1}{2}\right)$

41. $3x + 4y = 3$
$6x - 2y = 1$
$\left(\frac{1}{3}, \frac{1}{2}\right)$

42. $3x + 3y = 1$
$2x + 4y = 2$
$\left(-\frac{1}{3}, \frac{2}{3}\right)$

43. $5x - 2y = \frac{9}{5}$
$3x + 4y = -1$
$\left(\frac{1}{5}, -\frac{2}{5}\right)$

44. $2x + 3y = -\frac{1}{12}$
$5x + 4y = \frac{2}{3}$
$\left(\frac{1}{3}, -\frac{1}{4}\right)$

Skillscan (Section 3.4)

Solve each of the following equations.

a. $2x + 3(x + 1) = 13$ 2

b. $3(y - 1) + 4y = 18$ 3

c. $x + 2(3x - 5) = 25$ 5

d. $3x - 2(x - 7) = 12$ -2

e. $2(3y + 1) - 4y = 7$ $\frac{5}{2}$

f. $3x - 2(3x - 2) = 5$ $-\frac{1}{3}$

Answers

1. $(5, 1)$ **3.** $(1, 4)$ **5.** $(2, 3)$ **7.** $(6, 2)$ **9.** $\left(-5, \frac{3}{2}\right)$ **11.** $(3, 2)$ **13.** Inconsistent **15.** $(4, 2)$ **17.** $(2, 1)$ **19.** $(3, 6)$ **21.** $(2, -4)$ **23.** $\left(3, \frac{13}{2}\right)$ **25.** Dependent **27.** $(5, -3)$ **29.** $(6, 3)$ **31.** $(-8, -2)$ **33.** $\left(2, \frac{3}{2}\right)$ **35.** $(3, 0)$ **37.** $(4, 0)$ **39.** $\left(\frac{1}{2}, -\frac{1}{2}\right)$ **41.** $\left(\frac{1}{3}, \frac{1}{2}\right)$ **43.** $\left(\frac{1}{5}, -\frac{2}{5}\right)$ **a.** 2 **b.** 3 **c.** 5 **d.** -2 **e.** $\frac{5}{2}$ **f.** $-\frac{1}{3}$

8.3 Systems of Linear Equations: Solving by Substitution

OBJECTIVE
To find the solution of a system of equations by substitution.

In Sections 8.1 and 8.2, we looked at graphing and addition as methods of solving linear systems. A third method is called *solution by substitution.*

The substitution method uses the fact that if two expressions are equal, one may be substituted for the other in any equation. The following example illustrates the process.

Example 1

Solve by substitution.

$$x + y = 12 \qquad (1)$$
$$y = 3x \qquad (2)$$

Notice that equation (2) says that y and $3x$ name the same quantity. So we may substitute $3x$ for y in equation (1). We then have

Replace y with $3x$ in equation (1).

$$x + 3x = 12$$
$$4x = 12$$
$$x = 3$$

The resulting equation contains only the variable x, so substitution is just another way of eliminating one of the variables from our system.

We can now substitute 3 for x in equation (1) to find the corresponding y coordinate of the solution.

$$3 + y = 12$$
$$y = 9$$

So (3, 9) is the solution.

This last step is identical to the one you saw in Section 8.2. As before, you can substitute the known coordinate value back into either of the original equations to find the value of the remaining variable. The check is also identical.

CHECK YOURSELF 1

Solve by substitution.

$$x - y = 9$$
$$y = 4x$$

The same technique can be readily used any time one of the equations is *already solved* for x or for y. The following example illustrates.

Example 2

Solve by substitution.

$$2x + 3y = 3 \quad (1)$$
$$y = 2x - 7 \quad (2)$$

Since equation (2) tells us that y is $2x - 7$, we can replace y with $2x - 7$ in equation (1). This gives

y is now eliminated from the equation, and we can proceed to solve for x.

$$2x + 3(2x - 7) = 3$$
$$2x + 6x - 21 = 3$$
$$8x = 24$$
$$x = 3$$

We now know that 3 is the x coordinate for the solution. So substituting 3 for x in equation (2), we have

$$y = 2 \cdot 3 - 7$$
$$= 6 - 7$$
$$= -1$$

$(3, -1)$ is the solution. Once again you should verify this result.

CHECK YOURSELF 2

Solve by substitution.

$$2x - 3y = 6$$
$$x = 4y - 2$$

As we have seen, the substitution method works very well when one of the given equations is already solved for x or for y. It is also useful if you can readily solve for x or for y in one of the equations.

Example 3

Solve by substitution.

$$x - 2y = 5 \qquad (1)$$
$$3x + y = 8 \qquad (2)$$

Neither equation is solved for a variable. That is easily handled in this case. Solving for x in equation (1), we have

$$x = 2y + 5$$

Now substitute $2y + 5$ for x in equation (2).

$$\begin{aligned} 3(\overset{x}{2y + 5}) + y &= 8 \\ 6y + 15 + y &= 8 \\ 7y &= -7 \\ y &= -1 \end{aligned}$$

Substituting -1 for y in equation (2) yields

$$\begin{aligned} 3x + (-1) &= 8 \\ 3x &= 9 \\ x &= 3 \end{aligned}$$

So $(3, -1)$ is the solution.

You should check this result by substituting 3 for x and -1 for y in the equations of the original system. There are always other methods of solution. For example, you could solve for y in equation (2). This would give $y = -3x + 8$, and that can be substituted back into equation (1) to solve for x. Try it for yourself. You'll get the same result either way.

CHECK YOURSELF 3

Solve by substitution.

$$3x - y = 5$$
$$x + 4y = 6$$

Inconsistent systems and dependent systems will show up in a fashion similar to that which we saw in Section 8.2. The following example illustrates.

Example 4

(*a*) $4x - 2y = 6 \qquad (1)$

$y = 2x - 3 \qquad (2)$

From equation (2) we can substitute $2x - 3$ for y in equation (1).

$$4x - 2(2x - 3) = 6$$
$$4x - 4x + 6 = 6$$
$$6 = 6$$

Both variables have been eliminated, and we have the true statement $6 = 6$.

Recall from the last section that a true statement tells us that the lines coincide. We call this system dependent. There are an infinite number of solutions.

(*b*) $3x - 6y = 9$ (3)

$x = 2y + 2$ (4)

Substitute $2y + 2$ for x in equation (3).

$$3(2y + 2) - 6y = 9$$
$$6y + 6 - 6y = 9$$
$$6 = 9$$

This time we have a false statement.

This means that the system is *inconsistent* and that the graphs of the two equations are parallel lines. There is no solution.

CHECK YOURSELF 4

Indicate whether the systems are inconsistent (no solution) or dependent (an infinite number of solutions).

1. $5x + 15y = 10$
$x = -3y + 1$

2. $12x - 4y = 8$
$y = 3x - 2$

The following summarizes our work of this section.

TO SOLVE A SYSTEM OF LINEAR EQUATIONS BY SUBSTITUTION

STEP 1 Solve one of the given equations for x or y. If this is already done, go on to step 2.

STEP 2 Substitute this expression for x or for y into the other equation.

STEP 3 Solve the resulting equation for the remaining variable.

STEP 4 Substitute the known value into either of the original equations to find the value of the second variable.

STEP 5 Check your solution in both of the original equations.

Strategies in Solving Systems of Equations

You have now seen three different ways to solve systems of linear equations: by graphing, adding, and substitution. The natural question is which method to use in a given situation.

Graphing is the least exact of the methods, and solutions may have to be estimated.

The algebraic methods, addition and substitution, give exact solutions, and both will work for any system of linear equations. In fact, you may have noticed that several examples in this section could just as easily have been solved by adding (Example 3, for instance).

The choice of which of the algebraic methods, substitution or addition, to use is yours and depends largely on the given system. Here are some guidelines designed to help you choose an appropriate method for solving a linear system.

1. If one of the equations is already solved for x (or for y), then substitution is the preferred method.
2. If the coefficients of x (or of y) are the same, or opposites, in the two equations, then addition is the preferred method.
3. If solving for x (or for y) in either of the given equations will result in fractional coefficients, then addition is the preferred method.

CHECK YOURSELF 5

Select the most appropriate method for solving each of the following systems.

1. $2x + 5y = 3$
$8x - 5y = -13$

2. $4x - 3y = 2$
$y = 3x - 4$

3. $3x - 5y = 2$
$x = 3y - 2$

4. $5x - 2y = 19$
$4x + 6y = 38$

CHECK YOURSELF ANSWERS

1. $(-3, -12)$. **2.** $(6, 2)$. **3.** $(2, 1)$. **4.** (1) Inconsistent; (2) dependent.
5. (1) Addition; (2) substitution; (3) substitution; (4) addition.

8.3 Exercises

Solve each of the following systems by substitution.

1. $x + y = 10$
$y = 4x$
(2, 8)

2. $x - y = 4$
$x = 3y$
(6, 2)

3. $2x - y = 10$
$x = -2y$
(4, −2)

4. $x + 3y = 10$
$y = 3x$
(1, 3)

5. $3x + 2y = 12$
$y = 3x$
$\left(\frac{4}{3}, 4\right)$

6. $4x - 3y = 24$
$y = -4x$
$\left(\frac{3}{2}, -6\right)$

7. $x + y = 5$
$y = x - 3$
(4, 1)

8. $x + y = 9$
$x = y + 3$
(6, 3)

9. $x - y = 4$
$x = 2y - 2$
(10, 6)

10. $x - y = 7$
$y = 2x - 12$
(5, −2)

11. $2x + y = 7$
$y = x - 8$
(5, −3)

12. $3x - y = -15$
$x = y - 7$
(−4, 3)

13. $2x - 5y = 10$
$x = y + 8$
(10, 2)

14. $4x - 3y = 0$
$y = x + 1$
(3, 4)

15. $3x + 4y = 9$
$y = 3x + 1$
$\left(\frac{1}{3}, 2\right)$

16. $5x - 2y = -5$
$y = 5x + 3$
$\left(-\frac{1}{5}, 2\right)$

17. $3x - 18y = 4$
$x = 6y + 2$
Inconsistent

18. $4x + 5y = 6$
$y = 2x - 10$
(4, −2)

19. $5x - 3y = 6$
$y = 3x - 6$
(3, 3)

20. $8x - 4y = 16$
$y = 2x - 4$
Dependent

21. $8x - 5y = 16$
$y = 4x - 5$
$\left(\frac{3}{4}, -2\right)$

22. $6x - 5y = 27$
$x = 5y + 2$
$\left(5, \frac{3}{5}\right)$

23. $x + 3y = 7$
$x - y = 3$
(4, 1)

24. $2x - y = -4$
$x + y = -5$
(−3, −2)

25. $6x - 3y = 9$
$-2x + y = -3$
Dependent

26. $5x - 6y = 21$
$x - 2y = 5$
(3, −1)

27. $x - 7y = 3$
$2x - 5y = 15$
(10, 1)

28. $4x - 12y = 5$
$-x + 3y = -1$
Inconsistent

29. $4x + 3y = -11$
$5x + y = -11$
(−2, −1)

30. $5x - 4y = 5$
$4x - y = -7$
(−3, −5)

Solve each of the following systems by using either addition or substitution. If a unique solution does not exist, state whether the system is dependent or inconsistent.

31. $2x + 3y = -6$
$x = 3y + 6$
(0, −2)

32. $7x + 3y = 31$
$y = -2x + 9$
(4, 1)

33. $2x - y = 1$
$-2x + 3y = 5$
(2, 3)

34. $x + 3y = 12$
$2x - 3y = 6$
(6, 2)

35. $6x + 2y = 4$
$y = -3x + 2$
Dependent

36. $3x - 2y = 15$
$-x + 5y = -5$
(5, 0)

37. $x + 2y = -2$
$3x + 2y = -12$
$\left(-5, \frac{3}{2}\right)$

38. $10x + 2y = 7$
$y = -5x + 3$
Inconsistent

39. $2x - 3y = 14$
$4x + 5y = -5$
$\left(\frac{5}{2}, -3\right)$

40. $2x + 3y = 1$

$5x + 3y = 16$

(5, −3)

41. $4x - 2y = 0$

$x = \frac{3}{2}$

$\left(\frac{3}{2}, 3\right)$

42. $4x - 3y = \frac{11}{2}$

$y = -\frac{3}{2}$

$\left(\frac{1}{4}, -\frac{3}{2}\right)$

Skillscan (Section 7.3)

Write an equation that describes the following relationships.

a. The sum of n and d is 55.
$n + d = 55$

b. The value of L is 3 more than twice W.
$L = 2W + 3$

c. Twice the sum of x and y is 36.
$2(x + y) = 36$

d. 3 times the sum of m and n is 72.
$3(m + n) = 72$

e. 4 times p plus 3 times q is 880.
$4p + 3q = 880$

f. The sum of 9 times x and 7 times y is 3100.
$9x + 7y = 3100$

g. The larger number (y) is 8 less than twice the smaller number (x).
$y = 2x - 8$

h. The larger number (m) is 2 more than 3 times the smaller number (n).
$m = 3n + 2$

Answers

1. (2, 8) **3.** (4, −2) **5.** $\left(\frac{4}{3}, 4\right)$ **7.** (4, 1) **9.** (10, 6) **11.** (5, −3) **13.** (10, 2) **15.** $\left(\frac{1}{3}, 2\right)$ **17.** Inconsistent **19.** (3, 3) **21.** $\left(\frac{3}{4}, -2\right)$ **23.** (4, 1) **25.** Dependent **27.** (10, 1) **29.** (−2, −1) **31.** (0, −2) **33.** (2, 3) **35.** Dependent **37.** $\left(-5, \frac{3}{2}\right)$ **39.** $\left(\frac{5}{2}, -3\right)$ **41.** $\left(\frac{3}{2}, 3\right)$ **a.** $n + d = 55$ **b.** $L = 2W + 3$ **c.** $2(x + y) = 36$ **d.** $3(m + n) = 72$ **e.** $4p + 3q = 880$ **f.** $9x + 7y = 3100$ **g.** $y = 2x - 8$ **h.** $m = 3n + 2$

8.4 Systems of Linear Equations: Applications

OBJECTIVE

To solve applied or word problems using systems of equations.

In Chapter 3 we solved word problems by using equations in a single variable. Remember that all the unknowns in the problem had to be expressed in terms of that single variable.

Now that you have the background to use two equations in two variables to solve word problems, let's see how they can be applied. The five steps for solving word problems stay the same (in fact, we give them again for reference in our first example). Many students find that using two equations and two variables makes writing the necessary equations much easier.

Often problems can be solved by using either one or two variables. We'll start with an example done both ways so that you can compare.

Example 1

Here are the steps using a single variable:

The sum of two numbers is 25. If the second number is 5 less than twice the first number, what are the two numbers?

1. Read the problem carefully. What do you want to find?
2. Assign variables to the unknown quantities.

Step 1 You want to find the two unknown numbers.

Step 2 Let x = the first number.

Then $2x - 5$ is the second number.

Twice (points to $2x$); 5 less than (points to $- 5$)

Step 3

3. Write the equation for the solution.

$$x + (2x - 5) = 25$$

The sum of the numbers is 25.

Step 4

4. Solve the equation.

$$x + (2x - 5) = 25$$
$$3x - 5 = 25$$
$$3x = 30$$
$$x = 10$$

x → 10 is the first number.

$2x - 5$ → 15 is the second number.

Step 5

5. Verify your result by returning to the original problem.

$$10 + 15 = 25$$

The solution checks.

CHECK YOURSELF 1

The sum of two numbers is 28. The second number is 4 more than twice the first number. Find the two numbers.

Use an equation in a single variable for the solution.

Now, to compare the two approaches, let's look again at the problem of Example 1. This time we'll use two equations in two variables.

Example 2

The sum of two numbers is 25. If the second number is 5 less than twice the first number, what are the two numbers?

1. What do you want to find?

Step 1 You want to find the two unknown numbers.

2. Assign variables. This time we use two letters, x and y.

Step 2 Let x = the first number and y = the second number.

3. Write equations for the solution. Here two equations are needed because we have introduced two variables.

Step 3

$$\underbrace{x + y}_{\text{The sum}} = \underbrace{25}_{\text{is 25}}$$

$$\underbrace{y}_{\text{The second number}} = \underbrace{2x - 5}_{\text{is 5 less than twice the first}}$$

4. Solve the system of equations.

Step 4

$$x + y = 25 \qquad (1)$$
$$y = 2x - 5 \qquad (2)$$

We have used the substitution method because equation (2) is already solved for y.

Substitute $2x - 5$ for y in equation (1).

$$x + (2x - 5) = 25$$
$$3x - 5 = 25$$
$$x = 10$$

From equation (1),

$$10 + y = 25$$
$$y = 15$$

The two numbers are 10 and 15.

5. Check the result.

Step 5 The sum of the numbers is 25. The second number, 15, is 5 less than twice the first number, 10. The solution checks.

You should go back now and compare our two approaches to this problem. See which you prefer.

CHECK YOURSELF 2

Solve the problem of Check Yourself 1, this time using two variables for the solution.

There are some problems that cannot be readily solved by using a single variable, but can be solved rather easily in two variables. The following example illustrates.

Example 3

Ryan bought eight pens and seven pencils and paid a total of $7.40. Ashleigh purchased two pens and 10 pencils and paid $3.50. Find the cost for a single pen and a single pencil.

Step 1 You want to find the cost of a single pen and the cost of a single pencil.

Step 2 Let x be the cost of a pen and y be the cost of a pencil.

Step 3 Write the two necessary equations.

In the first equation, $8x$ is the total cost of the pens, $7y$ the total cost of the pencils. The second equation is formed in a similar fashion.

$$8x + 7y = 7.40 \qquad (1)$$

$$2x + 10y = 3.50 \qquad (2)$$

Step 4 Solve the system formed in step 3. We will multiply equation (2) by -4. Adding will then eliminate the variable x.

$$\begin{aligned} 8x + 7y &= 7.40 \\ -8x - 40y &= -14.00 \end{aligned}$$

Now adding the equations, we have

$$-33y = -6.60$$

or

$$y = 0.20$$

Substituting 0.20 for y in equation (1), we have

$$\begin{aligned} 8x + 7(0.20) &= 7.40 \\ 8x + 1.40 &= 7.40 \\ 8x &= 6.00 \end{aligned}$$

or

$$x = 0.75$$

Step 5 From the results of step 4 we see that the pens are 75¢ each and that the pencils are 20¢ each.

To check these solutions replace x with 0.75 and y with 0.20 in equation (1).

$$8(0.75) + 7(0.20) = 7.40$$
$$6.00 + 1.40 = 7.40$$
$$7.40 = 7.40 \quad \text{(True)}$$

We will leave it to the reader to check these values in equation (2).

CHECK YOURSELF 3

Annette bought three digital tapes and two compact discs on sale for $66. At the same sale, Terry bought three digital tapes and four compact discs for $96. Find the individual price for a tape and a disc.

Example 4

An 18-ft board is cut into two pieces, one of which is 4 ft longer than the other. How long is each piece?

Step 1 You want to find the two lengths.

Step 2 Let x be the length of the longer piece and y the length of the shorter piece. You should always draw a sketch of the problem whenever it is appropriate.

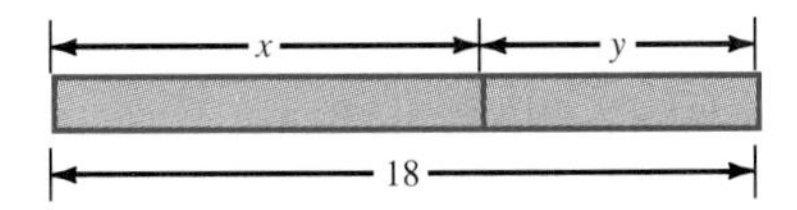

Step 3 Write the equations for the solution.

$$x + y = 18$$

The total length is 18.

$$x - y = 4$$

The difference in lengths is 4.

Our second equation could also be written as

$x = y + 4$

in which case you would probably use substitution for the solution.

Step 4 To solve the system, add:

$$\begin{aligned} x + y &= 18 \quad &(1) \\ x - y &= 4 \quad &(2) \\ \hline 2x \quad &= 22 \\ x &= 11 \end{aligned}$$

Replace x with 11 in equation (1).

$$11 + y = 18$$
$$y = 7$$

The longer piece has length 11 ft, the shorter piece 7 ft.

Step 5 We'll leave it to you to check this result in the original problem.

CHECK YOURSELF 4

A 20-ft board is cut into two pieces, one of which is 6 ft longer than the other. How long is each piece?

Sketches will always be helpful in solving applications from geometry. Let's look at such an example.

Example 5

The length of a rectangle is 3 m more than twice its width. If the perimeter of the rectangle is 42 m, find the dimensions of the rectangle.

Step 1 You want to find the dimensions (length and width) of the rectangle.

We have used x and y as our two variables in the previous examples. Use whatever letters you want. The process is the same and sometimes it helps you remember what letter stands for what. Here L = length and W = width.

Step 2 Let L be the length of the rectangle and W the width. Now draw a sketch of the problem.

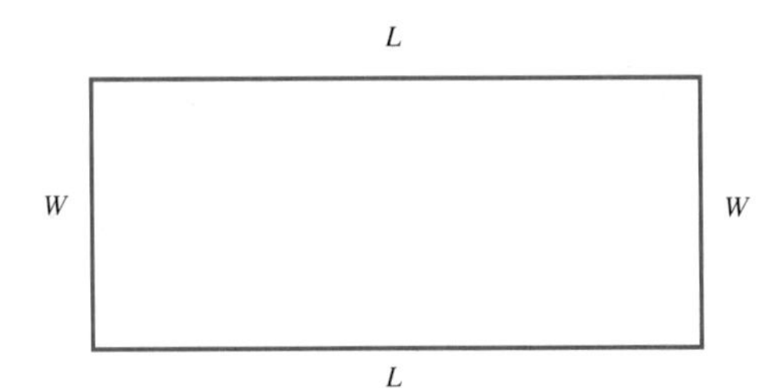

Step 3 Write the equations for the solution.

$$L = \underbrace{2W + 3}$$

↑ 3 more than twice the width

$$\underbrace{2L + 2W} = 42$$

↑ The perimeter

Step 4 Solve the system:

$$L = 2W + 3 \quad (1)$$
$$2L + 2W = 42 \quad (2)$$

From equation (1) we can substitute $2W + 3$ for L in equation (2).

$$2(2W + 3) + 2W = 42$$
$$4W + 6 + 2W = 42$$
$$6W = 36$$
$$W = 6$$

Replace W with 6 in equation (1) to find L.

$$L = 2 \cdot 6 + 3$$
$$= 12 + 3$$
$$= 15$$

The length is 15 m, the width is 6 m.

Step 5 Check these results. The perimeter is $2 \cdot 15 + 2 \cdot 6$ m, or 42 m. The length (15 m) is 3 m more than twice the width (6 m).

CHECK YOURSELF 5

The length of the two equal legs of an isosceles triangle is 5 in less than the length of the base. If the perimeter of the triangle is 50 in, find the lengths of the legs and the base.

Using two equations in two variables may also help you in writing equations to solve mixture problems.

Example 6

Winnifred has collected \$4.50 in nickels and dimes. If she has 55 coins, how many of each kind of coin does she have?

Step 1 You want to find the number of nickels and the number of dimes.

Step 2 Let

n = the number of nickels

d = the number of dimes

Again we choose appropriate variables—n for nickels, d for dimes.

Step 3 Write the equations for the solution.

$$n + d = 55$$

There are 55 coins in all.

Remember: The value of a number of coins is the value per coin times the number of coins: $5n$, $10d$, etc.

$$5n + 10d = 450$$

Value of nickels — Value of dimes — Total value (in cents)

Step 4 We now have the system

$$n + d = 55 \quad (1)$$
$$5n + 10d = 450 \quad (2)$$

Let's solve this system by addition. Multiply equation (1) by -5. Adding will then eliminate the variable n.

$$\begin{aligned} -5n - 5d &= -275 \\ 5n + 10d &= 450 \\ \hline 5d &= 175 \\ d &= 35 \end{aligned}$$

We now substitute d for 35 in equation (1).

$$\begin{aligned} n + 35 &= 55 \\ n &= 20 \end{aligned}$$

There are 20 nickels and 35 dimes.

Step 5 We'll leave it to you to check this result. Just verify that the value of these coins is \$4.50.

CHECK YOURSELF 6

Tickets for a play cost \$8 or \$6. If 350 tickets were sold in all and receipts were \$2500, how many of each price ticket were sold?

We can also solve mixture problems which involve percentages by using two equations in two unknowns. Look at the following example.

Example 7

In a chemistry lab are a 20% acid solution and a 60% acid solution. How many milliliters of each should be mixed to produce 200 milliliters (mL) of a 44% acid solution?

Step 1 You need to know the amount of each solution to use.

Step 2 Let:

x = the amount of 20% acid solution
y = the amount of 60% acid solution

Step 3 A drawing will help. Note that a 20% acid solution is 20% acid and 80% water.

20% x mL + 60% y mL = 44% 200 mL

We can write equations from the total amount of the solution, here 200 mL, and from the amount of acid in that solution. Many students find a table helpful in organizing the information at this point. Here, for example, we might have

Note: The amount of acid is the amount of solution times the percent acid (as a decimal). That is the key to forming the third column of our table.

	AMOUNT OF SOLUTION	% ACID	AMOUNT OF ACID
	x	0.20	$0.20x$
	y	0.60	$0.60y$
TOTALS	200	0.44	(0.44)(200)

Now we are ready to form our system.

The total amount of the solution from the first column of our table.

The amount of acid from the third column of our table.

$$x + y = 200 \qquad (1)$$

$$0.20x + 0.60y = 0.44\ (200) \qquad (2)$$

Acid in 20% solution; Acid in 60% solution; Acid in mixture

Step 4 If we multiply equation (2) by 100 to clear of decimals, we have

$$\begin{aligned} x + \quad y &= 200 \\ 20x + 60y &= 8800 \end{aligned}$$

To complete the solution, multiply the first equation by -20 and add. We can then solve for y.

$$\begin{aligned} -20x - 20y &= -4000 \\ 20x + 60y &= 8800 \\ \hline 40y &= 4800 \\ y &= 120 \end{aligned}$$

Substituting 120 for y in equation (1), we have

$$x + 120 = 200$$
$$x = 80$$

The amounts to be mixed are 80 mL (20% acid solution) and 120 mL (60% acid solution).

Step 5 You can check this solution by verifying that the amount of acid from the 20% solution and the amount from the 60% solution are equal to the amount of acid in the mixture.

CHECK YOURSELF 7

You have a 30% alcohol solution and a 50% alcohol solution. How much of each solution should be combined to make 400 mL of a 45% alcohol solution?

A related kind of application involves interest. The key equation involves the *principal* (the amount invested), the annual *interest rate,* the *time* (in years) that the money is invested, and the amount of *interest* you receive.

$$I = P \cdot r \cdot t$$

Interest Principal Rate Time

For 1 year we have

$I = P \cdot r$ since $t = 1$

Example 8

Jeremy inherits $20,000 and invests part of the money in bonds with an interest rate of 11 percent. The remainder of the money is in savings at a 9 percent rate. What amount has he invested at each rate if he receives $2040 in interest for 1 year?

Step 1 You want to find the amounts invested at 11 percent and at 9 percent.

Step 2 Let x = the amount invested at 11 percent and y = the amount invested at 9 percent. Once again you may find a table helpful at this point. Here we might have

Note: $I = P \cdot r$—interest equals principal times rate—is the key to forming the third column of our table.

	PRINCIPAL	RATE	INTEREST
	x	11%	$0.11x$
	y	9%	$0.09y$
TOTALS	20,000		2040

Step 3 Form the equations for the solution, using the first and second columns of the above table.

$$x + y = 20{,}000$$

He has $20,000 invested in all.

$$0.11x + 0.09y = 2040$$

The interest at 11% (rate · principal) — The interest at 9% — The total interest

Step 4 To solve the following system, use addition.

$$x + y = 20{,}000 \quad (1)$$
$$0.11x + 0.09y = 2{,}040 \quad (2)$$

To do this, multiply both sides of equation (1) by −9. Multiplying both sides of equation (2) by 100 will clear decimals. Adding the resulting equations will eliminate y.

$$\begin{aligned} -9x - 9y &= -180{,}000 \\ 11x + 9y &= 204{,}000 \\ \hline 2x &= 24{,}000 \\ x &= 12{,}000 \end{aligned}$$

Now substitute 12,000 for x in equation (1) and solve for y.

$$\begin{aligned} 12{,}000 + y &= 20{,}000 \\ y &= 8{,}000 \end{aligned}$$

Jeremy has $12,000 invested at 11 percent and $8000 invested at 9 percent.

Step 5 To check, the interest at 11 percent is ($12,000)(0.11), or $1320. The interest at 9 percent is ($8000)(0.09), or $720. The total interest is $2040, and the solution is verified.

CHECK YOURSELF 8

Jan has $2000 more invested in a stock that pays 9 percent interest than in a savings account paying 8 percent. If her total interest for 1 year is \$860, how much does she have invested at each rate?

Another group of applications is called *motion problems*. They involve a distance traveled, the rate, and the time of the travel. Recall from Section 4.7 that these quantities are related by the equation

$$d = r \cdot t$$

Distance ↗ Rate ↑ Time ↖

Our final example shows the use of this equation to form a system of equations to solve a motion problem.

Example 9

A boat can travel 36 mi downstream in 2 h. In coming back upstream, the trip takes 3 h. What is the rate of the boat in still water? What is the rate of the current?

Step 1 You want to find the two rates (of the boat and the current).

Step 2 Let

$x =$ the rate of the boat in still water

$y =$ the rate of the current.

Step 3 To write the equations, think about the following: What is the effect of the current? Suppose the boat's rate in still water is 10 mi/h and the current is 2 mi/h.

The current *increases* the rate *downstream* to 12 mi/h (10 + 2). The current *decreases* the rate *upstream* to 8 mi/h (10 − 2). So here the rate downstream will be $x + y$, and the rate upstream will be $x - y$. At this point a table of information can again be helpful.

	DISTANCE	RATE	TIME
DOWNSTREAM	36	$x + y$	2
UPSTREAM	36	$x - y$	3

From the relationship $d = r \cdot t$ we can now use our table to write the system

From line 1 of our table $\quad 36 = 2(x + y)$

From line 2 $\quad 36 = 3(x - y)$

Step 4 Removing the parentheses in the equations of step 3, we have

$$2x + 2y = 36$$
$$3x - 3y = 36$$

By either of our earlier methods, this system gives values of 15 for x and 3 for y.

The rate in still water is 15 mi/h, and the rate of the current is 3 mi/h. We will leave the check to you.

CHECK YOURSELF 9

A plane flies 480 mi with the wind in 4 h. In returning against the wind, the trip takes 6 h. What is the rate of the plane in still air? What was the rate of the wind?

CHECK YOURSELF ANSWERS

1. The numbers are 8 and 20. **2.** The numbers are 8 and 20.
3. Tape $12; disc: $15. **4.** 7 ft and 13 ft.
5. The legs have length 15 in; the base is 20 in.
6. 150 $6 tickets and 200 $8 tickets were sold.
7. 100 mL (30%), 300 mL (50%).
8. $4000 at 8 percent, $6000 at 9 percent.
9. Plane's rate in still air, 100 mi/h; wind's rate, 20 mi/h.

8.4 Exercises

Solve each of the following problems. Be sure to show the equations used for the solution.

Find the two numbers in Exercises 1 to 8.

1. The sum of the numbers is 40. Their difference is 8. 24, 16

2. The sum of the numbers is 100. The second is 3 times the first. 25, 75

3. The sum of the numbers is 70. The second is 10 more than 3 times the first. 15, 55

4. The sum of the numbers is 56. The second is 4 less than twice the first. 20, 36

5. The difference of the numbers is 4. The larger is 8 less than twice the smaller.
16, 12

6. The difference of the numbers is 22. The larger is 2 more than 3 times the smaller.
32, 10

7. One number is 18 more than another, and the sum of the smaller number and twice the larger number is 45. 3, 21

8. One number is 5 times another. The larger number is 9 more than twice the smaller.
3, 15

9. Eight eagle stamps and two raccoon stamps cost \$2.80. Three eagle stamps and four raccoon stamps cost \$2.35. Find the cost of each kind of stamp. 25¢, 40¢

10. Alice bought four chocolate bars and a pack of gum and paid \$2.75. Sue bought two chocolate bars and three packs of gum and paid \$2.25. Find the cost of each. 60¢, 35¢

11. Jim bought 5 Red Delicious apples and 4 Granny Smith apples at a cost of \$4.81. Lynn bought one of each of the two types at a cost of \$1.08. Find the cost for each kind of apple. 49¢, 59¢

12. Four single-sided disks and two double-sided disks cost a total of \$5.10. Two single-sided and four double-sided disks cost \$5.40. Find the individual cost for each. 80¢, 95¢

13. A 30-m rope is cut into two pieces so that one piece is 6 m longer than the other. How long is each piece? 18 m, 12 m

14. An 18-ft board is cut into two pieces, one of which is twice as long as the other. How long is each piece? 12 ft, 6 ft

15. Two packages together weigh 32 kilograms (kg). The smaller package weighs 6 kg less than the larger. How much does each package weigh? 13 kg, 19 kg

16. A washer-dryer combination costs \$400. If the washer costs \$40 more than the dryer, what does each appliance cost separately? \$220, \$180

17. In a town election, the winning candidate had 220 more votes than the loser. If 810 votes were cast in all, how many votes did each candidate receive? 515, 295

18. An office desk and chair together cost \$250. If the desk cost \$20 less than twice as much as the chair, what did each cost? \$160, \$90

19. The length of a rectangle is 2 in more than twice its width. If the perimeter of the rectangle is 34 in, find the dimensions of the rectangle. 5 in by 12 in

20. The perimeter of an isosceles triangle is 37 in. The lengths of the two equal legs are 6 in less than 3 times the length of the base. Find the lengths of the three sides.
7 in, 15 in, 15 in

21. Jill has \$3.50 in nickels and dimes. If she has 50 coins, how many of each type of coin does she have? 30 nickels, 20 dimes

22. Richard has 22 coins with a total value of \$4. If the coins are all quarters and dimes, how many of each type of coin does he have? 10 dimes, 12 quarters

23. Theater tickets are \$4 for general admission and \$3 for students. During one evening 240 tickets were sold, and the receipts were \$880. How many of each kind of ticket were sold? 160 general, 80 student

24. 400 tickets were sold for a concert. The receipts from ticket sales were \$3100, and the ticket prices were \$7 and \$9. How many of each price ticket were sold?
250 at \$7, 150 at \$9

25. A coffee merchant has coffee beans which sell for \$3 per pound and \$5 per pound. The two types are to be mixed to create 100 lb of a mixture that will sell for \$4.50 per pound. How much of each type of bean should be used in the mixture? 25 lb at \$3, 75 lb at \$5

26. Peanuts are selling for \$2 per pound, and cashews are selling for \$5 per pound. How much of each type of nut would be needed to create 20 lb of a mixture that would sell for \$2.75 per pound? 15 lb of peanuts, 5 lb of cashews

27. A chemist has a 25% and a 50% acid solution. How much of each solution should be used to form 200 milliliters (mL) of a 35% acid solution? 120 mL of 25%, 80 mL of 50%

28. A pharmacist wishes to prepare 150 mL of a 20% alcohol solution. She has a 30% solution and a 15% solution in her stock. How much of each should be used in forming the desired mixture? 50 mL of 30%, 100 mL of 15%

29. You have two alcohol solutions, one a 15% solution and one a 45% solution. How much of each solution should be used to obtain 300 mL of a 25% solution?
200 mL of 15%, 100 mL of 45%

30. If you combine a 10% acid solution and a 50% acid solution, how much of each should be used to make 40 centiliters (cL) of a 40% acid solution? 10 cL of 10%, 30 cL of 50%

31. Roger has a total of \$12,000 invested in two accounts. One account pays 8 percent, and the other 9 percent. If his interest for 1 year is \$1010, how much does he have invested at each rate? \$7000 at 8%, \$5000 at 9%

32. Toni invests a part of \$8000 in bonds paying 12 percent interest. The remainder is in a savings account at 8 percent. If she receives \$840 in interest for 1 year, how much does she have invested at each rate? \$3000 at 8%, \$5000 at 12%

33. John has \$2000 more invested in an account paying 10 percent interest than in a second account that pays 8 percent. If his interest is \$920 for 1 year, what does he have invested at each rate? \$6000 at 10%, \$4000 at 8%

34. Jill invested twice as much in an account paying 6 percent than she did in one paying 7 percent. She received \$950 in interest for 1 year. What amount did she have invested at each rate? \$10,000 at 6%, \$5000 at 7%

35. David was able to row 16 mi downstream in 2 h. Returning upstream, he took 4 h to make the trip. How fast can he row in still water? What was the rate of the current? 6 mi/h, 2 mi/h

36. A plane flies 450 mi with the wind in 3 h. Flying back against the wind, the plane takes 5 h to make the trip. What was the rate of the plane in still air? What was the rate of the wind? 120 mi/h, 30 mi/h

37. An airliner made a trip of 1800 mi in 3 h flying east across the country with the jetstream directly behind it. The return trip, against the jetstream, took 4 h. Find the speed of the plane in still air and the speed of the jetstream. 525 mi/h, 75 mi/h

38. A boat traveled 60 mi upstream in 5 h. Returning downstream with the current, the boat took 3 h to make the trip. What was the boat's speed in still water, and what was the speed of the river's current? 16 mi/h, 4 mi/h

Each of the following applications can be solved by the use of a system of linear equations. Match the application with the system on the right that could be used for its solution.

39. One number is 4 less than 3 times another. If the sum of the numbers is 36, what are the two numbers? *d*

40. Suppose that a movie theater sold 300 adult and student tickets for a showing with a revenue of \$1440. If the adult tickets were \$6 and the student tickets \$4, how many of each type of ticket were sold? *f*

41. The length of a rectangle is 3 cm more than twice its width. If the perimeter of the rectangle is 36 cm, find the dimensions of the rectangle. *g*

42. An order of 12 dozen roller-ball pens and 5 dozen ballpoint pens cost \$116. A later order for 8 dozen roller-ball pens and 12 dozen ballpoint pens cost \$112. What was the cost of 1 dozen of each of the pens? *a*

(*a*) $12x + 5y = 116$
$8x + 12y = 112$

(*b*) $x + y = 8000$
$0.06x + 0.09y = 600$

(*c*) $x + y = 200$
$0.20x + 0.60y = 90$

(*d*) $x + y = 36$
$y = 3x - 4$

(*e*) $2(x + y) = 36$
$3(x - y) = 36$

(*f*) $x + y = 300$
$6x + 4y = 1440$

(*g*) $L = 2W + 3$
$2L + 2W = 36$

(*h*) $x + y = 140$
$2x + 5.5y = 420$

43. A candy merchant wishes to mix peanuts selling at $2 per pound with cashews selling for $5.50 per pound to form 140 lb of a mixed-nut blend which will sell for $3 per pound. What amount of each type of nut should be used? *h*

44. Donald has investments totaling $8000 in two accounts, one a savings account paying 6 percent interest and the other a bond paying 9 percent. If the annual interest from the two investments was $600, how much did he have invested at each rate? *b*

45. A chemist wants to combine a 20% alcohol solution with a 60% solution to form 200 mL of a 45% solution. How much of each of the solutions should be used to form the mixture? *c*

46. Mark was able to make a downstream trip of 36 mi in 2 h. In returning upstream, it took 3 h to make the trip. How fast can his boat travel in still water? What was the rate of the river's current? *e*

Answers

1. 24, 16 **3.** 15, 55 **5.** 16, 12 **7.** 3, 21 **9.** 25¢, 40¢ **11.** 49¢, 59¢ **13.** 18 m, 12 m **15.** 13 kg, 19 kg **17.** 515, 295 **19.** 5 in by 12 in **21.** 30 nickels, 20 dimes **23.** 160 general, 80 student **25.** 25 lb at $3, 75 lb at $5 **27.** 120 mL of 25%, 80 mL of 50% **29.** 200 mL of 15%, 100 mL of 45% **31.** $7000 at 8%, $5000 at 9% **33.** $6000 at 10%, $4000 at 8% **35.** 6 mi/h, 2 mi/h **37.** 525 mi/h, 75 mi/h **39.** *d* **41.** *g* **43.** *h* **45.** *c*

Summary

Systems of Linear Equations [8.1] to [8.3]

A System of Equations Two or more equations considered together.

Solution The solution of a system of two equations in two unknowns is an ordered pair that satisfies each equation of the system.

Solving by Graphing

1. Graph both equations on the same coordinate system.
2. The system may have
   (*a*) *One solution.* The lines intersect at one point (a consistent system). The solution is the ordered pair corresponding to that point.
   (*b*) *No solution.* The lines are parallel (an inconsistent system).
   (*c*) *Infinitely many solutions.* The two equations have the same graph (a dependent system). Any ordered pair corresponding to a point on the line is a solution.

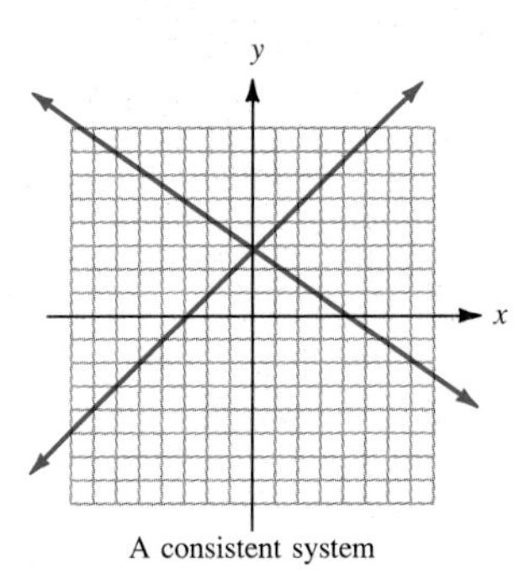

A consistent system

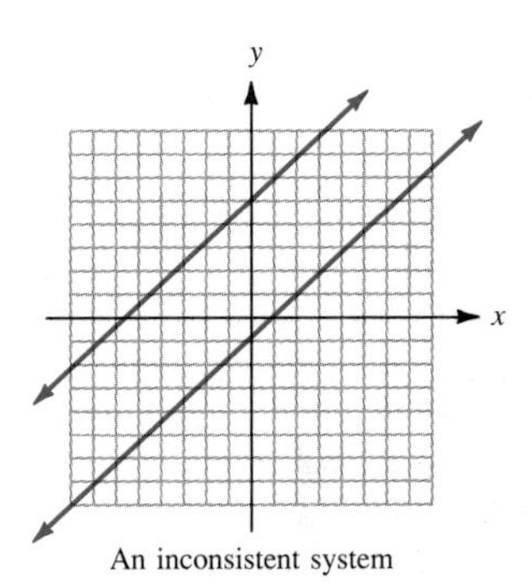

An inconsistent system

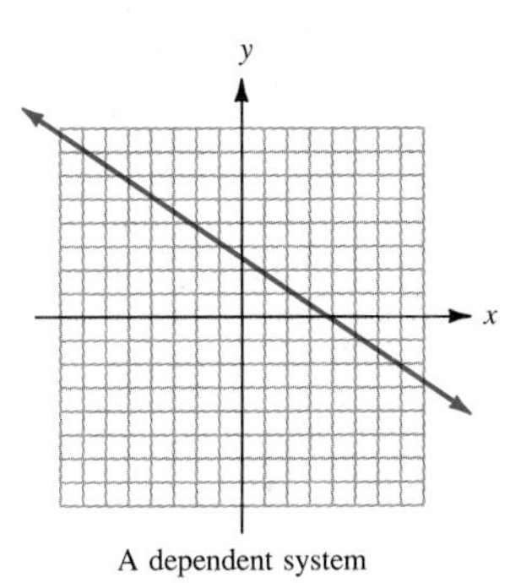

A dependent system

Solving by Adding

1. Multiply both sides of one or both equations by nonzero numbers to form an equivalent system in which the coefficients of one of the variables are opposites, if necessary.

   $2x - y = 4$ (1)
   $3x + 2y = 13$ (2)
   Multiply equation (1) by 2.
   $4x - 2y = 8$
   $3x + 2y = 13$

2. Add the equations of the new system.

   Add.
   $7x = 21$

3. Solve the resulting equation for the remaining variable.

   $x = 3$

4. Substitute the value found in step 3 into either of the original equations to find the value of the second variable.
5. Check your solution in both of the original equations.

   In equation (1),
   $2 \cdot 3 - y = 4$
   $y = 2$
   (3, 2) is the solution.

Solving by Substitution

1. Solve one of the given equations for x or for y. If this is already done, go on to step 2.

$$x - 2y = 3 \quad (1)$$
$$2x + 3y = 13 \quad (2)$$

From equation (1),

$$x = 2y + 3$$

2. Substitute this expression for x or for y into the other equation.

Substitute in equation (2):

$$2(2y + 3) + 3y = 13$$

3. Solve the resulting equation for the remaining variable.

$$y = 1$$

Steps 4 and 5 are the same as above.

Continue as before.

Solving Word Problems by Using Systems of Equations [8.4]

Applying Systems of Equations Often word problems can be solved by using two variables and two equations to represent the unknowns and the given relationships in the problem.

The Solution Steps

1. Read the problem carefully. Then reread it to decide what you are asked to find.
2. Choose letters to represent the unknowns.
3. Translate the problem to the language of algebra to form a system of equations.
4. Solve the system.
5. Verify your solution in the original problem.

Summary Exercises Chapter 8

This summary exercise set is provided to give you practice with each of the objectives of the chapter. Each exercise is keyed to the appropriate chapter section. The answers are provided in the instructor's manual. Your instructor will give you guidelines on how to best use these exercises in your instructional setting.

[8.1] Solve each of the following systems by graphing.

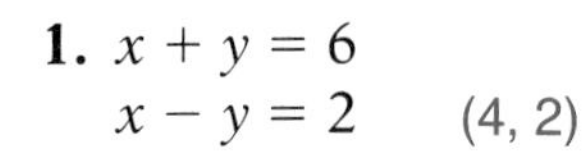

1. $x + y = 6$
$x - y = 2$ (4, 2)

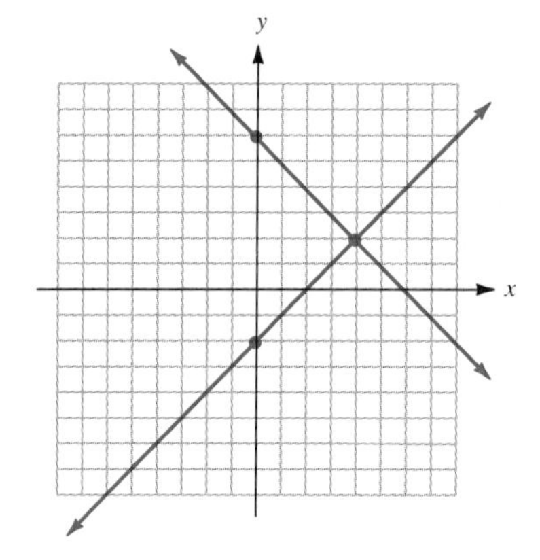

2. $x - y = 8$
$2x + y = 7$ (5, −3)

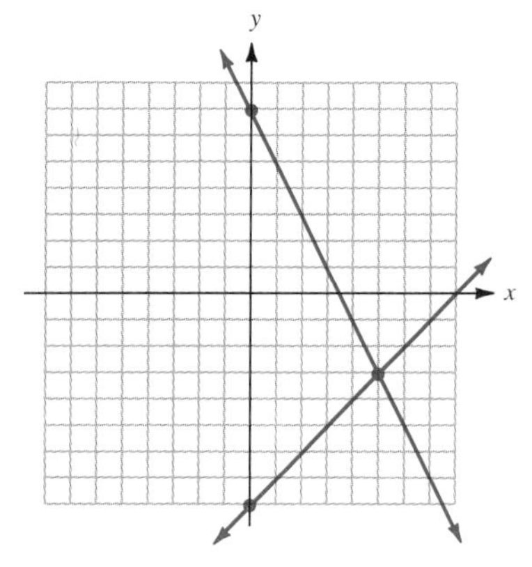

3. $x + 2y = 4$
$x + 2y = 6$ Inconsistent

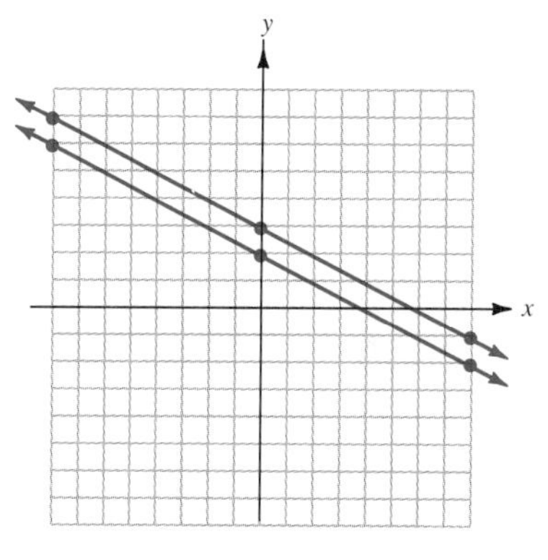

4. $2x - y = 8$
$y = 2$ (5, 2)

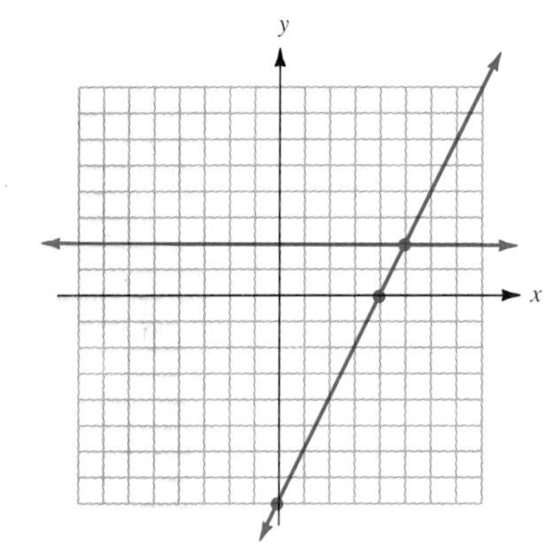

5. $2x - 4y = 8$
$x - 2y = 4$ Dependent

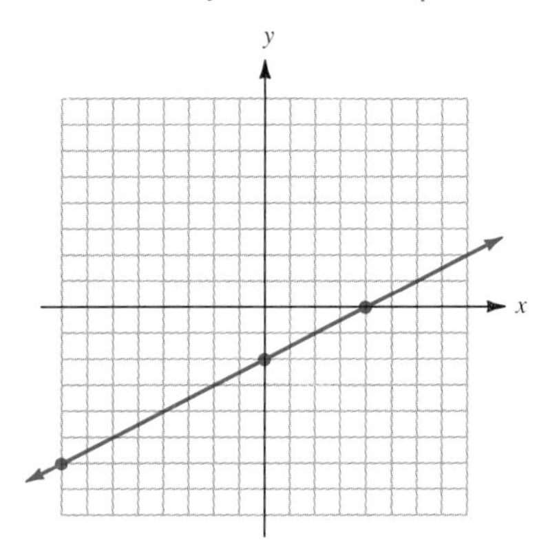

6. $3x + 2y = 6$
$4x - y = 8$ (2, 0)

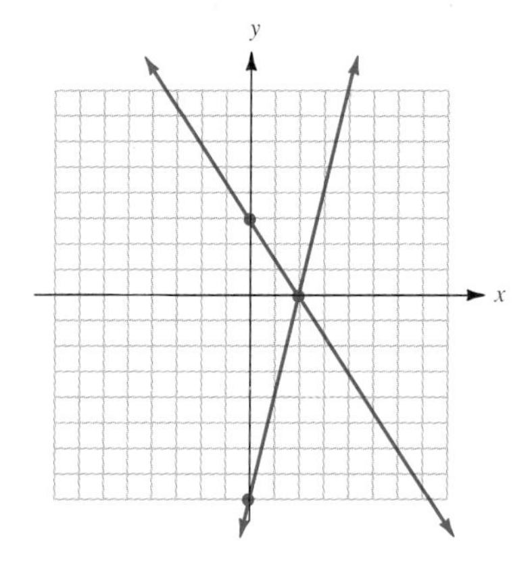

[8.2] Solve each of the following systems by adding.

7. $x + y = 8$
$x - y = 2$
(5, 3)

8. $-x - y = 4$
$x - y = -8$
(−6, 2)

9. $2x - 3y = 16$
$5x + 3y = 19$
(5, −2)

10. $2x + y = 7$
$3x - y = 3$
(2, 3)

11. $3x - 5y = 14$
$3x + 2y = 7$
(3, −1)

12. $2x - 4y = 8$
$x - 2y = 4$
Dependent

13. $4x - 3y = -22$
$4x + 5y = -6$
$(-4, 2)$

14. $5x - 2y = 17$
$3x - 2y = 9$
$\left(4, \frac{3}{2}\right)$

15. $4x - 3y = 10$
$2x - 3y = 6$
$\left(2, -\frac{2}{3}\right)$

16. $2x + 3y = -10$
$-2x + 5y = 10$
$(-5, 0)$

17. $3x + 2y = 3$
$6x + 4y = 5$
Inconsistent

18. $3x - 2y = 23$
$x + 5y = -15$
$(5, -4)$

19. $5x - 2y = -1$
$10x + 3y = 12$
$\left(\frac{3}{5}, 2\right)$

20. $x - 3y = 9$
$5x - 15y = 45$
Dependent

21. $2x - 3y = 18$
$5x - 6y = 42$
$(6, -2)$

22. $3x + 7y = 1$
$4x - 5y = 30$
$(5, -2)$

23. $5x - 4y = 12$
$3x + 5y = 22$
$(4, 2)$

24. $6x + 5y = -6$
$9x - 2y = 10$
$\left(\frac{2}{3}, -2\right)$

25. $4x - 3y = 7$
$-8x + 6y = -10$
Inconsistent

26. $3x + 2y = 8$
$-x - 5y = -20$
$(0, 4)$

27. $3x - 5y = -14$
$6x + 3y = -2$
$\left(-\frac{4}{3}, 2\right)$

[8.3] Solve each of the following systems by substitution.

28. $x + 2y = 10$
$y = 2x$
$(2, 4)$

29. $x - y = 10$
$x = -4y$
$(8, -2)$

30. $2x - y = 10$
$x = 3y$
$(6, 2)$

31. $2x + 3y = 2$
$y = x - 6$
$(4, -2)$

32. $4x + 2y = 4$
$y = 2 - 2x$
Dependent

33. $x + 5y = 20$
$x = y + 2$
$(5, 3)$

34. $6x + y = 2$
$y = 3x - 4$
$\left(\frac{2}{3}, -2\right)$

35. $2x + 6y = 10$
$x = 6 - 3y$
Inconsistent

36. $2x + y = 9$
$x - 3y = 22$
$(7, -5)$

37. $x - 3y = 17$
$2x + y = 6$
$(5, -4)$

38. $2x + 3y = 4$
$y = 2$
$(-1, 2)$

39. $4x - 5y = -2$
$x = -3$
$(-3, -2)$

40. $-6x + 3y = -4$
$y = -\frac{2}{3}$
$\left(\frac{1}{3}, -\frac{2}{3}\right)$

41. $5x - 2y = -15$
$y = 2x + 6$
$(-3, 0)$

42. $3x + y = 15$
$x = 2y + 5$
$(5, 0)$

[8.3] Solve each of the following systems by either addition or substitution.

43. $x - 4y = 0$
$4x + y = 34$
$(8, 2)$

44. $2x + y = 0$
$y = -2x$
$(2, -4)$

45. $3x - 3y = 30$
$x = -2y - 8$
$(4, -6)$

46. $5x + 4y = 40$
$x + 2y = 11$
$\left(6, \frac{5}{2}\right)$

47. $x - 6y = -8$
$2x + 3y = 4$
$\left(0, \frac{4}{3}\right)$

48. $4x - 3y = 9$
$2x + y = 12$
$\left(\frac{9}{2}, 3\right)$

49. $9x + y = 9$
$x + 3y = 14$
$\left(\frac{1}{2}, \frac{9}{2}\right)$

50. $3x - 2y = 8$
$-6x + 4y = -16$
Dependent

51. $3x - 2y = 8$
$2x - 3y = 7$
(2, −1)

[8.4] Solve the following problems. Be sure to show the equations used.

52. The sum of two numbers is 40. If their difference is 10, find the two numbers. 25, 15

53. The sum of two numbers is 17. If the second number is 1 more than 3 times the first, what are the two numbers? 4, 13

54. The difference of two numbers is 8. The larger number is 2 less than twice the smaller. Find the numbers. 10, 18

55. Five writing tablets and three pencils cost \$8.25. Two tablets and two pencils cost \$3.50. Find the cost for each item. \$1.50, 25¢

56. A cable which is 200 ft long is cut into two pieces so that one piece is 12 ft longer than the other. How long is each piece? 94 ft, 106 ft

57. An amplifier and a pair of speakers cost \$925. If the amplifier costs \$75 more than the speakers, what does each cost? \$425, \$500

58. A sofa and chair cost \$850 as a set. If the sofa costs \$100 more than twice as much as the chair, what is the cost of each? \$250, \$600

59. The length of a rectangle is 4 cm more than its width. If the perimeter of the rectangle is 64 cm, find the dimensions of the rectangle. 14 cm, 18 cm

60. The perimeter of an isosceles triangle is 29 in. The lengths of the two equal legs are 2 in more than twice the length of the base. Find the lengths of the three sides.
5 in, 12 in, 12 in

61. Robert has 30 coins with a value of \$5.50. If they are all nickels and quarters, how many of each kind of coin does he have? 10 nickels, 20 quarters

62. Tickets for a concert sold for \$11 and \$8. If 600 tickets were sold for one evening and the receipts were \$5550, how many of each kind of ticket were sold? 250 at \$11, 350 at \$8

63. A laboratory has a 20% acid solution and a 50% acid solution. How much of each should be used to produce 600 mL of a 40% acid solution? 200 mL of 20%, 400 mL of 50%

64. A service station wishes to mix 40 L of a 78% antifreeze solution. How many liters of a 75% solution and a 90% solution should be used in forming the mixture?
32 L of 75%, 8 L of 90%

65. Martha has \$18,000 invested. Part of the money is invested in a bond which yields 11 percent interest. The remainder is in her savings account, which pays 7 percent. If she earns \$1660 in interest for 1 year, how much does she have invested at each rate?
\$10,000 at 11 percent, \$8000 at 7 percent

66. A boat travels the 24 mi upstream in 3 h. It then takes the same 3 h to go 36 mi downstream. Find the speed of the boat in still water and the speed of the current.
Boat 10 mi/h, current 2 mi/h

67. A plane flying with the wind makes a trip of 2200 mi in 4 h. Returning against the wind, it can travel only 1800 mi in 4 h. What is the plane's rate in still air? What is the wind speed? Plane 500 mi/h, wind 50 mi/h

Self-Test for Chapter Eight

The purpose of this self-test is to help you check your progress and to review for a chapter test in class. Allow yourself about an hour to take the test. When you are done, check your answers in the back of the book. If you missed any problems, be sure to go back and review the appropriate sections in the chapter and the exercises that are provided.

Solve each of the following systems by graphing.

1. $x + y = 5$
$x - y = 3$

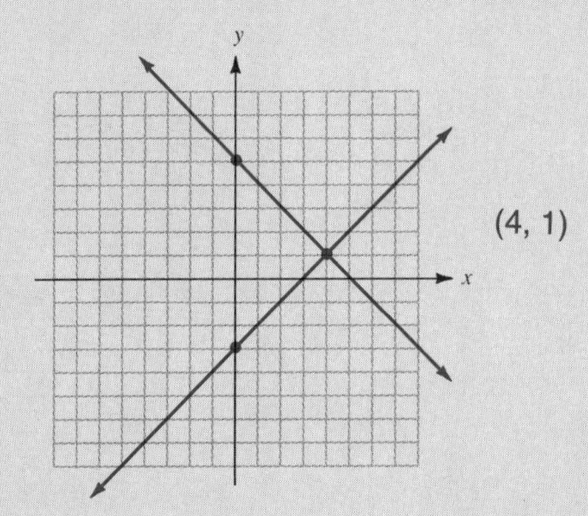

(4, 1)

2. $x + 2y = 8$
$x - y = 2$

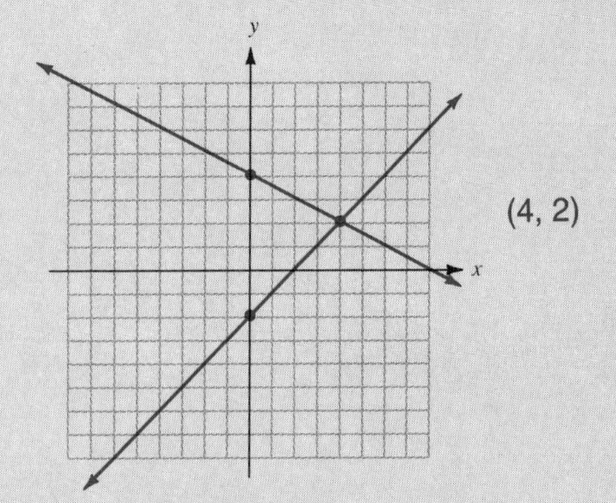

(4, 2)

3. $x - 3y = 3$
$x - 3y = 6$

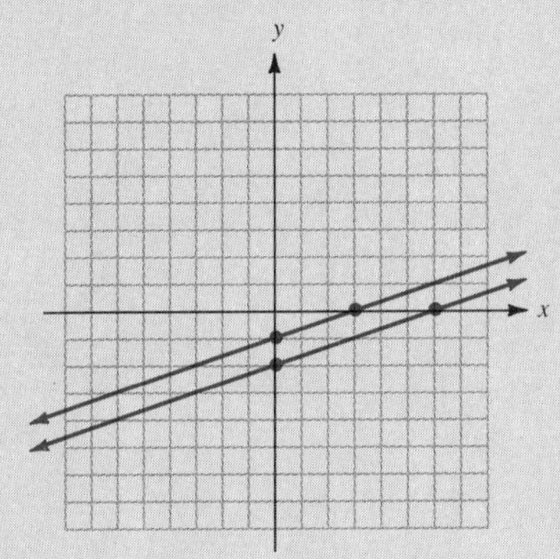

Inconsistent

4. $4x - y = 4$
$x - 2y = -6$

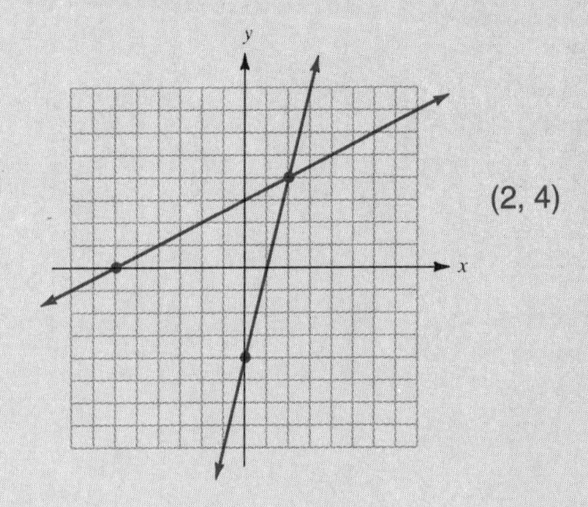

(2, 4)

Solve each of the following systems by adding.

5. $x + y = 5$
$x - y = 3$

(4, 1)

6. $x + 2y = 8$
$x - y = 2$

(4, 2)

7. $3x + y = 6$
$-3x + 2y = 3$

(1, 3)

8. $3x + 2y = 11$
$5x + 2y = 15$

$\left(2, \frac{5}{2}\right)$

9. $3x - 6y = 12$
$x - 2y = 4$

Dependent

10. $4x + y = 2$
$8x - 3y = 9$

$\left(\frac{3}{4}, -1\right)$

11. $2x - 5y = 2$
$3x + 4y = 26$

(6, 2)

12. $x + 3y = 6$
$3x + 9y = 9$

Inconsistent

Solve each of the following systems by substitution.

13. $x + y = 8$
$y = 3x$
(2, 6)

14. $x - y = 9$
$x = -2y$
(6, −3)

15. $2x - y = 10$
$x = y + 4$
(6, 2)

16. $x - 3y = -7$
$y = x - 1$
(5, 4)

17. $3x + y = -6$
$y = 2x + 9$
(−3, 3)

18. $4x + 2y = 8$
$y = 3 - 2x$
Inconsistent

19. $5x + y = 10$
$x + 2y = -7$
(3, −5)

20. $3x - 2y = 5$
$2x + y = 8$
(3, 2)

Solve the following problems. Be sure to show the equations used.

21. The sum of two numbers is 30, and their difference is 6. Find the two numbers. 12, 18

22. A rope 50 m long is cut into two pieces so that one piece is 8 m longer than the other. How long is each piece? 21 m, 29 m

23. The length of a rectangle is 4 in less than twice its width. If the perimeter of the rectangle is 64 in, what are the dimensions of the rectangle? 12 in by 20 in

24. Murray has 30 coins with a value of \$5.70. If the coins are all dimes and quarters, how many of each coin does he have? 12 dimes, 18 quarters

25. Jackson was able to travel 36 mi downstream in 2 hours. In returning upstream, it took 3 hours to make the trip. How fast can his boat travel in still water? What was the rate of the river current? Boat 15 mi/h, current 3 mi/h

Cumulative Test for Chapters 7 and 8

This test is provided to help you in the process of review of the previous chapters. Answers are provided in the back of the book. If you missed any problems, be sure to go back and review the appropriate chapter sections.

Determine which of the ordered pairs are solutions for the given equations.

1. $x + y = 8$ $\quad (4, 4), (8, 0), (5, 2), (3, 5)$ $\quad$ (4, 4), (8, 0), (3, 5)

2. $3x - y = 6$ $\quad (0, -6), \left(\frac{1}{3}, 5\right), (2, 0), (1, -3)$ $\quad$ (0, −6), (2, 0), (1, −3)

Complete the ordered pairs so that each pair is a solution for the given equation.

3. $2x - y = 8$ $\quad (4, \ ), (\ , -8), \left(\frac{5}{2}, \ \right), (\ , -6)$ $\quad$ 0, 0, −3, 1

4. $3x + 4y = 12$ $\quad (4, \ ), (8, \ ), (\ , 3), (\ , 6)$ $\quad$ 0, −3, 0, −4

Find four solutions for each of the given equations.

5. $3x - y = 6$
(2, 0), (1, −3), (0, −6), (−1, −9)

6. $x + 5y = 10$
(0, 2), (5, 1), (10, 0), (15, −1)

Plot points with the given coordinates.

7. $A(0, -7)$

8. $B(4, -3)$

9. $C(-2, -5)$

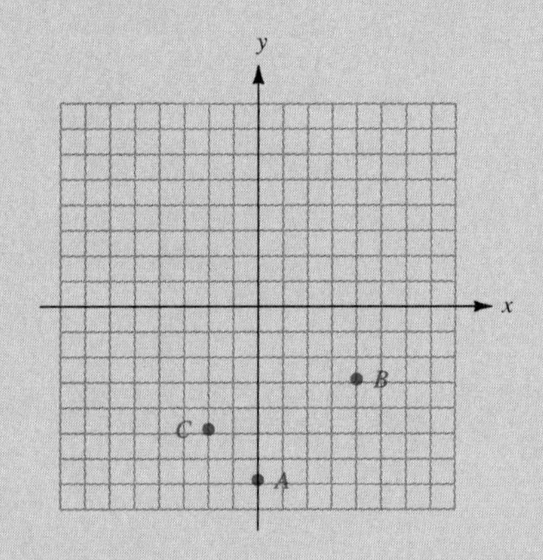

Graph each of the following equations.

10. $x - y = 5$

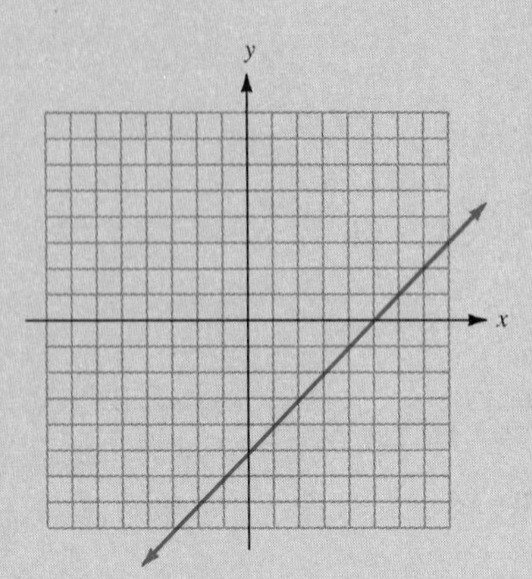

11. $y = \frac{2}{3}x + 3$

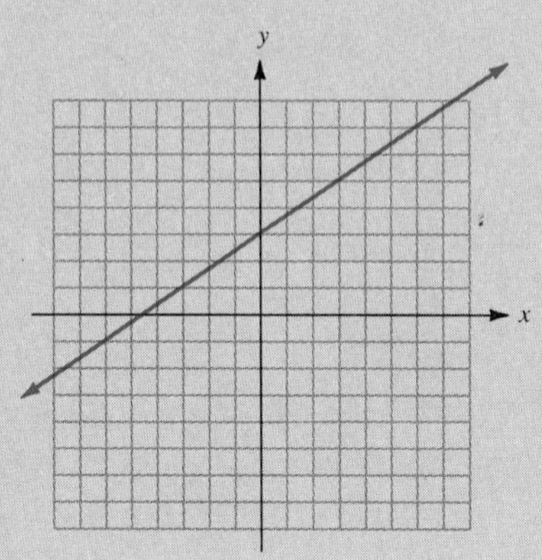

12. $x + 2y = 6$

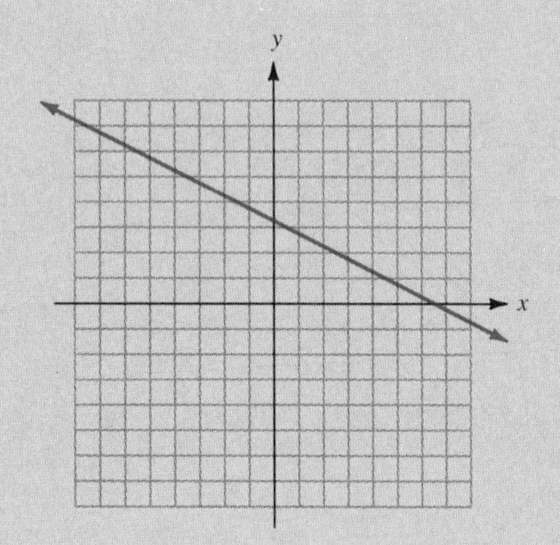

13. $2x - 5y = 10$

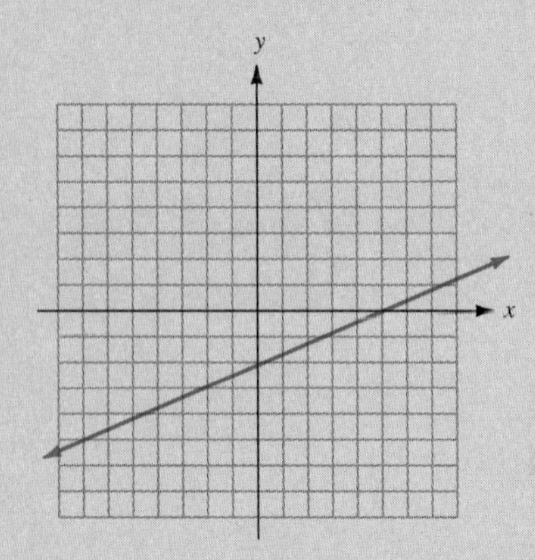

14. $y = -5$

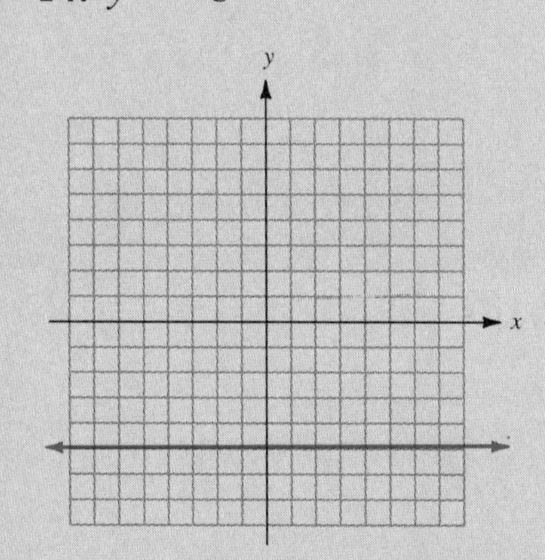

Find the slope of the line through the following pairs of points.

15. $(-2, -3)$ and $(5, 7)$ $\quad \frac{10}{7}$

16. $(-3, 4)$ and $(2, -3)$ $\quad -\frac{7}{5}$

Find the slope and y intercept of the lines described by each of the following equations.

17. $y = -3x + 7$ $\quad m = -3, b = 7$

18. $5x - 3y = 15$ $\quad m = \frac{5}{3}, b = -5$

Given the slope and y intercept for each of the following lines, write the equation of the line. Then graph the line.

19. $m = 2, b = -5$ $\quad y = 2x - 5$

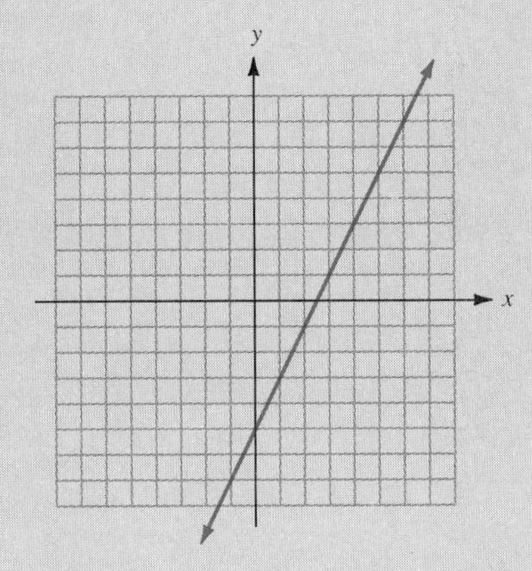

20. $m = -\dfrac{3}{2}, b = 5$ $\quad y = -\dfrac{3}{2}x + 5$

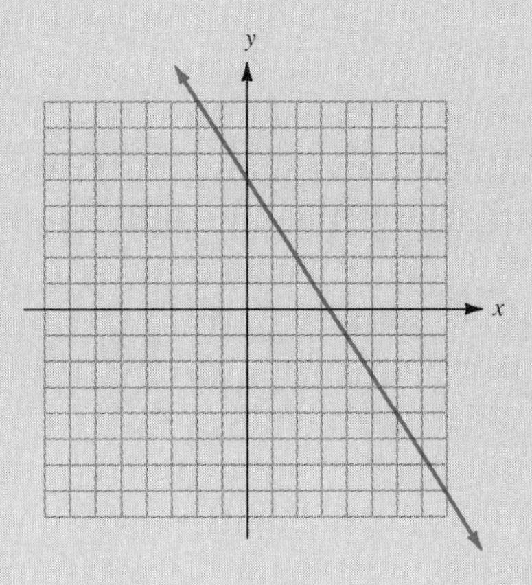

Graph each of the following inequalities.

21. $x + 2y < 6$

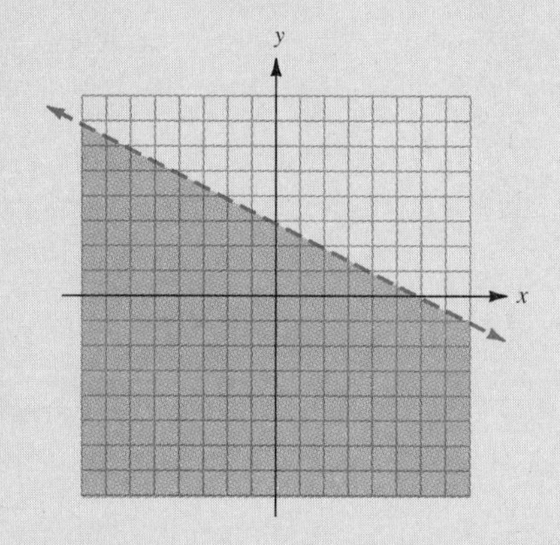

22. $3x - 4y \geq 12$

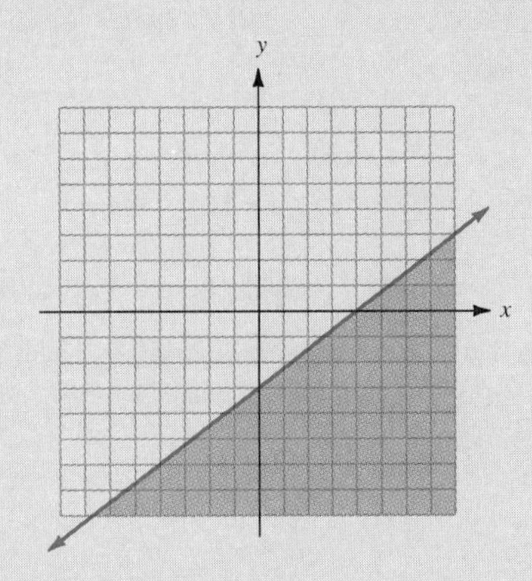

Solve each of the following systems by graphing.

23. $x - y = 2$
$x + 3y = 6$ $\quad (3, 1)$

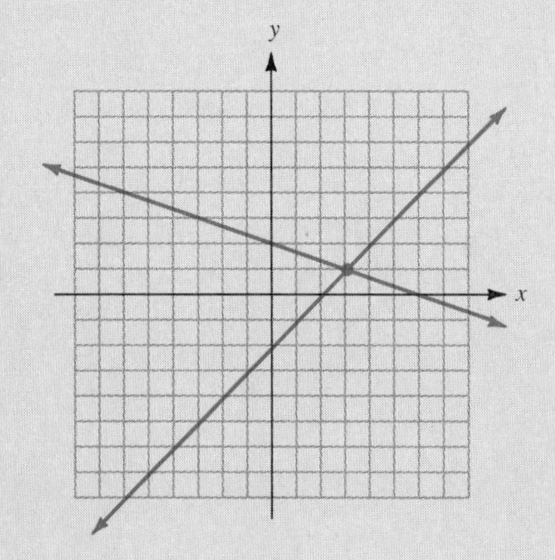

24. $3x + 2y = 6$
$x + 2y = -2$ $\quad (4, -3)$

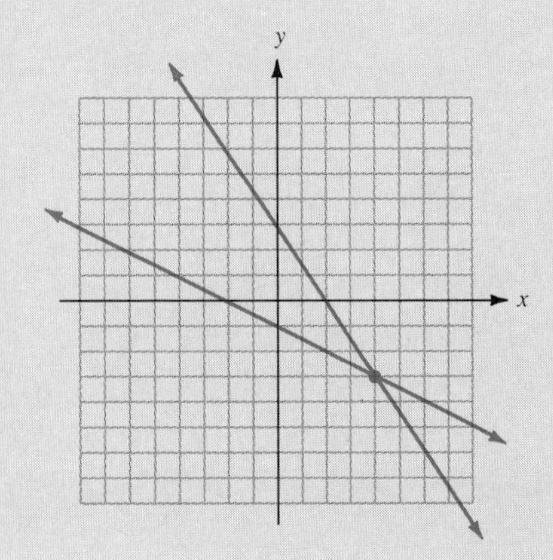

Solve each of the following systems. If a unique solution does not exist, state whether the system is inconsistent or dependent.

25. $2x - 3y = 6$
$x - 3y = 2$
$\left(4, \frac{2}{3}\right)$

26. $2x + y = 4$
$y = 2x - 8$
$(3, -2)$

27. $5x + 2y = 30$
$x - 4y = 17$
$\left(7, -\frac{5}{2}\right)$

28. $2x - 6y = 8$
$x = 3y + 4$
Dependent

29. $4x - 5y = 20$
$2x + 3y = 10$
$(5, 0)$

30. $5x - 4y = -7$
$3x + 5y = -19$
$(-3, -2)$

31. $4x + 2y = 11$
$2x + y = 5$
Inconsistent

32. $4x - 3y = 7$
$6x + 6y = 7$
$\left(\frac{3}{2}, -\frac{1}{3}\right)$

33. $5x - 3y = 11$
$y = 10x - 7$
$\left(\frac{2}{5}, -3\right)$

34. $2x - 5y = -17$
$x = \frac{3}{2}$ $\left(\frac{3}{2}, 4\right)$

Solve each of the following applications. Be sure to show the system of equations used for your solution.

35. One number is 4 less than 5 times another. If the sum of the numbers is 26, what are the two numbers? 5, 21

36. Cynthia bought five blank VHS tapes and four cassette tapes for \$28.50. Charlie bought four VHS tapes and two cassette tapes for \$21.00. Find the cost of each type of tape. \$4.50, \$1.50

37. Receipts for a concert, attended by 450 people, were \$2775. If reserved-seat tickets were \$7 and general-admission tickets were \$4, how many of each type of ticket were sold? 325 at \$7, 125 at \$4

38. Anthony invested part of his \$12,000 inheritance in a bond paying 9 percent and the other part in a savings account paying 6 percent. If his interest from the two investments was \$930 in the first year, how much did he have invested at each rate? \$5000 at 6%, \$7000 at 9%

39. A chemist has a 30% acid solution and a 60% acid solution already prepared. How much of each of the two solutions should be mixed in order to form 300mL of a 50% acid solution? 100 mL of 30%, 200 mL of 60%

40. Andrew was able to travel 75 mi downstream in 3 h. Returning upstream, he took 5 h to make the trip. How fast can his boat travel in still water, and what was the rate of the current? Boat 20 mi/h, current 5 mi/h

Chapter Nine

Exponents and Radicals

9.1 Extending the Properties of Exponents

OBJECTIVE
To simplify expressions involving exponents by using the five properties of exponents.

In Section 1.4, we introduced the first two properties of exponents. We stated there that the exponent notation indicates repeated multiplication and that the exponent tells us how many times the base is to be used as a factor.

Exponent (pointing to 5); Base (pointing to 3)

$$3^5 = \underbrace{3 \cdot 3 \cdot 3 \cdot 3 \cdot 3}_{\text{Five factors}} = 243$$

Our first property for exponents allowed us to multiply expressions to the same base.

PROPERTY 1 FOR EXPONENTS

For any real number a and positive integers m and n,

$$a^m \cdot a^n = a^{m+n}$$

In words, to multiply expressions with the same base, keep the base and add the exponents.

Example 1

Multiply.

(*a*) $x^6 \cdot x^7 = x^{6+7} = x^{13}$ Add the exponents.
(*b*) $2^3 \cdot 2^5 = 2^{3+5} = 2^8$ The base in the product is still 2.

CHECK YOURSELF 1

Multiply.

1. $y^5 \cdot y^6$ **2.** $3^2 \cdot 3^4$

Our second property allowed us to divide expressions with the same base.

PROPERTY 2 FOR EXPONENTS

For any real number *a*, where *a* is not equal to zero, and positive integers *m* and *n*, where *m* is greater than *n*,

$$\frac{a^m}{a^n} = a^{m-n}$$

In words, to divide expressions with the same base, keep the base and subtract the exponents.

Example 2

Divide.

(*a*) $\frac{y^8}{y^4} = y^{8-4} = y^4$ Subtract the exponents.

(*b*) $\frac{3^6}{3^3} = 3^{6-3} = 3^3$ The base in the quotient is still 3.

CHECK YOURSELF 2

Divide.

1. $\frac{b^9}{b^5}$ **2.** $\frac{4^5}{4^2}$

In this section we want to expand our work with exponents to introduce three further properties. Consider the following expression:

Note that this means that the base, x^2, is used as a factor *four* times.

$$(x^2)^4$$

We can write this as

Using property 1, we can *add* the exponents for the result x^8. However, note also that $2 + 2 + 2 + 2 = 2 \times 4 = 8$.

$$(x^2)^4 = x^2 \cdot x^2 \cdot x^2 \cdot x^2 = x^8$$

and this leads us to our third property for exponents.

PROPERTY 3 FOR EXPONENTS

For any real number a and positive integers m and n,

$(a^m)^n = a^{m \cdot n}$

In words, to raise a power to a power, keep the base and multiply the exponents.

The use of this new property is illustrated in our next example.

Example 3

Simplify each expression.

(*a*) $(x^4)^5 = x^{4 \cdot 5} = x^{20}$ Multiply the exponents.
(*b*) $(2^3)^4 = 2^{3 \cdot 4} = 2^{12}$

Be Careful! Be sure to distinguish between the correct use of Property 1 and Property 3.

$(x^4)^5 = x^{4 \cdot 5} = x^{20}$ This is a *power* raised to a *power*, so multiply the exponents.

but

$x^4 \cdot x^5 = x^{4+5} = x^9$ This is the product where each factor has the same base so *add* the exponents.

CHECK YOURSELF 3

Simplify each expression.

1. $(m^5)^6$ **2.** $(m^5)(m^6)$
3. $(3^2)^4$ **4.** $(3^2)(3^4)$

Suppose we now have a product raised to a power. Consider an expression such as

Here the base is $3x$.

$(3x)^4$

We know that

$$(3x)^4 = (3x)(3x)(3x)(3x)$$
$$= (3 \cdot 3 \cdot 3 \cdot 3)(x \cdot x \cdot x \cdot x)$$
$$= 3^4 \cdot x^4 = 81x^4$$

Here we have applied the commutative and associative laws.

Note that the power, here 4, has been applied to each factor, 3 and x.

In general, we have

PROPERTY 4 FOR EXPONENTS

For any real numbers a and b and positive integer m,

$(ab)^m = a^m b^m$

In words, to raise a product to a power, raise each factor to that same power.

The use of this property is shown in the following example.

Example 4

Simplify each expression.

Note that $(2x)^5$ and $2x^5$ are entirely different expressions. For $(2x)^5$, the base is $2x$ so we raise each factor to the fifth power. For $2x^5$, the base is x and so the exponent applies only to x.

(a) $(2x)^5 = 2^5 \cdot x^5 = 32x^5$
(b) $(3ab)^4 = 3^4 \cdot a^4 \cdot b^4 = 81a^4b^4$
(c) $5(2r)^3 = 5 \cdot 2^3 \cdot r^3 = 40r^3$

CHECK YOURSELF 4

Simplify each expression.

1. $(3y)^4$ **2.** $(2mn)^6$
3. $3(4x)^2$

We may have to use more than one of our properties in simplifying an expression involving exponents. Consider our next example.

Example 5

Simplify each expression.

To help you understand each step of the simplification we will provide reference to the property being applied. Make a list of the properties now to help you as you work through the remainder of this and the next section.

(a) $(r^4s^3)^3 = (r^4)^3 \cdot (s^3)^3$ Property 4
$= r^{12}s^9$ Property 3

(b) $(3x^2)^2 \cdot (2x^3)^3$
$= 3^2(x^2)^2 \cdot 2^3 \cdot (x^3)^3$ Property 4
$= 9x^4 \cdot 8x^9$ Property 3
$= 72x^{13}$ Multiply the coefficients and apply property 1.

(c) $\dfrac{(a^3)^5}{a^4} = \dfrac{a^{15}}{a^4}$ Property 3
$= a^{11}$ Property 2

CHECK YOURSELF 5

Simplify each expression.

1. $(m^5n^2)^3$ **2.** $(2p)^4(4p^2)^2$ **3.** $\dfrac{(s^4)^3}{s^5}$

We have one final exponent property to develop. Suppose now that we have a quotient raised to a power. Consider the following expression.

$$\left(\frac{x}{3}\right)^3$$

We know that

$$\left(\frac{x}{3}\right)^3 = \frac{x}{3} \cdot \frac{x}{3} \cdot \frac{x}{3} = \frac{x \cdot x \cdot x}{3 \cdot 3 \cdot 3} = \frac{x^3}{3^3}$$

Note that the power, here 3, has been applied to the numerator x and to the denominator 3. In general, we have

PROPERTY 5 FOR EXPONENTS

For any real numbers a and b, where b is not equal to zero, and positive integer m,

$$\left(\frac{a}{b}\right)^m = \frac{a^m}{b^m}$$

In words, to raise a quotient to a power, raise the numerator and denominator to that same power.

Our final example illustrates the use of this property. Again note that the other properties may also have to be applied in simplifying an expression.

Example 6

Simplify each expression.

(a) $\left(\dfrac{3}{4}\right)^3 = \dfrac{3^3}{4^3} = \dfrac{27}{64}$ Property 5

(b) $\left(\dfrac{x^3}{y^2}\right)^4 = \dfrac{(x^3)^4}{(y^2)^4}$ Property 5

$= \dfrac{x^{12}}{y^8}$ Property 3

$$(c)\ \left(\frac{r^2s^3}{t^4}\right)^2 = \frac{(r^2s^3)^2}{(t^4)^2} \qquad \text{Property 5}$$

$$= \frac{(r^2)^2(s^3)^2}{(t^4)^2} \qquad \text{Property 4}$$

$$= \frac{r^4s^6}{t^8} \qquad \text{Property 3}$$

CHECK YOURSELF 6

Simplify each expression.

1. $\left(\frac{2}{3}\right)^4$ **2.** $\left(\frac{m^3}{n^4}\right)^5$ **3.** $\left(\frac{a^2b^3}{c^5}\right)^2$

The following chart summarizes the five properties of exponents that have been discussed in this section.

GENERAL FORM		EXAMPLE
I. $a^m a^n = a^{m+n}$		$x^2 \cdot x^3 = x^5$
II. $\frac{a^m}{a^n} = a^{m-n}$	$(m > n)$	$\frac{5^7}{5^3} = 5^4$
III. $(a^m)^n = a^{mn}$		$(z^5)^4 = z^{20}$
IV. $(ab)^m = a^m b^m$		$(4x)^3 = 4^3x^3 = 64x^3$
V. $\left(\frac{a}{b}\right)^m = \frac{a^m}{b^m}$		$\left(\frac{2}{3}\right)^6 = \frac{2^6}{3^6} = \frac{64}{729}$

CHECK YOURSELF ANSWERS

1. (1) y^{11}; (2) 3^6. **2.** (1) b^4; (2) 4^3.
3. (1) m^{30}; (2) m^{11}; (3) 3^8; (4) 3^6. **4.** (1) $81y^4$; (2) $64m^6n^6$; (3) $48x^2$.
5. (1) $m^{15}n^6$; (2) $256p^8$; (3) s^7 **6.** (1) $\frac{16}{81}$; (2) $\frac{m^{15}}{n^{20}}$; (3) $\frac{a^4b^6}{c^{10}}$.

9.1 Exercises

Use Properties 1 and 2 for exponents to simplify each of the following expressions.

1. $x^3 \cdot x^5$ x^8 **2.** $a^7 \cdot a^6$ a^{13} **3.** $3^3 \cdot 3^4$ 3^7 **4.** $5^2 \cdot 5^3$ 5^5

5. $y^3 \cdot y^2 \cdot y^4$ y^9 **6.** $b^5 \cdot b^4 \cdot b$ b^{10} **7.** $2^3 \cdot 2^2 \cdot 2$ 2^6 **8.** $3^6 \cdot 3^4 \cdot 3^2$ 3^{12}

9. $\dfrac{x^9}{x^4}$ x^5 **10.** $\dfrac{y^{10}}{y^3}$ y^7 **11.** $\dfrac{5^5}{5^2}$ 5^3 **12.** $\dfrac{7^9}{7^6}$ 7^3

Use Property 3 for exponents to simplify each of the following expressions.

13. $(x^2)^3$ x^6 **14.** $(a^5)^3$ a^{15} **15.** $(m^4)^4$ m^{16} **16.** $(p^7)^2$ p^{14}

17. $(2^4)^2$ 2^8 **18.** $(3^3)^2$ 3^6 **19.** $(5^3)^5$ 5^{15} **20.** $(7^2)^4$ 7^8

Use Properties 4 and 5 for exponents to simplify each of the following expressions.

21. $(3x)^3$ $27x^3$ **22.** $(4m)^2$ $16m^2$ **23.** $(2xy)^4$ $16x^4y^4$ **24.** $(5pq)^3$ $125p^3q^3$

25. $5(3ab)^3$ $135a^3b^3$ **26.** $4(2rs)^4$ $64r^4s^4$ **27.** $\left(\dfrac{3}{4}\right)^2$ $\dfrac{9}{16}$ **28.** $\left(\dfrac{2}{3}\right)^3$ $\dfrac{8}{27}$

29. $\left(\dfrac{x}{5}\right)^3$ $\dfrac{x^3}{125}$ **30.** $\left(\dfrac{a}{2}\right)^5$ $\dfrac{a^5}{32}$

Use the properties for exponents to simplify each of the following expressions.

31. $(2x^3)^3$ $8x^9$ **32.** $(3y^5)^2$ $9y^{10}$ **33.** $(a^4b^3)^4$ $a^{16}b^{12}$

34. $(p^2q^5)^3$ p^6q^{15} **35.** $(3x^3y)^2$ $9x^6y^2$ **36.** $(5m^2n^2)^3$ $125m^6n^6$

37. $(2m^3)^2(m^2)^3$ $4m^{12}$ **38.** $(y^3)^5(3y^2)^2$ $9y^{19}$ **39.** $\dfrac{(x^4)^3}{x^2}$ x^{10}

40. $\dfrac{(m^5)^3}{m^6}$ m^9 **41.** $\dfrac{(s^3)^2(s^2)^3}{(s^5)^2}$ s^2 **42.** $\dfrac{(y^5)^3(y^3)^2}{(y^4)^4}$ y^5

43. $\left(\dfrac{m^3}{n^2}\right)^3$ $\dfrac{m^9}{n^6}$ **44.** $\left(\dfrac{a^4}{b^3}\right)^4$ $\dfrac{a^{16}}{b^{12}}$ **45.** $\left(\dfrac{a^3b^2}{c^4}\right)^2$ $\dfrac{a^6b^4}{c^8}$

46. $\left(\dfrac{x^5y^2}{z^4}\right)^3$ $\dfrac{x^{15}y^6}{z^{12}}$

47. Write x^{12} as a power of x^2. $(x^2)^6$

48. Write y^{15} as a power of y^3. $(y^3)^5$

49. Write a^{16} as a power of a^2. $(a^2)^8$

50. Write m^{20} as a power of m^5. $(m^5)^4$

Skillscan (Section 4.2)

Reduce each of the following fractions to simplest form.

a. $\frac{m^3}{m^5}$ $\frac{1}{m^2}$ **b.** $\frac{x^7}{x^{10}}$ $\frac{1}{x^3}$ **c.** $\frac{a^3}{a^9}$ $\frac{1}{a^6}$ **d.** $\frac{y^4}{y^8}$ $\frac{1}{y^4}$

e. $\frac{x^3}{x^3}$ 1 **f.** $\frac{b^5}{b^5}$ 1 **g.** $\frac{s^7}{s^7}$ 1 **h.** $\frac{r^{10}}{r^{10}}$ 1

Answers

1. x^8 **3.** 3^7 **5.** y^9 **7.** 2^6 **9.** x^5 **11.** 5^3 **13.** x^6 **15.** m^{16} **17.** 2^8 **19.** 5^{15} **21.** $27x^3$ **23.** $16x^4y^4$ **25.** $135a^3b^3$ **27.** $\frac{9}{16}$ **29.** $\frac{x^3}{125}$ **31.** $8x^9$ **33.** $a^{16}b^{12}$ **35.** $9x^6y^2$ **37.** $4m^{12}$ **39.** x^{10} **41.** s^2 **43.** $\frac{m^9}{n^6}$ **45.** $\frac{a^6b^4}{c^8}$ **47.** $(x^2)^6$ **49.** $(a^2)^8$ **a.** $\frac{1}{m^2}$ **b.** $\frac{1}{x^3}$ **c.** $\frac{1}{a^6}$ **d.** $\frac{1}{y^4}$ **e.** 1 **f.** 1 **g.** 1 **h.** 1

9.2 Zero and Negative Exponents

1. To evaluate expressions involving zero or negative exponents.
2. To simplify expressions involving zero or negative exponents.

In the last section, we continued our discussion of the properties of exponents with the introduction of rules for raising expressions involving powers, products, and quotients to a power. We now want to extend our exponent notation to include 0 and negative integers as exponents.

First, what do we do with x^0? It will help to look at a problem that gives us x^0 as a result. What if the numerator and denominator of a fraction have the same base raised to the same power and we extend our division rule? For example,

By Property 2,

$$\frac{a^m}{a^n} = a^{m-n}$$

where $m > n$. Here m and n are *both* 5 and $m = n$.

$$\frac{a^5}{a^5} = a^{5-5} = a^0 \quad (1)$$

But from our experience with fractions we know that

$$\frac{a^5}{a^5} = 1 \quad (2)$$

By comparing Equations (1) and (2), it seems reasonable to make the following definition:

> For any number a, $a \neq 0$,
>
> $a^0 = 1$
>
> In words, an expression raised to the zero power is always 1 as long as the expression itself is not equal to 0.

The following example illustrates the use of this definition.

Example 1

(*a*) $5^0 = 1$
(*b*) $27^0 = 1$
(*c*) $(x^2y)^0 = 1$ if $x \neq 0$ and $y \neq 0$
(*d*) $6x^0 = 6 \cdot 1 = 6$ if $x \neq 0$

Be Careful! In part (*d*) the zero exponent applies only to the x and *not* to the factor 6, since the base is x.

CHECK YOURSELF 1

Evaluate. Assume all variables are nonzero.

1. 7^0 **2.** $(-8)^0$ **3.** $(xy^3)^0$ **4.** $3x^0$

Our second property for exponents will also allow us to define what is meant by a negative exponent. Suppose that the exponent in the denominator is *greater than* the exponent in the numerator. Consider the expression $\frac{x^2}{x^5}$.

Our previous work with fractions tells us that

Divide numerator and denominator by the two common factors of x

$$\frac{x^2}{x^5} = \frac{x \cdot x}{x \cdot x \cdot x \cdot x \cdot x} = \frac{1}{x^3} \quad (1)$$

However, if we extend the second property to let n be greater than m, we have

Remember:

$$\frac{a^m}{a^n} = a^{m-n}$$

$$\frac{x^2}{x^5} = x^{2-5} = x^{-3} \quad (2)$$

Now, by comparing equations (1) and (2), it seems reasonable to define x^{-3} as $\frac{1}{x^3}$.

In general, we have this result:

John Wallis (1616–1703), an English mathematician, was the first to fully discuss the meaning of 0 and negative exponents.

For any number a, $a \neq 0$, and any positive integer n,

$$a^{-n} = \frac{1}{a^n}$$

Example 2

(a) $x^{-4} = \frac{1}{x^4}$

Negative exponent in the numerator

Positive exponent in the denominator

(b) $m^{-7} = \frac{1}{m^7}$

(c) $3^{-2} = \frac{1}{3^2}$ or $\frac{1}{9}$

(d) $10^{-3} = \frac{1}{10^3}$ or $\frac{1}{1000}$

(e) $2x^{-3} = 2 \cdot \frac{1}{x^3} = \frac{2}{x^3}$

The -3 exponent applies only to the x, since x is the base.

(f) $\frac{a^5}{a^9} = a^{5-9} = a^{-4} = \frac{1}{a^4}$

CHECK YOURSELF 2

Write, using positive exponents.

1. a^{-10} **2.** 4^{-3} **3.** $3x^{-2}$ **4.** $\frac{x^5}{x^8}$

We can now allow negative integers as exponents in our first property for exponents. Consider our next example.

Example 3

Simplify each expression.

$a^m \cdot a^n = a^{m+n}$ for *any* integers m and n. So add the exponents.

(a) $x^5x^{-2} = x^{5+(-2)} = x^3$

Note: An alternative approach would be

By definition

$x^{-2} = \frac{1}{x^2}$

$$x^5x^{-2} = x^5 \cdot \frac{1}{x^2} = \frac{x^5}{x^2} = x^3$$

(*b*) $a^7a^{-5} = a^{7+(-5)} = a^2$

(*c*) $y^5y^{-9} = y^{5+(-9)} = y^{-4} = \frac{1}{y^4}$

CHECK YOURSELF 3

Simplify.

1. x^7x^{-2} **2.** b^3b^{-8}

Our final example shows that all the exponent properties introduced in the last section can be extended to allow negative exponents.

Example 4

Simplify each expression.

Property 2

(*a*) $\frac{m^{-3}}{m^4} = m^{-3-4}$

$= m^{-7} = \frac{1}{m^7}$

Apply Property 2 to each variable.

(*b*) $\frac{a^{-2}b^6}{a^5b^{-4}} = a^{-2-5}b^{6-(-4)}$

$= a^{-7}b^{10} = \frac{b^{10}}{a^7}$

Definition of the negative exponent

(*c*) $(2x^4)^{-3} = \frac{1}{(2x^4)^3}$

Property 4

$= \frac{1}{2^3(x^4)^3}$

Property 3

$= \frac{1}{8x^{12}}$

Property 3

(*d*) $\frac{(y^{-2})^4}{(y^3)^{-2}} = \frac{y^{-8}}{y^{-6}}$

Property 2

$= y^{-8-(-6)}$

$= y^{-2} = \frac{1}{y^2}$

CHECK YOURSELF 4

Simplify each expression.

1. $\frac{x^5}{x^{-3}}$ **2.** $\frac{m^3n^{-5}}{m^{-2}n^3}$ **3.** $(3a^3)^{-4}$ **4.** $\frac{(r^3)^{-2}}{(r^{-4})^2}$

CHECK YOURSELF ANSWERS

1. (1) 1; (2) 1; (3) 1; (4) 3. **2.** (1) $\frac{1}{a^{10}}$; (2) $\frac{1}{4^3}$ or $\frac{1}{64}$; (3) $\frac{3}{x^2}$; (4) $\frac{1}{x^3}$.
3. (1) x^5; (2) $\frac{1}{b^5}$. **4.** (1) x^8; (2) $\frac{m^5}{n^8}$; (3) $\frac{1}{81a^{12}}$; (4) r^2.

9.2 Exercises

Evaluate (assume the variables are nonzero).

1. 4^0 1 **2.** $(-7)^0$ 1 **3.** $(-29)^0$ 1 **4.** 75^0 1

5. $(x^3y^2)^0$ 1 **6.** $7m^0$ 7 **7.** $9x^0$ 9 **8.** $(3a^2b^6)^0$ 1

9. $(-2p^5q^7)^0$ 1 **10.** $-5x^0$ −5

Write each of the following expressions using positive exponents; simplify where possible.

11. b^{-8} $\frac{1}{b^8}$ **12.** p^{-12} $\frac{1}{p^{12}}$ **13.** 3^{-4} $\frac{1}{81}$ **14.** 2^{-5} $\frac{1}{32}$

15. 5^{-2} $\frac{1}{25}$ **16.** 4^{-3} $\frac{1}{64}$ **17.** 10^{-4} $\frac{1}{10,000}$ **18.** 10^{-5} $\frac{1}{100,000}$

19. $5x^{-1}$ $\frac{5}{x}$ **20.** $3a^{-2}$ $\frac{3}{a^2}$ **21.** $(5x)^{-1}$ $\frac{1}{5x}$ **22.** $(3a)^{-2}$ $\frac{1}{9a^2}$

23. $-2x^{-5}$ $-\frac{2}{x^5}$ **24.** $3x^{-4}$ $\frac{3}{x^4}$ **25.** $(-2x)^{-5}$ $-\frac{1}{32x^5}$ **26.** $(3x)^{-4}$ $\frac{1}{81x^4}$

Use Properties 1 and 2 to simplify each of the following expressions. Write your answers with positive exponents only.

27. a^5a^3 a^8 **28.** m^5m^7 m^{12} **29.** x^9x^{-3} x^6 **30.** $a^{10}a^{-6}$ a^4

31. b^5b^{-9} $\frac{1}{b^4}$ **32.** y^3y^{-10} $\frac{1}{y^7}$ **33.** x^0x^5 x^5 **34.** $r^{-3}r^0$ $\frac{1}{r^3}$

35. $\frac{a^8}{a^5}$ a^3 **36.** $\frac{m^9}{m^4}$ m^5 **37.** $\frac{x^7}{x^9}$ $\frac{1}{x^2}$ **38.** $\frac{a^3}{a^{10}}$ $\frac{1}{a^7}$

39. $\frac{r^{-3}}{r^5}$ $\frac{1}{r^8}$ **40.** $\frac{x^3}{x^{-5}}$ x^8 **41.** $\frac{x^{-4}}{x^{-5}}$ x **42.** $\frac{p^{-6}}{p^{-3}}$ $\frac{1}{p^3}$

Simplify each of the following expressions. Write your answers with positive exponents only.

43. $\frac{m^5n^{-3}}{m^{-4}n^5}$ $\frac{m^9}{n^8}$ **44.** $\frac{p^{-3}q^{-2}}{p^4q^{-3}}$ $\frac{q}{p^7}$ **45.** $(2a^{-3})^4$ $\frac{16}{a^{12}}$ **46.** $(3x^2)^{-3}$ $\frac{1}{27x^6}$

47. $(x^{-2}y^3)^{-2}$ $\frac{x^4}{y^6}$ **48.** $(a^5b^{-3})^{-3}$ $\frac{b^9}{a^{15}}$ **49.** $\frac{(r^{-2})^3}{r^{-4}}$ $\frac{1}{r^2}$ **50.** $\frac{(y^3)^{-4}}{y^{-6}}$ $\frac{1}{y^6}$

51. $\frac{(x^{-3})^3}{(x^4)^{-2}}$ $\frac{1}{x}$ **52.** $\frac{(m^4)^{-3}}{(m^{-2})^4}$ $\frac{1}{m^4}$ **53.** $\frac{(a^{-3})^2(a^4)}{(a^{-3})^{-3}}$ $\frac{1}{a^{11}}$ **54.** $\frac{(x^2)^{-3}(x^{-2})}{(x^2)^{-4}}$ 1

Skillscan (Section 2.4)

Evaluate each expression.

a. 3^2 9 **b.** $(-3)^2$ 9 **c.** 5^3 125 **d.** $(-5)^3$ −125

e. 2^4 16 **f.** $(-2)^4$ 16 **g.** 3^5 243 **h.** $(-3)^5$ −243

Answers

1. 1 **3.** 1 **5.** 1 **7.** 9 **9.** 1 **11.** $\frac{1}{b^8}$ **13.** $\frac{1}{3^4}$ or $\frac{1}{81}$ **15.** $\frac{1}{5^2}$ or $\frac{1}{25}$ **17.** $\frac{1}{10^4}$ or $\frac{1}{10{,}000}$
19. $\frac{5}{x}$ **21.** $\frac{1}{5x}$ **23.** $-\frac{2}{x^5}$ **25.** $-\frac{1}{32x^5}$ **27.** a^8 **29.** x^6 **31.** $\frac{1}{b^4}$ **33.** x^5 **35.** a^3 **37.** $\frac{1}{x^2}$
39. $\frac{1}{r^8}$ **41.** x **43.** $\frac{m^9}{n^8}$ **45.** $\frac{16}{a^{12}}$ **47.** $\frac{x^4}{y^6}$ **49.** $\frac{1}{r^2}$ **51.** $\frac{1}{x}$ **53.** $\frac{1}{a^{11}}$ **a.** 9 **b.** 9 **c.** 125
d. −125 **e.** 16 **f.** 16 **g.** 243 **h.** −243

9.3 Roots and Radicals

OBJECTIVES

1. To use the radical notation to represent roots.
2. To distinguish between rational and irrational numbers.

In the last section, we extended the properties of exponents to include the entire set of integers. Over the next four sections, we will be working with a new notation that "reverses" the process of raising to a power.

From our work in Section 1.5, we know that when we have a statement such as

$$x^2 = 9$$

it is read as "x squared equals 9."

Here we are concerned with the relationship between the variable x and the number 9. We call that relationship the *square root* and say, equivalently, that "x is the square root of 9."

We know from experience that x must be 3 [since $3^2 = 9$] or -3 [since $(-3)^2 = 9$]. We see that 9 has two square roots, 3 and -3. In fact, every positive number will have *two* square roots. In general, if $x^2 = a$, we call x the *square root* of a.

The symbol $\sqrt{\ }$ first appeared in print in 1525. In Latin, *radix* means "root," and this was contracted to a small *r*. The present symbol may have been used because it resembled the manuscript form of that small *r*.

We are now ready for our new notation. The symbol $\sqrt{\ }$ is called a *radical sign*. We saw above that 3 was the positive square root of 9. We also call 3 the *principal square root* of 9 and can write

$$\sqrt{9} = 3$$

to indicate that 3 is the principal square root of 9.

To summarize:

$\sqrt{9}$ asks, "What number must we square (or multiply by itself) to get 9?"

$\sqrt{9}$ is read the "*square root*" of 9, and since $3^2 = 9$,

$$\sqrt{9} = 3$$

This leads us to the following definition.

DEFINITION

$\sqrt{a}$ is the positive (or *principal*) square root of a. It is the positive number whose square is a.

This definition is illustrated in our first example.

Example 1

Find the following square roots.

(*a*) $\sqrt{49} = 7$ Since 7 is the positive number, we must square to get 49.

(*b*) $\sqrt{81} = 9$

(*c*) $\sqrt{\frac{4}{9}} = \frac{2}{3}$ Since $\frac{2}{3}$ is the positive number, we must square to get $\frac{4}{9}$.

CHECK YOURSELF 1

Find the following square roots.

1. $\sqrt{64}$ **2.** $\sqrt{144}$ **3.** $\sqrt{\frac{16}{25}}$

Note: When you use the radical sign, you will get only the *positive square root:*

$\sqrt{25} = 5$

Each positive number has two square roots. For instance, 25 has square roots of 5 and -5 because

$5^2 = 25$ and $(-5)^2 = 25$

If you want to indicate the negative square root, you must use a negative sign in front of the radical.

$-\sqrt{25} = -5$

Example 2

Find the following square roots.

(*a*) $\sqrt{100} = 10$ The principal root

(*b*) $-\sqrt{100} = -10$ The negative square root

(*c*) $-\sqrt{\frac{9}{16}} = -\frac{3}{4}$

CHECK YOURSELF 2

Find the following square roots.

1. $\sqrt{16}$ **2.** $-\sqrt{16}$ **3.** $-\sqrt{\frac{16}{25}}$

The square roots of negative numbers are *not* real numbers. For instance, $\sqrt{-9}$ is *not* a real number because there is *no* real number x such that

$x^2 = -9$

Be Careful! Do not confuse

$-\sqrt{9}$ with $\sqrt{-9}$.

The expression $-\sqrt{9}$ is -3, while $\sqrt{-9}$ is not a real number.

The following example summarizes our discussion thus far.

Example 3

(*a*) $\sqrt{36} = 6$
(*b*) $\sqrt{121} = 11$
(*c*) $-\sqrt{64} = -8$
(*d*) $\sqrt{-64}$ is not a real number
(*e*) $\sqrt{0} = 0$ (Because $0 \cdot 0 = 0$)

CHECK YOURSELF 3

Evaluate if possible.

1. $\sqrt{81}$ **2.** $\sqrt{49}$ **3.** $-\sqrt{49}$ **4.** $\sqrt{-49}$

As we mentioned earlier, finding the square root of a number is the reverse of squaring a number. We can extend that idea to work with other roots of numbers. For instance, the *cube root* of a number is the number we must cube (or raise to the third power) to get that number.

The cube root of 8 is 2 since $2^3 = 8$, and we write

$\sqrt[3]{8}$ is read "the cube root of 8."

$$\sqrt[3]{8} = 2$$

We give special names to the parts of a radical expression. These are summarized as follows.

The index for $\sqrt[3]{a}$ is 3.

The index of 2 for square roots is generally not written.

We understand that $\sqrt{a}$ is the principal square root of *a*.

Every radical expression contains three parts as shown below. The principal *n*th root of *a* is written as

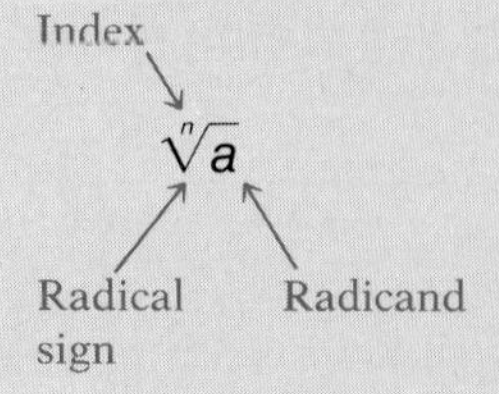

To illustrate, the *cube root* of 64 is written

Index of 3 ⟶ $\sqrt[3]{64}$

and it represents the number we must cube (or raise to the third power) to get 64.

$\sqrt[3]{64} = 4$

because $4^3 = 64$.

Index of 4 → $\sqrt[4]{81}$

is the *fourth root* of 81, and it represents the number we must raise to the fourth power to get 81.

$\sqrt[4]{81} = 3$

because $3^4 = 81$.

We can find roots of negative numbers as long as the index is *odd* (3, 5, etc.). For example,

$\sqrt[3]{-64} = -4$

because $(-4)^3 = -64$.

If the index is *even* (2, 4, etc.), roots of negative numbers are *not* real numbers.

$\sqrt[4]{-16}$

The *even power* of a real number is always *positive* or *zero.*

is not a real number because there is no real number x such that $x^4 = -16$.

The following table shows the most common roots.

It would be helpful for your work in the remainder of this chapter to memorize these roots.

SQUARE ROOTS		CUBE ROOTS	FOURTH ROOTS
$\sqrt{1} = 1$	$\sqrt{49} = 7$	$\sqrt[3]{1} = 1$	$\sqrt[4]{1} = 1$
$\sqrt{4} = 2$	$\sqrt{64} = 8$	$\sqrt[3]{8} = 2$	$\sqrt[4]{16} = 2$
$\sqrt{9} = 3$	$\sqrt{81} = 9$	$\sqrt[3]{27} = 3$	$\sqrt[4]{81} = 3$
$\sqrt{16} = 4$	$\sqrt{100} = 10$	$\sqrt[3]{64} = 4$	$\sqrt[4]{256} = 4$
$\sqrt{25} = 5$	$\sqrt{121} = 11$	$\sqrt[3]{125} = 5$	$\sqrt[4]{625} = 5$
$\sqrt{36} = 6$	$\sqrt{144} = 12$		

You can use the table in the following example which summarizes the discussion above.

Example 4

(*a*) $\sqrt[3]{125} = 5$ because $5^3 = 125$.
(*b*) $\sqrt[5]{32} = 2$ because $2^5 = 32$.
(*c*) $\sqrt[3]{-125} = -5$ because $(-5)^3 = -125$.

The cube root of a negative number will be negative

The fourth root of a negative number is not a real number.

(*d*) $\sqrt[4]{-81}$ is not a real number.

CHECK YOURSELF 4

Evaluate if possible.

1. $\sqrt[3]{64}$ **2.** $\sqrt[4]{16}$ **3.** $\sqrt[4]{-256}$ **4.** $\sqrt[3]{-8}$

The radical notation allows us to distinguish between two important types of numbers.

First, a *rational number* can be represented by a fraction whose numerator and denominator are integers and whose denominator is nonzero. The form of a rational number is

$$\frac{a}{b} \quad \text{where } a \text{ and } b \text{ are integers and } b \neq 0$$

Some examples of rational numbers are $\frac{-2}{3}$, $\frac{4}{5}$, 2, and $\frac{7}{3}$. Certain square roots are rational numbers also. For example

Note that each radicand is a *perfect square integer* (that is, an integer which is the square of another integer).

$$\sqrt{4} \qquad \sqrt{25} \qquad \text{and} \qquad \sqrt{64}$$

represent the rational numbers 2, 5, and 8, respectively.

There are numbers that represent a point on the number line but that *cannot* be written as the ratio of two integers. These are called *irrational numbers.* For example, the square root of any positive number which is not itself a perfect square is an irrational number. The expressions

This is because the radicands are *not* perfect squares.

$$\sqrt{2} \qquad \sqrt{3} \qquad \text{and} \qquad \sqrt{5}$$

represent irrational numbers.

Note: The fact that the square root of 2 is irrational will be proved in later mathematics courses and was known to Greek mathematicians over 2000 years ago.

Example 5

(*a*) $\sqrt{7}$, $\sqrt{10}$, $\sqrt{17}$, and $\sqrt{\frac{2}{3}}$ are irrational numbers.

(*b*) $\sqrt{16}$ and $\sqrt{25}$ are rational numbers because 16 and 25 are perfect squares.

(*c*) $\sqrt{\frac{4}{9}}$ is rational because $\sqrt{\frac{4}{9}} = \frac{2}{3}$, a ratio of two integers.

CHECK YOURSELF 5

Which of the following numbers are rational and which are irrational?

1. $\sqrt{26}$ **2.** $\sqrt{49}$ **3.** $\sqrt{\dfrac{6}{7}}$

4. $\sqrt{100}$ **5.** $\sqrt{105}$ **6.** $\sqrt{\dfrac{16}{9}}$

The decimal representation of a rational number always terminates or repeats. For instance

$\frac{3}{8} = 0.375$

$\frac{5}{11} = 0.454545\ldots$

or 0.45

An important fact about the irrational numbers is that their decimal representations are always *nonterminating* and *nonrepeating*. We can therefore only approximate irrational numbers with a terminating decimal. A table of roots can be found at the end of this text or a calculator can be used to find roots. However, note that the values found for the irrational roots are only approximations. For instance, $\sqrt{2}$ is approximately 1.414 (to three decimal places), and we can write

$$\sqrt{2} \approx 1.414$$

Read "approximately equal to"

With a calculator we find that

1.414 is an approximation to the number whose square is 2.

$$(1.414)^2 = 1.999 \qquad \text{(to three decimal places)}$$

The set of all rational numbers and the set of all irrational numbers considered together form the set of *real numbers*. The real numbers will represent every point that can be pictured on the number line. Some examples are shown below.

For this reason we will refer to the number line as the *real number line.*

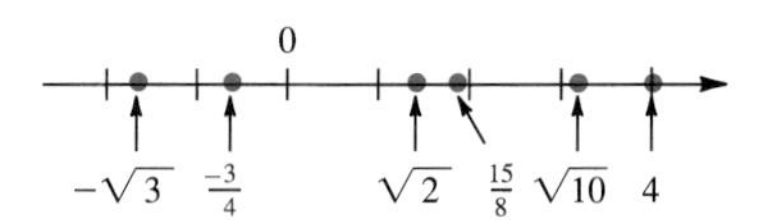

The following diagram will summarize the relationships among the various numeric sets that have been introduced in this section.

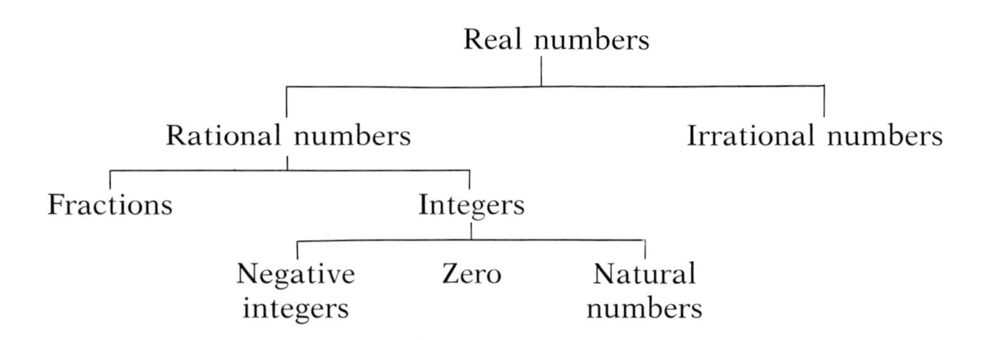

We conclude our work in this section by developing a general result that we will need later. Let's start by looking at two numerical examples.

$$\sqrt{2^2} = \sqrt{4} = 2 \qquad (1)$$

$$\sqrt{(-2)^2} = \sqrt{4} = 2 \qquad \text{since } (-2)^2 = 4 \qquad (2)$$

Consider the value of $\sqrt{x^2}$ where x is positive or negative.

This is because the principal square root of a number is always positive or zero.

In (1) where $x = 2$: In (2) where $x = -2$:

$\sqrt{2^2} = 2$ $\qquad \sqrt{(-2)^2} \neq -2$

Here $\sqrt{(-2)^2} = -(-2) = 2$

Comparing the results above, we see that $\sqrt{x^2}$ is x if x is positive (or 0) and $\sqrt{x^2}$ is $-x$ if x is negative. We can write

$$\sqrt{x^2} = \begin{cases} x, & \text{where } x \geq 0 \\ -x, & \text{where } x < 0 \end{cases}$$

From your earlier work with absolute values you will remember that

$$|x| = \begin{cases} x, & \text{where } x \geq 0 \\ -x, & \text{where } x < 0 \end{cases}$$

and we can summarize the discussion by writing

$$\sqrt{x^2} = |x| \qquad \text{for any real number } x$$

Example 6

Evaluate (*a*) $\sqrt{5^2}$ and (*b*) $\sqrt{(-4)^2}$.

(*a*) $\sqrt{5^2} = 5$

(*b*) $\sqrt{(-4)^2} = |-4| = 4$

Note: Alternatively in (*b*), we could write

$$\sqrt{(-4)^2} = \sqrt{16} = 4$$

CHECK YOURSELF 6

Evaluate.

1. $\sqrt{6^2}$ **2.** $\sqrt{(-6)^2}$

Note: The case for roots with indices that are odd does *not* require the use of absolute value, as illustrated in the last example. For instance,

$$\sqrt[3]{3^3} = \sqrt[3]{27} = 3$$
$$\sqrt[3]{(-3)^3} = \sqrt[3]{-27} = -3$$

and we see that

$$\sqrt[n]{x^n} = x \qquad \text{where } n \text{ is odd}$$

To summarize, we can write

$$\sqrt[n]{x^n} = \begin{cases} |x| & \text{where } n \text{ is even} \\ x & \text{where } n \text{ is odd} \end{cases}$$

Example 7

Evaluate (*a*) $\sqrt{(-4)^2}$, (*b*) $\sqrt[3]{5^3}$, (*c*) $\sqrt[3]{(-5)^3}$, and (*d*) $\sqrt[4]{(-2)^4}$.

(*a*) $\sqrt{(-4)^2} = |-4| = 4$ — $\sqrt{x} = |x|$

(*b*) $\sqrt[3]{5^3} = 5$ — $\sqrt[3]{x} = x$ (the index is odd)

(*c*) $\sqrt[3]{(-5)^3} = -5$

(*d*) $\sqrt[4]{(-2)^4} = |-2| = 2$ — $\sqrt[4]{x^4} = |x|$ (the index is even)

CHECK YOURSELF 7

Evaluate.

1. $\sqrt[3]{4^3}$ **2.** $\sqrt[4]{(-3)^4}$ **3.** $\sqrt[5]{(-2)^5}$

CHECK YOURSELF ANSWERS

1. (1) 8; (2) 12; (3) $\frac{4}{5}$. **2.** (1) 4; (2) −4; (3) $-\frac{4}{5}$.
3. (1) 9; (2) 7; (3) −7; (4) not a real number.
4. (1) 4; (2) 2; (3) not a real number; (4) −2.
5. (1) Irrational; (2) rational (since $\sqrt{49} = 7$); (3) irrational; (4) rational (since $\sqrt{100} = 10$); (5) irrational; (6) rational $\left(\text{since } \sqrt{\frac{16}{9}} = \frac{4}{3}\right)$.
6. (1) 6; (2) 6. **7.** (1) 4; (2) 3; (3) −2.

9.3 Exercises

Evaluate if possible.

1. $\sqrt{16}$ 4 **2.** $\sqrt{121}$ 11 **3.** $\sqrt{400}$ 20 **4.** $\sqrt{64}$ 8

5. $-\sqrt{100}$
-10

6. $\sqrt{-100}$
Not a real number

7. $\sqrt{-81}$
Not a real number

8. $-\sqrt{81}$
-9

9. $\sqrt{\frac{16}{9}}$ $\frac{4}{3}$

10. $-\sqrt{\frac{1}{25}}$ $-\frac{1}{5}$

11. $\sqrt{-\frac{4}{5}}$
Not a real number

12. $\sqrt{\frac{4}{25}}$ $\frac{2}{5}$

13. $\sqrt[3]{27}$ 3

14. $\sqrt[4]{81}$ 3

15. $\sqrt[3]{-27}$ -3

16. $\sqrt[4]{-16}$
Not a real number

17. $\sqrt[4]{-81}$
Not a real number

18. $-\sqrt[3]{64}$ -4

19. $-\sqrt[3]{27}$ -3

20. $-\sqrt[3]{-8}$ 2

21. $\sqrt[4]{625}$ 5

22. $\sqrt[3]{1000}$ 10

23. $\sqrt[3]{\frac{1}{27}}$ $\frac{1}{3}$

24. $\sqrt[3]{-\frac{8}{27}}$ $-\frac{2}{3}$

Which of the following roots are rational numbers and which are irrational numbers?

25. $\sqrt{19}$
Irrational

26. $\sqrt{36}$
Rational

27. $\sqrt{100}$
Rational

28. $\sqrt{7}$
Irrational

29. $\sqrt[3]{9}$
Irrational

30. $\sqrt[3]{8}$
Rational

31. $\sqrt[4]{16}$
Rational

32. $\sqrt{\frac{4}{9}}$
Rational

33. $\sqrt{\frac{4}{7}}$
Irrational

34. $\sqrt[3]{5}$
Irrational

35. $\sqrt[3]{-27}$
Rational

36. $-\sqrt[4]{81}$
Rational

Evaluate each of the following expressions.

37. $\sqrt{5^5}$ 5

38. $\sqrt{(-5)^2}$ 5

39. $\sqrt[3]{4^3}$ 4

40. $\sqrt[4]{(-3)^4}$ 3

41. $\sqrt[4]{2^4}$ 2

42. $\sqrt[3]{(-5)^3}$ -5

43. $\sqrt[5]{3^5}$ 3

44. $\sqrt[5]{(-2)^5}$ -2

Skillscan (Section 1.5)
Find each of the following products.

a. $(4x^2)(2x)$
$8x^3$

b. $(9a^4)(5a)$
$45a^5$

c. $(16m^2)(3m)$
$48m^3$

d. $(8b^3)(2b)$
$16b^4$

e. $(27p^6)(3p)$
$81p^7$

f. $(81s^4)(s^3)$
$81s^7$

g. $(100y^4)(2y)$
$200y^5$

h. $(49m^6)(2m)$
$98m^7$

Answers

1. 4 **3.** 20 **5.** -10 **7.** Not a real number **9.** $\frac{4}{3}$ **11.** Not a real number **13.** 3 **15.** -3
17. Not a real number **19.** -3 **21.** 5 **23.** $\frac{1}{3}$ **25.** Irrational **27.** Rational **29.** Irrational
31. Rational **33.** Irrational **35.** Rational **37.** 5 **39.** 4 **41.** 2 **43.** 3 **a.** $8x^3$ **b.** $45a^5$
c. $48m^3$ **d.** $16b^4$ **e.** $81p^7$ **f.** $81s^7$ **g.** $200y^5$ **h.** $98m^7$

9.4 Simplifying Radical Expressions

OBJECTIVE
To simplify expressions involving radicals.

In Section 9.3, we introduced the radical notation. For most applications, we will want to make sure that all radical expressions are in *simplest form*. The following three conditions must be satisfied.

> An expression involving square roots is in *simplest form* if
>
> **1.** There are no perfect-square factors in a radical.
> **2.** No fraction appears inside a radical.
> **3.** No radical appears in the denominator.

For instance, considering condition 1,

$\sqrt{17}$ is in simplest form since 17 has *no* perfect square factors.

while

$\sqrt{12}$ is *not* in simplest form

because

$\sqrt{12} = \sqrt{4 \cdot 3}$

A perfect square

To simplify radical expressions, we'll need to develop two important properties. First, look at the following expressions:

$$\sqrt{4 \cdot 9} = \sqrt{36} = 6$$
$$\sqrt{4} \cdot \sqrt{9} = 2 \cdot 3 = 6$$

Since this tells us that $\sqrt{4 \cdot 9} = \sqrt{4} \cdot \sqrt{9}$, the following general rule for radicals is suggested.

PROPERTY 1 FOR RADICALS

For any positive real numbers a and b,

$$\sqrt{ab} = \sqrt{a} \cdot \sqrt{b}$$

In words, the square root of a product is the product of the square roots.

Let's see how this property is applied in simplifying expressions when radicals are involved.

Example 1

(*a*) To simplify $\sqrt{12}$, look for perfect-square factors of 12.

Perfect-square factors are 1, 4, 9, 16, 25, 36, 49, 64, 81, 100, and so on.

$$\sqrt{12} = \sqrt{4 \cdot 3}$$

A perfect square (4)

Apply Property 1.

$$= \sqrt{4} \cdot \sqrt{3}$$
$$= 2\sqrt{3}$$

Note that we have removed the perfect-square factor from inside the radical, so the expression is in simplest form.

(*b*) To simplify $\sqrt{45}$, write

$$\sqrt{45} = \sqrt{9 \cdot 5}$$

A perfect square (9)

$$= \sqrt{9} \cdot \sqrt{5}$$
$$= 3\sqrt{5}$$

It would not have helped to write

$$\sqrt{45} = \sqrt{15 \cdot 3}$$

since neither factor is a perfect square.

(*c*) To simplify $\sqrt{72}$, write

We look for the *largest* perfect square factor, here 36.

$$\sqrt{72} = \sqrt{36 \cdot 2}$$

A perfect square (36)

Then apply Property 1.

$$= \sqrt{36} \cdot \sqrt{2}$$
$$= 6\sqrt{2}$$

(*d*) To simplify $5\sqrt{18}$, write

$$5\sqrt{18} = 5\sqrt{9 \cdot 2}$$

A perfect square (9)

$$= 5 \cdot \sqrt{9} \cdot \sqrt{2} = 5 \cdot 3\sqrt{2}$$
$$= 15\sqrt{2}$$

Be Careful! Even though

$\sqrt{a \cdot b} = \sqrt{a} \cdot \sqrt{b}$

$\sqrt{a + b}$ is *not the same* as $\sqrt{a} + \sqrt{b}$

Let $a = 4$ and $b = 9$, and substitute.

$\sqrt{a + b} = \sqrt{4 + 9} = \sqrt{13}$

$\sqrt{a} + \sqrt{b} = \sqrt{4} + \sqrt{9} = 2 + 3 = 5$

Since $\sqrt{13} \neq 5$, we see that the expressions $\sqrt{a + b}$ and $\sqrt{a} + \sqrt{b}$ are not in general the same.

CHECK YOURSELF 1

Simplify.

1. $\sqrt{20}$ **2.** $\sqrt{75}$ **3.** $\sqrt{98}$

In our remaining work with radicals we will assume that all variables represent positive real numbers.

The process is the same if variables are involved in a radical expression.

Example 2

(*a*) $\sqrt{x^3} = \sqrt{x^2 \cdot x}$

A perfect square (x^2)

$= \sqrt{x^2} \cdot \sqrt{x}$

$= x\sqrt{x}$

By our first rule for radicals.

Note: $\sqrt{x^2} = x$ (as long as x is positive).

(*b*) $\sqrt{4b^3} = \sqrt{4 \cdot b^2 \cdot b}$

Perfect squares (4, b^2)

$= \sqrt{4b^2} \cdot \sqrt{b}$

$= 2b\sqrt{b}$

(*c*) $\sqrt{18a^5} = \sqrt{9 \cdot a^4 \cdot 2a}$

Perfect squares (9, a^4)

$= \sqrt{9a^4} \cdot \sqrt{2a}$

$= 3a^2\sqrt{2a}$

Note that we want the perfect-square factor to have the largest possible even exponent, here 4. Keep in mind that

$a^2 \cdot a^2 = a^4$

CHECK YOURSELF 2

Simplify.

1. $\sqrt{9x^3}$ **2.** $\sqrt{27m^3}$ **3.** $\sqrt{50b^5}$

To develop a second property for radicals, look at the following expressions:

$$\sqrt{\frac{16}{4}} = \sqrt{4} = 2$$

$$\frac{\sqrt{16}}{\sqrt{4}} = \frac{4}{2} = 2$$

Since $\sqrt{\frac{16}{4}} = \frac{\sqrt{16}}{\sqrt{4}}$, a second general rule for radicals is suggested.

PROPERTY 2 FOR RADICALS

For any positive real numbers a and b,

$$\sqrt{\frac{a}{b}} = \frac{\sqrt{a}}{\sqrt{b}}$$

In words, the square root of a quotient is the quotient of the square roots.

This property is used in a fashion similar to the first in simplifying radical expressions. Remember that our second condition for a radical expression to be in simplest form states that no fraction should appear inside a radical. The next example illustrates how expressions that violate that condition are simplified.

Example 3

Write each expression in simplest form.

Apply Property 2 to write the numerator and denominator as separate radicals.

(a) $\sqrt{\frac{9}{4}} = \frac{\sqrt{9}}{\sqrt{4}}$ Remove any perfect squares from the radical.

$= \frac{3}{2}$

Apply Property 2.

(b) $\sqrt{\frac{2}{25}} = \frac{\sqrt{2}}{\sqrt{25}}$

$= \frac{\sqrt{2}}{5}$

Apply Property 2.

$$(c)\ \sqrt{\frac{8x^2}{9}} = \frac{\sqrt{8x^2}}{\sqrt{9}}$$

Factor $8x^2$ as $4x^2 \cdot 2x$.

$$= \frac{\sqrt{4x^2 \cdot 2}}{3}$$

Apply Property 1 in the numerator.

$$= \frac{\sqrt{4x^2} \cdot \sqrt{2}}{3}$$

$$= \frac{2x\sqrt{2}}{3}$$

CHECK YOURSELF 3

Simplify.

1. $\sqrt{\frac{25}{16}}$ 2. $\sqrt{\frac{7}{9}}$ 3. $\sqrt{\frac{12x^2}{49}}$

In our previous examples, the denominator of the fraction appearing in the radical was a perfect square, and we were able to write each expression in simplest radical form by removing that perfect square from the denominator.

If the denominator of the fraction in the radical is *not* a perfect square, we can still apply Property 2 for radicals. As we will see in our next example, the third condition for a radical to be in simplest form is then violated, and a new technique is necessary.

Example 4

Write each expression in simplest form.

We begin by applying Property 2.

$$(a)\ \sqrt{\frac{1}{3}} = \frac{\sqrt{1}}{\sqrt{3}} = \frac{1}{\sqrt{3}}$$

Do you see that $\frac{1}{\sqrt{3}}$ is still not in simplest form because of the radical in the denominator? To solve this problem, we multiply the numerator and denominator by $\sqrt{3}$. Note that the denominator will become

In general,

$\sqrt{a} \cdot \sqrt{a} = a$

where a is a positive real number.

$$\sqrt{3} \cdot \sqrt{3} = \sqrt{9} = 3$$

We then have

$$\frac{1}{\sqrt{3}} = \frac{1 \cdot \sqrt{3}}{\sqrt{3} \cdot \sqrt{3}} = \frac{\sqrt{3}}{3}$$

The expression $\frac{\sqrt{3}}{3}$ is now in simplest form since all three of our conditions are satisfied.

Note:

$\sqrt{2} \cdot \sqrt{5} = \sqrt{2 \cdot 5} = \sqrt{10}$

$\sqrt{5} \cdot \sqrt{5} = 5$

(*b*) $$\sqrt{\frac{2}{5}} = \frac{\sqrt{2}}{\sqrt{5}} = \frac{\sqrt{2} \cdot \sqrt{5}}{\sqrt{5} \cdot \sqrt{5}} = \frac{\sqrt{10}}{5}$$

and the expression is in simplest form since again our three conditions are satisfied.

We multiply numerator and denominator by $\sqrt{7}$ to "clear" the denominator of the radical.

(*c*) $$\sqrt{\frac{3x}{7}} = \frac{\sqrt{3x}}{\sqrt{7}} = \frac{\sqrt{3x} \cdot \sqrt{7}}{\sqrt{7} \cdot \sqrt{7}} = \frac{\sqrt{21x}}{7}$$

The expression is in simplest form.

CHECK YOURSELF 4

Simplify.

1. $\sqrt{\frac{1}{2}}$ **2.** $\sqrt{\frac{2}{3}}$ **3.** $\sqrt{\frac{2y}{5}}$

Both of the properties of radicals given in this section are true for cube roots, fourth roots, and so on. Here we have limited ourselves to simplifying expressions involving square roots.

CHECK YOURSELF ANSWERS

1. (1) $2\sqrt{5}$; (2) $5\sqrt{3}$; (3) $7\sqrt{2}$.
2. (1) $3x\sqrt{x}$; (2) $3m\sqrt{3m}$; (3) $5b^2\sqrt{2b}$.
3. (1) $\frac{5}{4}$; (2) $\frac{\sqrt{7}}{3}$; (3) $\frac{2x\sqrt{3}}{7}$.
4. (1) $\frac{\sqrt{2}}{2}$; (2) $\frac{\sqrt{6}}{3}$; (3) $\frac{\sqrt{10y}}{5}$.

9.4 Exercises

Use Property 1 to simplify each of the following radical expressions. Assume that all variables represent positive real numbers.

1. $\sqrt{18}$ $3\sqrt{2}$

2. $\sqrt{50}$ $5\sqrt{2}$

3. $\sqrt{28}$ $2\sqrt{7}$

4. $\sqrt{108}$ $6\sqrt{3}$

5. $\sqrt{45}$ $3\sqrt{5}$

6. $\sqrt{80}$ $4\sqrt{5}$

7. $\sqrt{48}$ $4\sqrt{3}$

8. $\sqrt{125}$ $5\sqrt{5}$

9. $\sqrt{200}$ $10\sqrt{2}$

10. $\sqrt{96}$ $4\sqrt{6}$

11. $\sqrt{147}$ $7\sqrt{3}$

12. $\sqrt{300}$ $10\sqrt{3}$

13. $3\sqrt{12}$ $6\sqrt{3}$

14. $5\sqrt{24}$ $10\sqrt{6}$

15. $\sqrt{5x^2}$ $x\sqrt{5}$

16. $\sqrt{7a^2}$ $a\sqrt{7}$

17. $\sqrt{3y^4}$ $y^2\sqrt{3}$

18. $\sqrt{10x^6}$ $x^3\sqrt{10}$

19. $\sqrt{2r^3}$ $r\sqrt{2r}$

20. $\sqrt{5a^5}$ $a^2\sqrt{5a}$

21. $\sqrt{27b^2}$ $3b\sqrt{3}$

22. $\sqrt{98m^4}$ $7m^2\sqrt{2}$

23. $\sqrt{24x^4}$ $2x^2\sqrt{6}$

24. $\sqrt{72x^3}$ $6x\sqrt{2x}$

25. $\sqrt{54a^5}$ $3a^2\sqrt{6a}$

26. $\sqrt{200y^6}$ $10y^3\sqrt{2}$

27. $\sqrt{x^3y^2}$ $xy\sqrt{x}$

28. $\sqrt{a^2b^5}$ $ab^2\sqrt{b}$

Use Property 2 to simplify each of the following radical expressions.

29. $\sqrt{\frac{4}{25}}$ $\frac{2}{5}$

30. $\sqrt{\frac{64}{9}}$ $\frac{8}{3}$

31. $\sqrt{\frac{9}{16}}$ $\frac{3}{4}$

32. $\sqrt{\frac{49}{25}}$ $\frac{7}{5}$

33. $\sqrt{\frac{3}{4}}$ $\frac{\sqrt{3}}{2}$

34. $\sqrt{\frac{5}{9}}$ $\frac{\sqrt{5}}{3}$

35. $\sqrt{\frac{5}{36}}$ $\frac{\sqrt{5}}{6}$

36. $\sqrt{\frac{10}{49}}$ $\frac{\sqrt{10}}{7}$

Use the properties for radicals to simplify each of the following expressions. Assume that all variables represent positive real numbers.

37. $\sqrt{\frac{8a^2}{25}}$ $\frac{2a\sqrt{2}}{5}$

38. $\sqrt{\frac{12y^2}{49}}$ $\frac{2y\sqrt{3}}{7}$

39. $\sqrt{\frac{1}{5}}$ $\frac{\sqrt{5}}{5}$

40. $\sqrt{\frac{1}{7}}$ $\frac{\sqrt{7}}{7}$

41. $\sqrt{\frac{3}{2}}$ $\frac{\sqrt{6}}{2}$

42. $\sqrt{\frac{5}{3}}$ $\frac{\sqrt{15}}{3}$

43. $\sqrt{\frac{3a}{5}}$ $\frac{\sqrt{15a}}{5}$

44. $\sqrt{\frac{2x}{7}}$ $\frac{\sqrt{14x}}{7}$

45. $\sqrt{\frac{2x^2}{3}}$ $\frac{x\sqrt{6}}{3}$ 46. $\sqrt{\frac{5m^2}{2}}$ $\frac{m\sqrt{10}}{2}$ 47. $\sqrt{\frac{8s^3}{7}}$ $\frac{2s\sqrt{14s}}{7}$ 48. $\sqrt{\frac{12x^3}{5}}$ $\frac{2x\sqrt{15x}}{5}$

Skillscan (Section 1.4)
Use the distributive property to combine the like terms in each of the following expressions.

a. $5x + 6x$ $11x$ **b.** $8a - 3a$ $5a$ **c.** $10y - 12y$ $-2y$ **d.** $7m + 10m$ $17m$

e. $9a + 7a - 12a$ $4a$ **f.** $5s - 8s + 4s$ s **g.** $12m + 3n - 6m$ $6m + 3n$ **h.** $8x + 5y - 4x$ $4x + 5y$

1. $3\sqrt{2}$ 3. $2\sqrt{7}$ 5. $3\sqrt{5}$ 7. $4\sqrt{3}$ 9. $10\sqrt{2}$ 11. $7\sqrt{3}$ 13. $6\sqrt{3}$ 15. $x\sqrt{5}$ 17. $y^2\sqrt{3}$ 19. $r\sqrt{2r}$ 21. $3b\sqrt{3}$ 23. $2x^2\sqrt{6}$ 25. $3a^2\sqrt{6a}$ 27. $xy\sqrt{x}$ 29. $\frac{2}{5}$ 31. $\frac{3}{4}$ 33. $\frac{\sqrt{3}}{2}$ 35. $\frac{\sqrt{5}}{6}$ 37. $\frac{2a\sqrt{2}}{5}$ 39. $\frac{\sqrt{5}}{5}$ 41. $\frac{\sqrt{6}}{2}$ 43. $\frac{\sqrt{15a}}{5}$ 45. $\frac{x\sqrt{6}}{3}$ 47. $\frac{2s\sqrt{14s}}{7}$ a. $11x$ b. $5a$ c. $-2y$ d. $17m$ e. $4a$ f. s g. $6m + 3n$ h. $4x + 5y$

9.5 Adding and Subtracting Radicals

OBJECTIVE
To add or subtract expressions involving radicals.

Like radicals have the same index and the same radicand (the expression inside the radical).

$2\sqrt{3}$ and $5\sqrt{3}$ are like radicals.

$\sqrt{2}$ and $\sqrt{5}$ are not like radicals—they have different radicands.

"Indices" is the plural of index.

$\sqrt{2}$ and $\sqrt[3]{2}$ are not like radicals—they have different indices (2 and 3, representing a square root and a cube root).

Like radicals can be added (or subtracted) in the same way as like terms. We apply the distributive property and then combine the coefficients:

$$2x^2 + 3x^2 = (2 + 3)x^2 = 5x^2$$

and

$$2\sqrt{5} + 3\sqrt{5} = (2 + 3)\sqrt{5} = 5\sqrt{5}$$

Example 1

Simplify each expression.

Apply the distributive property, then combine the coefficients.

(*a*) $5\sqrt{2} + 3\sqrt{2} = (5 + 3)\sqrt{2} = 8\sqrt{2}$

(*b*) $7\sqrt{5} - 2\sqrt{5} = (7 - 2)\sqrt{5} = 5\sqrt{5}$

(*c*) $8\sqrt{7} - \sqrt{7} + 2\sqrt{7} = (8 - 1 + 2)\sqrt{7}$

$= 9\sqrt{7}$

CHECK YOURSELF 1

Simplify.

1. $2\sqrt{5} + 7\sqrt{5}$ **2.** $9\sqrt{7} - \sqrt{7}$ **3.** $5\sqrt{3} - 2\sqrt{3} + \sqrt{3}$

If a sum or difference involves terms that are *not* like radicals, we may be able to combine terms after simplifying the radicals according to our earlier methods.

Example 2

Simplify each expression.

(*a*) $3\sqrt{2} + \sqrt{8}$

We do not have like radicals, but we can simplify $\sqrt{8}$.

Remember that

$$\sqrt{8} = \sqrt{4 \cdot 2} = 2\sqrt{2}$$

so

$$3\sqrt{2} + \sqrt{8} = 3\sqrt{2} + \overbrace{2\sqrt{2}}^{\sqrt{8}}$$
$$= (3 + 2)\sqrt{2}$$
$$= 5\sqrt{2}$$

Simplify $\sqrt{12}$.

The radicals can now be combined. Do you see why?

(*b*) $5\sqrt{3} + \sqrt{12} = 5\sqrt{3} + \sqrt{4 \cdot 3}$

$= 5\sqrt{3} + \sqrt{4}\,\sqrt{3}$

$= 5\sqrt{3} + 2\sqrt{3}$

$= (5 + 2)\sqrt{3}$

$= 7\sqrt{3}$

Simplify both terms.

(*c*) $\sqrt{50} - 2\sqrt{18} = \sqrt{25 \cdot 2} - 2\sqrt{9 \cdot 2}$

$= \sqrt{25}\,\sqrt{2} - 2\sqrt{9}\,\sqrt{2}$

$= 5\sqrt{2} - 2 \cdot 3\,\sqrt{2}$

$= 5\sqrt{2} - 6\sqrt{2}$

$= (5 - 6)\,\sqrt{2}$

$= -\sqrt{2}$

(*d*) $3\sqrt{20} - 2\sqrt{45} + 4\sqrt{5} = 3\sqrt{4 \cdot 5} - 2\sqrt{9 \cdot 5} + 4\sqrt{5}$
$= 3\sqrt{4}\,\sqrt{5} - 2\sqrt{9}\,\sqrt{5} + 4\sqrt{5}$
$= 3 \cdot 2\sqrt{5} - 2 \cdot 3\sqrt{5} + 4\sqrt{5}$
$= 6\sqrt{5} - 6\sqrt{5} + 4\sqrt{5}$
$= (6 - 6 + 4)\,\sqrt{5}$
$= 4\,\sqrt{5}$

CHECK YOURSELF 2

Simplify.

1. $\sqrt{2} + \sqrt{18}$ **2.** $5\sqrt{3} - \sqrt{27}$ **3.** $\sqrt{80} - 2\sqrt{5} + 4\sqrt{20}$

There is an important point to keep in mind when radical expressions are combined. Combining terms requires the use of the distributive property, and that property can be applied *only* with like radicals. The following example illustrates.

Example 3

We begin by simplifying *each* term as before.

Simplify $3\sqrt{50} - 2\sqrt{48} - \sqrt{72}$.

$3\sqrt{50} - 2\sqrt{48} - \sqrt{72}$
$= 3\sqrt{25 \cdot 2} - 2\sqrt{16 \cdot 3} - \sqrt{36 \cdot 2}$
$= 3\sqrt{25}\,\sqrt{2} - 2\sqrt{16}\,\sqrt{3} - \sqrt{36}\,\sqrt{2}$
$= 3 \cdot 5\,\sqrt{2} - 2 \cdot 4\,\sqrt{3} - 6\sqrt{2}$
$= 15\sqrt{2} - 8\sqrt{3} - 6\sqrt{2}$
$= 9\sqrt{2} - 8\sqrt{3}$

Now note that *only* the first and third terms are *like radicals* and can be combined.

CHECK YOURSELF 3

Simplify $2\sqrt{80} + \sqrt{54} - 3\sqrt{45}$.

If variables are involved in radical expressions, the process of combining terms proceeds in a fashion similar to that shown in previous examples. Consider our final example. We again assume that all variables represent positive real numbers.

Simplify each expression.

Since like radicals are involved, we apply the distributive property and combine terms as before.

(*a*) $5\sqrt{3x} - 2\sqrt{3x}$
$= (5 - 2)\,\sqrt{3x} = 3\sqrt{3x}$

Simplify the first term.

The radicals can now be combined.

(*b*) $2\sqrt{3a^3} + 5a\sqrt{3a}$
$= 2\sqrt{a^2 \cdot 3a} + 5a\sqrt{3a}$
$= 2\sqrt{a^2} \cdot \sqrt{3a} + 5a\sqrt{3a}$
$= 2a\sqrt{3a} + 5a\sqrt{3a}$
$= (2a + 5a)\sqrt{3a} = 7a\sqrt{3a}$

CHECK YOURSELF 4

Simplify each expression.

1. $2\sqrt{7y} + 3\sqrt{7y}$ **2.** $\sqrt{20a^2} - a\sqrt{45}$

CHECK YOURSELF ANSWERS

1. (1) $9\sqrt{5}$; (2) $8\sqrt{7}$; (3) $4\sqrt{3}$. **2.** (1) $4\sqrt{2}$; (2) $2\sqrt{3}$; (3) $10\sqrt{5}$.
3. $3\sqrt{6} - \sqrt{5}$. **4.** (1) $5\sqrt{7y}$; (2) $-a\sqrt{5}$.

9.5 Exercises

Simplify by combining like terms.

1. $\sqrt{2} + 5\sqrt{2}$
$6\sqrt{2}$

2. $4\sqrt{3} + 2\sqrt{3}$
$6\sqrt{3}$

3. $9\sqrt{7} - 2\sqrt{7}$
$7\sqrt{7}$

4. $7\sqrt{2} - 3\sqrt{3}$
Cannot be simplified

5. $5\sqrt{7} + 3\sqrt{6}$
Cannot be simplified

6. $3\sqrt{5} - 5\sqrt{5}$
$-2\sqrt{5}$

7. $2\sqrt{3} - 5\sqrt{3}$
$-3\sqrt{3}$

8. $2\sqrt{11} + 5\sqrt{11}$
$7\sqrt{11}$

9. $2\sqrt{3x} + 5\sqrt{3x}$
$7\sqrt{3x}$

10. $7\sqrt{2a} - 3\sqrt{2a}$
$4\sqrt{2a}$

11. $2\sqrt{3} + \sqrt{3} + 3\sqrt{3}$
$6\sqrt{3}$

12. $3\sqrt{5} + 2\sqrt{5} + \sqrt{5}$
$6\sqrt{5}$

13. $5\sqrt{7} - 2\sqrt{7} + \sqrt{7}$
$4\sqrt{7}$

14. $3\sqrt{10} - 2\sqrt{10} + \sqrt{10}$
$2\sqrt{10}$

15. $2\sqrt{5x} + 5\sqrt{5x} - 2\sqrt{5x}$
$5\sqrt{5x}$

16. $5\sqrt{3b} - 2\sqrt{3b} + 4\sqrt{3b}$
$7\sqrt{3b}$

17. $2\sqrt{3} + \sqrt{12}$
$4\sqrt{3}$

18. $5\sqrt{2} + \sqrt{18}$
$8\sqrt{2}$

19. $\sqrt{20} - \sqrt{5}$
$\sqrt{5}$

20. $\sqrt{98} - 3\sqrt{2}$
$4\sqrt{2}$

21. $2\sqrt{6} - \sqrt{54}$
$-\sqrt{6}$

22. $2\sqrt{3} - \sqrt{27}$
$-\sqrt{3}$

23. $\sqrt{72} + \sqrt{50}$
$11\sqrt{2}$

24. $\sqrt{27} - \sqrt{12}$
$\sqrt{3}$

25. $3\sqrt{12} - \sqrt{48}$
$2\sqrt{3}$

26. $5\sqrt{8} + 2\sqrt{18}$
$16\sqrt{2}$

27. $4\sqrt{20} - 2\sqrt{45}$
$2\sqrt{5}$

28. $2\sqrt{72} - \sqrt{200}$
$2\sqrt{2}$

29. $\sqrt{12} + \sqrt{27} - \sqrt{3}$
$4\sqrt{3}$

30. $\sqrt{50} + \sqrt{32} - \sqrt{8}$
$7\sqrt{2}$

31. $3\sqrt{24} - \sqrt{54} + \sqrt{6}$
$4\sqrt{6}$

32. $\sqrt{63} - 2\sqrt{28} + 5\sqrt{7}$
$4\sqrt{7}$

33. $2\sqrt{50} + 3\sqrt{18} - \sqrt{32}$
$15\sqrt{2}$

34. $3\sqrt{27} + 4\sqrt{12} - \sqrt{300}$
$7\sqrt{3}$

35. $a\sqrt{27} - 2\sqrt{3a^2}$
$a\sqrt{3}$

36. $5\sqrt{2y^2} - 3y\sqrt{8}$
$-y\sqrt{2}$

37. $5\sqrt{3x^3} + 2\sqrt{27x}$
$(5x + 6)\sqrt{3x}$

38. $7\sqrt{2a^3} - \sqrt{8a}$
$(7a - 2)\sqrt{2a}$

39. $\sqrt{6} - \sqrt{\frac{2}{3}}$
$\frac{2\sqrt{6}}{3}$

40. $\sqrt{15} + \sqrt{\frac{3}{5}}$
$\frac{6\sqrt{15}}{5}$

Skillscan (Section 4.3)

Perform the indicated multiplication.

a. $2(x + 5)$
$2x + 10$

b. $3(a - 3)$
$3a - 9$

c. $m(m - 8)$
$m^2 - 8m$

d. $y(y + 7)$
$y^2 + 7y$

e. $(w + 2)(w - 2)$
$w^2 - 4$

f. $(x - 3)(x + 3)$
$x^2 - 9$

g. $(x + y)(x + y)$
$x^2 + 2xy + y^2$

h. $(b - 7)(b - 7)$
$b^2 - 14b + 49$

Answers

1. $6\sqrt{2}$ **3.** $7\sqrt{7}$ **5.** Cannot be simplified **7.** $-3\sqrt{3}$ **9.** $7\sqrt{3x}$ **11.** $6\sqrt{3}$ **13.** $4\sqrt{7}$ **15.** $5\sqrt{5x}$ **17.** $4\sqrt{3}$ **19.** $\sqrt{5}$ **21.** $-\sqrt{6}$ **23.** $11\sqrt{2}$ **25.** $2\sqrt{3}$ **27.** $2\sqrt{5}$ **29.** $4\sqrt{3}$ **31.** $4\sqrt{6}$ **33.** $15\sqrt{2}$ **35.** $a\sqrt{3}$ **37.** $(5x + 6)\sqrt{3x}$ **39.** $\frac{2\sqrt{6}}{3}$ **a.** $2x + 10$ **b.** $3a - 9$ **c.** $m^2 - 8m$ **d.** $y^2 + 7y$ **e.** $w^2 - 4$ **f.** $x^2 - 9$ **g.** $x^2 + 2xy + y^2$ **h.** $b^2 - 14b + 49$

9.6 Multiplying and Dividing Radicals

OBJECTIVE

To multiply or divide expressions involving radicals.

In Section 9.4 we stated the first property for radicals:

$\sqrt{ab} = \sqrt{a} \cdot \sqrt{b}$ where a and b are any positive real numbers

That property has been used to simplify radical expressions up to this point. Suppose now that we want to find a product, such as $\sqrt{3} \cdot \sqrt{5}$.

We can use our first radical rule in the opposite manner.

The product of square roots is equal to the square root of the product of the radicands.

$\sqrt{a} \cdot \sqrt{b} = \sqrt{ab}$

so

$\sqrt{3} \cdot \sqrt{5} = \sqrt{3 \cdot 5} = \sqrt{15}$

Example 1

(*a*) $\sqrt{2}\,\sqrt{7} = \sqrt{2 \cdot 7} = \sqrt{14}$

(*b*) $\sqrt{5}\,\sqrt{3x} = \sqrt{5 \cdot 3x} = \sqrt{15x}$

(*c*) $\sqrt{5}\,\sqrt{3}\,\sqrt{2} = \sqrt{5 \cdot 3 \cdot 2} = \sqrt{30}$

CHECK YOURSELF 1

Multiply.

1. $\sqrt{3}\,\sqrt{10}$ **2.** $\sqrt{5}\,\sqrt{7a}$ **3.** $\sqrt{2}\,\sqrt{7}\,\sqrt{5}$

We may have to simplify after multiplying, as the following examples illustrate.

Example 2

(*a*) $\sqrt{5}\sqrt{10} = \sqrt{5 \cdot 10} = \sqrt{50}$
$= \sqrt{25 \cdot 2} = 5\sqrt{2}$

(*b*) $\sqrt{12}\,\sqrt{6} = \sqrt{12 \cdot 6} = \sqrt{72}$
$= \sqrt{36 \cdot 2} = \sqrt{36}\,\sqrt{2} = 6\sqrt{2}$

Note: An alternative approach would be to simplify $\sqrt{12}$ first.

$$\begin{aligned}\sqrt{12}\,\sqrt{6} = 2\sqrt{3}\,\sqrt{6} &= 2\sqrt{18}\\ &= 2\sqrt{9 \cdot 2} = 2\sqrt{9}\,\sqrt{2}\\ &= 2 \cdot 3\sqrt{2} = 6\sqrt{2}\end{aligned}$$

(*c*) $\sqrt{10x}\,\sqrt{2x} = \sqrt{20x^2} = \sqrt{4x^2 \cdot 5}$
$= \sqrt{4x^2}\,\sqrt{5} = 2x\sqrt{5}$

CHECK YOURSELF 2

Simplify.

1. $\sqrt{3}\,\sqrt{6}$ **2.** $\sqrt{3} \cdot \sqrt{18}$ **3.** $\sqrt{8a}\,\sqrt{3a}$

If coefficients are involved in a product, we can use the commutative and associative properties to change the order and grouping of the factors. This is illustrated in our next example.

Example 3

Multiply.

Note: In practice, it is not necessary to show the intermediate steps.

$$(2\sqrt{5})(3\sqrt{6}) = (2 \cdot 3)(\sqrt{5}\,\sqrt{6})$$
$$= 6\sqrt{5 \cdot 6}$$
$$= 6\sqrt{30}$$

CHECK YOURSELF 3

Multiply $(3\sqrt{7})(5\sqrt{3})$.

The distributive property can also be applied in multiplying radical expressions. Consider the following.

Example 4

Multiply.

(*a*) $\sqrt{3}\,(\sqrt{2} + \sqrt{3})$

$= \sqrt{3} \cdot \sqrt{2} + \sqrt{3} \cdot \sqrt{3}$ The distributive property

$= \sqrt{6} + 3$ Multiply the radicals.

(*b*) $\sqrt{5}\,(2\sqrt{6} + 3\sqrt{3})$

$= \sqrt{5} \cdot 2\sqrt{6} + \sqrt{5} \cdot 3\sqrt{3}$ The distributive property

$= 2 \cdot \sqrt{5} \cdot \sqrt{6} + 3 \cdot \sqrt{5} \cdot \sqrt{3}$ The commutative property

$= 2\sqrt{30} + 3\sqrt{15}$

CHECK YOURSELF 4

Multiply.

1. $\sqrt{5}\,(\sqrt{6} + \sqrt{5})$ **2.** $\sqrt{3}\,(2\sqrt{5} + 3\sqrt{2})$

Our earlier FOIL pattern for multiplying binomials can also be applied in multiplying radical expressions. This is shown in our next example.

Example 5

Multiply.

(*a*) $(\sqrt{3} + 2)\,(\sqrt{3} + 5)$

$= \sqrt{3} \cdot \sqrt{3} + 5\sqrt{3} + 2\sqrt{3} + 2 \cdot 5$

$= 3 + 5\sqrt{3} + 2\sqrt{3} + 10$ Combine like terms.

$= 13 + 7\sqrt{3}$

Be Careful! This result *cannot* be further simplified: 13 and $7\sqrt{3}$ are *not* like terms.

Note: You can use the pattern $(a + b)(a - b) = a^2 - b^2$ where $a = \sqrt{7}$ and $b = 2$, for the same result. $\sqrt{7} + 2$ and $\sqrt{7} - 2$ are called *conjugates* of each other. Note that their product is the rational number 3. The product of conjugates will *always be rational*.

(*b*) $(\sqrt{7} + 2)(\sqrt{7} - 2)$
$= \sqrt{7} \cdot \sqrt{7} - 2\sqrt{7} + 2\sqrt{7} - 4$
$= 7 - 4 = 3$

(*c*) $(\sqrt{3} + 5)^2$
$= (\sqrt{3} + 5)(\sqrt{3} + 5)$
$= \sqrt{3}\,\sqrt{3} + 5\sqrt{3} + 5\sqrt{3} + 5 \cdot 5$
$= 3 + 5\sqrt{3} + 5\sqrt{3} + 25$
$= 28 + 10\sqrt{3}$

CHECK YOURSELF 5

Multiply.

1. $(\sqrt{5} + 3)(\sqrt{5} - 2)$
2. $(\sqrt{3} + 4)(\sqrt{3} - 4)$
3. $(\sqrt{2} - 3)^2$

We can also use our second property for radicals in the opposite manner.

The quotient of square roots is equal to the square root of the quotient of the radicands.

$$\frac{\sqrt{a}}{\sqrt{b}} = \sqrt{\frac{a}{b}}$$

One use of this property to divide radical expressions is illustrated in our next example.

Example 6

The clue to recognizing when to use this approach is in noting that 48 is divisible by 3.

(*a*) $\dfrac{\sqrt{48}}{\sqrt{3}} = \sqrt{\dfrac{48}{3}} = \sqrt{16} = 4$

(*b*) $\dfrac{\sqrt{200}}{\sqrt{2}} = \sqrt{\dfrac{200}{2}} = \sqrt{100} = 10$

(*c*) $\dfrac{\sqrt{125x^2}}{\sqrt{5}} = \sqrt{\dfrac{125x^2}{5}} = \sqrt{25x^2} = 5x$

CHECK YOURSELF 6

Simplify.

1. $\dfrac{\sqrt{75}}{\sqrt{3}}$

2. $\dfrac{\sqrt{81s^2}}{\sqrt{9}}$

Another approach is called for if the quotient cannot be simplified as in the previous examples. Consider the following.

Example 7

(*a*) Simplify $\frac{\sqrt{3}}{\sqrt{5}}$.

Note that this approach to simplifying a quotient is identical to our earlier work in Section 9.4.

Multiply the numerator and denominator by $\sqrt{5}$. The value of the fraction is unchanged, and we will have a rational number in the denominator.

$$\frac{\sqrt{3}}{\sqrt{5}} = \frac{\sqrt{3} \cdot \sqrt{5}}{\sqrt{5} \cdot \sqrt{5}} = \frac{\sqrt{15}}{\sqrt{25}} = \frac{\sqrt{15}}{5}$$

The expression is now in simplest radical form (there is no radical in the denominator). This process is called *rationalizing the denominator*.

(*b*) Simplify $\frac{2}{\sqrt{7}}$.

Multiply numerator and denominator by $\sqrt{7}$.

$$\frac{2}{\sqrt{7}} = \frac{2\sqrt{7}}{\sqrt{7}\sqrt{7}} = \frac{2\sqrt{7}}{\sqrt{49}} = \frac{2\sqrt{7}}{7}$$

(*c*) Simplify $\frac{5}{\sqrt{10}}$.

$$\frac{5}{\sqrt{10}} = \frac{5\sqrt{10}}{\sqrt{10}\sqrt{10}}$$

Divide numerator and denominator by 5 to simplify.

$$= \frac{5\sqrt{10}}{10} = \frac{\sqrt{10}}{2}$$

(*d*) $$\frac{\sqrt{3x^3y^2}}{\sqrt{5}} = \frac{\sqrt{3x^3y^2} \cdot \sqrt{5}}{\sqrt{5} \cdot \sqrt{5}}$$

$$= \frac{\sqrt{15x^3y^2}}{5}$$

We must now simplify the radical in the numerator.

$$= \frac{\sqrt{x^2y^2 \cdot 15x}}{5}$$

Apply Property 1.

$$= \frac{\sqrt{x^2y^2} \cdot \sqrt{15x}}{5}$$

$$= \frac{xy\sqrt{15x}}{5}$$

CHECK YOURSELF 7

Simplify by rationalizing the denominators.

1. $\dfrac{3}{\sqrt{6}}$ **2.** $\dfrac{\sqrt{10}}{\sqrt{3}}$ **3.** $\dfrac{\sqrt{2a^2b}}{\sqrt{5}}$

We will now look at another type of division problem involving radical expressions. Here the divisor (the denominator) will be a binomial. Simplifying such an expression will use the idea of conjugates introduced in Example 5 earlier in this section. The following example illustrates.

Example 8

Rationalize the denominator in the expression

$$\frac{5}{\sqrt{7}+2}$$

Recall that $\sqrt{7}+2$ and $\sqrt{7}-2$ are conjugates of each other and that the product of conjugates will *always be a rational number.* Therefore, to rationalize the denominator in this expression, we can multiply numerator and denominator by $\sqrt{7}-2$.

If an expression involves a binomial with a radical in the denominator, multiply the numerator and denominator by the *conjugate* of the denominator to rationalize.

See Example 5(*b*) for the details of the multiplication in the denominator.

$$\frac{5}{\sqrt{7}+2} = \frac{5(\sqrt{7}-2)}{(\sqrt{7}+2)(\sqrt{7}-2)}$$

$$= \frac{5\sqrt{7}-5\cdot 2}{3}$$

$$= \frac{5\sqrt{7}-10}{3}$$

The quotient is now in simplest form.

CHECK YOURSELF 8

Rationalize the denominator of the expression $\dfrac{7}{\sqrt{10}-2}$.

There is one final quotient form which you may encounter in simplifying expressions such as that of the last example, and it will be extremely important in our work with quadratic equations in the next chapter. This form is shown in our final example.

Example 9

Simplify the expression

$$\frac{3 + \sqrt{72}}{3}$$

First, we must simplify the radical in the numerator.

$$\frac{3 + \sqrt{72}}{3} = \frac{3 + \sqrt{36 \cdot 2}}{3}$$

Use Property 1 to simplify $\sqrt{72}$.

$$= \frac{3 + \sqrt{36} \cdot \sqrt{2}}{3} = \frac{3 + 6\sqrt{2}}{3}$$

$$= \frac{3(1 + 2\sqrt{2})}{3} = 1 + 2\sqrt{2}$$

Factor the numerator—then divide by the *common* factor of 3.

Be Careful! Students are sometimes tempted to write

$$\frac{\cancel{3} + 6\sqrt{2}}{\cancel{3}} = 1 + 6\sqrt{2}$$

This is *not* correct. We must divide *both terms* of the numerator by the common factor.

CHECK YOURSELF 9

Simplify $\frac{15 + \sqrt{75}}{5}$.

CHECK YOURSELF ANSWERS

1. (1) $\sqrt{30}$; (2) $\sqrt{35a}$; (3) $\sqrt{70}$. **2.** (1) $3\sqrt{2}$; (2) $3\sqrt{6}$; (3) $2a\sqrt{6}$. **3.** $15\sqrt{21}$. **4.** (1) $\sqrt{30} + 5$; (2) $2\sqrt{15} + 3\sqrt{6}$. **5.** (1) $-1 + \sqrt{5}$; (2) -13; (3) $11 - 6\sqrt{2}$. **6.** (1) 5; (2) $3s$. **7.** (1) $\frac{\sqrt{6}}{2}$; (2) $\frac{\sqrt{30}}{3}$; (3) $\frac{a\sqrt{10b}}{5}$. **8.** $\frac{7\sqrt{10} + 14}{6}$. **9.** $3 + \sqrt{3}$.

9.6 Exercises

Perform the indicated multiplication. Then simplify each radical expression.

1. $\sqrt{7}\sqrt{5}$
$\sqrt{35}$

2. $\sqrt{3}\sqrt{7}$
$\sqrt{21}$

3. $\sqrt{5}\sqrt{11}$
$\sqrt{55}$

4. $\sqrt{13}\sqrt{5}$
$\sqrt{65}$

5. $\sqrt{3}\sqrt{10m}$
$\sqrt{30m}$

6. $\sqrt{7a}\sqrt{13}$
$\sqrt{91a}$

7. $\sqrt{2x}\sqrt{15}$
$\sqrt{30x}$

8. $\sqrt{17}\sqrt{2b}$
$\sqrt{34b}$

9. $\sqrt{3}\sqrt{7}\sqrt{2}$
$\sqrt{42}$

10. $\sqrt{5}\sqrt{7}\sqrt{3}$
$\sqrt{105}$

11. $\sqrt{3}\sqrt{12}$
6

12. $\sqrt{7}\sqrt{7}$
7

13. $\sqrt{10}\sqrt{10}$
10

14. $\sqrt{5}\sqrt{15}$
$5\sqrt{3}$

15. $\sqrt{18}\sqrt{6}$
$6\sqrt{3}$

16. $\sqrt{8}\sqrt{10}$
$4\sqrt{5}$

17. $\sqrt{2x}\sqrt{6x}$
$2x\sqrt{3}$

18. $\sqrt{3a}\sqrt{15a}$
$3a\sqrt{5}$

19. $2\sqrt{3}\sqrt{7}$
$2\sqrt{21}$

20. $3\sqrt{2}\sqrt{5}$
$3\sqrt{10}$

21. $(3\sqrt{3})(5\sqrt{7})$
$15\sqrt{21}$

22. $(2\sqrt{5})(3\sqrt{11})$
$6\sqrt{55}$

23. $(3\sqrt{5})(2\sqrt{10})$
$30\sqrt{2}$

24. $(4\sqrt{3})(3\sqrt{6})$
$36\sqrt{2}$

25. $\sqrt{5}(\sqrt{2}+\sqrt{5})$
$\sqrt{10}+5$

26. $\sqrt{3}(\sqrt{5}-\sqrt{3})$
$\sqrt{15}-3$

27. $\sqrt{3}(2\sqrt{5}-3\sqrt{3})$
$2\sqrt{15}-9$

28. $\sqrt{7}(2\sqrt{3}+3\sqrt{7})$
$2\sqrt{21}+21$

29. $(\sqrt{3}+5)(\sqrt{3}+3)$
$18+8\sqrt{3}$

30. $(\sqrt{5}-2)(\sqrt{5}-1)$
$7-3\sqrt{5}$

31. $(\sqrt{5}-1)(\sqrt{5}+3)$
$2+2\sqrt{5}$

32. $(\sqrt{2}+3)(\sqrt{2}-7)$
$-19-4\sqrt{2}$

33. $(\sqrt{5}-2)(\sqrt{5}+2)$
1

34. $(\sqrt{7}+5)(\sqrt{7}-5)$
-18

35. $(\sqrt{10}+5)(\sqrt{10}-5)$
-15

36. $(\sqrt{11}-3)(\sqrt{11}+3)$
2

37. $(\sqrt{x}+3)(\sqrt{x}-3)$
$x-9$

38. $(\sqrt{a}-4)(\sqrt{a}+4)$
$a-16$

39. $(\sqrt{3}+2)^2$
$7+4\sqrt{3}$

40. $(\sqrt{5}-3)^2$
$14-6\sqrt{5}$

41. $(\sqrt{y}-5)^2$
$y-10\sqrt{y}+25$

42. $(\sqrt{x}+4)^2$
$x+8\sqrt{x}+16$

Perform the indicated division. Rationalize the denominator if necessary. Then simplify each radical expression.

43. $\dfrac{\sqrt{98}}{\sqrt{2}}$
7

44. $\dfrac{\sqrt{108}}{\sqrt{3}}$
6

45. $\dfrac{\sqrt{72a^2}}{\sqrt{2}}$
$6a$

46. $\dfrac{\sqrt{48m^2}}{\sqrt{3}}$
$4m$

47. $\dfrac{\sqrt{5}}{\sqrt{2}}$
$\dfrac{\sqrt{10}}{2}$

48. $\dfrac{\sqrt{3}}{\sqrt{5}}$
$\dfrac{\sqrt{15}}{5}$

49. $\dfrac{3}{\sqrt{7}}$
$\dfrac{3\sqrt{7}}{7}$

50. $\dfrac{2}{\sqrt{5}}$
$\dfrac{2\sqrt{5}}{5}$

51. $\dfrac{4}{\sqrt{6}}$
$\dfrac{2\sqrt{6}}{3}$

52. $\dfrac{2}{\sqrt{10}}$
$\dfrac{\sqrt{10}}{5}$

53. $\dfrac{2}{\sqrt{8}}$
$\dfrac{\sqrt{2}}{2}$

54. $\dfrac{3}{\sqrt{27}}$
$\dfrac{\sqrt{3}}{3}$

55. $\dfrac{\sqrt{2a^2b}}{\sqrt{3}}$ $\dfrac{a\sqrt{6b}}{3}$

56. $\dfrac{\sqrt{5x^3y^2}}{\sqrt{2}}$ $\dfrac{xy\sqrt{10x}}{2}$

57. $\dfrac{\sqrt{5x^4y}}{\sqrt{7}}$ $\dfrac{x^2\sqrt{35y}}{7}$

58. $\dfrac{\sqrt{7m^2n^3}}{\sqrt{3}}$ $\dfrac{mn\sqrt{21n}}{3}$

59. $\dfrac{3}{\sqrt{6}+2}$ $\dfrac{3\sqrt{6}-6}{2}$

60. $\dfrac{4}{\sqrt{7}-2}$ $\dfrac{4\sqrt{7}+8}{3}$

61. $\dfrac{12}{\sqrt{10}-2}$ $2\sqrt{10}+4$

62. $\dfrac{6}{\sqrt{11}-3}$ $3\sqrt{11}+9$

63. $\dfrac{4}{\sqrt{5}+\sqrt{3}}$ $2\sqrt{5}-2\sqrt{3}$

64. $\dfrac{10}{\sqrt{7}-\sqrt{2}}$ $2\sqrt{7}+2\sqrt{2}$

65. $\dfrac{6+\sqrt{18}}{3}$ $2+\sqrt{2}$

66. $\dfrac{6-\sqrt{20}}{2}$ $3-\sqrt{5}$

67. $\dfrac{15-\sqrt{75}}{5}$ $3-\sqrt{3}$

68. $\dfrac{8+\sqrt{48}}{4}$ $2+\sqrt{3}$

Answers

1. $\sqrt{35}$ **3.** $\sqrt{55}$ **5.** $\sqrt{30m}$ **7.** $\sqrt{30x}$ **9.** $\sqrt{42}$ **11.** 6 **13.** 10 **15.** $6\sqrt{3}$ **17.** $2x\sqrt{3}$ **19.** $2\sqrt{21}$ **21.** $15\sqrt{21}$ **23.** $30\sqrt{2}$ **25.** $\sqrt{10}+5$ **27.** $2\sqrt{15}-9$ **29.** $18+8\sqrt{3}$ **31.** $2+2\sqrt{5}$ **33.** 1 **35.** -15 **37.** $x-9$ **39.** $7+4\sqrt{3}$ **41.** $y-10\sqrt{y}+25$ **43.** 7 **45.** $6a$ **47.** $\dfrac{\sqrt{10}}{2}$ **49.** $\dfrac{3\sqrt{7}}{7}$ **51.** $\dfrac{2\sqrt{6}}{3}$ **53.** $\dfrac{\sqrt{2}}{2}$ **55.** $\dfrac{a\sqrt{6b}}{3}$ **57.** $\dfrac{x^2\sqrt{35y}}{7}$ **59.** $\dfrac{3\sqrt{6}-6}{2}$ **61.** $2\sqrt{10}+4$ **63.** $2\sqrt{5}-2\sqrt{3}$ **65.** $2+\sqrt{2}$ **67.** $3-\sqrt{3}$

Summary

Properties of Exponents [9.1]

For any real numbers a and b and integers m and n:

Property 1 $a^m \cdot a^n = a^{m+n}$ $x^5 \cdot x^7 = x^{5+7} = x^{12}$

Property 2 $\dfrac{a^m}{a^n} = a^{m-n}$ $(a \neq 0)$ $\dfrac{x^7}{x^5} = x^{7-5} = x^2$

Property 3 $(a^m)^n = a^{m \cdot n}$ $(x^5)^3 = x^{5\cdot 3} = x^{15}$

Property 4 $(ab)^m = a^m b^m$ $(2xy)^3 = 2^3x^3y^3 = 8x^3y^3$

Property 5 $\left(\dfrac{a}{b}\right)^m = \dfrac{a^m}{b^m}$ $(b \neq 0)$ $\left(\dfrac{x^2}{3}\right)^2 = \dfrac{(x^2)^2}{3^2} = \dfrac{x^4}{9}$

Zero and Negative Exponents [9.2]

Zero as an Exponent $a^0 = 1 \quad (a \neq 0)$

$5^0 = 1$
$(-3)^0 = 1$
$2x^0 = 2 \cdot 1 = 2$

Negative Integers as Exponents $a^{-n} = \frac{1}{a^n} \quad (a \neq 0)$

$x^{-3} = \frac{1}{x^3}$

$3x^{-5} = \frac{3}{x^5}$

Radicals [9.3]

Square Roots $\sqrt{x}$ is the principal (or positive) square root of x. It is the positive number we must square to get x.

$\sqrt{49} = 7$

$-\sqrt{x}$ is the negative square root of x.

$-\sqrt{49} = -7$

The square root of a negative number is not a real number.

$\sqrt{-49}$ is not a real number.

Other Roots $\sqrt[3]{x}$ is the cube root of x.

$\sqrt[3]{64} = 4$ because $4^3 = 64$.

$\sqrt[4]{x}$ is the fourth root of x.

$\sqrt[4]{81} = 3$ because $3^4 = 81$.

Rational and Irrational Numbers Rational numbers can be expressed as the quotient of two integers with a nonzero denominator.

$\frac{2}{3}, \frac{-7}{12}, 5, \sqrt{36}$, and $\sqrt[3]{64}$ are rational numbers.

Irrational numbers cannot be expressed as the quotient of two integers.

$\sqrt{5}, \sqrt{37}$, and $\sqrt[3]{65}$ are irrational numbers.

Real Numbers The real numbers are the set of rational numbers and the set of irrational numbers considered together.

Definitions $\sqrt{x^2} = |x|$ for any real number x

$\sqrt[3]{x^3} = x$ for any real number x

$\sqrt{5^2} = 5 \qquad \sqrt{(-3)^2} = 3$

$\sqrt[3]{2^3} = 2 \qquad \sqrt[3]{(-3)^3} = -3$

Simplifying Radical Expressions [9.4]

> An expression involving square roots is in *simplest form* if
>
> **1.** There are no perfect-square factors in a radical.
> **2.** No fraction appears inside a radical.
> **3.** No radical appears in the denominator.

To simplify a radical expression, use one of the following properties. The square root of a product is the product of the square roots.

$$\sqrt{ab} = \sqrt{a} \cdot \sqrt{b}$$

$$\begin{aligned}\sqrt{40} &= \sqrt{4 \cdot 10} \\ &= \sqrt{4} \cdot \sqrt{10} \\ &= 2\sqrt{10}\end{aligned}$$

$$\begin{aligned}\sqrt{12x^3} &= \sqrt{4x^2 \cdot 3x} \\ &= \sqrt{4x^2}\,\sqrt{3x} \\ &= 2x\sqrt{3x}\end{aligned}$$

The square root of a quotient is the quotient of the square roots.

$$\sqrt{\frac{a}{b}} = \frac{\sqrt{a}}{\sqrt{b}}$$

$$\sqrt{\frac{5}{16}} = \frac{\sqrt{5}}{\sqrt{16}} = \frac{\sqrt{5}}{4}$$

$$\sqrt{\frac{2y}{3}} = \frac{\sqrt{2y}}{\sqrt{3}} = \frac{\sqrt{2y} \cdot \sqrt{3}}{\sqrt{3} \cdot \sqrt{3}} = \frac{\sqrt{6y}}{\sqrt{9}} = \frac{\sqrt{6y}}{3}$$

Adding and Subtracting Radicals [9.5]

Like radicals have the same index and the same radicand (the expression inside the radical).

$3\sqrt{5}$ and $2\sqrt{5}$ are like radicals.

Like radicals can be added (or subtracted) in the same way as like terms. Apply the distributive law and combine the coefficients.

$$2\sqrt{3} + 3\sqrt{3} = (2 + 3)\sqrt{3} = 5\sqrt{3}$$

$$5\sqrt{7} - 2\sqrt{7} = (5 - 2)\sqrt{7} = 3\sqrt{7}$$

Certain expressions can be combined after one or more of the terms involving radicals are simplified.

$$\sqrt{12} + \sqrt{3} = 2\sqrt{3} + \sqrt{3} = (2 + 1)\sqrt{3} = 3\sqrt{3}$$

Multiplying and Dividing Radicals [9.6]

Multiplying To multiply radical expressions, use the first radical law in the following way:

$$\sqrt{a}\sqrt{b} = \sqrt{ab}$$

$$\sqrt{7}\sqrt{5} = \sqrt{7 \cdot 5} = \sqrt{35}$$

$$\sqrt{6}\sqrt{15} = \sqrt{6 \cdot 15} = \sqrt{90} = \sqrt{9 \cdot 10} = 3\sqrt{10}$$

The distributive property can also be applied in multiplying radical expressions.

$$\sqrt{5}(\sqrt{3} + 2\sqrt{5}) = \sqrt{5} \cdot \sqrt{3} + \sqrt{5} \cdot 2\sqrt{5} = \sqrt{15} + 10$$

The FOIL pattern allows us to find the product of binomial radical expressions.

$$(\sqrt{5} + 2)(\sqrt{5} - 1) = \sqrt{5} \cdot \sqrt{5} - \sqrt{5} + 2\sqrt{5} - 2 = 3 + \sqrt{5}$$

$$(\sqrt{10} + 3)(\sqrt{10} - 3) = 10 - 9 = 1$$

Dividing To divide radical expressions, use the second radical law in the following way:

$$\frac{\sqrt{a}}{\sqrt{b}} = \sqrt{\frac{a}{b}}$$

$$\frac{\sqrt{50}}{\sqrt{2}} = \sqrt{\frac{50}{2}} = \sqrt{25} = 5$$

When necessary, multiply the numerator and denominator of the expression by the same root to rationalize the denominator.

$$\frac{\sqrt{3}}{\sqrt{5}} = \frac{\sqrt{3} \cdot \sqrt{5}}{\sqrt{5} \cdot \sqrt{5}} = \frac{\sqrt{15}}{\sqrt{25}} = \frac{\sqrt{15}}{5}$$

If the divisor (the denominator) is a binomial, multiply the numerator and denominator by the *conjugate* of the denominator.

$\sqrt{5} - 2$ is the conjugate of $\sqrt{5} + 2$.

To simplify,

$$\frac{3}{\sqrt{5} - 2} = \frac{3(\sqrt{5} + 2)}{(\sqrt{5} - 2)(\sqrt{5} + 2)} = \frac{3(\sqrt{5} + 2)}{5 - 4} = 3\sqrt{5} + 6$$

Summary Exercises Chapter 9

This summary exercise set is provided to give you practice with each of the objectives of the chapter. Each exercise is keyed to the appropriate chapter section. The answers are provided in the instructor's manual. Your instructor will give you guidelines on how to best use these exercises in your instructional setting.

[9.1] Use the properties of exponents to simplify each of the following expressions. Assume that all variables represent nonzero real numbers.

1. $2^3 \cdot 2^4$ 2^7

2. $\frac{x^{12}}{x^8}$ x^4

3. $(a^4)^3$ a^{12}

4. $(3xy)^4$ $81x^4y^4$

5. $(r^4s^5)^3$ $r^{12}s^{15}$

6. $(3x^2)^2(x^3)^3$ $9x^{13}$

7. $\frac{(b^4)^4}{b^{10}}$ b^6

8. $\left(\frac{x^5}{y^3}\right)^2$ $\frac{x^{10}}{y^6}$

[9.2] Evaluate each of the following expressions.

9. 5^0 1

10. $(5m)^0$ 1

11. $5m^0$ 5

12. $(4x^3y)^0$ 1

[9.2] Write, using positive exponents.

13. a^{-7} $\frac{1}{a^7}$

14. 2^{-4} $\frac{1}{16}$

15. 10^{-3} $\frac{1}{1000}$

16. $4x^{-3}$ $\frac{4}{x^3}$

17. $\frac{a^5}{a^8}$ $\frac{1}{a^3}$

18. p^6p^{-8} $\frac{1}{p^2}$

19. $\frac{m^{-3}}{m^{-8}}$ m^5

20. $\frac{x^3y^{-4}}{x^{-4}y^3}$ $\frac{x^7}{y^7}$

21. $(3s^{-2})^3$ $\frac{27}{s^6}$

22. $\frac{(y^3)^{-4}}{(y^{-2})^{-3}}$ $\frac{1}{y^{18}}$

[9.3] Evaluate if possible.

23. $\sqrt{81}$ 9

24. $-\sqrt{49}$ -7

25. $\sqrt{-49}$ Not a real number

26. $\sqrt[3]{64}$ 4

27. $\sqrt[3]{-64}$ -4

28. $\sqrt[4]{81}$ 3

29. $\sqrt[4]{-81}$ Not a real number

[9.4] Simplify each of the following radical expressions. Assume that all variables represent positive real numbers.

30. $\sqrt{50}$ $5\sqrt{2}$

31. $\sqrt{45}$ $3\sqrt{5}$

32. $\sqrt{7a^3}$ $a\sqrt{7a}$

33. $\sqrt{20x^4}$ $2x^2\sqrt{5}$

34. $\sqrt{49m^5}$ $7m^2\sqrt{m}$

35. $\sqrt{200b^3}$ $10b\sqrt{2b}$

36. $\sqrt{147r^3s^2}$ $7rs\sqrt{3r}$

37. $\sqrt{108a^2b^5}$ $6ab^2\sqrt{3b}$

38. $\sqrt{\dfrac{10}{81}}$ $\dfrac{\sqrt{10}}{9}$

39. $\sqrt{\dfrac{18x^2}{25}}$ $\dfrac{3x\sqrt{2}}{5}$

40. $\sqrt{\dfrac{12m^5}{49}}$ $\dfrac{2m^2\sqrt{3m}}{7}$

41. $\sqrt{\dfrac{3}{7}}$ $\dfrac{\sqrt{21}}{7}$

42. $\sqrt{\dfrac{3a}{2}}$ $\dfrac{\sqrt{6a}}{2}$

43. $\sqrt{\dfrac{8x^2}{7}}$ $\dfrac{2x\sqrt{14}}{7}$

[9.5] Simplify by combining like terms.

44. $\sqrt{3} + 4\sqrt{3}$ $5\sqrt{3}$

45. $9\sqrt{5} - 3\sqrt{5}$ $6\sqrt{5}$

46. $3\sqrt{2} + 2\sqrt{3}$ Cannot be simplified

47. $3\sqrt{3a} - \sqrt{3a}$ $2\sqrt{3a}$

48. $7\sqrt{6} - 2\sqrt{6} + \sqrt{6}$ $6\sqrt{6}$

49. $5\sqrt{3} + \sqrt{12}$ $7\sqrt{3}$

50. $3\sqrt{18} - 5\sqrt{2}$ $4\sqrt{2}$

51. $\sqrt{32} - \sqrt{18}$ $\sqrt{2}$

52. $\sqrt{27} - \sqrt{3} + 2\sqrt{12}$ $6\sqrt{3}$

53. $\sqrt{8} + 2\sqrt{27} - \sqrt{75}$ $2\sqrt{2} + \sqrt{3}$

54. $x\sqrt{18} - 3\sqrt{8x^2}$ $-3x\sqrt{2}$

[9.6] Simplify each radical expression.

55. $\sqrt{6}\sqrt{5}$ $\sqrt{30}$

56. $\sqrt{3}\sqrt{6}$ $3\sqrt{2}$

57. $\sqrt{3x}\sqrt{2}$ $\sqrt{6x}$

58. $\sqrt{2}\sqrt{8}\sqrt{3}$ $4\sqrt{3}$

59. $\sqrt{5a} \cdot \sqrt{10a}$ $5a\sqrt{2}$

60. $\sqrt{2}(\sqrt{3} + \sqrt{5})$ $\sqrt{6} + \sqrt{10}$

61. $\sqrt{7}(2\sqrt{3} - 3\sqrt{7})$ $2\sqrt{21} - 21$

62. $(\sqrt{3} + 5)(\sqrt{3} - 3)$ $-12 + 2\sqrt{3}$

63. $(\sqrt{15} - 3)(\sqrt{15} + 3)$ 6

64. $(\sqrt{2} + 3)^2$ $11 + 6\sqrt{2}$

65. $\dfrac{\sqrt{6}}{\sqrt{5}}$ $\dfrac{\sqrt{30}}{5}$

66. $\dfrac{4}{\sqrt{10}}$ $\dfrac{2\sqrt{10}}{5}$

67. $\dfrac{\sqrt{7x^3}}{\sqrt{3}}$ $\dfrac{x\sqrt{21x}}{3}$

68. $\dfrac{5}{\sqrt{11} - 2}$ $\dfrac{5\sqrt{11} + 10}{7}$

69. $\dfrac{6}{\sqrt{7} + 2}$ $2\sqrt{7} - 4$

70. $\dfrac{9 - \sqrt{108}}{3}$ $3 - 2\sqrt{3}$

Self-Test for Chapter Nine

The purpose of this self-test is to help you check your progress and to review for a chapter test in class. Allow yourself about an hour to take the test. When you are done, check your answers in the back of the book. If you missed any problems, be sure to go back and review the appropriate sections in the chapter and the exercises that are provided.

Use the properties of exponents to simplify each of the following expressions. Assume that all variables represent nonzero real numbers.

1. $(x^5)^3$ x^{15}

2. $(4a^2b)^2$ $16a^4b^2$

3. $\frac{(y^5)^2}{y^7}$ y^3

4. $\left(\frac{m^5}{n^2}\right)^3$ $\frac{m^{15}}{n^6}$

Evaluate (assume the variables are nonzero).

5. 9^0 1

6. $7x^0$ 7

Write using positive exponents.

7. x^{-7} $\frac{1}{x^7}$

8. $5a^{-4}$ $\frac{5}{a^4}$

9. x^5x^{-7} $\frac{1}{x^2}$

10. $\frac{m^{-4}}{m^4}$ $\frac{1}{m^8}$

Evaluate if possible.

11. $\sqrt{121}$ 11

12. $\sqrt[3]{27}$ 3

13. $\sqrt{-144}$ Not a real number

Simplify each of the following radical expressions.

14. $\sqrt{75}$ $5\sqrt{3}$

15. $\sqrt{24a^3}$ $2a\sqrt{6a}$

16. $\sqrt{\frac{16}{25}}$ $\frac{4}{5}$

17. $\sqrt{\frac{5}{9}}$ $\frac{\sqrt{5}}{3}$

Simplify by combining like terms.

18. $2\sqrt{10} - 3\sqrt{10} + 5\sqrt{10}$ $4\sqrt{10}$

19. $\sqrt{3} + \sqrt{12}$ $3\sqrt{3}$

20. $3\sqrt{8} - \sqrt{18}$ $3\sqrt{2}$

Simplify each of the following radical expressions.

21. $\sqrt{3x} \cdot \sqrt{6x}$ $3x\sqrt{2}$

22. $\sqrt{3}(\sqrt{7} - 2\sqrt{3})$ $\sqrt{21} - 6$

23. $(\sqrt{5} + 3)(\sqrt{5} + 2)$ $11 + 5\sqrt{5}$

24. $\dfrac{\sqrt{7}}{\sqrt{2}}$ $\dfrac{\sqrt{14}}{2}$

25. $\dfrac{4}{\sqrt{11} + 3}$ $2\sqrt{11} - 6$

Chapter Ten

Quadratic Equations

10.1 More on Quadratic Equations

OBJECTIVES

1. To solve quadratic equations of the form

$ax^2 = k$

2. To solve quadratic equations of the form

$(x - h)^2 = k$

We now have more tools for solving quadratic equations. In this section and the next we will be using the ideas of Sections 9.3 and 9.4 to extend our solution techniques.

In Section 5.5 we identified all equations of the form

$$ax^2 + bx + c = 0$$

as quadratic equations in standard form. In that section, we discussed solving these equations whenever the quadratic member was factorable. In this chapter, we want to extend our equation-solving techniques so that we can find solutions for all such quadratic equations.

Let's first review the factoring method of solution that we introduced in Chapter 5.

Example 1

Solve each quadratic equation by factoring.

(a) $x^2 = -7x - 12$

First, we write the equation in standard form.

$$x^2 + 7x + 12 = 0$$

Add $7x$ and 12 to both sides of the equation. The equation must be *set equal to 0.*

Once the equation is in standard form, we can factor the quadratic member.

$$(x + 3)(x + 4) = 0$$

Finally, using the zero product rule, we solve the equations $x + 3 = 0$ and $x + 4 = 0$ as follows:

These solutions can be checked as before by substitution into the original equation.

$x = -3 \quad \text{or} \quad x = -4$

(*b*) $x^2 = 16$

Again, we write the equation in standard form.

Here we factor the quadratic member of the equation as a difference of squares.

$x^2 - 16 = 0$

Factoring, we have

$(x + 4)(x - 4) = 0$

and finally the solutions are

$x = -4 \qquad \text{or} \qquad x = 4$

CHECK YOURSELF 1

Solve each of the following quadratic equations.

1. $x^2 - 4x = 45$ **2.** $x^2 = 25$

Approaches other than factoring can also be taken for the solution of certain quadratic equations. For instance, the equation of Example 1*b* could have been solved by an alternative method. Let's return to that equation.

Beginning with

$x^2 = 16$

we can take the square root of each side, to write

$\sqrt{x^2} = \sqrt{16}$

From Section 9.3, we know that this is equivalent to

$$\sqrt{x^2} = 4 \tag{1}$$

or

Recall that by definition

$\sqrt{x^2} = |x|$

$$|x| = 4 \tag{2}$$

Values for x of 4 or -4 will both satisfy equation (2), and so we have the two solutions

$x = 4 \qquad \text{or} \qquad x = -4$

We usually write the solutions as

$x = \pm 4$

$x = \pm 4$ is simply a convenient "shorthand" for indicating the two solutions and we generally will go directly to this form.

Let's look at two more equations solved by this method in the next example.

Example 2

Solve each of the following equations by the square root method shown above.

(*a*) $x^2 = 9$

By taking the square root of each side, we have

$x = \pm\sqrt{9}$

or

$x = \pm 3$

(*b*) $x^2 = 5$

Again, we take the square root of each side to write our two solutions as

$x = \pm\sqrt{5}$

CHECK YOURSELF 2

Solve.

1. $x^2 = 100$ **2.** $x^2 = 15$

You may have to add or subtract on both sides of the equation to write an equation in the form of those in the previous example. Our next example illustrates.

Example 3

Solve $x^2 - 8 = 0$.

First, add 8 to both sides of the equation. We have

$x^2 = 8$

Now take the square root of both sides.

$x = \pm\sqrt{8}$

Recall that

$\sqrt{8} = \sqrt{4 \cdot 2}$
$= \sqrt{4} \cdot \sqrt{2}$
$= 2\sqrt{2}$

Normally the solution should be written in the simplest form. In this case we would have

$$x = \pm 2\sqrt{2}$$

CHECK YOURSELF 3

Solve.

1. $x^2 - 18 = 0$ **2.** $x^2 + 1 = 7$

Equations of the form $(x - h)^2 = k$ can also be solved by taking the square root of both sides of the equation. Consider the next example.

Example 4

Solve $(x - 1)^2 = 6$.

Again, take the square root of both sides of the equation.

$$x - 1 = \pm\sqrt{6}$$

Now add 1 to both sides of the equation to isolate x.

$$x = 1 \pm \sqrt{6}$$

CHECK YOURSELF 4

Solve $(x + 2)^2 = 12$.

In the form

$ax^2 = k$

a is the coefficient of x^2 and k is some number.

To solve a quadratic equation of the form $ax^2 = k$, divide both sides of the equation by a as the first step, when $a \neq 1$. This is shown in our final example.

Example 5

Solve $4x^2 = 3$.

Divide both sides of the equation by 4.

$$x^2 = \frac{3}{4}$$

Now take the square root of both sides.

$$x = \pm\sqrt{\frac{3}{4}}$$

Recall that

$$\sqrt{\frac{3}{4}} = \frac{\sqrt{3}}{\sqrt{4}} = \frac{\sqrt{3}}{2}$$

Again write your result in the simplest form, so

$$x = \pm\frac{\sqrt{3}}{2}$$

CHECK YOURSELF 5

Solve $9x^2 = 5$.

What about an equation such as the following?

$$x^2 + 5 = 0$$

If we apply the above methods, we first subtract 5 from both sides, to write

$$x^2 = -5$$

Taking the square root of both sides gives

$$x = \pm\sqrt{-5}$$

But we know there are no square roots of -5 in the real numbers, so this equation has *no real number* solutions. You'll work with this type of equation in your next algebra course.

CHECK YOURSELF ANSWERS

1. (1) $-5, 9$; (2) $-5, 5$. **2.** (1) ± 10; (2) $\pm\sqrt{15}$. **3.** (1) $\pm 3\sqrt{2}$; (2) $\pm\sqrt{6}$. **4.** $-2 \pm 2\sqrt{3}$. **5.** $\pm\frac{\sqrt{5}}{3}$.

10.1 Exercises

Solve each of the equations for x.

1. $x^2 = 3$
$\pm\sqrt{3}$

2. $x^2 = 11$
$\pm\sqrt{11}$

3. $x^2 = 29$
$\pm\sqrt{29}$

4. $x^2 = 37$
$\pm\sqrt{37}$

5. $x^2 - 7 = 0$
$\pm\sqrt{7}$

6. $x^2 - 13 = 0$
$\pm\sqrt{13}$

7. $x^2 - 20 = 0$
$\pm 2\sqrt{5}$

8. $x^2 = 28$
$\pm 2\sqrt{7}$

9. $x^2 = 40$
$\pm 2\sqrt{10}$

10. $x^2 - 54 = 0$
$\pm 3\sqrt{6}$

11. $(x - 1)^2 = 5$
$1 \pm \sqrt{5}$

12. $(x - 3)^2 = 10$
$3 \pm \sqrt{10}$

13. $(x + 1)^2 = 12$
$-1 \pm 2\sqrt{3}$

14. $(x + 2)^2 = 32$
$-2 \pm 4\sqrt{2}$

15. $(x - 3)^2 = 24$
$3 \pm 2\sqrt{6}$

16. $(x - 5)^2 = 27$
$5 \pm 3\sqrt{3}$

17. $(x + 5)^2 = 25$
$-10, 0$

18. $(x + 2)^2 = 16$
$-6, 2$

19. $x^2 + 3 = 12$
± 3

20. $x^2 - 7 = 18$
± 5

21. $x^2 + 5 = 8$
$\pm\sqrt{3}$

22. $x^2 - 4 = 17$
$\pm\sqrt{21}$

23. $x^2 - 2 = 16$
$\pm 3\sqrt{2}$

24. $x^2 + 6 = 30$
$\pm 2\sqrt{6}$

25. $9x^2 = 25$
$\pm\frac{5}{3}$

26. $16x^2 = 9$
$\pm\frac{3}{4}$

27. $36x^2 = 5$
$\frac{\pm\sqrt{5}}{6}$

28. $25x^2 = 7$
$\frac{\pm\sqrt{7}}{5}$

29. $5x^2 = 3$
$\frac{\pm\sqrt{15}}{5}$

30. $2x^2 = 7$
$\frac{\pm\sqrt{14}}{2}$

31. $x^2 - 2x + 1 = 7$
(*Hint:* Factor the left-hand side.)
$1 \pm \sqrt{7}$

32. $x^2 + 4x + 4 = 7$
(*Hint:* Factor the left-hand side.)
$-2 \pm \sqrt{7}$

Skillscan (Section 4.4)

Multiply each of the following expressions.

a. $(x + 1)^2$
$x^2 + 2x + 1$

b. $(x + 5)^2$
$x^2 + 10x + 25$

c. $(x - 2)^2$
$x^2 - 4x + 4$

d. $(x - 7)^2$
$x^2 - 14x + 49$

e. $(x + 4)^2$
$x^2 + 8x + 16$

f. $(x - 3)^2$
$x^2 - 6x + 9$

g. $(2x + 5)^2$
$4x^2 + 20x + 25$

h. $(2x - 1)^2$
$4x^2 - 4x + 1$

Answers

1. $\pm\sqrt{3}$ **3.** $\pm\sqrt{29}$ **5.** $\pm\sqrt{7}$ **7.** $\pm 2\sqrt{5}$ **9.** $\pm 2\sqrt{10}$ **11.** $1 \pm \sqrt{5}$ **13.** $-1 \pm 2\sqrt{3}$ **15.** $3 \pm 2\sqrt{6}$ **17.** $-10, 0$ **19.** ± 3 **21.** $\pm\sqrt{3}$ **23.** $\pm 3\sqrt{2}$ **25.** $\pm\frac{5}{3}$ **27.** $\pm\frac{\sqrt{5}}{6}$ **29.** $\pm\frac{\sqrt{15}}{5}$ **31.** $1 \pm \sqrt{7}$

a. $x^2 + 2x + 1$ **b.** $x^2 + 10x + 25$ **c.** $x^2 - 4x + 4$ **d.** $x^2 - 14x + 49$ **e.** $x^2 + 8x + 16$ **f.** $x^2 - 6x + 9$ **g.** $4x^2 + 20x + 25$ **h.** $4x^2 - 4x + 1$

10.2 Completing the Square

OBJECTIVE

To solve a quadratic equation by completing the square.

We can solve a quadratic equation like

$$x^2 - 2x + 1 = 5$$

very easily if we notice that the expression on the left is a perfect-square trinomial. Factoring, we have

$$(x - 1)^2 = 5$$

so

$$x - 1 = \pm\sqrt{5} \qquad \text{or} \qquad x = 1 \pm \sqrt{5}$$

The solutions for the original equation are then $1 + \sqrt{5}$ and $1 - \sqrt{5}$.

It is true that every quadratic equation can be written in the form above (with a perfect-square trinomial on the left). That is the basis for the *completing-the-square* method for solving quadratic equations.

First, let's look at two perfect-square trinomials.

$$x^2 + 6x + 9 = (x + 3)^2 \qquad (1)$$

$$x^2 - 8x + 16 = (x - 4)^2 \qquad (2)$$

There is an important relationship between the coefficient of the middle term (the x term) and the constant.

In equation (1),

$$\left(\frac{1}{2} \cdot 6\right)^2 = 3^2 = 9$$

The x coefficient — The constant

In equation (2),

$$\left[\frac{1}{2}(-8)\right]^2 = (-4)^2 = 16$$

The x coefficient — The constant

It is always true, that in a perfect-square trinomial with a coefficient of 1 for x^2, the square of one-half of the x coefficient is equal to the constant term.

Example 1

(*a*) Find the term that should be added to $x^2 + 4x$ so that the expression is a perfect-square trinomial.

To complete the square of $x^2 + 4x$, add the square of one-half of 4 (the x coefficient).

$$x^2 + 4x + \left(\frac{1}{2} \cdot 4\right)^2$$

or $\quad x^2 + 4x + 2^2 \quad$ or $\quad x^2 + 4x + 4$

The trinomial $x^2 + 4x + 4$ is a perfect square because

$$x^2 + 4x + 4 = (x + 2)^2$$

(*b*) Find the term that should be added to $x^2 - 10x$ so that the expression is a perfect-square trinomial.

To complete the square of $x^2 - 10x$, add the square of one-half of -10 (the x coefficient).

$$x^2 - 10x + \left[\frac{1}{2}(-10)\right]^2$$

or $\quad x^2 - 10x + (-5)^2 \quad$ or $\quad x^2 - 10x + 25$

Check for yourself, by factoring, that this is a perfect-square trinomial.

CHECK YOURSELF 1

Complete the square and factor.

1. $x^2 + 2x$ **2.** $x^2 - 12x$

We can now use the above process along with the solution methods of Section 10.1 to solve a quadratic equation.

Example 2

Solve $x^2 + 4x - 2 = 0$ by completing the square.

Add 2 to both sides to remove -2 from the left side.

$$x^2 + 4x = 2$$

We find the term needed to complete the square by squaring one-half of the x coefficient.

$$\left(\frac{1}{2} \cdot 4\right)^2 = 2^2 = 4$$

We now add 4 to both sides of the equation.

This *completes the square* on the left.

$$x^2 + 4x + 4 = 2 + 4$$

Now factor on the left and simplify on the right.

$$(x + 2)^2 = 6$$

Now solving as before we have

$$x + 2 = \pm\sqrt{6}$$
$$x = -2 \pm \sqrt{6}$$

CHECK YOURSELF 2

Solve by completing the square.

$x^2 + 6x - 4 = 0$

For the completing the square method to work, the coefficient of x^2 must be 1. The following example illustrates the solution process when that coefficient is not equal to 1.

Example 3

Solve $2x^2 - 4x - 5 = 0$ by completing the square.

$$2x^2 - 4x - 5 = 0$$

Add 5 to both sides.

$$2x^2 - 4x = 5$$

Since the coefficient of x^2 is not 1 (here it is 2), divide every term by 2.

$$x^2 - 2x = \frac{5}{2}$$

We now complete the square and solve as before.

$$x^2 - 2x + 1 = \frac{5}{2} + 1$$

$$(x - 1)^2 = \frac{7}{2}$$

$$x - 1 = \pm\sqrt{\frac{7}{2}}$$

Simplify the radical on the right.

$$x - 1 = \pm\frac{\sqrt{14}}{2}$$

$$x = 1 \pm \frac{\sqrt{14}}{2}$$

or

We have combined the terms on the right with the common denominator of 2.

$$x = \frac{2 \pm \sqrt{14}}{2}$$

CHECK YOURSELF 3

Solve by completing the square.

$3x^2 - 6x + 2 = 0$

Let's summarize by listing the steps to solve a quadratic equation by completing the square.

SOLVING A QUADRATIC EQUATION BY COMPLETING THE SQUARE

STEP 1 Write the equation in the form

$ax^2 + bx = k$

so that the variable terms are on the left side and the constant is on the right side.

STEP 2 If the leading coefficient (of x^2) is not 1, divide both sides of the equation by that coefficient.

STEP 3 Add the square of one-half the coefficient of x to both sides of the equation.

STEP 4 The left side of the equation is now a perfect-square trinomial. Factor and solve as before.

CHECK YOURSELF ANSWERS

1. (1) $x^2 + 2x + 1 = (x + 1)^2$; (2) $x^2 - 12x + 36 = (x - 6)^2$.
2. $-3 \pm \sqrt{13}$. **3.** $\frac{3 \pm \sqrt{3}}{3}$.

10.2 Exercises

Determine whether the following trinomials are perfect squares.

1. $x^2 + 4x + 4$ Yes
2. $x^2 + 6x - 9$ No
3. $x^2 - 10x - 25$ No
4. $x^2 - 12x + 36$ Yes
5. $x^2 - 18x + 81$ Yes
6. $x^2 - 24x + 48$ No

Find the constant term that should be added to make each of the following expressions perfect-square trinomials.

7. $x^2 + 6x$ 9
8. $x^2 - 8x$ 16
9. $x^2 - 10x$ 25
10. $x^2 + 5x$ $\frac{25}{4}$
11. $x^2 + 9x$ $\frac{81}{4}$
12. $x^2 - 20x$ 100

Solve each of the quadratic equations by completing the square.

13. $x^2 + 4x - 12 = 0$
$-6, 2$

14. $x^2 - 6x + 8 = 0$
$2, 4$

15. $x^2 - 2x - 5 = 0$
$1 \pm \sqrt{6}$

16. $x^2 + 4x - 7 = 0$
$-2 \pm \sqrt{11}$

17. $x^2 + 3x - 27 = 0$
$\frac{-3 \pm 3\sqrt{13}}{2}$

18. $x^2 + 5x - 3 = 0$
$\frac{-5 \pm \sqrt{37}}{2}$

19. $x^2 + 6x - 1 = 0$
$-3 \pm \sqrt{10}$

20. $x^2 + 4x - 4 = 0$
$-2 \pm 2\sqrt{2}$

21. $x^2 - 5x + 6 = 0$
2, 3

22. $x^2 - 6x - 3 = 0$
$3 \pm 2\sqrt{3}$

23. $x^2 + 6x - 5 = 0$
$-3 \pm \sqrt{14}$

24. $x^2 - 2x = 1$
$1 \pm \sqrt{2}$

25. $x^2 = 8x + 3$

26. $x^2 = 2 - 5x$
$\frac{-5 \pm \sqrt{33}}{2}$

27. $x^2 = 3 - x$
$\frac{-1 \pm \sqrt{13}}{2}$

28. $x^2 = x + 1$
$\frac{1 \pm \sqrt{5}}{2}$

29. $2x^2 - 6x + 1 = 0$
$\frac{3 \pm \sqrt{7}}{2}$

30. $2x^2 + 10x + 11 = 0$
$\frac{-5 \pm \sqrt{3}}{2}$

31. $2x^2 - 4x + 1 = 0$
$\frac{2 \pm \sqrt{2}}{2}$

32. $2x^2 - 8x + 5 = 0$
$\frac{4 \pm \sqrt{6}}{2}$

33. $4x^2 - 2x - 1 = 0$
$\frac{1 \pm \sqrt{5}}{4}$

34. $3x^2 - x - 2 = 0$
$-\frac{2}{3}, 1$

Skillscan (Section 2.6)

Evaluate the expression $b^2 - 4ac$ for each set of values.

a. $a = 1, b = 1, c = -3$
13

b. $a = 1, b = -1, c = -1$
5

c. $a = 1, b = -8, c = -3$
76

d. $a = 1, b = -2, c = -1$
8

e. $a = -2, b = 4, c = -2$
0

f. $a = 2, b = -3, c = 4$
-23

Answers

1. Yes **3.** No **5.** Yes **7.** 9 **9.** 25 **11.** $\frac{81}{4}$ **13.** $-6, 2$ **15.** $1 \pm \sqrt{6}$ **17.** $\frac{-3 \pm 3\sqrt{13}}{2}$
19. $-3 \pm \sqrt{10}$ **21.** 2, 3 **23.** $-3 \pm \sqrt{14}$ **25.** $4 \pm \sqrt{19}$ **27.** $\frac{-1 \pm \sqrt{13}}{2}$ **29.** $\frac{3 \pm \sqrt{7}}{2}$
31. $\frac{2 \pm \sqrt{2}}{2}$ **33.** $\frac{1 \pm \sqrt{5}}{4}$ **a.** 13 **b.** 5 **c.** 76 **d.** 8 **e.** 0 **f.** -23

10.3 The Quadratic Formula

OBJECTIVE
To solve a quadratic equation by using the quadratic formula.

We are now ready to derive and use the *quadratic formula,* which will allow us to solve all quadratic equations. We will derive the formula by using the method of completing the square.

The derivation of the quadratic formula is shown below.

To use the quadratic formula, the quadratic equation you want to solve must be in *standard form*. That form is

$$ax^2 + bx + c = 0 \qquad \text{where } a \neq 0$$

Example 1

(*a*) $2x^2 - 5x + 3 = 0$ is in standard form.

$a = 2 \qquad b = -5 \qquad$ and $\qquad c = 3$

(*b*) $5x^2 + 3x = 5$ is *not* in standard form. Rewrite the equation by subtracting 5 from both sides.

$5x^2 + 3x - 5 = 0$ is in standard form.

$a = 5 \qquad b = 3 \qquad$ and $\qquad c = -5$

CHECK YOURSELF 1

Rewrite each quadratic in standard form.

1. $x^2 - 3x = 5$ **2.** $3x^2 = 7 - 2x$

TO DERIVE THE QUADRATIC FORMULA

Let $ax^2 + bx + c = 0$, where $a \neq 0$.

$ax^2 + bx = -c$ Subtract c from both sides.

$x^2 + \frac{b}{a}x = -\frac{c}{a}$ Divide both sides by a.

This is the completing-the-square step that makes the left-hand side a perfect square.

$x^2 + \frac{b}{a}x + \frac{b^2}{4a^2} = \frac{b^2}{4a^2} - \frac{c}{a}$ Add $\frac{b^2}{4a^2}$ to both sides.

$\left(x + \frac{b}{2a}\right)^2 = \frac{b^2 - 4ac}{4a^2}$ Factor on the left, and add the fractions on the right.

$x + \frac{b}{2a} = \pm\sqrt{\frac{b^2 - 4ac}{4a^2}}$ Take the square root of both sides.

$x + \frac{b}{2a} = \pm\frac{\sqrt{b^2 - 4ac}}{2a}$ Simplify the radical on the right.

$x = -\frac{b}{2a} \pm \frac{\sqrt{b^2 - 4ac}}{2a}$ Subtract $\frac{b}{2a}$ from both sides.

$$\boxed{x = \frac{-b \pm \sqrt{b^2 - 4ac}}{2a}}$$ THE QUADRATIC FORMULA

Example 2

Solve $x^2 - 5x + 4 = 0$ by formula.

The equation is in standard form, so first identify a, b, and c.

$$x^2 - 5x + 4 = 0$$

$a = 1$ $b = -5$ $c = 4$

We now substitute the values for a, b, and c into the formula.

$$x = \frac{-b \pm \sqrt{b^2 - 4ac}}{2a}$$

$$= \frac{-(-5) \pm \sqrt{(-5)^2 - 4(1)(4)}}{2(1)}$$

Simplify the expression.

$$= \frac{5 \pm \sqrt{25 - 16}}{2}$$

$$= \frac{5 \pm \sqrt{9}}{2}$$

$$= \frac{5 \pm 3}{2}$$

Now,

$$x = \frac{5 + 3}{2} \quad \text{or} \quad x = \frac{5 - 3}{2}$$

$$= 4 \qquad\qquad = 1$$

The solutions are 4 and 1.

Note: These results could also have been found by factoring the original equation. You should check that for yourself.

CHECK YOURSELF 2

Solve $x^2 - 2x - 8 = 0$ by formula. Check your result by factoring.

Of course, the main use of the formula is to solve equations that cannot be factored.

Example 3

Solve $2x^2 = x + 4$ by formula.

First, the equation *must be written* in standard form to find a, b, and c.

$$2x^2 - x - 4 = 0$$

$a = 2 \quad b = -1 \quad c = -4$

Substitute the values for a, b, and c into the formula.

$$x = \frac{-b \pm \sqrt{b^2 - 4ac}}{2a}$$

$$= \frac{-(-1) \pm \sqrt{(-1)^2 - 4(2)(-4)}}{2(2)}$$

$$= \frac{1 \pm \sqrt{1 + 32}}{4}$$

$$= \frac{1 \pm \sqrt{33}}{4}$$

CHECK YOURSELF 3

Solve $3x^2 = 3x + 4$ by formula.

Example 4

Solve $x^2 - 2x = 4$ by formula.

In standard form, the equation is

$$x^2 - 2x - 4 = 0$$

$a = 1 \quad b = -2 \quad c = -4$

Again substitute the values into the quadratic formula.

$$x = \frac{-(-2) \pm \sqrt{(-2)^2 - 4(1)(-4)}}{2(1)}$$

$$= \frac{2 \pm \sqrt{20}}{2}$$

You should always write your solution in simplest form.

Since 20 has a perfect square factor,

$\sqrt{20} = \sqrt{4 \cdot 5}$
$= 2\sqrt{5}$

Now factor the numerator and divide by the common factor of 2.

$$x = \frac{2 \pm 2\sqrt{5}}{2}$$

$$= \frac{2(1 \pm \sqrt{5})}{2}$$

$$= 1 \pm \sqrt{5}$$

CHECK YOURSELF 4

Solve $3x^2 = 2x + 4$ by formula.

There is another point that should be made about the use of the quadratic formula. This is illustrated in our next example.

Example 5

Solve $3x^2 - 6x - 3 = 0$ by formula. Since the equation is in standard form, we could use

$a = 3 \qquad b = -6 \qquad \text{and} \qquad c = -3$

in the quadratic formula. There is, however, a better approach.

Note the common factor of 3 in the quadratic member of the original equation. Factoring we have

$$3(x^2 - 2x - 1) = 0$$

and dividing both sides of the equation by 3 gives

$$x^2 - 2x - 1 = 0$$

The advantage to this approach is that these values will require much less simplification after substituting into the quadratic formula.

Now let $a = 1$, $b = -2$, and $c = -1$. Then

$$x = \frac{-(-2) \pm \sqrt{(-2)^2 - 4(1)(-1)}}{2 \cdot 1}$$

$$= \frac{2 \pm \sqrt{8}}{2}$$

$$= \frac{2 \pm 2\sqrt{2}}{2}$$

$$= \frac{2(1 \pm \sqrt{2})}{2}$$

$$= 1 \pm \sqrt{2}$$

CHECK YOURSELF 5

Solve $4x^2 - 20x = 12$ by formula.

In applications that lead to quadratic equations, you may want to find approximate values for the solutions.

Example 6

Solve $x^2 - 5x + 5 = 0$ by formula, and write your solutions in approximate decimal form.

Substituting $a = 1$, $b = -5$, and $c = 5$ gives

$$x = \frac{-(-5) \pm \sqrt{(-5)^2 - 4(1)(5)}}{2(1)}$$

$$= \frac{5 \pm \sqrt{5}}{2}$$

From the table of square roots in Appendix 2 or a calculator we have, $\sqrt{5} \approx 2.236$, so

$$x \approx \frac{5 + 2.236}{2} \quad \text{or} \quad x \approx \frac{5 - 2.236}{2}$$

$$= \frac{7.236}{2} \qquad\qquad = \frac{2.764}{2}$$

$$= 3.618 \qquad\qquad = 1.382$$

CHECK YOURSELF 6

Solve $x^2 - 3x - 5 = 0$ by formula, and approximate the solutions in decimal form.

You may be wondering whether the quadratic formula can be used to solve all quadratic equations. It can, but not all quadratic equations will have real solutions, as the next example shows.

Example 7

Solve $x^2 - 3x + 5 = 0$ by formula.

Substituting $a = 1$, $b = -3$, and $c = 5$, we have

$$x = \frac{-(-3) \pm \sqrt{(-3)^2 - 4(1)(5)}}{2(1)}$$

$$= \frac{3 \pm \sqrt{-11}}{2}$$

In this case there are no real number solutions because of the negative number in the radical.

CHECK YOURSELF 7

Solve $x^2 - 3x = -3$ by formula.

Let's review the steps used for solving equations by the use of the quadratic formula.

SOLVING EQUATIONS WITH THE QUADRATIC FORMULA

STEP 1 Rewrite the equation in standard form.

$ax^2 + bx + c = 0$

STEP 2 If a common factor exists, divide both sides of the equation by that common factor.

STEP 3 Identify the coefficients a, b, and c.

STEP 4 Substitute values for a, b, and c into the formula

$$x = \frac{-b \pm \sqrt{b^2 - 4ac}}{2a}$$

STEP 5 Simplify the right side of the expression formed in step 4 to write the solutions for the original equation.

Often applied problems will lead to quadratic equations that must be solved by the methods of this or the previous section. Our final example illustrates such an application.

Example 8

The word "numerically" is used because we cannot compare units of area to units of length.

The perimeter of a square is numerically 6 less than its area. Find the length of one side of the square.

Step 1 You want to find the length of one side of the square.

Step 2 Let x represent the length of one side. A sketch of the problem will help.

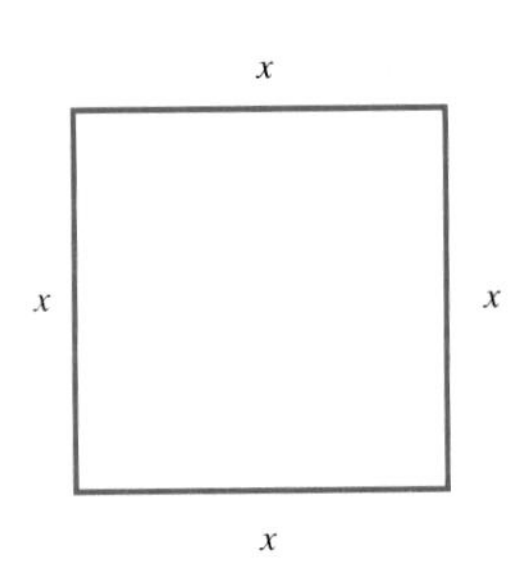

Step 3 The perimeter is $4x$. The area is x^2. So

$$4x = x^2 - 6$$

The perimeter — The area — 6 less than

Step 4 Writing the equation in standard form, we solve as before.

$$x^2 - 4x - 6 = 0$$

has solutions

$$x = 2 + \sqrt{10} \approx 5.162$$

$$x = 2 - \sqrt{10} \approx -1.162$$

We reject the solution—1.162; a length can't be negative.

The length of a side is approximately 5.162.

Step 5 If you have a calculator handy, you can easily check this result. Letting x be 5.162, find the perimeter and area to verify (approximately) the result in the original problem.

CHECK YOURSELF 8

The perimeter of a square is numerically 2 more than its area. Find the length of one side.

You have now studied four methods for solving quadratic equations:

1. Factoring
2. Extracting roots
3. Completing the square
4. The quadratic formula

The choice of which method to use depends largely on the equation you want to solve. Factoring is usually easiest and should be tried first.

Extracting roots is used only for equations in the particular form $(x - h)^2 = k$.

For this reason, it is important that you commit the quadratic formula to memory.

Both the completing-the-square method and the quadratic formula are applicable to all quadratic equations and can always be used. Many students seem to find the quadratic formula quicker and easier to apply.

CHECK YOURSELF ANSWERS

1. (1) $x^2 - 3x - 5 = 0$; (2) $3x^2 + 2x - 7 = 0$.
2. $x = 4, -2$.
3. $x = \frac{3 \pm \sqrt{57}}{6}$.
4. $x = \frac{1 \pm \sqrt{13}}{3}$.
5. $x = \frac{5 \pm \sqrt{37}}{2}$
6. $x \approx 4.193$ or -1.193
7. $\frac{3 \pm \sqrt{-3}}{2}$, no real solutions
8. Approximately 3.414 or 0.586

10.3 Exercises

Solve each of the following quadratic equations by formula.

1. $x^2 + 3x + 2 = 0$
   $-2, -1$
2. $x^2 - 4x + 3 = 0$
   1, 3
3. $x^2 - 6x + 5 = 0$
   1, 5
4. $x^2 - 7x + 6 = 0$
   1, 6
5. $3x^2 + 2x - 1 = 0$
   $-1, \frac{1}{3}$
6. $x^2 - 8x + 16 = 0$
   4

7. $x^2 + 5x = -4$

$-4, -1$

8. $4x^2 + 5x = 6$

$-2, \frac{3}{4}$

9. $x^2 = 6x - 9$

3

10. $2x^2 - 5x = 3$

$-\frac{1}{2}, 3$

11. $2x^2 - 3x - 7 = 0$

$\frac{3 \pm \sqrt{65}}{4}$

12. $x^2 - 5x + 2 = 0$

$\frac{5 \pm \sqrt{17}}{2}$

13. $x^2 + 2x - 4 = 0$

$-1 \pm \sqrt{5}$

14. $x^2 - 4x + 2 = 0$

$2 \pm \sqrt{2}$

15. $2x^2 - 3x = 3$

$\frac{3 \pm \sqrt{33}}{4}$

16. $3x^2 - 2x + 1 = 0$

No real solutions

17. $3x^2 - 2x = 6$

$\frac{1 \pm \sqrt{19}}{3}$

18. $4x^2 = 4x + 5$

$\frac{1 \pm \sqrt{6}}{2}$

19. $2x^2 + 2x + 1 = 0$

No real solutions

20. $3x^2 - x = 1$

$\frac{1 \pm \sqrt{13}}{6}$

21. $5x^2 = 8x - 2$

$\frac{4 \pm \sqrt{6}}{5}$

22. $5x^2 - 2 = 2x$

$\frac{1 \pm \sqrt{11}}{5}$

23. $2x^2 - 9 = 4x$

$\frac{2 \pm \sqrt{22}}{2}$

24. $3x^2 - 6x = 2$

$\frac{3 \pm \sqrt{15}}{3}$

25. $3x - 5 = \frac{1}{x}$

$\frac{5 \pm \sqrt{37}}{6}$

26. $x + 3 = \frac{1}{x}$

$\frac{-3 \pm \sqrt{13}}{2}$

27. $(x - 2)(x + 1) = 3$

$\frac{1 \pm \sqrt{21}}{2}$

28. $(x - 3)(x + 2) = 5$

$\frac{1 \pm 3\sqrt{5}}{2}$

Solve the following quadratic equations by factoring or by any of the techniques of this chapter.

29. $(x - 1)^2 = 7$

$1 \pm \sqrt{7}$

30. $(2x + 3)^2 = 5$

$\frac{-3 \pm \sqrt{5}}{2}$

31. $x^2 - 5x - 14 = 0$

$-2, 7$

32. $3x^2 + 2x - 1 = 0$

$-1, \frac{1}{3}$

33. $6x^2 - 23x + 10 = 0$

$\frac{1}{2}, \frac{10}{3}$

34. $x^2 + 7x - 18 = 0$

$-9, 2$

35. $2x^2 - 8x + 3 = 0$

$\frac{4 \pm \sqrt{10}}{2}$

36. $x^2 + 2x - 1 = 0$

$-1 \pm \sqrt{2}$

37. $x^2 - 8x + 16 = 9$

$1, 7$

38. $3x^2 + 6x + 1 = 1$

$-2, 0$

39. $2x^2 - 16x + 32 = 3$

$\frac{8 \pm \sqrt{6}}{2}$

40. $x^2 - 3x = 10$

$-2, 5$

Use your calculator or a table of square roots for the following problems.

41. The perimeter of a square is numerically 3 less than its area. Find the length of one side. Approximately 4.646

42. The perimeter of a square is numerically 1 more than its area. Find the length of one side. Approximately 0.268 or 3.732

43. A picture frame is 15 in by 12 in. The area of the picture that shows is 140 in^2. What is the width of the frame? Approximately 0.787 in

44. A garden area is 30 ft long by 20 ft wide. A path of uniform width is set around the edge. If the remaining garden area is 400 ft^2, what is the width of the path?
Approximately 2.192 ft

Skillscan (Section 9.3)
Find the value of the expression $\sqrt{a^2 + b^2}$ for the following pairs of values. Where necessary approximate the value using a calculator or our table of square roots.

a. $a = 3, b = 4$
5

b. $a = 5, b = 12$
13

c. $a = 5, b = 8$
9.434

d. $a = 6, b = 9$
10.817

e. $a = 7, b = 8$
10.630

f. $a = 1, b = 2$
2.236

g. $a = 7, b = 24$
25

h. $a = 10, b = 15$
18.028

Answers

1. $-2, -1$ **3.** $1, 5$ **5.** $-1, \frac{1}{3}$ **7.** $-4, -1$ **9.** 3 **11.** $\frac{3 \pm \sqrt{65}}{4}$ **13.** $-1 \pm \sqrt{5}$ **15.** $\frac{3 \pm \sqrt{33}}{4}$ **17.** $\frac{1 \pm \sqrt{19}}{3}$ **19.** No real solutions **21.** $\frac{4 \pm \sqrt{6}}{5}$ **23.** $\frac{2 \pm \sqrt{22}}{2}$ **25.** $\frac{5 \pm \sqrt{37}}{6}$ **27.** $\frac{1 \pm \sqrt{21}}{2}$ **29.** $1 \pm \sqrt{7}$ **31.** $-2, 7$ **33.** $\frac{1}{2}, \frac{10}{3}$ **35.** $\frac{4 \pm \sqrt{10}}{2}$ **37.** $1, 7$ **39.** $\frac{8 \pm \sqrt{6}}{2}$ **41.** Approximately 4.646 **43.** Approximately 0.787 in **a.** 5 **b.** 13 **c.** 9.434 **d.** 10.817 **e.** 10.630 **f.** 2.236 **g.** 25 **h.** 18.028

10.4 The Pythagorean Theorem

Pythagoras, born 572 B.C., was the founder of the Greek society, the Pythagoreans. Although the theorem bears Pythagoras' name, his own work on this theorem is uncertain as the Pythagoreans credited new discoveries to their founder.

OBJECTIVE
To apply the Pythagorean theorem in solving problems.

One very important application of our work with quadratic equations is called the *Pythagorean theorem* (named for the Greek mathematician Pythagoras). The theorem deals with *right triangles.* These are triangles with a right angle (a square corner). We give special names to the parts of a right triangle.

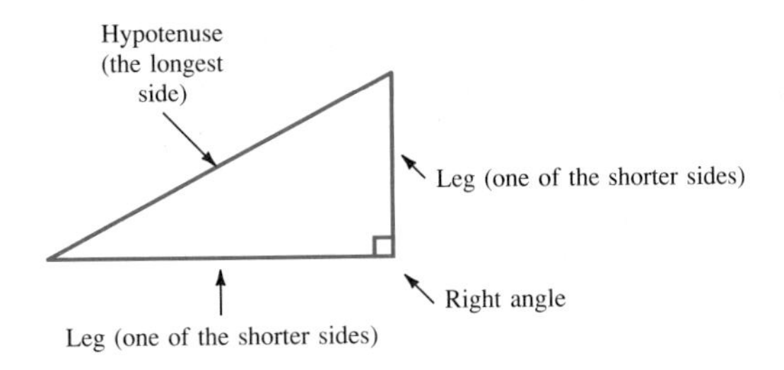

For every right triangle, the following is true.

THE PYTHAGOREAN THEOREM

The square of the length of the hypotenuse is equal to the sum of the squares of the lengths of the legs.

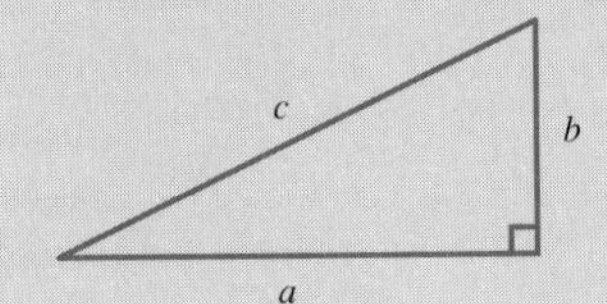

$$c^2 = a^2 + b^2$$

Example 1

(*a*) For the right triangle below,

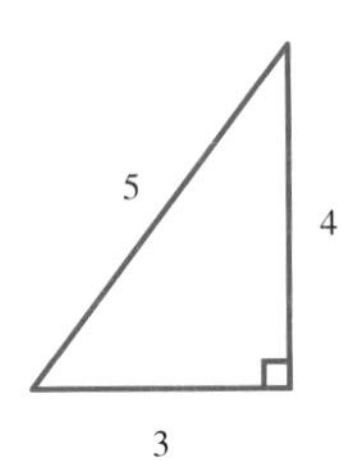

$$5^2 \stackrel{?}{=} 3^2 + 4^2$$
$$25 \stackrel{?}{=} 9 + 16$$
$$25 = 25$$

(*b*) For the right triangle below.

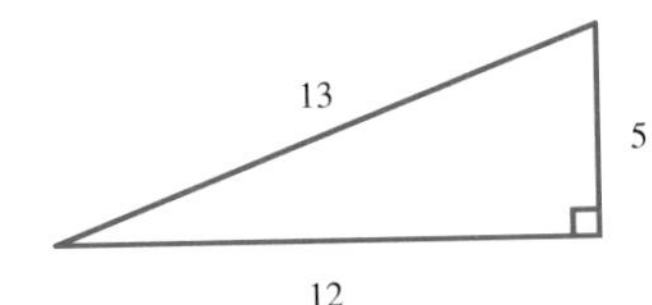

$$13^2 \stackrel{?}{=} 12^2 + 5^2$$
$$169 \stackrel{?}{=} 144 + 25$$
$$169 = 169$$

CHECK YOURSELF 1

Verify the Pythagorean theorem for the right triangle shown.

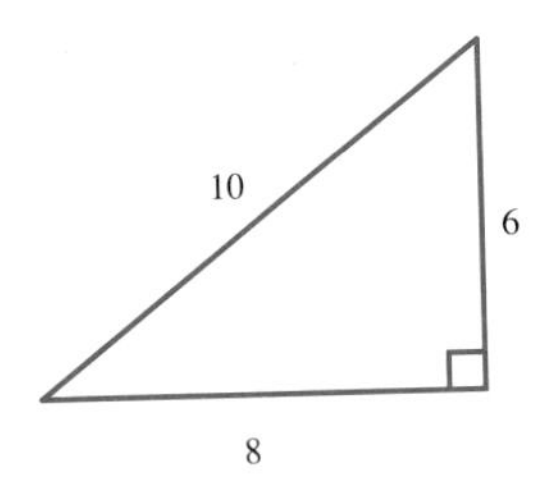

The Pythagorean theorem can be used to find the length of one side of a right triangle when the lengths of the two other sides are known.

Example 2

Find length x.

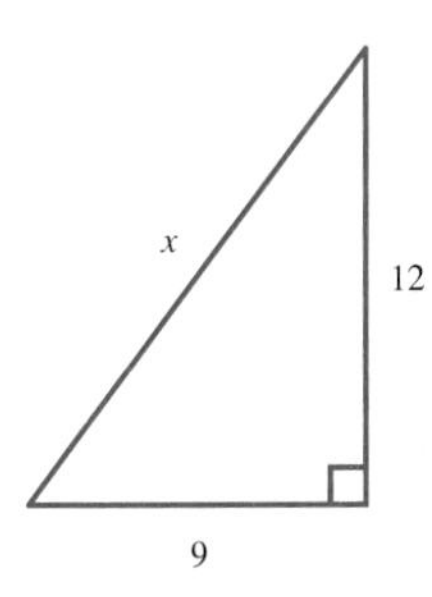

$$x^2 = 9^2 + 12^2$$
$$= 81 + 144$$
$$= 225$$

so

$x = 15$ or $x = -15$

We reject this solution because a length must be positive.

CHECK YOURSELF 2

Find length x.

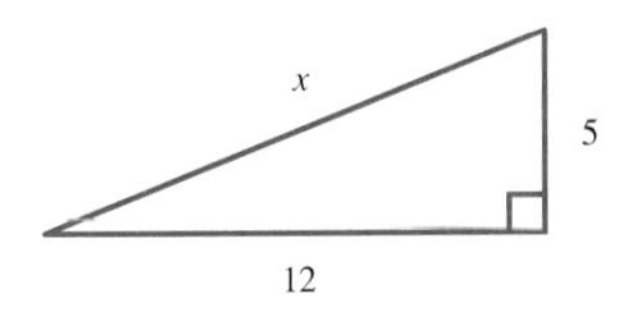

One or more of the lengths of the sides may be represented by an irrational number.

Example 3

Find length x.

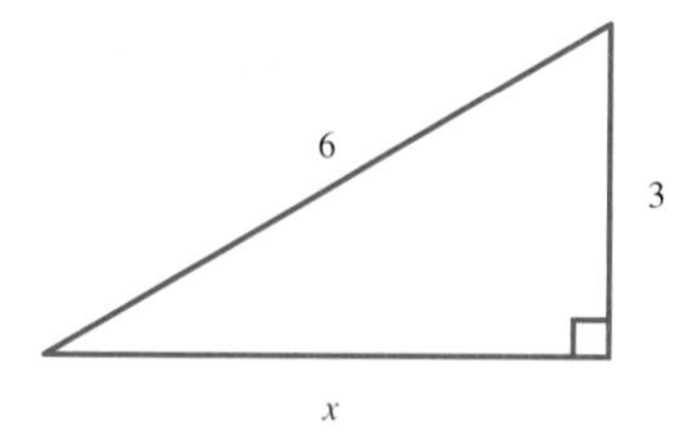

$$3^2 + x^2 = 6^2$$
$$9 + x^2 = 36$$
$$x^2 = 27$$

so

$x = \sqrt{27}$ or $3\sqrt{3}$

x is approximately 5.2.

Note: You can approximate $3\sqrt{3}$ (or $\sqrt{27}$) with the use of a calculator or by the table in Appendix 2.

CHECK YOURSELF 3

Find length x.

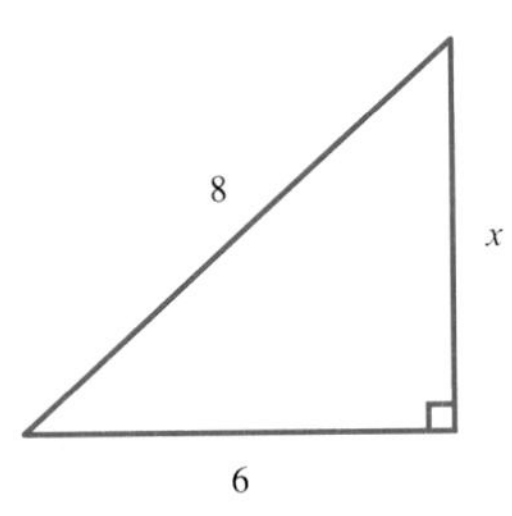

The Pythagorean theorem can be applied to solve a variety of geometric problems.

Example 4

Find the length of the diagonal of a rectangle whose length is 8 cm and whose width is 5 cm. Let x be the unknown length of the diagonal:

Always draw and label a sketch showing the information from a problem when geometric figures are involved.

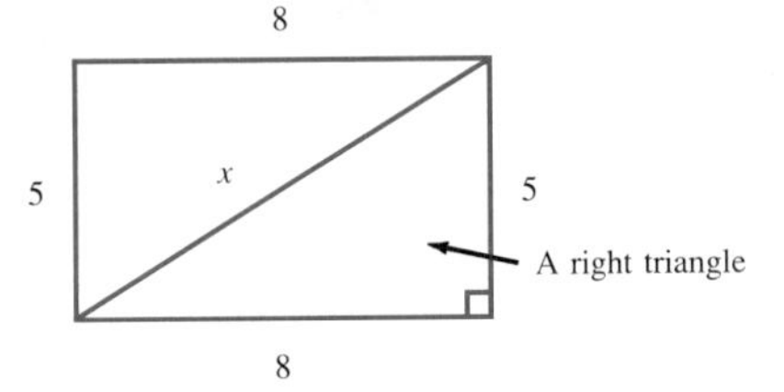

So

$$\begin{aligned} x^2 &= 5^2 + 8^2 \\ &= 25 + 64 \\ &= 89 \\ x &= \sqrt{89} \text{ cm} \end{aligned}$$

CHECK YOURSELF 4

The diagonal of a rectangle is 12 in, and its width is 6 in. Find its length.

Example 5

How long must a guy wire be to reach from the top of a 30-ft pole to a point on the ground 20 ft from the base of the pole?

Again be sure to draw a sketch of the problem.

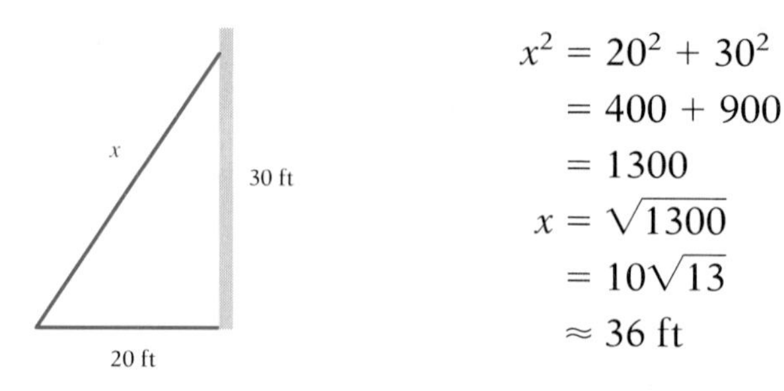

$$\begin{aligned} x^2 &= 20^2 + 30^2 \\ &= 400 + 900 \\ &= 1300 \\ x &= \sqrt{1300} \\ &= 10\sqrt{13} \\ &\approx 36 \text{ ft} \end{aligned}$$

CHECK YOURSELF 5

A 16-ft ladder leans against a wall with its base 4 ft from the wall. How far off the floor is the top of the ladder?

Example 6

The length of one leg of a right triangle is 2 cm more than the other. If the length of the hypotenuse is 6 cm, what are the lengths of the two legs?

Draw a sketch of the problem, labeling the known and unknown lengths. Here, if one leg is represented by x, the other must be represented by $x + 2$

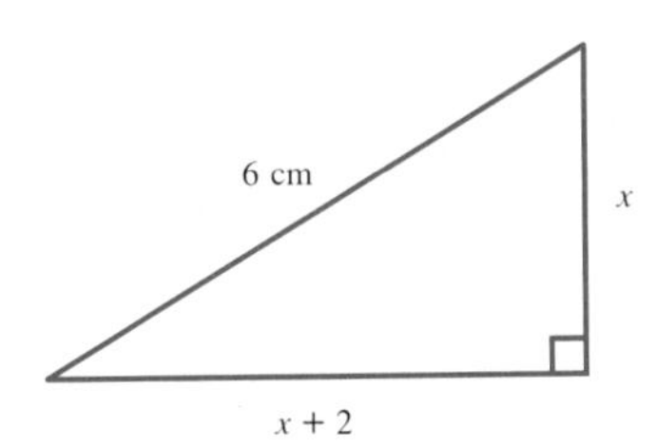

Use the Pythagorean theorem to form an equation.

The sum of the squares of the lengths of the unknown legs is equal to the square of the length of the hypotenuse.

$$\begin{aligned} x^2 + (x + 2)^2 &= 6^2 \\ x^2 + x^2 + 4x + 4 &= 36 \\ 2x^2 + 4x - 32 &= 0 \\ x^2 + 2x - 16 &= 0 \quad \text{Divide both sides by 2.} \end{aligned}$$

We apply the quadratic formula as before:

$$x = \frac{-2 \pm \sqrt{2^2 - 4(1)(-16)}}{2(1)}$$

$$= -1 \pm \sqrt{17}$$

Now,

$$x = -1 + \sqrt{17} \quad \text{or} \quad x = -1 - \sqrt{17}$$
$$\approx 3.123 \qquad\qquad \approx -5.123$$

Reject the negative solution in a geometric problem.

If $x \approx 3.123$, then $x + 2 \approx 5.123$. The lengths of the legs are approximately 3.123 and 5.123 cm.

CHECK YOURSELF 6

The length of one leg of a right triangle is 1 in more than the other. If the length of the hypotenuse is 3 in, what are the lengths of the legs?

CHECK YOURSELF ANSWERS

1. $10^2 \stackrel{?}{=} 8^2 + 6^2$; $100 \stackrel{?}{=} 64 + 36$; $100 = 100$.
2. 13.
3. $2\sqrt{7}$; or approximately 5.3.
4. Length: $6\sqrt{3}$ in.
5. The height is approximately 15.5 ft.
6. Approximately 1.561 and 2.561 in.

10.4 Exercises

Find the length x in each triangle.

1.

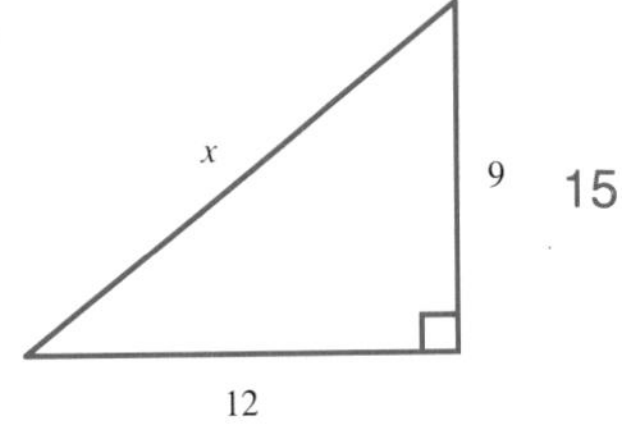

15

2.

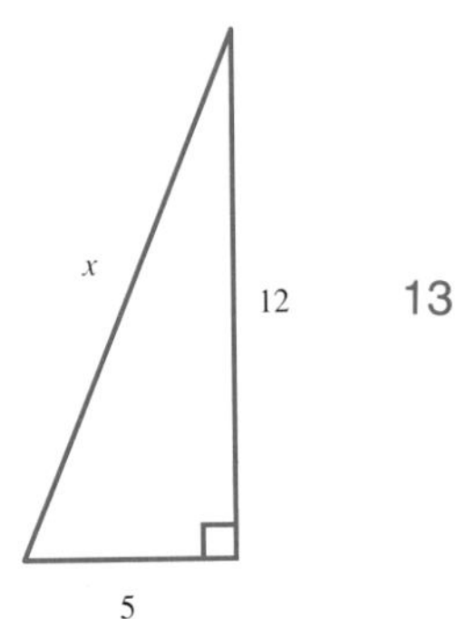

13

3.

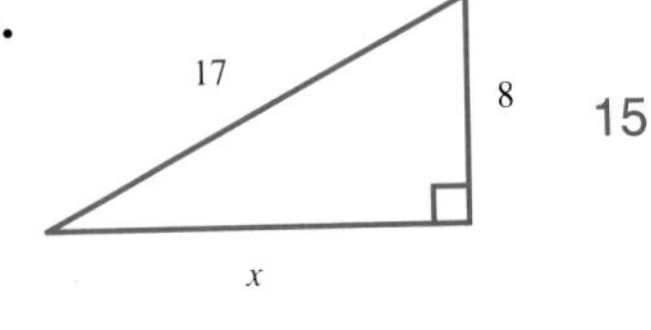

15

4.

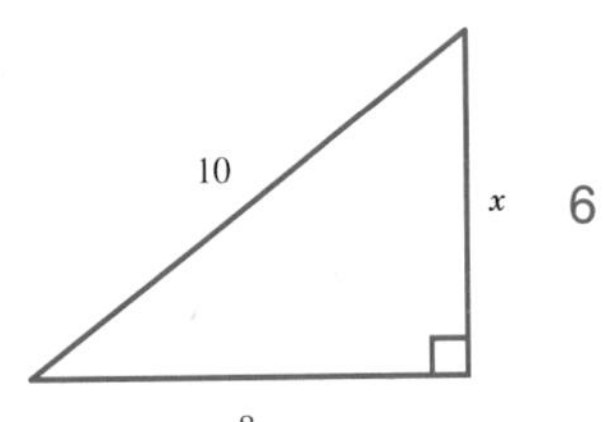

6

5.

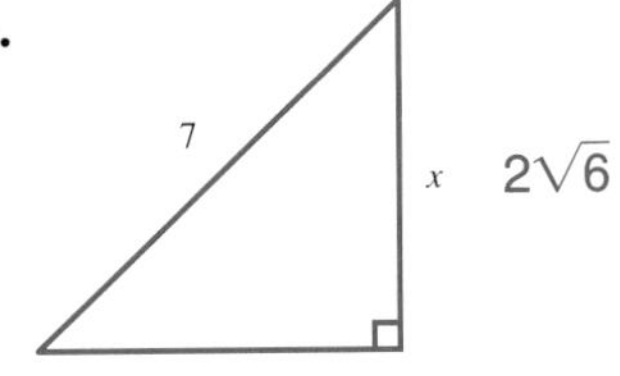

$2\sqrt{6}$

6.

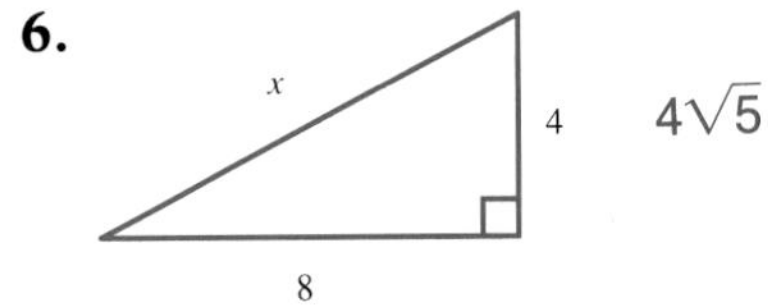

$4\sqrt{5}$

7. Find the length of the diagonal of a rectangle whose length is 10 cm and whose width is 7 cm. ≈12.207 cm

8. Find the length of the diagonal of a rectangle whose width is 5 in and whose length is 7 in. ≈8.602 in

9. Find the width of a rectangle whose diagonal is 12 ft and whose length is 10 ft. ≈6.633 ft

10. Find the length of a rectangle whose diagonal is 9 in and whose width is 6 in. ≈6.708 in

11. How long must a guy wire be to run from the top of a 20-ft pole to a point on the ground 8 ft from the base of the pole? ≈21.541 ft

12. The base of a 15-ft ladder is 5 ft away from a wall. How high from the floor is the top of the ladder? ≈14.142 ft

13. The length of one leg of a right triangle is 3 in more than the other. If the length of the hypotenuse is 8 in, what are the lengths of the two legs? ≈3.954 in, ≈6.954 in

14. The length of a rectangle is 1 cm longer than its width. If the diagonal of the rectangle is 4 cm, what are the dimensions (the length and width) of the rectangle? ≈2.284 cm, ≈3.284 cm

Skillscan (Section 7.3)

Graph each line after completing the table of values.

a. $y = 2x + 3$

x	y
0	3
1	5
2	7

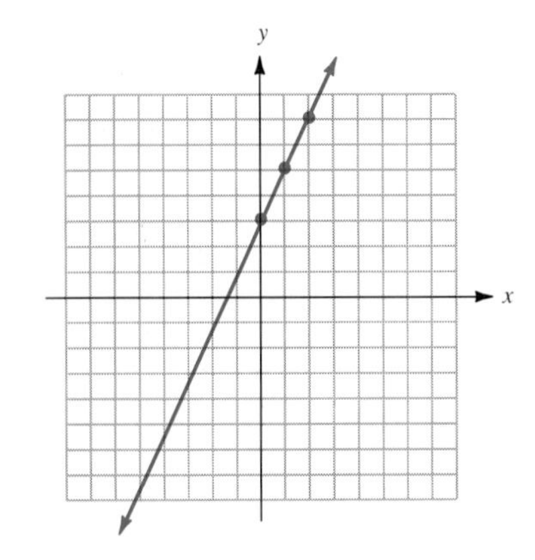

b. $y = -3x + 4$

x	y
0	4
1	1
2	−2

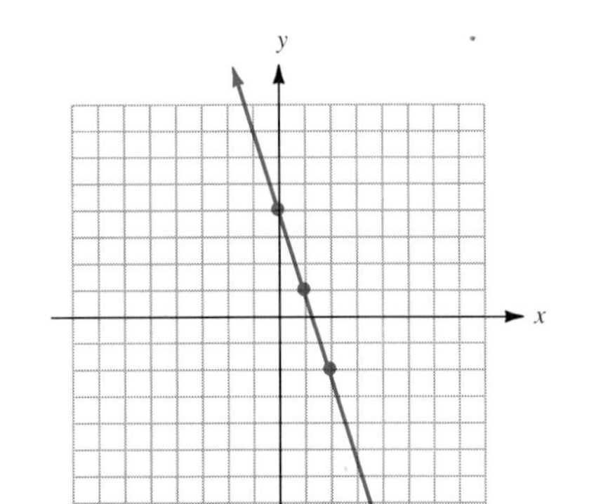

c. $y = \frac{2}{3}x - 1$

x	y
0	−1
3	1
6	3

d. $y = -\frac{3}{4}x + 3$

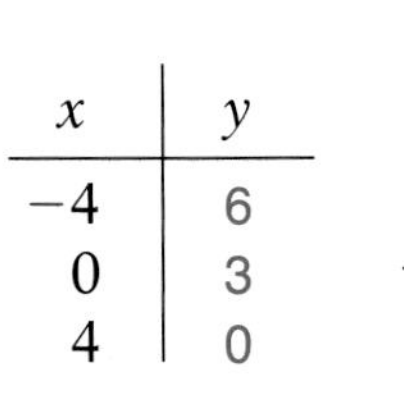

x	y
−4	6
0	3
4	0

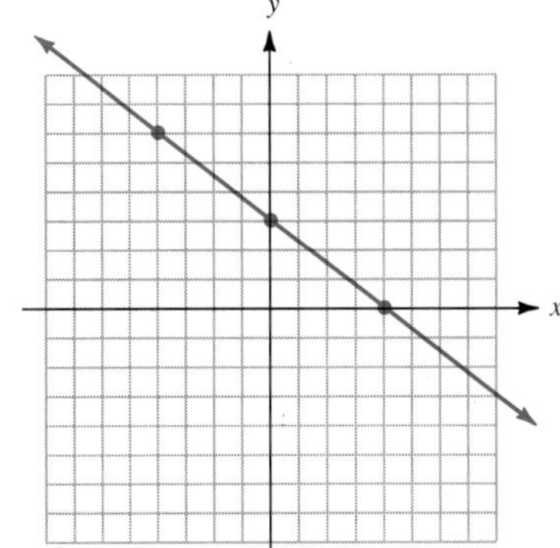

e. $2x - 3y = 6$

x	y
−3	−4
0	−2
3	0

f. $5x + 2y = 10$

x	y
0	5
2	0
4	−5

Answers

1. 15 **3.** 15 **5.** $2\sqrt{6}$

The following are approximations.

7. 12.207 cm **9.** 6.633 ft **11.** 21.541 ft **13.** 3.954 in, 6.954 in

a. $y = 2x + 3$

x	y
0	3
1	5
2	7

b. $y = -3x + 4$

x	y
0	4
1	1
2	−2

c. $y = \frac{2}{3}x - 1$

x	y
0	−1
3	1
6	3

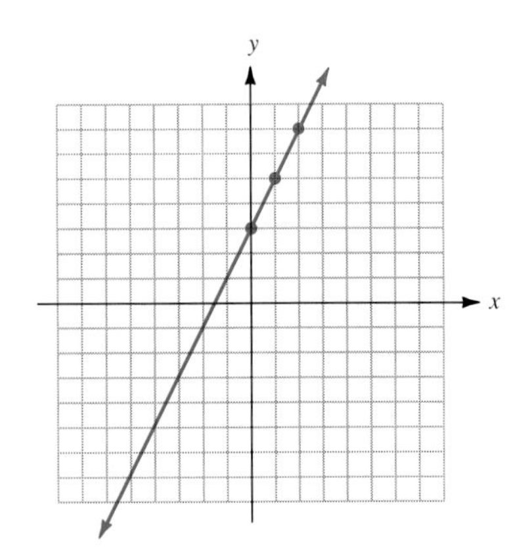

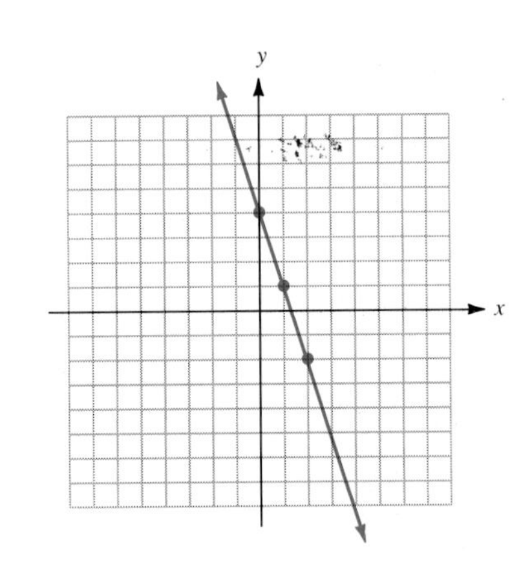

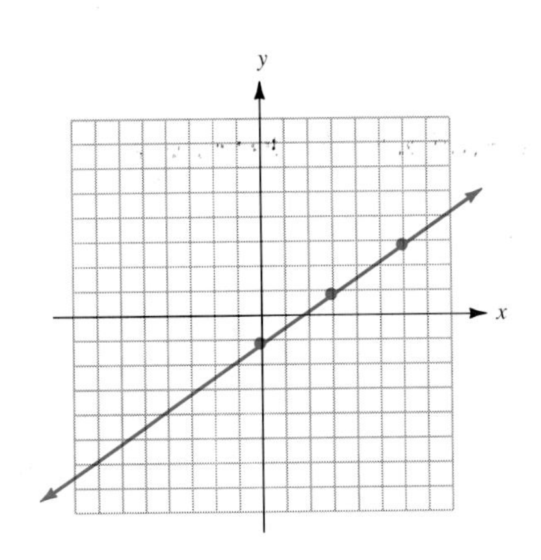

d. $y = -\frac{3}{4}x + 3$

x	y
−4	6
0	3
4	0

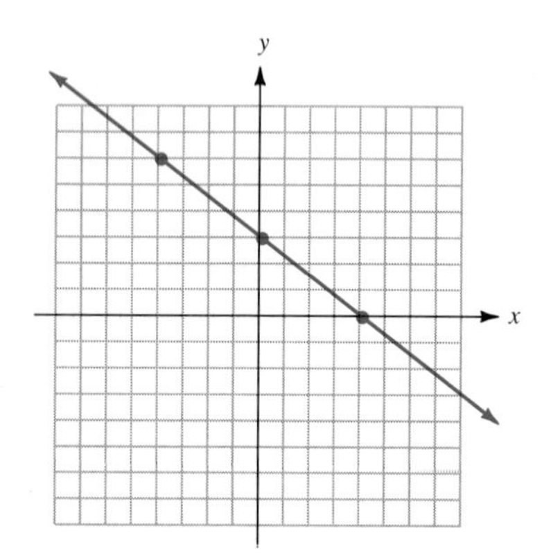

e. $2x - 3y = 6$

x	y
−3	−4
0	−2
3	0

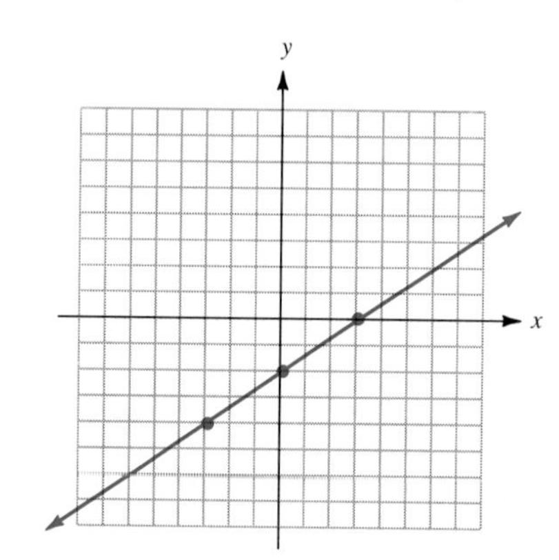

f. $5x + 2y = 10$

x	y
0	5
2	0
4	−5

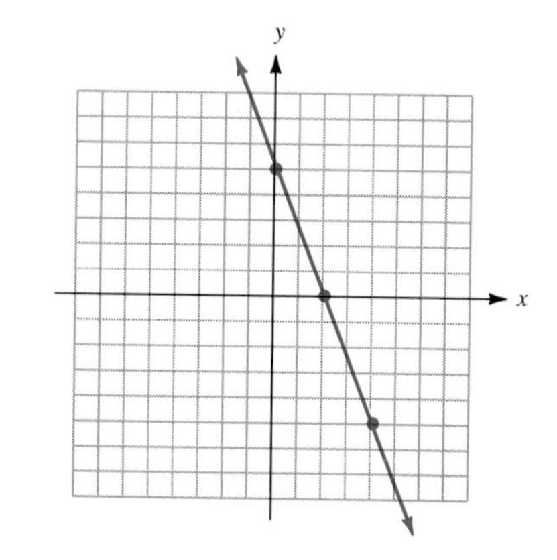

10.5 Graphing Quadratic Equations

OBJECTIVE
To graph a quadratic equation with the form

$y = ax^2 + bx + c \qquad a \neq 0$

In Section 7.3 you learned to graph first-degree equations. Similar methods will allow you to graph quadratic equations with the form

$$y = ax^2 + bx + c \qquad a \neq 0$$

The first thing you will notice is that the graph of an equation in this form is not a straight line. The graph is always a curve called a *parabola*.

Here are some examples:

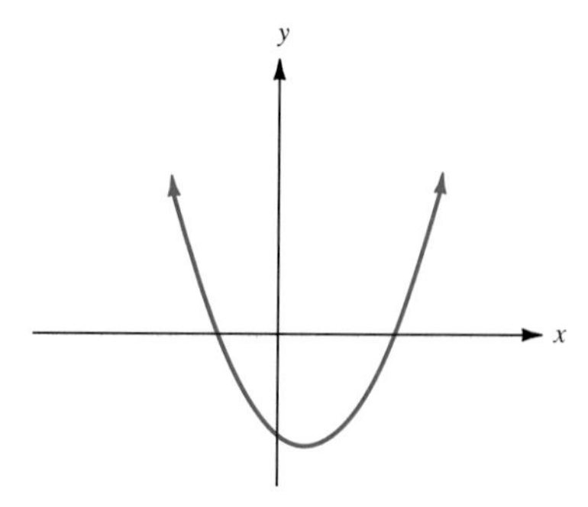

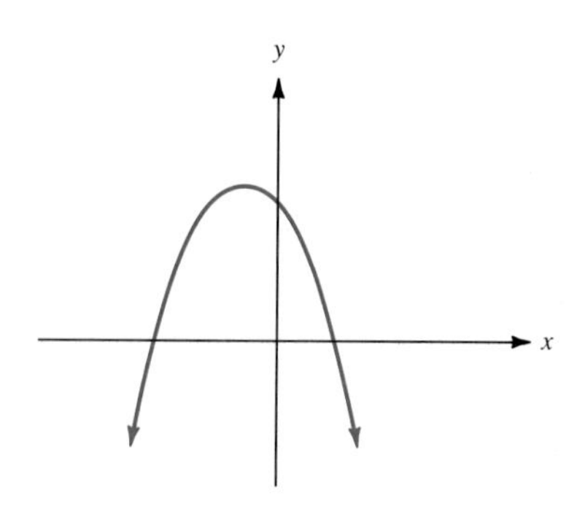

We can start as before by finding solutions for the equation. Choose any convenient values for x. Then use the given equation to compute the corresponding values for y.

Example 1

If $y = x^2$, complete the ordered pairs to form solutions. Then show these results in a table of values.

$(-2,\ ), (-1,\ ), (0,\ ), (1,\ ), (2,\ )$

For example, to complete the pair $(-2,\ )$, substitute -2 for x in the given equation.

$y = (-2)^2 = 4$

Remember that a solution is a pair of values that make the equation a true statement.

So $(-2, 4)$ is a solution.

Substituting the other values for x in the same manner, we have the following table of values for $y = x^2$:

x	y
-2	4
-1	1
0	0
1	1
2	4

CHECK YOURSELF 1

If $y = x^2 + 2$, complete the ordered pairs to form solutions and form a table of values.

$(-2,\ ), (-1,\ ), (0,\ ), (1,\ ), (2,\ )$

Example 1 (continued)

Plot the points from the table of values corresponding to $y = x^2$.

x	y
-2	4
-1	1
0	0
1	1
2	4

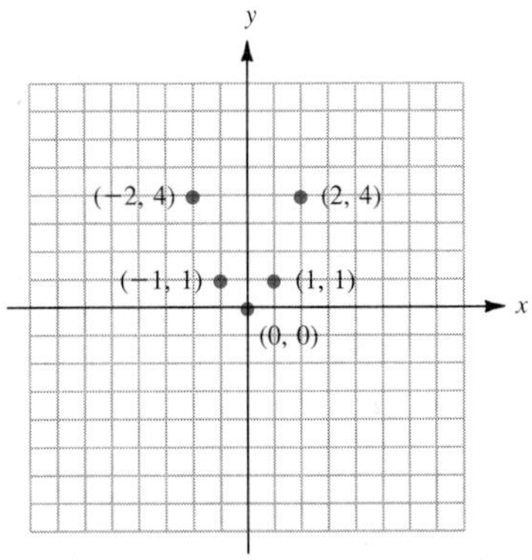

y
x
(−2, 4)
(2, 4)
(−1, 1)
(1, 1)
(0, 0)

CHECK YOURSELF 2

Plot the points from the table of values formed in Check Yourself 1.

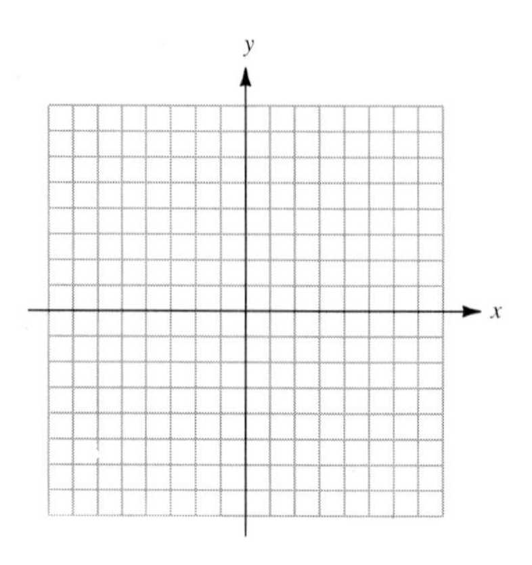

Example 1 (continued)

We can now draw a smooth curve between the points to form the graph of $y = x^2$.

As we mentioned earlier, the graph must be the curve called a parabola.

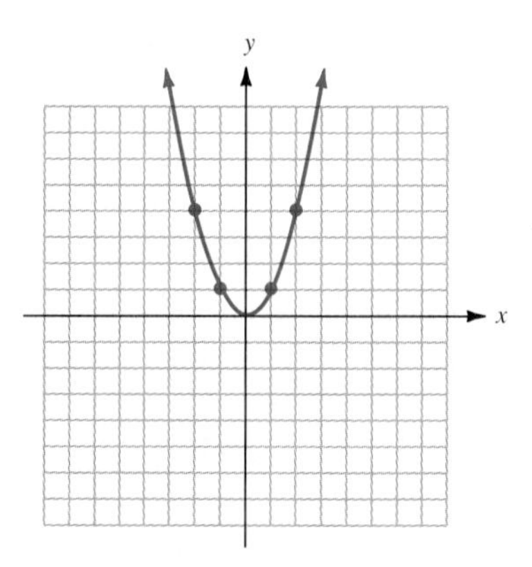

CHECK YOURSELF 3

Draw a smooth curve between the points plotted in the Check Yourself 2 exercise.

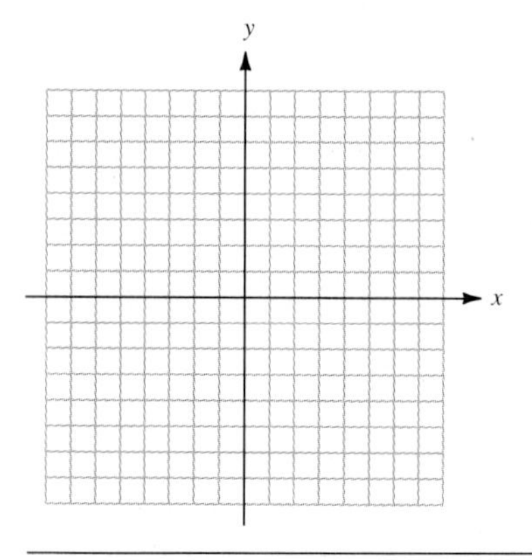

You can use any convenient values for x in forming your table of values. You should use as many pairs as are necessary to get the correct shape of the graph (a parabola).

Example 2

Graph $y = x^2 - 2x$. Use values of x between -1 and 3.

First, determine solutions for the equation.

For instance, if $x = -1$,

$$y = (-1)^2 - 2(-1)$$
$$= 1 + 2$$
$$= 3$$

and $(-1, 3)$ is a solution for the given equation.

Substituting the other values for x, we can form the table of values shown below. We then plot the corresponding points and draw a smooth curve to form our graph.

The graph of $y = x^2 - 2x$:

x	y
-1	3
0	0
1	-1
2	0
3	3

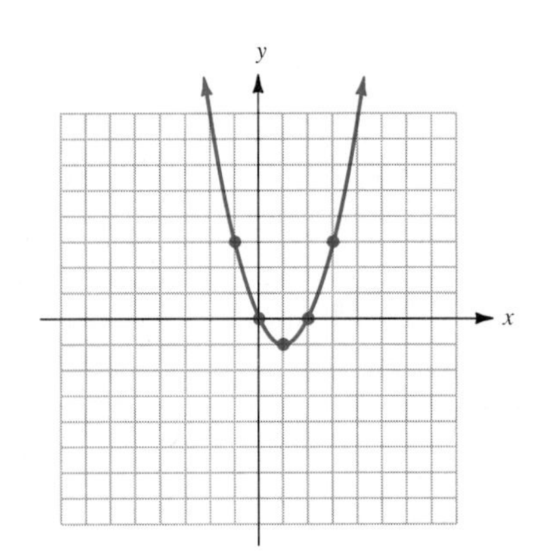

CHECK YOURSELF 4

Graph $y = x^2 + 4x$. Use values of x between -4 and 0.

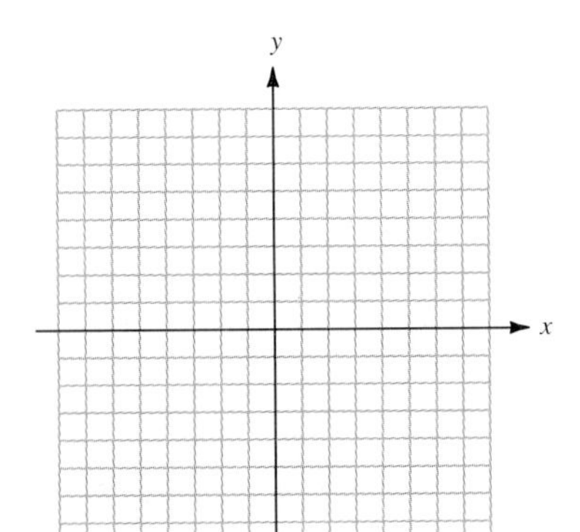

x	y

Example 3

Graph $y = x^2 - x - 2$. Use values of x between -2 and 3. We'll show the computation for two of the solutions.

If $x = -2$,

$$
\begin{aligned}
y &= (-2)^2 - (-2) - 2 \\
&= 4 + 2 - 2 \\
&= 4
\end{aligned}
$$

If $x = 3$,

$$
\begin{aligned}
y &= 3^2 - 3 - 2 \\
&= 9 - 3 - 2 \\
&= 4
\end{aligned}
$$

You should substitute the remaining values for x into the given equation to verify the other solutions shown in the table of values below.

x	y
-2	4
-1	0
0	-2
1	-2
2	0
3	4

The graph of $y = x^2 - x - 2$:

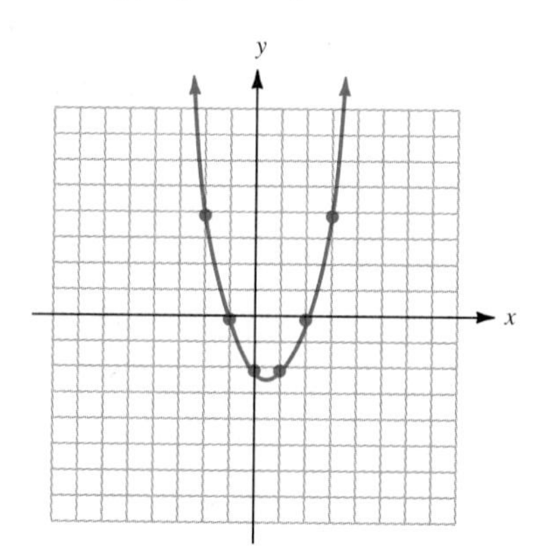

CHECK YOURSELF 5

Graph $y = x^2 - 4x + 3$. Use values of x between -1 and 4.

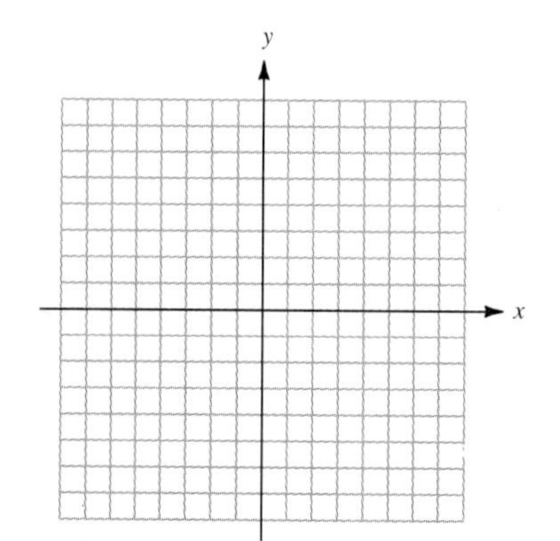

x	y

Example 4

Graph $y = -x^2 + 3$. Use x values between -2 and 2. Again we'll show two computations.

If $x = -2$,

Note: $-(-2)^2 = -4$

$$\begin{aligned} y &= -(-2)^2 + 3 \\ &= -4 + 3 \\ &= -1 \end{aligned}$$

If $x = 1$,

$$\begin{aligned} y &= -(1)^2 + 3 \\ &= -1 + 3 \\ &= 2 \end{aligned}$$

Verify the remainder of the solutions shown in the table of values below for yourself.

x	y
-2	-1
-1	2
0	3
1	2
2	-1

The graph of $y = -x^2 + 3$:

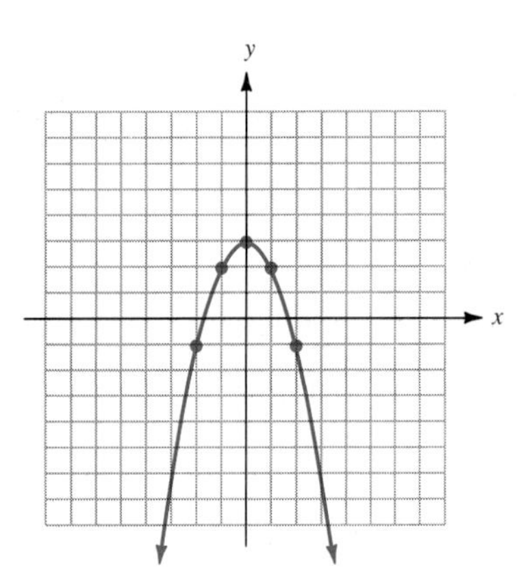

There is an important difference between this graph and the others we have seen. This time the parabola opens downward! Can you guess why? The answer is in the coefficient of the x^2 term.

If the coefficient of x^2 is *positive,* the parabola opens *upward.*

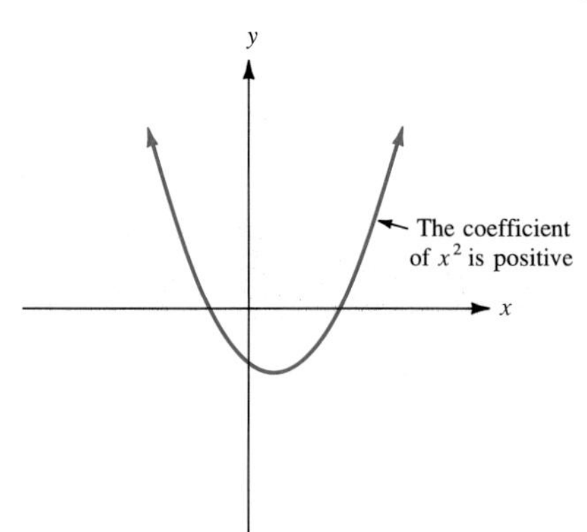

If the coefficient of x^2 is *negative*, the parabola opens *downward*.

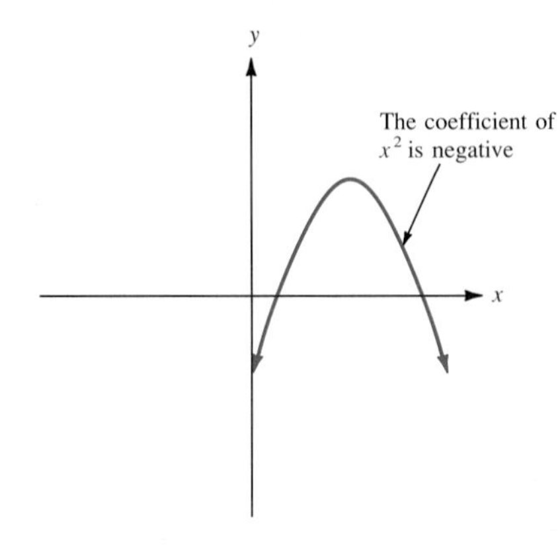

CHECK YOURSELF 6

Graph $y = -x^2 - 2x$. Use x values between -3 and 1.

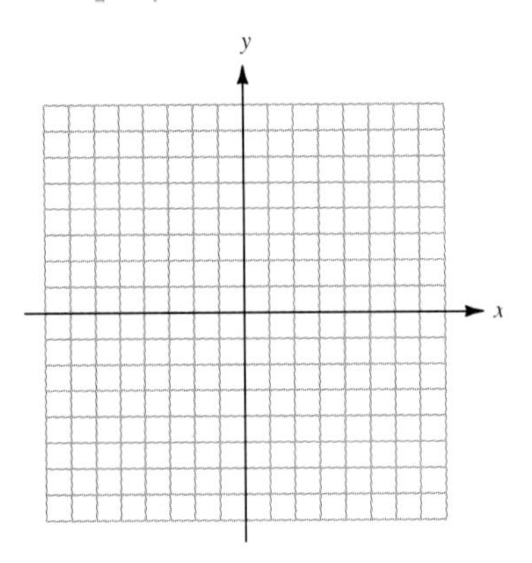

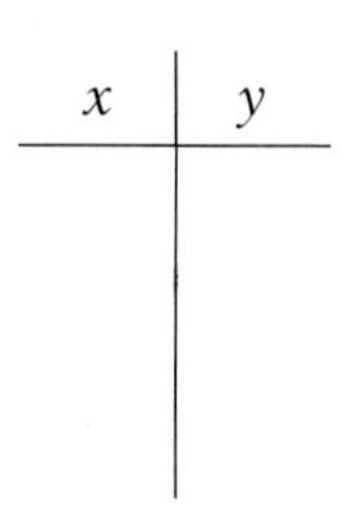

There are two other terms we would like to introduce before closing this section on graphing quadratic equations. As you may have noticed, all the parabolas that we graphed are symmetric about a vertical line. This is called the *axis of symmetry* for the parabola.

The point at which the parabola intersects that vertical line (this will be the lowest—or the highest—point on the parabola) is

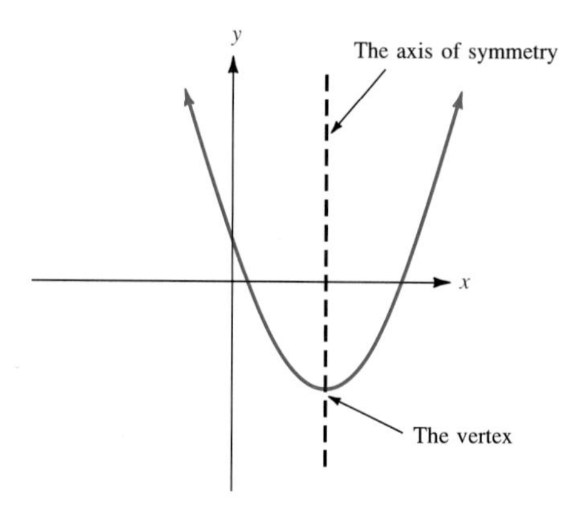

called the *vertex*. You'll learn more about finding the axis of symmetry and the vertex of a parabola in your next course in algebra.

CHECK YOURSELF 7

Refer to the parabola of Example 2 of this section. Can you tell by observation what the axis of symmetry is and where the vertex is located?

CHECK YOURSELF ANSWERS

1.

x	y
−2	6
−1	3
0	2
1	3
2	6

2.

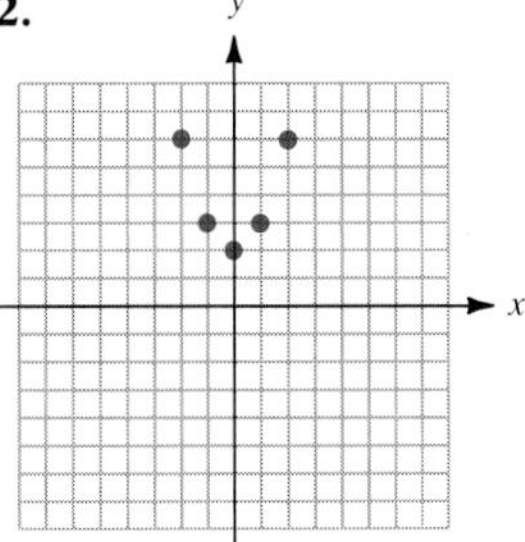

3. $y = x^2 + 2$

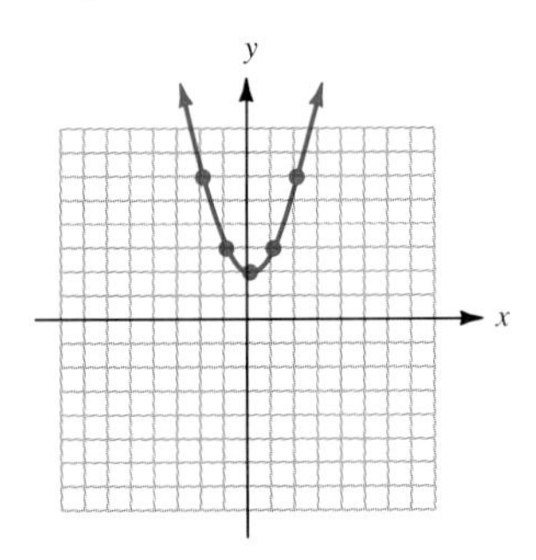

4. $y = x^2 + 4x$

x	y
−4	0
−3	−3
−2	−4
−1	−3
0	0

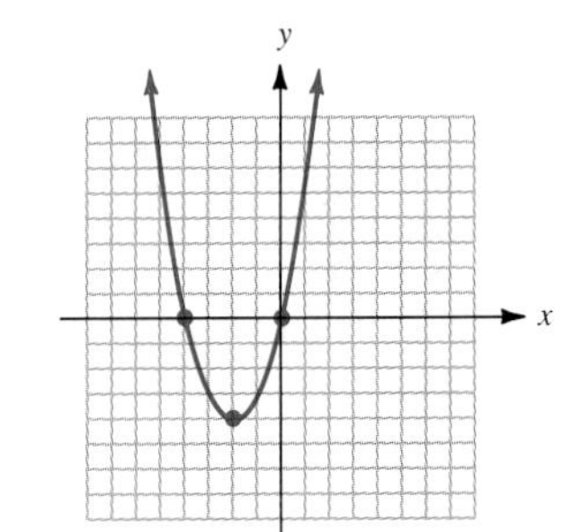

5. $y = x^2 - 4x + 3$

x	y
−1	8
0	3
1	0
2	−1
3	0
4	3

6. $y = -x^2 - 2x$

x	y
−3	−3
−2	0
−1	1
0	0
1	−3

7. The axis of symmetry is $x = 1$. The vertex has coordinates $(1, -1)$.

10.5 Exercises

Graph the following quadratic equations after completing the given table of values.

1. $y = x^2 + 1$

x	y
−2	5
−1	2
0	1
1	2
2	5

2. $y = x^2 - 2$

x	y
−2	2
−1	−1
0	−2
1	−1
2	2

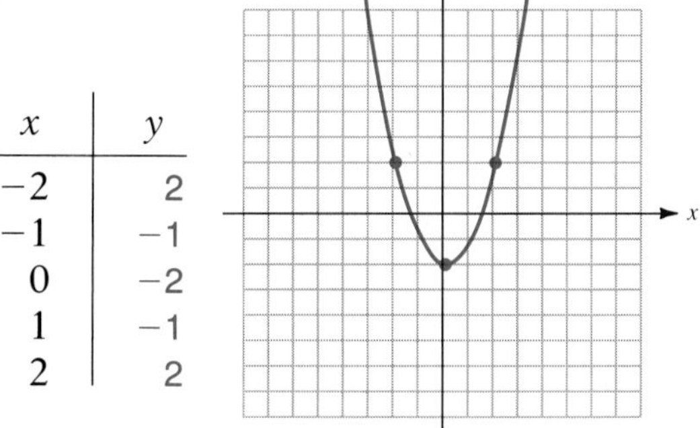

3. $y = x^2 - 4$

x	y
−2	0
−1	−3
0	−4
1	−3
2	0

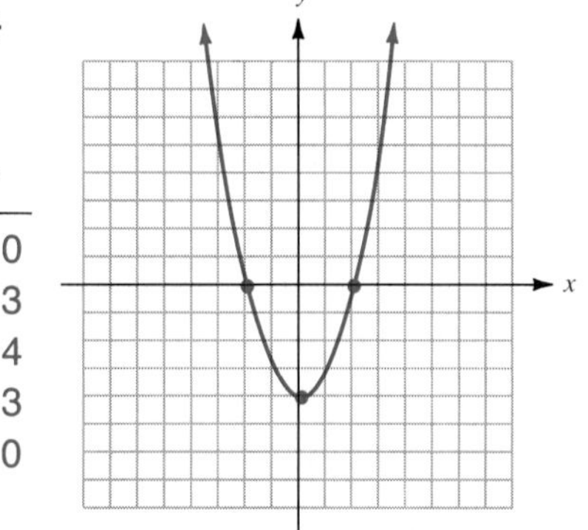

4. $y = x^2 + 3$

x	y
−2	7
−1	4
0	3
1	4
2	7

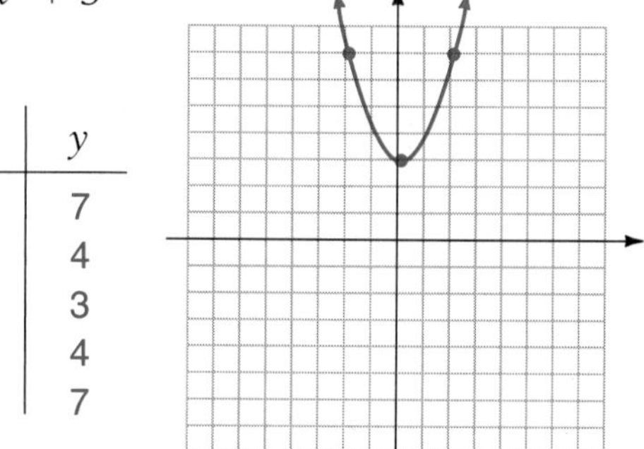

5. $y = x^2 - 4x$

x	y
0	0
1	−3
2	−4
3	−3
4	0

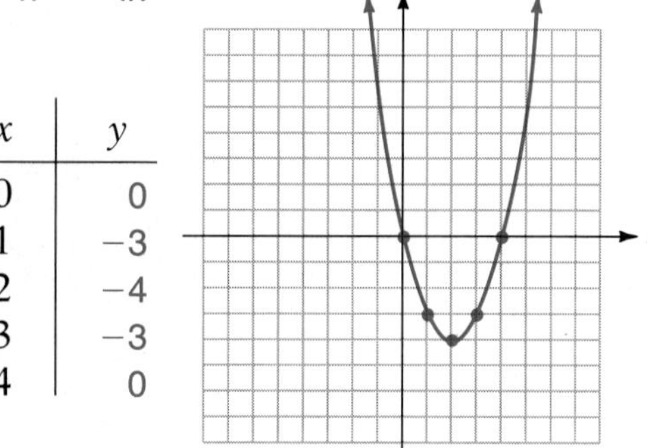

6. $y = x^2 + 2x$

x	y
−3	3
−2	0
−1	−1
0	0
1	3

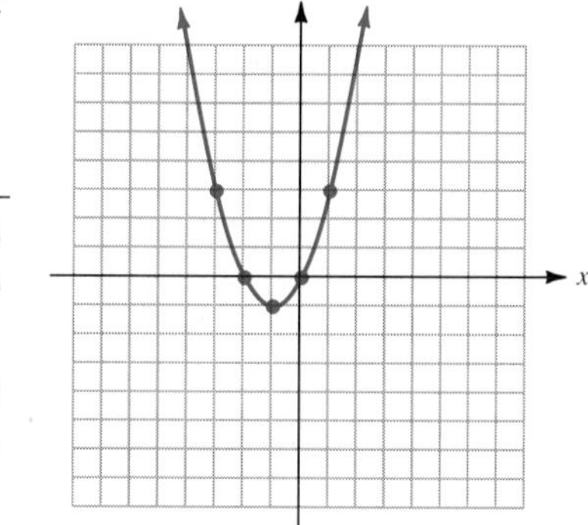

7. $y = x^2 + x$

x	y
−2	2
−1	0
0	0
1	2
2	6

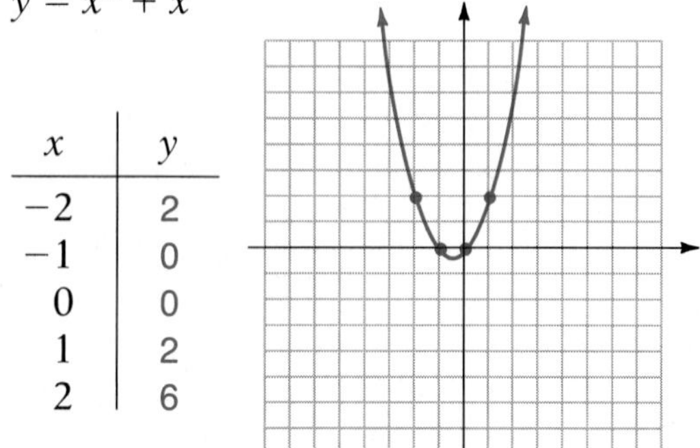

8. $y = x^2 - 3x$

x	y
−1	4
0	0
1	−2
2	−2
3	0

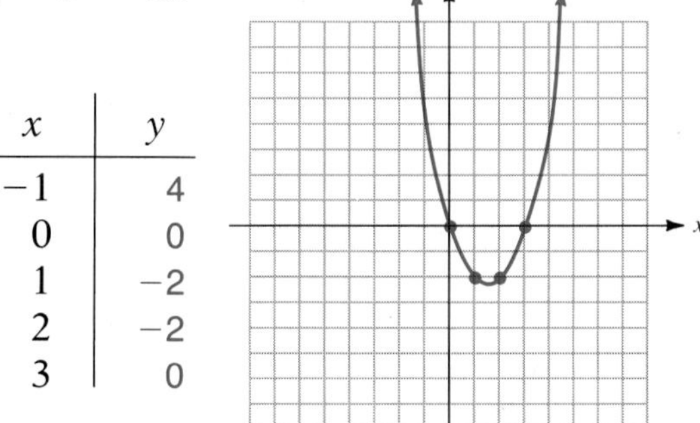

9. $y = x^2 - 2x - 3$

x	y
−1	0
−0	−3
1	−4
2	−3
3	0

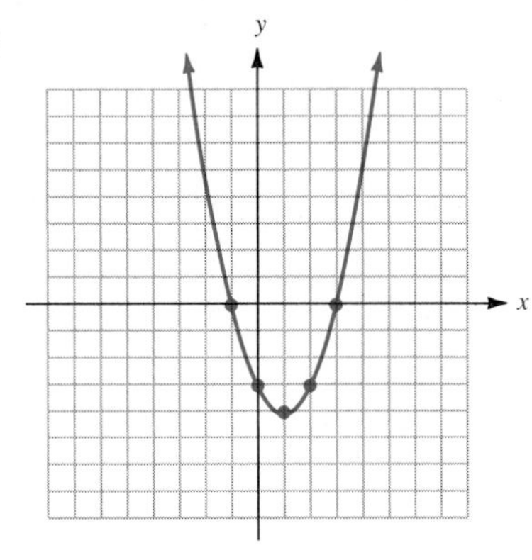

10. $y = x^2 - 5x + 6$

x	y
0	6
1	2
2	0
3	0
4	2

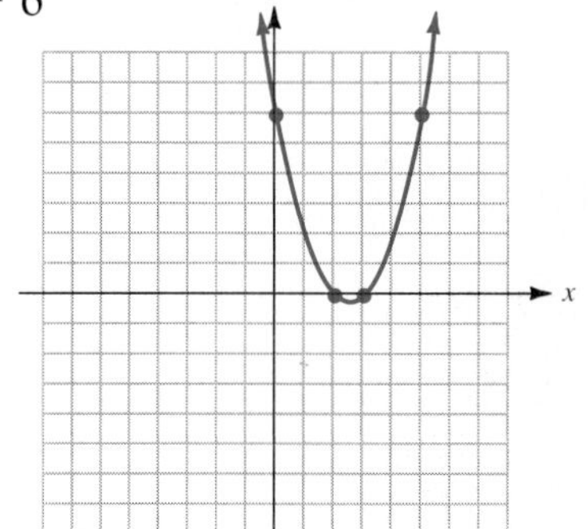

11. $y = x^2 - x - 6$

x	y
−1	−4
0	−6
1	−6
2	−4
3	0

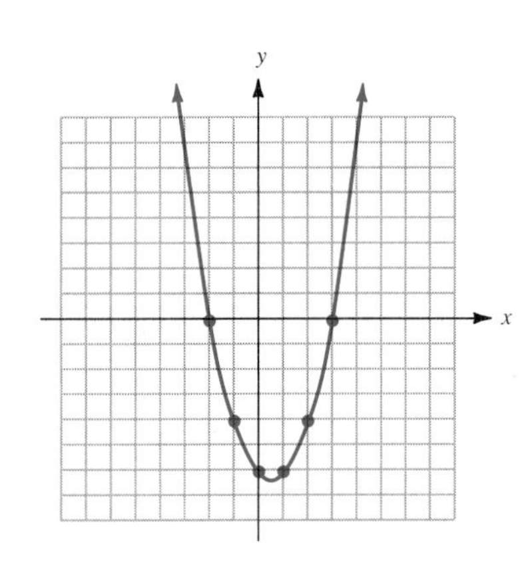

12. $y = x^2 + 3x - 4$

x	y
−4	0
−3	−4
−2	−6
−1	−6
0	−4

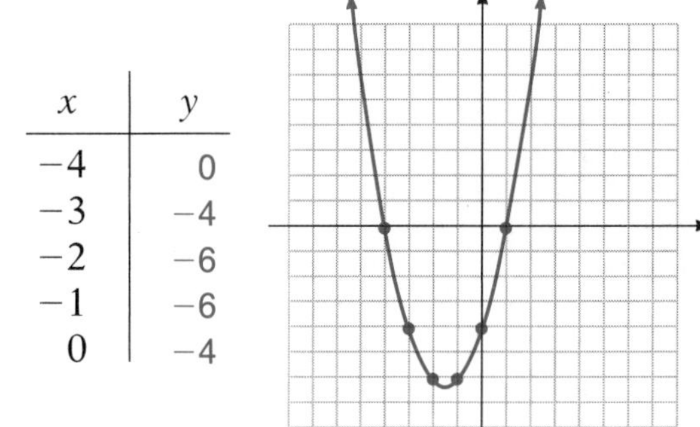

13. $y = -x^2 + 2$

x	y
−2	−2
−1	1
0	2
1	1
2	−2

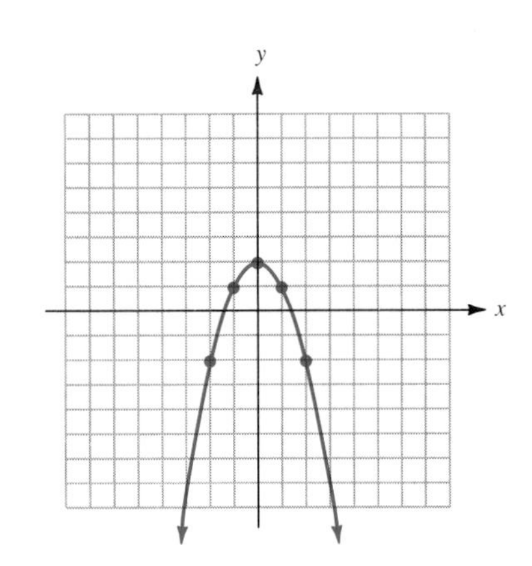

14. $y = -x^2 - 2$

x	y
−2	−6
−1	−3
0	−2
1	−3
2	−6

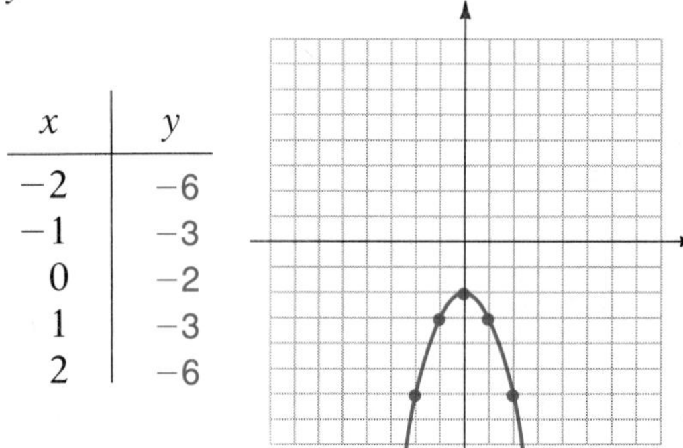

15. $y = -x^2 - 4x$

x	y
−4	0
−3	3
−2	4
−1	3
0	0

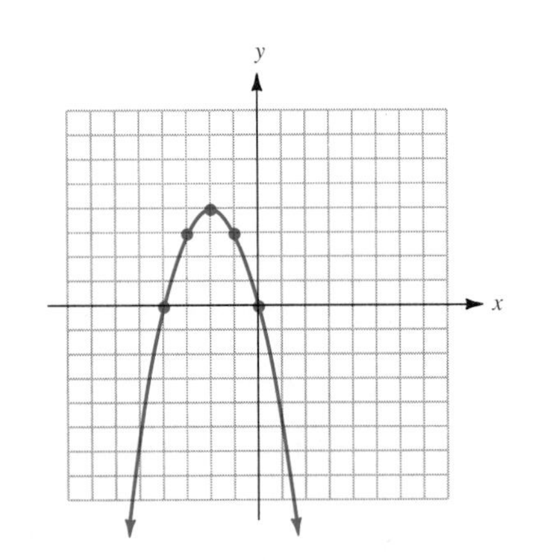

16. $y = -x^2 + 2x$

x	y
−1	−3
0	0
1	1
2	0
3	−3

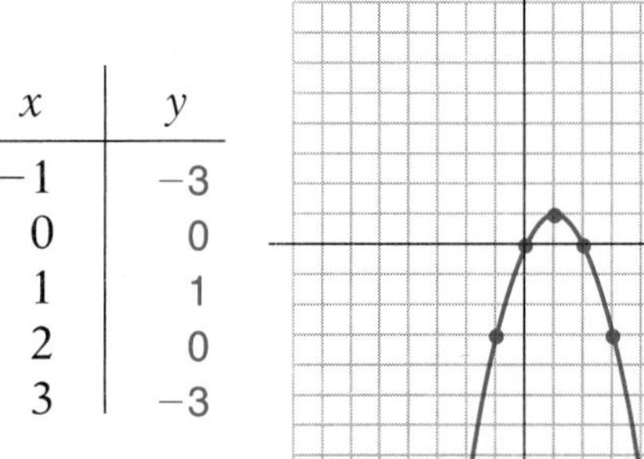

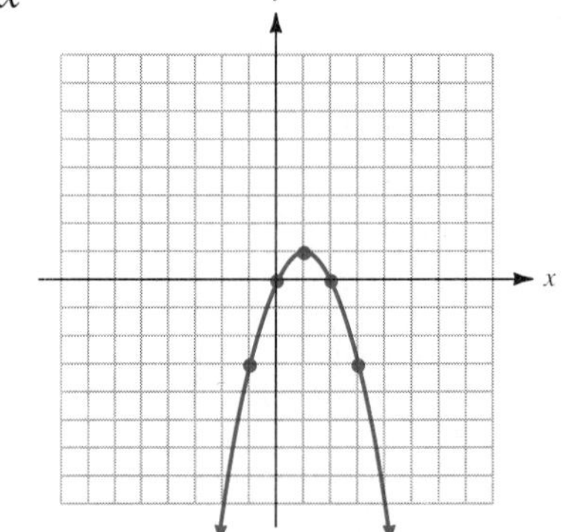

Match each graph with the correct equation at the right.

(*a*) $y = -x^2 + 1$

(*b*) $y = 2x$

(*c*) $y = x^2 - 4x$

(*d*) $y = -x + 1$

(*e*) $y = -x^2 + 3x$

(*f*) $y = x^2 + 1$

(*g*) $y = x + 1$

(*h*) $y = 2x^2$

17. 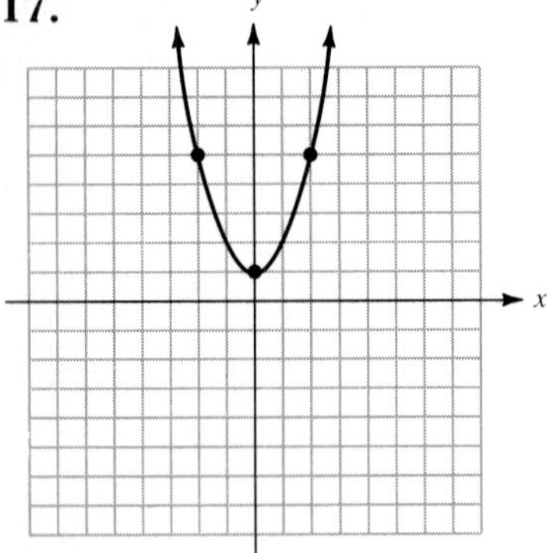

f

18. 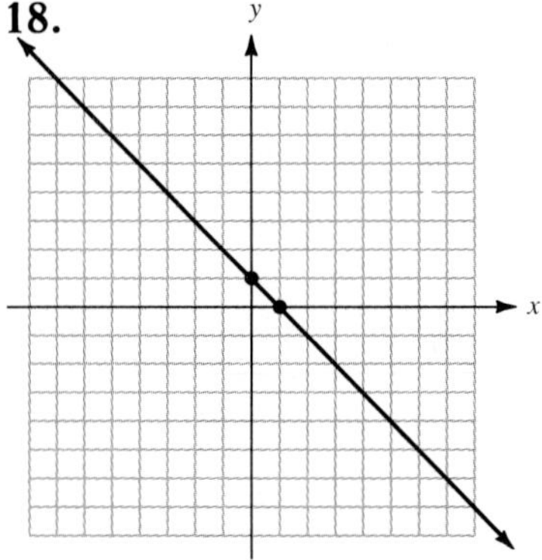

d

19. 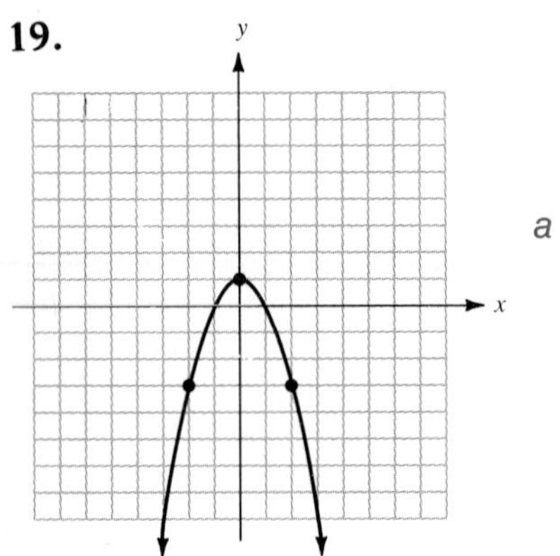

a

20. 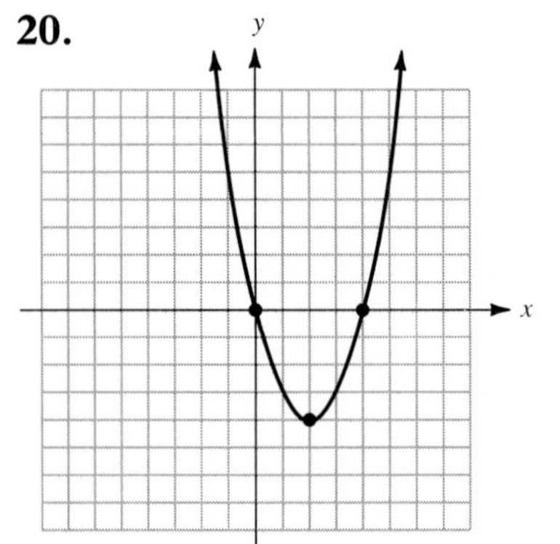

c

21. 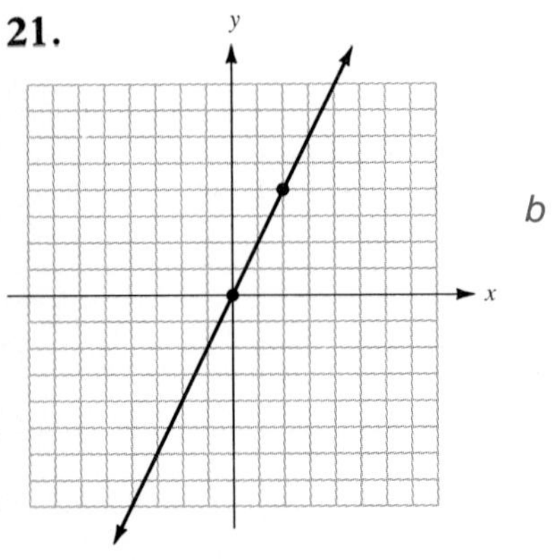

b

22. 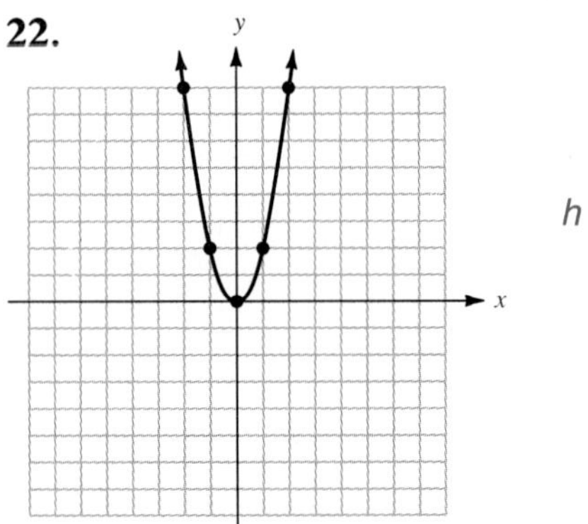

h

23. 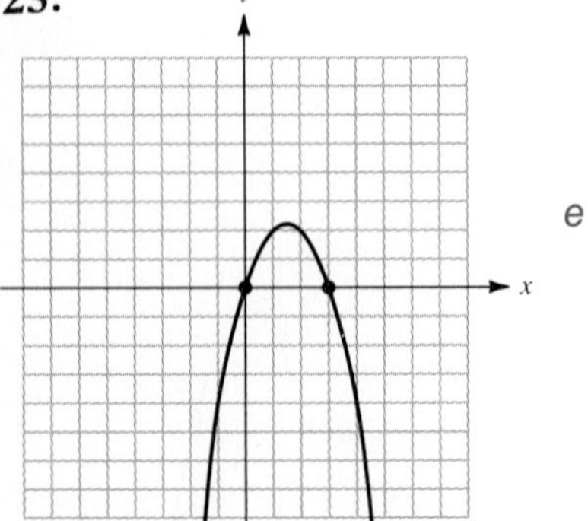

e

24. 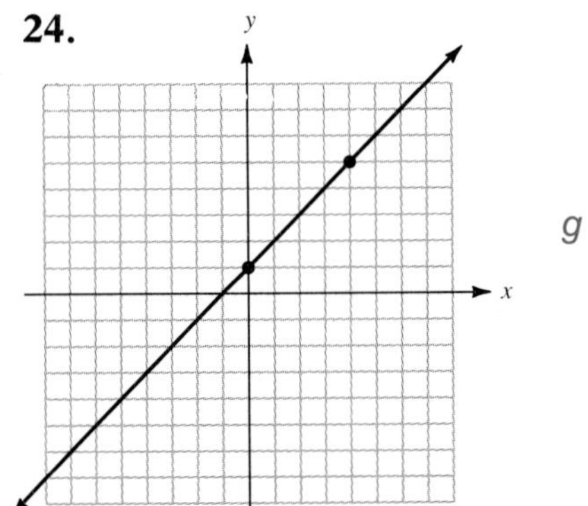

g

Answers

1. $y = x^2 + 1$

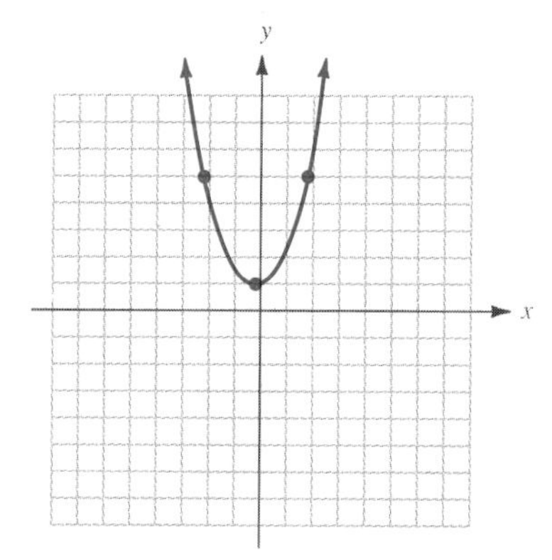

3. $y = x^2 - 4$

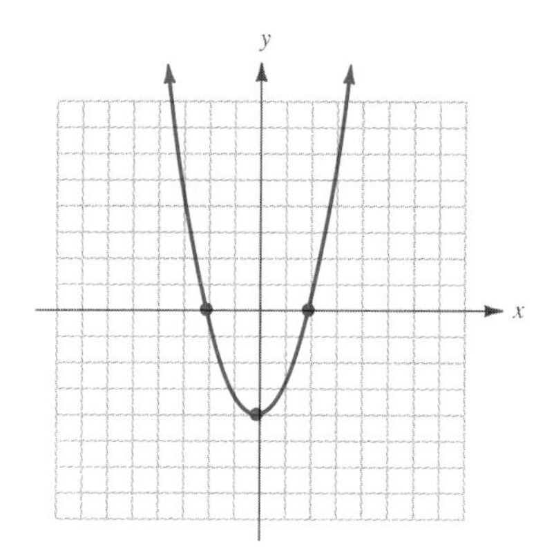

5. $y = x^2 - 4x$

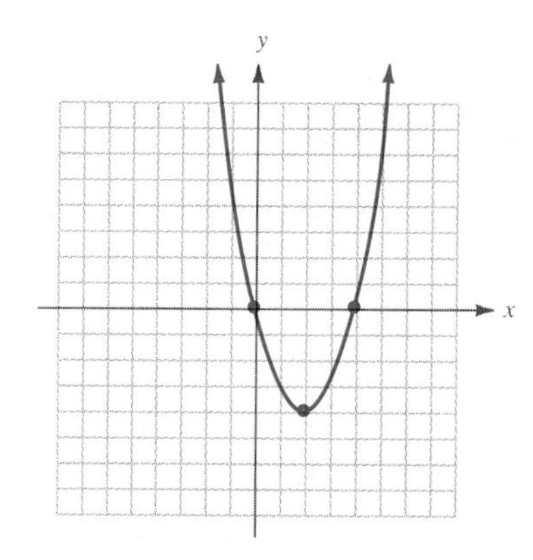

7. $y = x^2 + x$

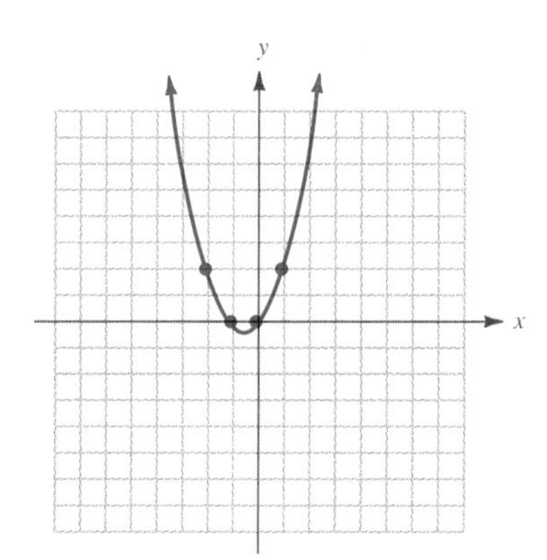

9. $y = x^2 - 2x - 3$

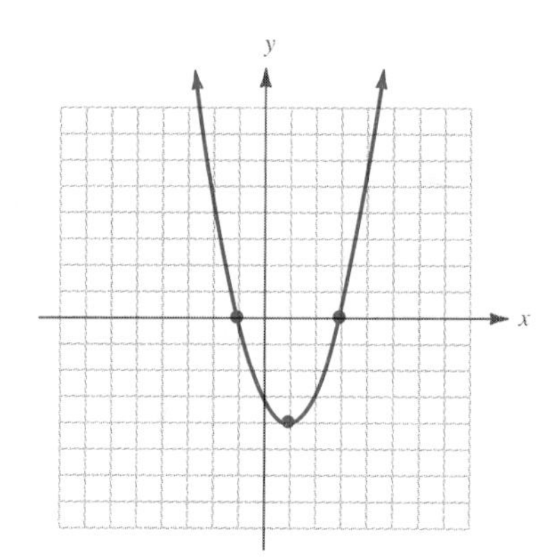

11. $y = x^2 - x - 6$

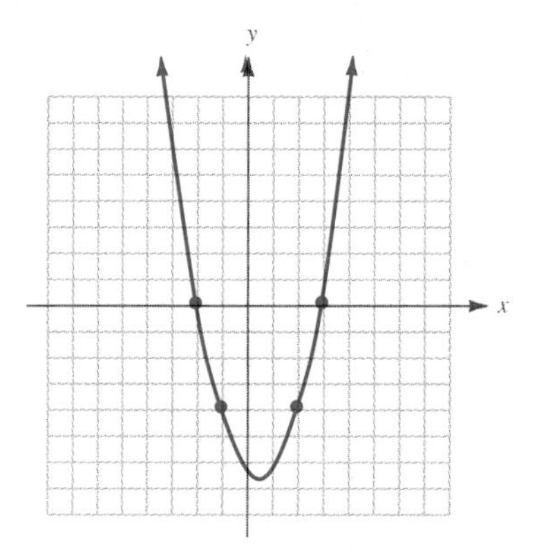

13. $y = -x^2 + 2$

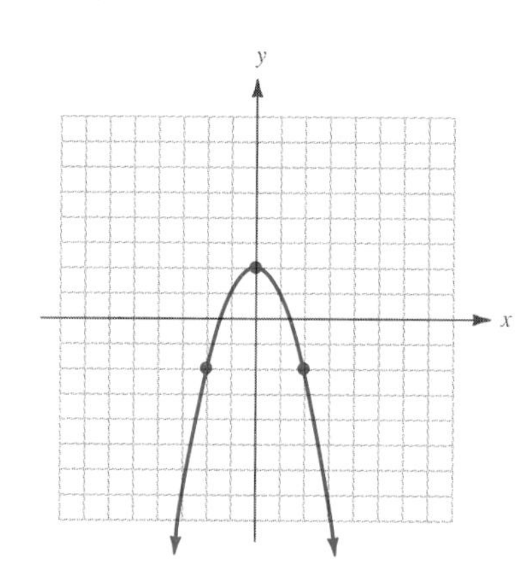

15. $y = -x^2 - 4x$

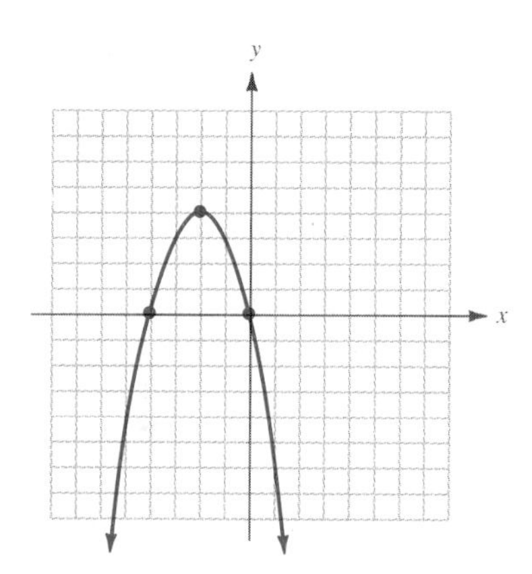

17. f

19. a

21. b

23. e

Summary

Solving Quadratic Equations [10.1] to [10.3]

Solving Equations of the Form

$ax^2 = k$

Divide both sides of the equation by a. The equation can then be solved by taking the square root of both sides.

$$x^2 = \frac{k}{a}$$

so

$$x = \sqrt{\frac{k}{a}} \quad \text{or} \quad x = -\sqrt{\frac{k}{a}}$$

Solve $4x^2 = 13$.

$$x^2 = \frac{13}{4}$$

$$x = \sqrt{\frac{13}{4}} \text{ or } -\sqrt{\frac{13}{4}}$$

Simplifying gives

$$x = \frac{\sqrt{13}}{2} \text{ or } -\frac{\sqrt{13}}{2}$$

Completing the Square To solve a quadratic equation by completing the square:

1. Write the equation in the form

   $ax^2 + bx = k$

   so that the variable terms are on the left side and the constant is on the right side.
2. If the leading coefficient (of x^2) is not 1, divide both sides by that coefficient.
3. Add the square of one-half the middle (x) coefficient to both sides of the equation.
4. The left side of the equation is now a perfect-square trinomial. Factor and solve as before.

To solve:

$$2x^2 + 2x - 1 = 0$$

$$2x^2 + 2x = 1$$

$$x^2 + x = \frac{1}{2}$$

$$x^2 + x + \left(\frac{1}{2}\right)^2 = \frac{1}{2} + \left(\frac{1}{2}\right)^2$$

or

$$x^2 + x + \frac{1}{4} = \frac{3}{4}$$

$$\left(x + \frac{1}{2}\right)^2 = \frac{3}{4}$$

$$x + \frac{1}{2} = \pm\sqrt{\frac{3}{4}}$$

or

$$x + \frac{1}{2} = \pm\frac{\sqrt{3}}{2}$$

$$x = \frac{-1 \pm \sqrt{3}}{2}$$

The Quadratic Formula To solve an equation by formula:

1. Rewrite the equation in standard form.

   $ax^2 + bx + c = 0$

2. If a common factor exists, divide both sides of the equation by that common factor.

To solve:

$x^2 - 2x = 4$

Write the equation as

$$x^2 - 2x - 4 = 0$$

$a = 1 \quad b = -2 \quad c = -4$

3. Identify the coefficients a, b, and c.
4. Substitute the values for a, b, and c into the quadratic formula.

$$x = \frac{-b \pm \sqrt{b^2 - 4ac}}{2a}$$

5. Simplify the right side of the expression formed in step 4 to write the solutions for the original equation.

$$x = \frac{-(-2) \pm \sqrt{(-2)^2 - 4(1)(-4)}}{2(1)}$$

$$= \frac{2 \pm \sqrt{20}}{2}$$

$$= \frac{2 \pm 2\sqrt{5}}{2} = \frac{2(1 \pm \sqrt{5})}{2}$$

$$= 1 \pm \sqrt{5}$$

The Pythagorean Theorem [10.4]

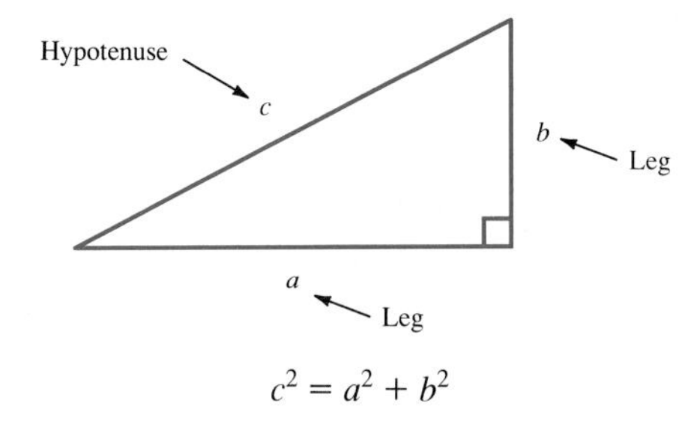

$c^2 = a^2 + b^2$

In words, the square of the length of the hypotenuse is equal to the sum of the squares of the lengths of the legs.

Find length x:

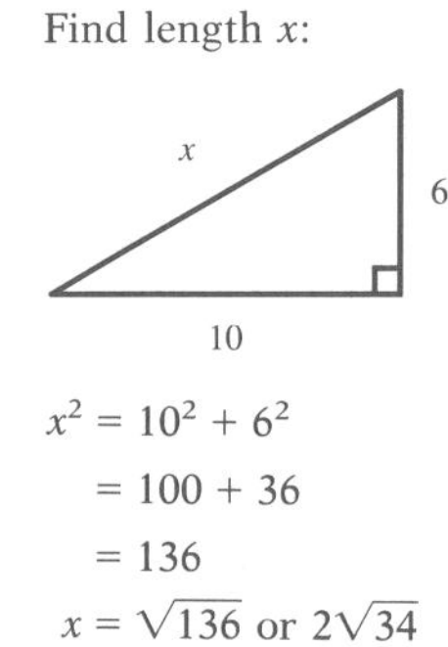

$$x^2 = 10^2 + 6^2$$
$$= 100 + 36$$
$$= 136$$
$$x = \sqrt{136} \text{ or } 2\sqrt{34}$$

Graphing Quadratic Equations [10.5]

To graph equations of the form

$y = ax^2 + bx + c$

1. Form a table of values by choosing convenient values for x and finding the corresponding values for y.
2. Plot the points from the table of values.
3. Draw a smooth curve between the points.

The graph of a quadratic equation will always be a parabola. The parabola opens upward if a, the coefficient of the x^2 term, is positive.

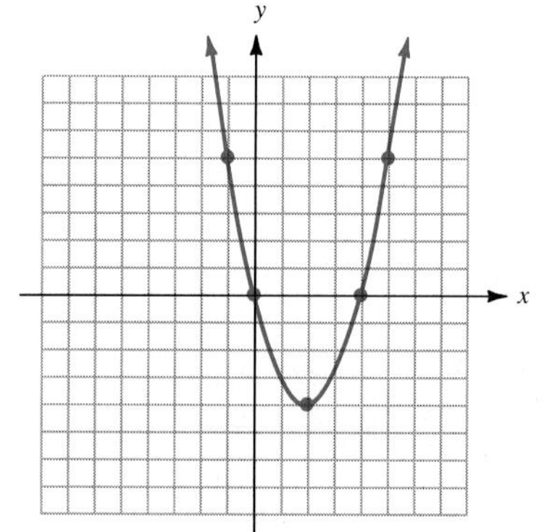

$y = x^2 - 4x$

x	y
−1	5
0	0
1	−3
2	−4
3	−3
4	0
5	5

The parabola opens downward if a, the coefficient of the x^2 term, is negative.

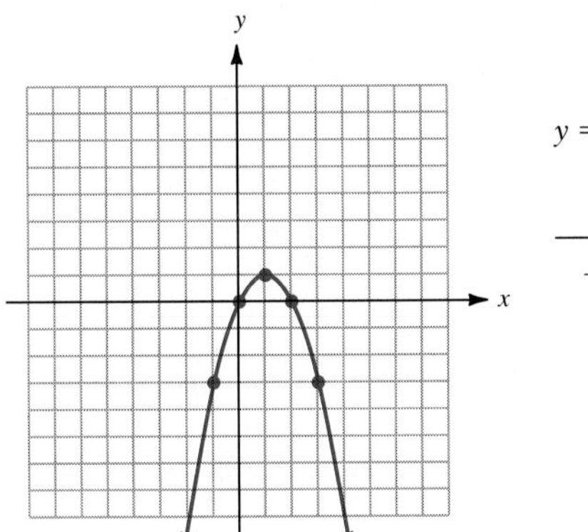

$y = -x^2 + 2x$

x	y
−1	−3
0	0
1	1
2	0
3	−3

Summary Exercises Chapter 10

This summary exercise set is provided to give you practice with each of the objectives of the chapter. Each exercise is keyed to the appropriate chapter section. The answers are provided in the instructor's manual. Your instructor will give you guidelines on how to best use these exercises in your instructional setting.

[10.1] Solve each of the equations for x by the square root method.

1. $x^2 = 10$
$\pm\sqrt{10}$

2. $x^2 = 48$
$\pm 4\sqrt{3}$

3. $x^2 - 20 = 0$
$\pm 2\sqrt{5}$

4. $x^2 + 2 = 8$
$\pm\sqrt{6}$

5. $(x - 1)^2 = 5$
$1 \pm \sqrt{5}$

6. $(x + 2)^2 = 8$
$-2 \pm 2\sqrt{2}$

7. $(x + 3)^2 = 5$
$-3 \pm \sqrt{5}$

8. $64x^2 = 25$
$\pm\frac{5}{8}$

9. $4x^2 = 27$
$\frac{\pm 3\sqrt{3}}{2}$

10. $9x^2 = 20$
$\frac{\pm 2\sqrt{5}}{3}$

11. $25x^2 = 7$
$\frac{\pm\sqrt{7}}{5}$

12. $7x^2 = 3$
$\frac{\pm\sqrt{21}}{7}$

[10.2] Solve each of the equations by completing the square.

13. $x^2 - 3x - 10 = 0$
$-2, 5$

14. $x^2 - 8x + 15 = 0$
$3, 5$

15. $x^2 - 5x + 2 = 0$
$\frac{5 \pm \sqrt{17}}{2}$

16. $x^2 - 2x - 2 = 0$
$1 \pm \sqrt{3}$

17. $x^2 - 4x - 4 = 0$
$2 \pm 2\sqrt{2}$

18. $x^2 + 3x = 7$
$\frac{-3 \pm \sqrt{37}}{2}$

19. $x^2 - 4x = -2$
$2 \pm \sqrt{2}$

20. $x^2 + 3x = 5$
$\frac{-3 \pm \sqrt{29}}{2}$

21. $x^2 - x = 7$
$\frac{1 \pm \sqrt{29}}{2}$

22. $2x^2 + 6x = 12$
$\frac{-3 \pm \sqrt{33}}{2}$

23. $2x^2 - 4x - 7 = 0$
$\frac{2 \pm 3\sqrt{2}}{2}$

24. $3x^2 + 5x + 1 = 0$
$\frac{-5 \pm \sqrt{13}}{6}$

[10.3] Solve each of the equations using the quadratic formula.

25. $x^2 - 5x - 14 = 0$ $\quad -2, 7$

26. $x^2 - 8x + 16 = 0$ $\quad 4$

27. $x^2 + 5x - 3 = 0$ $\quad \frac{-5 \pm \sqrt{37}}{2}$

28. $x^2 - 7x - 1 = 0$ $\quad \frac{7 \pm \sqrt{53}}{2}$

29. $x^2 - 6x + 1 = 0$ $\quad 3 \pm 2\sqrt{2}$

30. $x^2 - 3x + 5 = 0$ $\quad$ No real solution

31. $3x^2 - 4x = 2$ $\quad \frac{2 \pm \sqrt{10}}{3}$

32. $2x - 3 = \frac{3}{x}$ $\quad \frac{3 \pm \sqrt{33}}{4}$

33. $(x - 1)(x + 4) = 3$ $\quad \dfrac{-3 \pm \sqrt{37}}{2}$

34. $x^2 - 5x + 7 = 5$ $\quad \dfrac{5 \pm \sqrt{17}}{2}$

35. $2x^2 - 8x = 12$ $\quad 2 \pm \sqrt{10}$

36. $5x^2 = 15 - 15x$ $\quad \dfrac{-3 \pm \sqrt{21}}{2}$

Solve by factoring or by any of the methods of this chapter.

37. $5x^2 = 3x$ $\quad 0, \dfrac{3}{5}$

38. $(2x - 3)(x + 5) = -11$ $\quad -4, \dfrac{1}{2}$

39. $(x - 1)^2 = 10$ $\quad 1 \pm \sqrt{10}$

40. $2x^2 = 7$ $\quad \dfrac{\pm\sqrt{14}}{2}$

41. $2x^2 = 5x + 4$ $\quad \dfrac{5 \pm \sqrt{57}}{4}$

42. $2x^2 - 4x = 30$ $\quad 5, -3$

43. $2x^2 = 5x + 7$ $\quad \dfrac{7}{2}, -1$

44. $3x^2 - 4x = 2$ $\quad \dfrac{2 \pm \sqrt{10}}{3}$

45. $3x^2 + 6x - 15 = 0$ $\quad -1 \pm \sqrt{6}$

46. $x^2 - 3x = 2(x + 5)$ $\quad \dfrac{5 \pm \sqrt{65}}{2}$

47. $x - 2 = \dfrac{2}{x}$ $\quad 1 \pm \sqrt{3}$

48. The perimeter of a square is numerically 2 less than its area. Find the length of one side. (Approximate your answer to three decimal places, using a calculator or a table of square-roots.) 4.449

[10.4] Find length x in each triangle.

49.

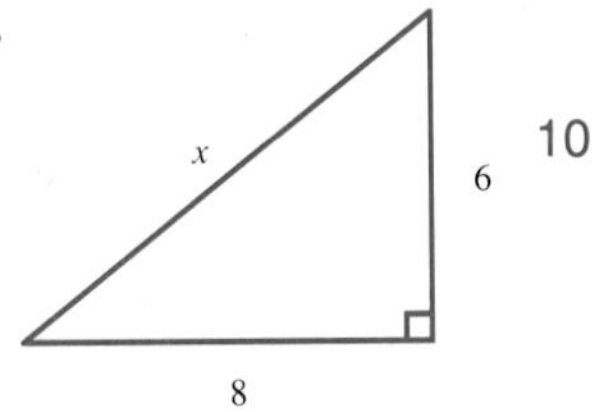

10

50.

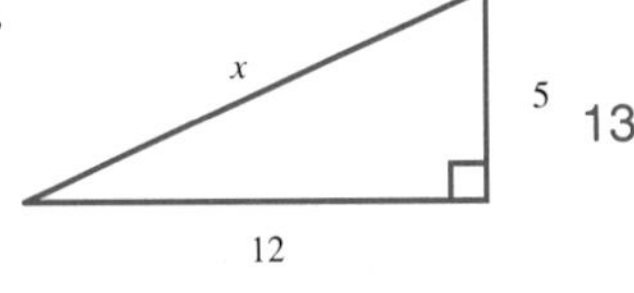

13

51.

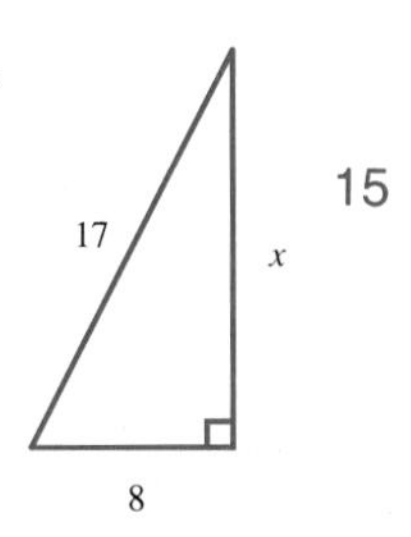

15

52.

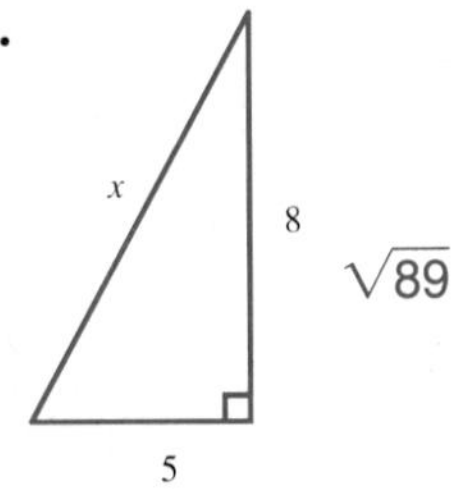

$\sqrt{89}$

53.

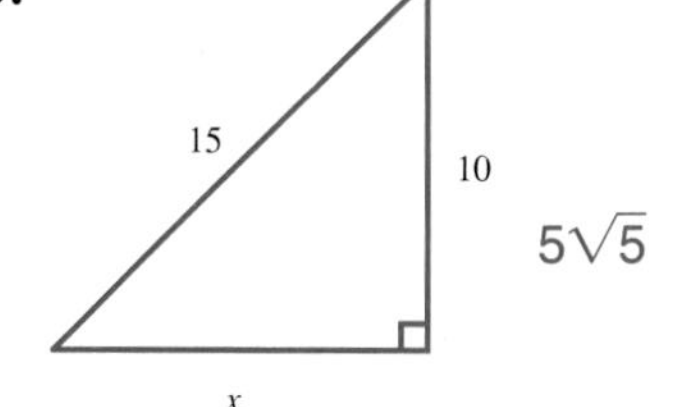

$5\sqrt{5}$

54.

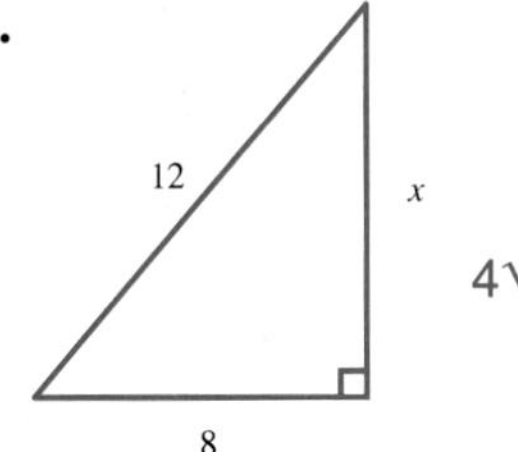

$4\sqrt{5}$

[10.4] Solve each of the following word problems. Approximate your answer to one decimal place where necessary.

55. Find the length of the diagonal of a rectangle whose length is 12 in and whose width is 9 in. 15 in

56. Find the length of a rectangle whose diagonal has a length of 10 cm and whose width is 8.7 cm

57. How long must a guy wire be to run from the top of an 18-ft pole to a point on level ground 16 ft away from the base of the pole? 24.1 ft

58. The length of one leg of a right triangle is 1 in more than the length of the other. If the length of the hypotenuse of the triangle is 6 in, what are the lengths of the two legs? 3.7 in, 4.7 in

[10.5] Graph each quadratic equation after completing the table of values.

59. $y = x^2 + 3$

x	y
−2	7
−1	4
0	3
1	4
2	7

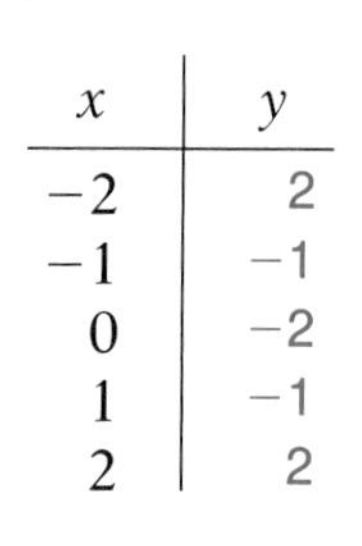

60. $y = x^2 - 2$

x	y
−2	2
−1	−1
0	−2
1	−1
2	2

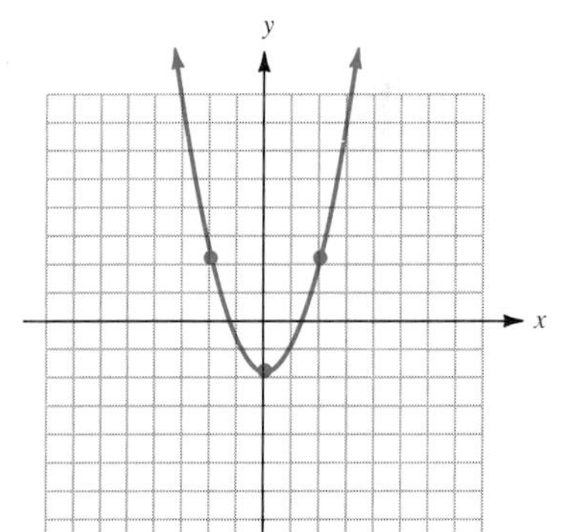

61. $y = x^2 - 3x$

x	y
−1	4
0	0
1	−2
2	−2
3	0

62. $y = x^2 + 4x$

x	y
−4	0
−3	−3
−2	−4
−1	−3
0	0

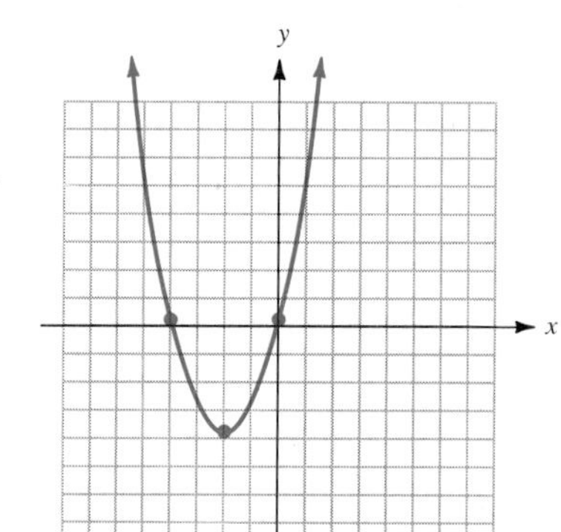

63. $y = x^2 - x - 2$

x	y
−1	0
0	−2
1	−2
2	0
3	4

64. $y = x^2 - 4x + 3$

x	y
0	3
1	0
2	−1
3	0
4	3

65. $y = x^2 + 2x - 3$

x	y
−3	0
−2	−3
−1	−4
0	−3
1	0

66. $y = 2x^2$

x	y
−2	8
−1	2
0	0
1	2
2	8

67. $y = 2x^2 - 3$

x	y
−2	5
−1	−1
0	−3
1	−1
2	5

68. $y = -x^2 + 3$

x	y
−2	−1
−1	2
0	3
1	2
2	−1

69. $y = -x^2 - 2$

x	y
−2	−6
−1	−3
0	−2
1	−3
2	−6

70. $y = -x^2 + 4x$

x	y
0	0
1	3
2	4
3	3
4	0

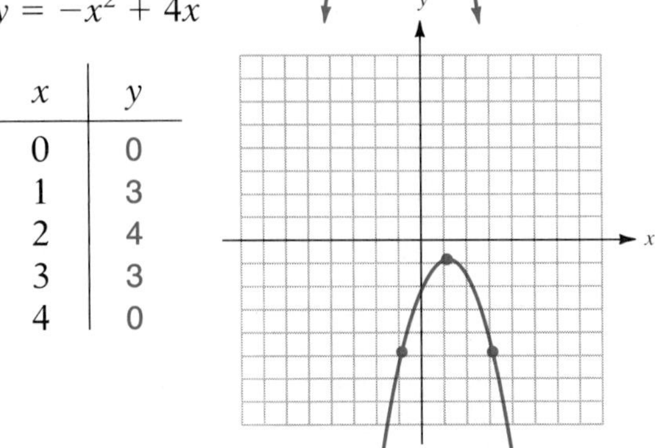

Self-Test for Chapter Ten

The purpose of this self-test is to help you check your progress and to review for a chapter test in class. Allow yourself about an hour to take the test. When you are done, check your answers in the back of the book. If you missed any problems, be sure to go back and review the appropriate sections in the chapter and the exercises that are provided.

Solve each of the equations for x.

1. $x^2 = 15$ $\pm\sqrt{15}$

2. $x^2 - 8 = 0$ $\pm 2\sqrt{2}$

3. $(x - 1)^2 = 7$ $1 \pm \sqrt{7}$

4. $9x^2 = 10$ $\dfrac{\pm\sqrt{10}}{3}$

Solve each of the equations by completing the square.

5. $x^2 - 2x - 8 = 0$ $4, -2$

6. $x^2 + 3x - 1 = 0$ $\dfrac{-3 \pm \sqrt{13}}{2}$

7. $x^2 + 2x - 5 = 0$ $-1 \pm \sqrt{6}$

8. $2x^2 - 5x + 1 = 0$ $\dfrac{5 \pm \sqrt{17}}{4}$

Solve each of the equations by using the quadratic formula.

9. $x^2 - 2x - 3 = 0$ $-1, 3$

10. $x^2 - 6x + 9 = 0$ 3

11. $x^2 - 5x = 2$ $\dfrac{5 \pm \sqrt{33}}{2}$

12. $2x^2 = 2x + 5$ $\dfrac{1 \pm \sqrt{11}}{2}$

13. $2x - 1 = \dfrac{4}{x}$ $\dfrac{1 \pm \sqrt{33}}{4}$

14. $(x - 1)(x + 3) = 2$ $-1 \pm \sqrt{6}$

Find length x in each triangle.

15.

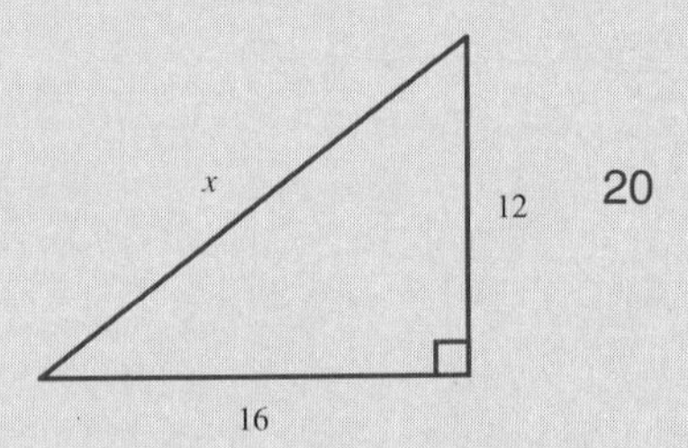

20

16.

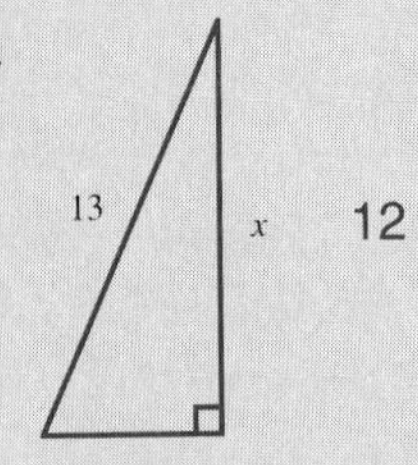

12

17.

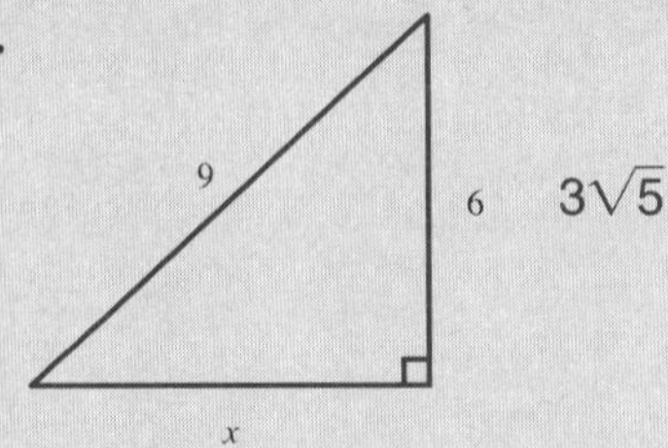

$3\sqrt{5}$

18.

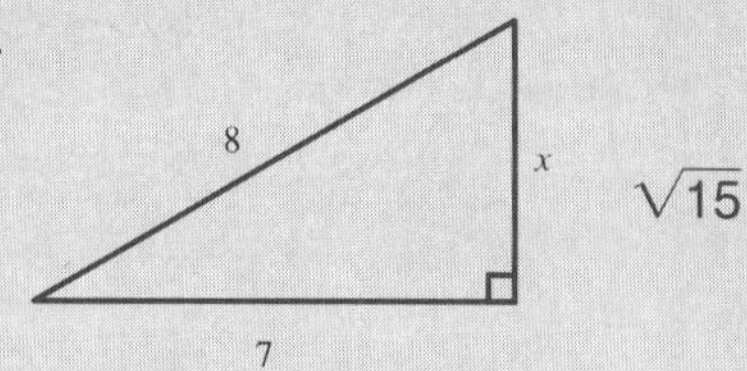

$\sqrt{15}$

Solve the following word problem.

19. If the length of the diagonal of a rectangle is 12 cm and the width of the rectangle is 7 cm, what is the length of the rectangle?
Approximately 9.747 cm

Graph each quadratic equation after completing the given table of values.

20. $y = x^2 + 4$

x	y
−2	8
−1	5
0	4
1	5
2	8

21. $y = x^2 - 2x$

x	y
−1	3
0	0
1	−1
2	0
3	3

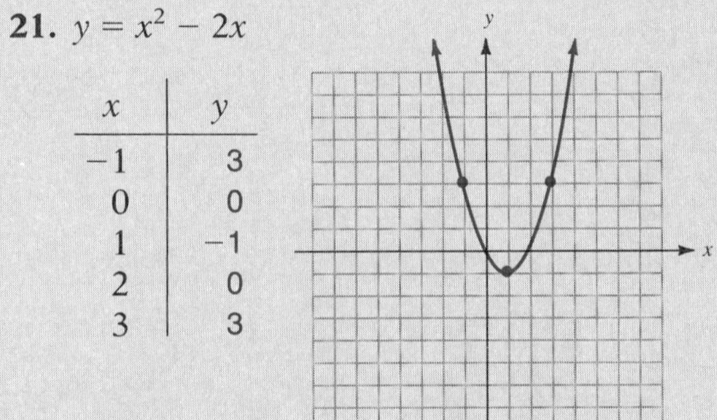

22. $y = x^2 - 3$

x	y
−2	1
−1	−2
0	−3
1	−2
2	1

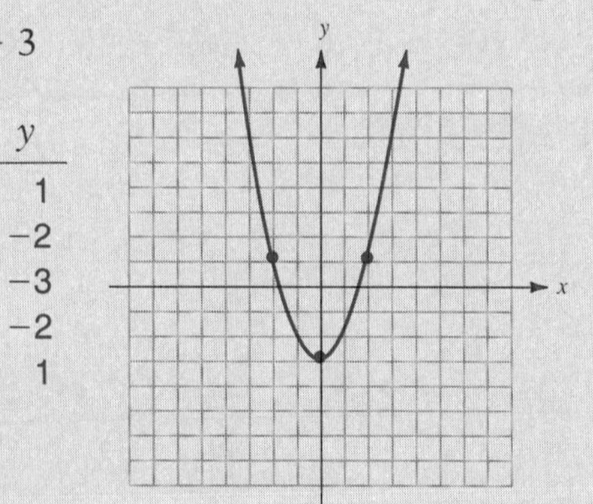

23. $y = x^2 + x - 2$

x	y
−2	0
−1	−2
0	−2
1	0
2	4

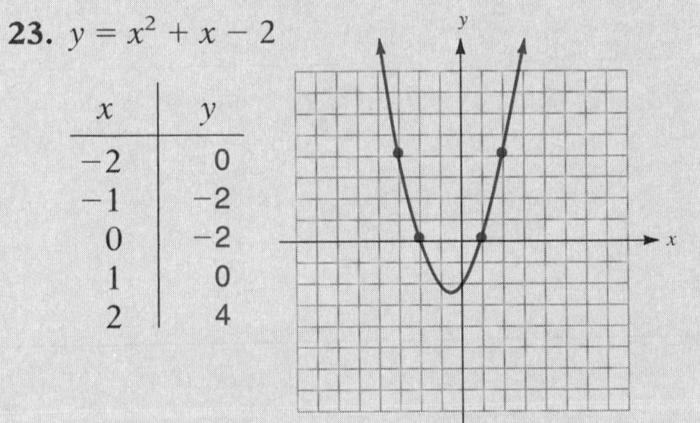

24. $y = -x^2 + 4$

x	y
−2	0
−1	3
0	4
1	3
2	0

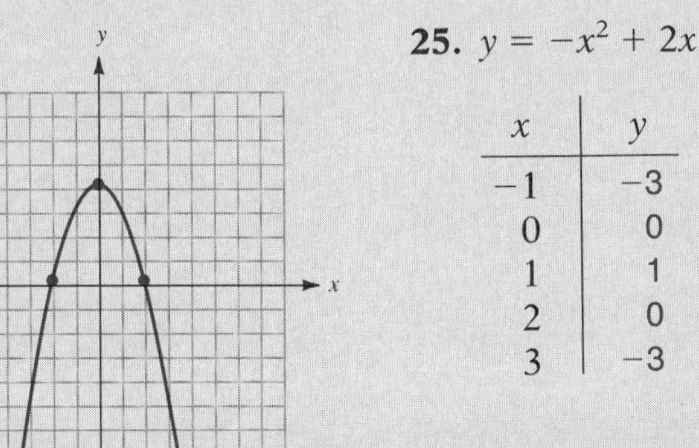

25. $y = -x^2 + 2x$

x	y
−1	−3
0	0
1	1
2	0
3	−3

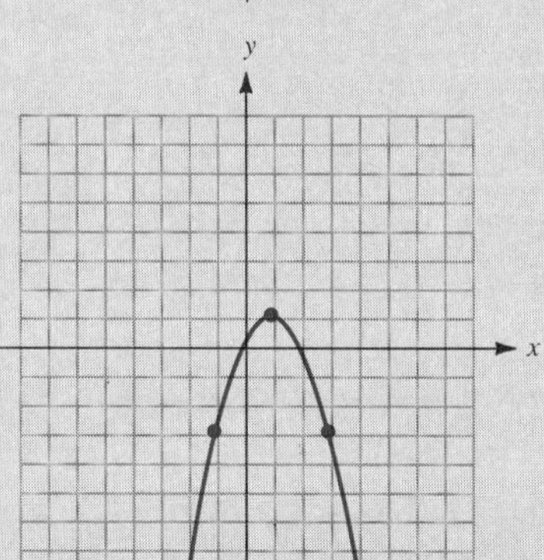

Cumulative Test for Chapters Nine and Ten

This test is provided to help you in the process of review of the previous chapters. Answers are provided in the back of the book. If you missed any problems, be sure to go back and review the appropriate chapter sections.

Use the properties of exponents to simplify each of the following expressions. Assume that all variables represent nonzero real numbers.

1. $(m^4)^5$ m^{20}

2. $(3xy^4)^2$ $9x^2y^8$

3. $(2a^3)^2(a^2)^4$ $4a^{14}$

4. $\left(\frac{r^4}{s^3}\right)^2$ $\frac{r^8}{s^6}$

Evaluate each of the following expressions.

5. $5x^0$ 5

6. $(5m^3n)^0$ 1

Simplify each expression, using positive exponents only.

7. x^{-8} $\frac{1}{x^8}$

8. $4a^{-5}$ $\frac{4}{a^5}$

9. r^4r^{-7} $\frac{1}{r^3}$

10. $\frac{y^{-5}}{y^5}$ $\frac{1}{y^{10}}$

Evaluate each root if possible.

11. $\sqrt{144}$ 12

12. $-\sqrt{144}$ -12

13. $\sqrt{-144}$ Not a real number

14. $\sqrt[3]{-27}$ -3

Simplify each of the following radical expressions. Assume that all variables represent positive real numbers.

15. $\sqrt{98x^4}$ $7x^2\sqrt{2}$

16. $\sqrt{150m^3n^2}$ $5mn\sqrt{6m}$

17. $\sqrt{\frac{12a^2}{25}}$ $\frac{2a\sqrt{3}}{5}$

18. $\sqrt{\frac{5y^3}{3}}$ $\frac{y\sqrt{15y}}{3}$

Simplify each of the following radical expressions by combining like terms.

19. $\sqrt{12} + 3\sqrt{27} - \sqrt{75}$ $6\sqrt{3}$

20. $a\sqrt{20} - 2\sqrt{45a^2}$ $-4a\sqrt{5}$

Simplify each of the following radical expressions.

21. $3\sqrt{2a} \cdot 5\sqrt{6a}$ $30a\sqrt{3}$

22. $\sqrt{5}\,(2\sqrt{3} - 3\sqrt{5})$ $2\sqrt{15} - 15$

23. $(\sqrt{2} - 5)(\sqrt{2} + 3)$ $-13 - 2\sqrt{2}$

24. $\dfrac{\sqrt{8x^3}}{\sqrt{3}}$

$\dfrac{2x\sqrt{6x}}{3}$

25. $\dfrac{12}{\sqrt{10} - 2}$

$2\sqrt{10} + 4$

Solve each of the following equations by factoring or by any of the methods of Chapter 10.

26. $x^2 - 72 = 0$

$\pm 6\sqrt{2}$

27. $x^2 + 6x - 3 = 0$

$-3 \pm 2\sqrt{3}$

28. $(x - 2)^2 = 7$

$2 \pm \sqrt{7}$

29. $3x^2 = 6x + 45$

$5, -3$

30. $2x^2 - 3x = 2(x + 1)$

$\dfrac{5 \pm \sqrt{41}}{4}$

31. $x^2 - 2x = 24$

$-4, 6$

32. $4x^2 - 10x = 8$

$\dfrac{5 \pm \sqrt{57}}{4}$

33. $2x - 2 = \dfrac{3}{x}$

$\dfrac{1 \pm \sqrt{7}}{2}$

Solve the following application. Approximate your answer to the nearest tenth of an inch.

34. The length of a rectangle is 2 in less than 3 times its width. If the area of the rectangle is 10 in^2, find the dimensions of the rectangle. 2.2 in, 4.6 in

Find length x in the triangle below, using the Pythagorean theorem.

35.

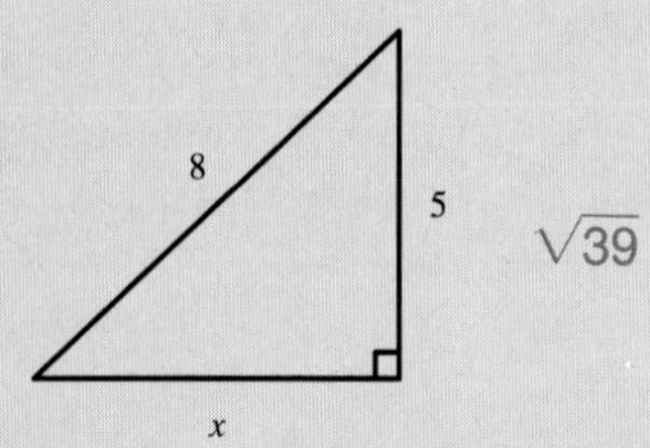

$\sqrt{39}$

Solve the following application by using the Pythagorean theorem. Approximate your answer to the nearest tenth of a foot.

36. The length of one leg of a right triangle is 2 ft more than the length of the other. If the length of the hypotenuse of the triangle is 6 ft, find the lengths of the two legs. Approximate your answer to the nearest hundredth of a foot. 3.12 ft, 5.12 ft

Graph each of the following quadratic equations after completing the table of values.

37. $y = x^2 - 2$

x	y
−2	2
−1	−1
0	−2
1	−1
2	2

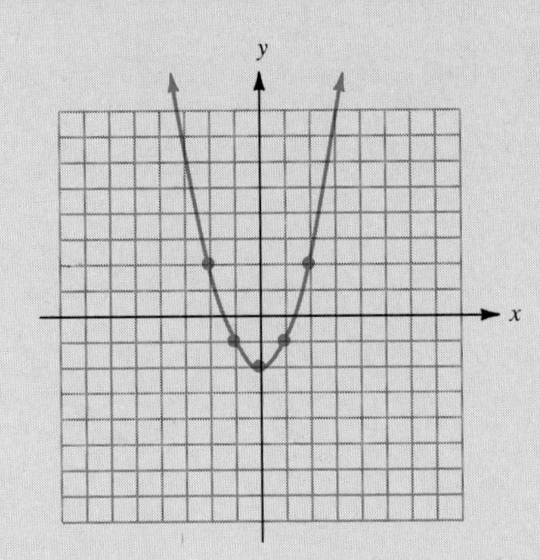

38. $y = x^2 - 4x$

x	y
0	0
1	−3
2	−4
3	−3
4	0

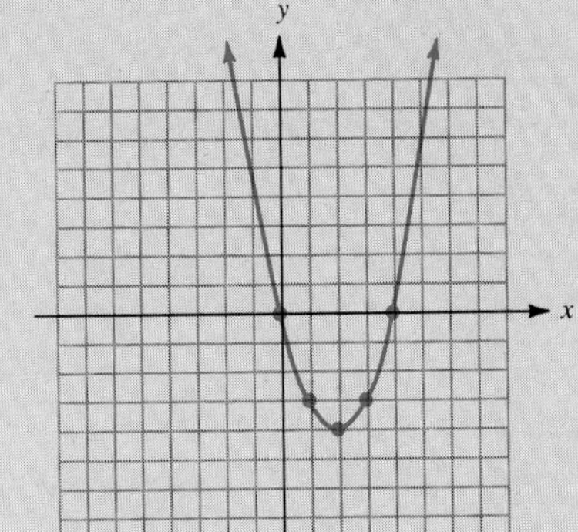

39. $y = x^2 - 3x + 2$

x	y
−1	6
0	2
1	0
2	0
3	2

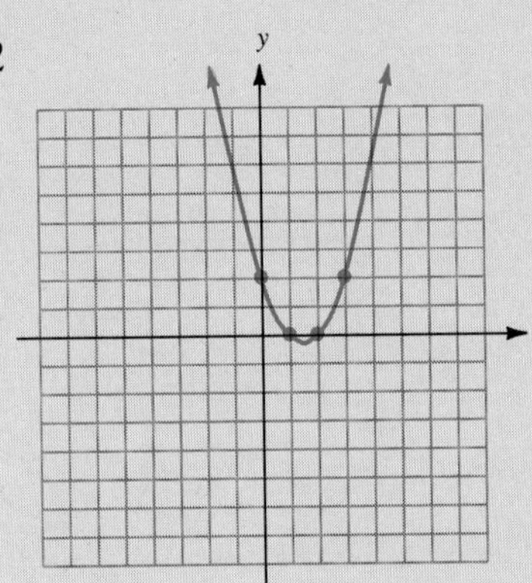

40. $y = -x^2 + 4$

x	y
−2	0
−1	3
0	4
1	3
2	0

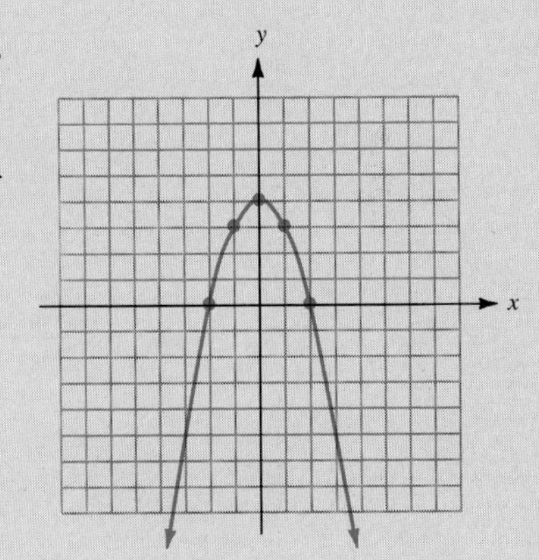

Appendix 1 The Arithmetic of Rational Numbers

This review is provided as an aid for students who wish to refresh their background in the arithmetic of fractions. You will find that much of what you will be asked to do with fractions or rational expressions in algebra has its basis in the methods you used in arithmetic.

Let's start the review with some terminology.

The Language of Fractions

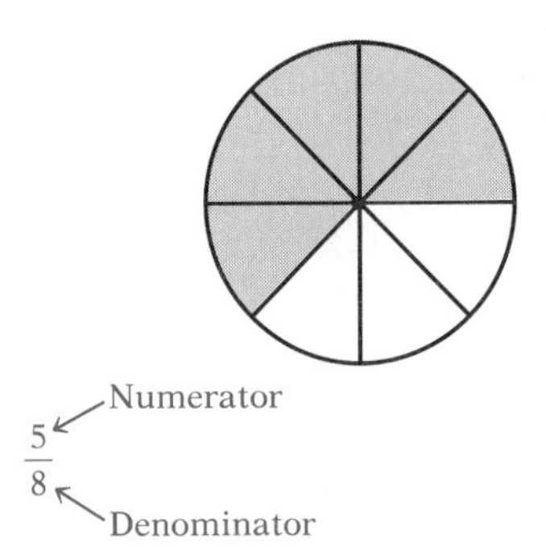

Fractions name a number of equal parts of a unit or whole. A fraction is written in the form $\frac{a}{b}$, where *a and* b *are whole numbers and* b *cannot be zero.*

$\frac{5}{8}$ ← Numerator (5), Denominator (8)

Denominator The number of equal parts the whole is divided into.

Numerator The number of equal parts of the whole that are used.

Proper Fraction A fraction whose numerator is less than its denominator. It names a number less than 1.

$\frac{2}{3}$ and $\frac{11}{15}$ are proper fractions.

Improper Fraction A fraction whose numerator is greater than or equal to its denominator. It names a number greater than or equal to 1.

$\frac{7}{5}$, $\frac{21}{20}$, and $\frac{8}{8}$ are improper fractions.

Mixed Number The sum of a whole number and a proper fraction.

$2\frac{1}{3}$ and $5\frac{7}{8}$ are mixed numbers. Note that $2\frac{1}{3}$ means $2 + \frac{1}{3}$.

Converting Mixed Numbers and Improper Fractions

To Change an Improper Fraction to a Mixed Number

1. Divide the numerator by the denominator. The quotient is the whole-number portion of the mixed number.
2. If there is a remainder, write the remainder over the original denominator. This gives the fractional portion of the mixed number.

Example 1

Change $\frac{22}{5}$ to a mixed number.

$$\begin{array}{r} 4 \\ 5\overline{)22} \\ \underline{20} \\ 2 \end{array}$$

Quotient (4)

Remainder (2)

$$\frac{22}{5} = 4\frac{2}{5}$$

To Change a Mixed Number to an Improper Fraction

1. Multiply the denominator of the fraction by the whole-number portion of the mixed number.
2. Add the numerator of the fraction to that product.
3. Write that sum over the original denominator to form the improper fraction.

Example 2

Change $5\frac{3}{4}$ to an improper fraction.

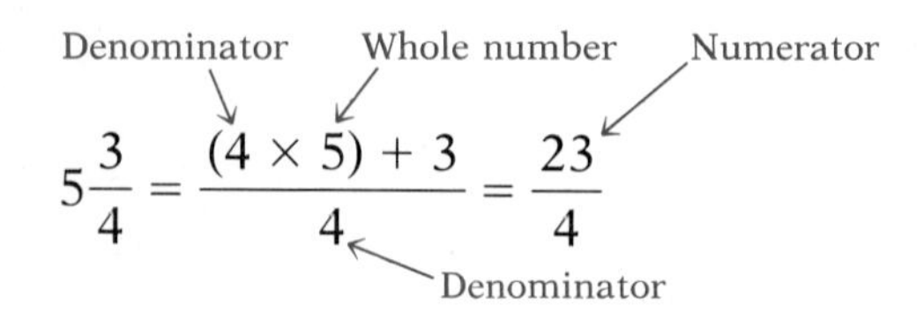

Equivalent Fractions

Equivalent Fractions Two fractions are equivalent (have equal value) if they are different names for the same number.

Cross Products

$$\frac{a}{b} = \frac{c}{d}$$

$a \times d$ and $b \times c$ are called the cross products.

If their cross products are equal, two fractions are equivalent.

$\frac{2}{3} = \frac{4}{6}$ because $2 \times 6 = 3 \times 4$.

The Fundamental Principle For the fraction $\frac{a}{b}$, and any nonzero number c,

$$\frac{a}{b} = \frac{a \times c}{b \times c}$$

$$\frac{1}{2} = \frac{1 \times 5}{2 \times 5} = \frac{5}{10}$$

$\frac{1}{2}$ and $\frac{5}{10}$ are equivalent fractions.

In words: We can multiply the numerator and denominator of a fraction by the same nonzero number. The result will be an equivalent fraction.

Simplest Form A fraction is in simplest form, or in lowest terms, if the numerator and denominator have no common factors other than 1. This means that the fraction has the smallest possible numerator and denominator.

$\frac{2}{3}$ is in simplest form.

$\frac{12}{18}$ is *not* in simplest form.

The numerator and denominator have the common factor 6.

To Write a Fraction in Simplest Form Divide the numerator and denominator by any common factors greater than 1 to write a fraction as an equivalent fraction in simplest form.

$$\frac{10}{15} = \frac{10 \div 5}{15 \div 5} = \frac{2}{3}$$

To Build a Fraction Multiply the numerator and denominator by any whole number greater than 1 to build a fraction to an equivalent fraction with a specified denominator.

$$\frac{3}{4} = \frac{3 \times 2}{4 \times 2} = \frac{6}{8}$$

Multiplying Fractions

To Multiply Fractions

1. Multiply numerator by numerator. This gives the numerator of the product.
2. Multiply denominator by denominator. This gives the denominator of the product.
3. Simplify the resulting fraction if possible.

Example 3

Multiply $\frac{5}{8} \times \frac{3}{7}$.

$$\frac{5}{8} \times \frac{3}{7} = \frac{5 \times 3}{8 \times 7} = \frac{15}{56}$$

In multiplying fractions, it is usually easiest to divide by any common factors in the numerator and denominator *before* multiplying.

Example 4

Multiply $\frac{7}{15} \times \frac{10}{21}$.

$$\frac{7}{15} \times \frac{10}{21} = \frac{7 \times 10}{15 \times 21} = \frac{\overset{1}{\cancel{7}} \times \overset{2}{\cancel{10}}}{\underset{3}{\cancel{15}} \times \underset{3}{\cancel{21}}} = \frac{2}{9}$$

We divide numerator and denominator by the common factors of 7 and 5.

Dividing Fractions

To Divide Fractions Invert the divisor and multiply.

Remember: The divisor *follows* the division sign. That is the fraction that is inverted or "turned."

Example 5

Divide $\frac{3}{7} \div \frac{4}{5}$.

Invert

$$\frac{3}{7} \div \frac{4}{5} = \frac{3}{7} \times \frac{5}{4} = \frac{15}{28}$$

You can divide by common factors *only* after the divisor has been inverted.

Example 6

Divide $\frac{3}{5} \div \frac{9}{10}$.

$$\frac{3}{5} \div \frac{9}{10} = \frac{3}{5} \times \frac{10}{9} = \frac{\overset{1}{\cancel{3}} \times \overset{2}{\cancel{10}}}{\underset{1}{\cancel{5}} \times \underset{3}{\cancel{9}}} = \frac{2}{3}$$

Note: Since a fraction bar also indicates division,

$$\frac{\frac{3}{5}}{\frac{9}{10}} = \frac{3}{5} \div \frac{9}{10}$$

and the quotient is simplified as in our example.

Multiplying or Dividing Mixed Numbers

To Multiply or Divide Mixed Numbers Convert any mixed or whole numbers to improper fractions. Then multiply or divide the fractions as before.

Example 7

Multiply $6\frac{2}{3} \times 3\frac{1}{5}$.

$$6\frac{2}{3} \times 3\frac{1}{5} = \frac{20}{3} \times \frac{16}{5} = \frac{\overset{4}{\cancel{20}} \times 16}{3 \times \underset{1}{\cancel{5}}} = \frac{64}{3} = 21\frac{1}{3}$$

Example 8

Divide $\frac{7}{8} \div 5\frac{1}{4}$.

$$\frac{7}{8} \div 5\frac{1}{4} = \frac{7}{8} \div \frac{21}{4} = \frac{\overset{1}{\cancel{7}}}{\underset{2}{\cancel{8}}} \times \frac{\overset{1}{\cancel{4}}}{\underset{3}{\cancel{21}}} = \frac{1}{6}$$

Finding the Least Common Denominator

To Find the LCD

1. Write the prime factorization for each of the denominators.
2. Find all the prime factors that appear in any one of the prime factorizations.
3. Form the product of those prime factors, using each factor the greatest number of times it occurs in any one factorization.

Example 9

Find the LCD of fractions with denominators 4, 6, and 15.

$$\begin{array}{rl} 4 = & 2 \times 2 \\ 6 = & 2 \qquad \times 3 \\ 15 = & \qquad\quad 3 \times 5 \\ \hline & 2 \times 2 \times 3 \times 5 \end{array}$$

The LCD = $2 \times 2 \times 3 \times 5$, or 60.

To form the LCD, use two factors of 2, one of 3, and one of 5.

Adding Fractions

To Add Like Fractions

1. Add the numerators.
2. Place the sum over the common denominator.
3. Simplify the resulting fraction if necessary.

Example 10

Add $\frac{5}{18} + \frac{7}{18}$.

$$\frac{5}{18} + \frac{7}{18} = \frac{12}{18} = \frac{\overset{2}{\cancel{12}}}{\underset{3}{\cancel{18}}} = \frac{2}{3}$$

To Add Unlike Fractions

1. Find the LCD of the fractions.
2. Change each fraction to equivalent fractions that have the LCD as a common denominator.
3. Add the resulting like fractions.

Example 11

Add $\frac{3}{4} + \frac{7}{10}$.

1. The LCD for 4 and 10 is 20.
2. $\frac{3}{4} = \frac{15}{20}$ (×5) $\qquad \frac{7}{10} = \frac{14}{20}$ (×2)
3. $\frac{3}{4} + \frac{7}{10} = \frac{15}{20} + \frac{14}{20} = \frac{29}{20} = 1\frac{9}{20}$

To convert to equivalent fractions with denominator 20, multiply numerator and denominator of the first fraction by 5, and of the second fraction by 2.

Subtracting Fractions

To Subtract Like Fractions

1. Subtract the numerators.
2. Place the difference over the common denominator.
3. Simplify the resulting fraction if necessary.

Example 12

$$\frac{17}{20} - \frac{7}{20} = \frac{10}{20} = \frac{\overset{1}{\cancel{10}}}{\underset{2}{\cancel{20}}} = \frac{1}{2}$$

To Subtract Unlike Fractions

1. Find the LCD of the fractions.
2. Change each fraction to equivalent fractions that have the LCD as a common denominator.
3. Subtract the resulting like fractions.

Example 13

Subtract $\frac{8}{9} - \frac{5}{6}$.

1. The LCD for 9 and 6 is 18.

2. $\frac{8}{9} = \frac{16}{18}$ (×2) $\qquad \frac{5}{6} = \frac{15}{18}$ (×3)

3. $\frac{8}{9} - \frac{5}{6} = \frac{16}{18} - \frac{15}{18} = \frac{1}{18}$

Adding or Subtracting Mixed Numbers

To Add or Subtract Mixed Numbers

1. Add or subtract the whole-number parts.
2. Add or subtract the fractional parts.
   Note: Subtracting may require renaming the first mixed number.
3. Combine the results as a mixed number.

Example 14

Add $1\frac{2}{3} + 2\frac{3}{4}$.

First note that the LCD for the fractional portions of the mixed numbers is 12. Then convert the fractional portions of the mixed numbers to fractions with that LCD.

$$1\frac{2}{3} + 2\frac{3}{4} = 1\frac{8}{12} + 2\frac{9}{12} = 3\frac{17}{12} = 4\frac{5}{12}$$

(1 + 2) $\left(\frac{8}{12} + \frac{9}{12}\right)$

Example 15

Subtract $5\frac{1}{2} - 3\frac{3}{4}$.

$$5\frac{1}{2} - 3\frac{3}{4} = 5\frac{2}{4} - 3\frac{3}{4}$$

Rename

$$= 4\frac{6}{4} - 3\frac{3}{4} = 1\frac{3}{4}$$

(4 − 3) $\left(\frac{6}{4} - \frac{3}{4}\right)$

To rename the first fraction, borrow 1 from 5 and think of that 1 as $\frac{4}{4}$.

Exercises A.1

1. Give the fractions that name the shaded portions of the following diagrams. Indicate the numerator and the denominator.

(*a*) 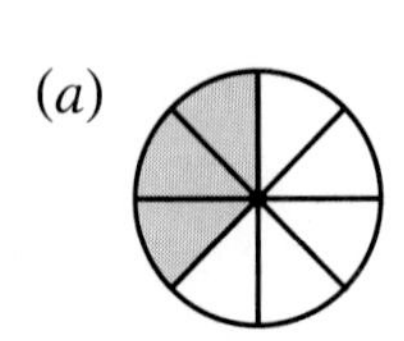

Fraction $\frac{3}{8}$

Numerator 3

Denominator 8

(*b*) 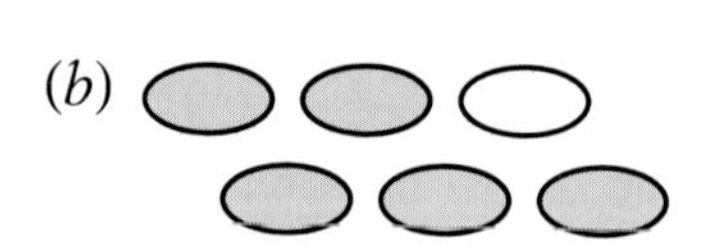

Fraction $\frac{5}{6}$

Numerator 5

Denominator 6

2. If your English class has 23 students and 13 are women:

(*a*) What fraction names the portion that are women? $\frac{13}{23}$

(*b*) What fraction names the portion that are not women? $\frac{10}{23}$

3. You are given the following group of numbers:

$$\frac{2}{3}, \frac{5}{4}, 2\frac{3}{7}, \frac{45}{8}, \frac{7}{7}, 3\frac{4}{5}, \frac{9}{1}, \frac{7}{10}, \frac{12}{5}, 5\frac{2}{9}$$

(*a*) List the proper fractions. $\frac{2}{3}, \frac{7}{10}$

(*b*) List the improper fractions. $\frac{5}{4}, \frac{45}{8}, \frac{7}{7}, \frac{9}{1}, \frac{12}{5}$

(*c*) List the mixed numbers. $2\frac{3}{7}, 3\frac{4}{5}, 5\frac{2}{9}$

4. Convert to mixed or whole numbers.

(*a*) $\frac{41}{6}$ $6\frac{5}{6}$

(*b*) $\frac{32}{8}$ 4

5. Convert to improper fractions.

(*a*) $3\frac{1}{6}$ $\frac{19}{6}$ (*b*) $4\frac{3}{8}$ $\frac{35}{8}$

6. Find out whether each pair of fractions is equivalent.

(*a*) $\frac{5}{8}, \frac{7}{12}$ not equivalent (*b*) $\frac{8}{15}, \frac{32}{60}$ equivalent

7. Reduce each fraction to lowest terms.

(*a*) $\frac{15}{18}$ $\frac{5}{6}$ (*b*) $\frac{24}{36}$ $\frac{2}{3}$ (*c*) $\frac{140}{180}$ $\frac{7}{9}$ (*d*) $\frac{210}{294}$ $\frac{5}{7}$

8. Find the missing numerators.

(*a*) $\frac{5}{8} = \frac{?}{24}$ 15 (*b*) $\frac{4}{5} = \frac{?}{40}$ 32

9. Arrange the fractions in order from smallest to largest.

(*a*) $\frac{5}{9}, \frac{4}{7}$ $\frac{5}{9}, \frac{4}{7}$ (*b*) $\frac{5}{6}, \frac{4}{5}, \frac{7}{10}$ $\frac{7}{10}, \frac{4}{5}, \frac{5}{6}$

10. Write as equivalent fractions with the LCD as a common denominator.

(*a*) $\frac{1}{6}, \frac{7}{8}$ $\frac{4}{24}, \frac{21}{24}$ (*b*) $\frac{3}{10}, \frac{5}{8}, \frac{7}{12}$ $\frac{36}{120}, \frac{75}{120}, \frac{70}{120}$

11. Multiply.

(*a*) $\frac{5}{8} \times \frac{3}{4}$ $\frac{15}{32}$ (*b*) $\frac{3}{5} \times \frac{4}{9}$ $\frac{4}{15}$ (*c*) $\frac{10}{27} \times \frac{9}{20}$ $\frac{1}{6}$ (*d*) $4 \times \frac{3}{8}$ $1\frac{1}{2}$

(*e*) $\frac{4}{7} \times 2\frac{3}{8}$ $1\frac{5}{14}$ (*f*) $5\frac{1}{3} \times 1\frac{4}{5}$ $9\frac{3}{5}$ (*g*) $2\frac{2}{5} \times 1\frac{7}{8}$ $4\frac{1}{2}$ (*h*) $1\frac{5}{12} \times 8$ $11\frac{1}{3}$

12. Divide.

(*a*) $\frac{3}{5} \div \frac{1}{4}$ $2\frac{2}{5}$ (*b*) $\frac{5}{12} \div \frac{5}{8}$ $\frac{2}{3}$ (*c*) $1\frac{7}{9} \div \frac{4}{9}$ 4 (*d*) $3\frac{3}{8} \div 2\frac{1}{4}$ $1\frac{1}{2}$ (*e*) $\frac{9}{10} \div 3$ $\frac{3}{10}$ (*f*) $4 \div 2\frac{2}{3}$ $1\frac{1}{2}$

13. A kitchen measures $5\frac{1}{3}$ yards (yd) by $4\frac{1}{4}$ yd. If you purchase linoleum which costs \$12 per square yard, what will it cost to cover the floor? \$272

14. If you drive 117 mi in $2\frac{1}{4}$ hours, what is your average speed? 52 mi/h

15. Add.

(*a*) $\frac{1}{8} + \frac{3}{8}$ $\frac{1}{2}$ (*b*) $\frac{6}{7} + \frac{5}{7}$ $1\frac{4}{7}$

16. Add.

(*a*) $\frac{1}{5} + \frac{3}{4}$ $\frac{19}{20}$ (*b*) $\frac{5}{8} + \frac{5}{6}$ $1\frac{11}{24}$ (*c*) $\frac{5}{18} + \frac{7}{12}$ $\frac{31}{36}$ (*d*) $\frac{3}{5} + \frac{1}{4} + \frac{5}{6}$ $1\frac{41}{60}$

17. Subtract.

(*a*) $\frac{5}{8} - \frac{3}{8}$ $\frac{1}{4}$ (*b*) $\frac{3}{5} - \frac{1}{6}$ $\frac{13}{30}$ (*c*) $\frac{7}{10} - \frac{7}{12}$ $\frac{7}{60}$ (*d*) $\frac{11}{20} - \frac{7}{25}$ $\frac{27}{100}$

18. Add or subtract as indicated.

(*a*) $5\frac{2}{5} + 4\frac{1}{5}$ $9\frac{3}{5}$ (*b*) $6\frac{5}{7} + 3\frac{4}{7}$ $10\frac{2}{7}$ (*c*) $4\frac{1}{8} + 3\frac{5}{12}$ $7\frac{13}{24}$

(*d*) $5\frac{7}{10} + 3\frac{11}{12}$ $9\frac{37}{60}$ (*e*) $2\frac{1}{2} + 3\frac{5}{6} + 4\frac{3}{8}$ $10\frac{17}{24}$ (*f*) $7\frac{7}{9} - 3\frac{4}{9}$ $4\frac{1}{3}$

(*g*) $5\frac{1}{7} - 3\frac{3}{7}$ $1\frac{5}{7}$ (*h*) $7\frac{1}{6} - 3\frac{1}{8}$ $4\frac{1}{24}$ (*i*) $6\frac{5}{12} - 3\frac{5}{8}$ $2\frac{19}{24}$

(*j*) $4 - 2\frac{2}{3}$ $1\frac{1}{3}$

19. Jan ran $3\frac{2}{3}$ mi on Monday, $1\frac{3}{4}$ mi on Wednesday, and $4\frac{1}{2}$ mi on Friday. How far did she run during the week? $9\frac{11}{12}$ mi

20. At the beginning of a year Miguel was $51\frac{3}{4}$ in tall. In June he measured $53\frac{1}{8}$ in. How much did he grow during that period? $1\frac{3}{8}$ in

21. Amelia buys an 8-yd roll of wallpaper on sale. After measuring, she finds that she needs the following amounts of the paper: $2\frac{1}{3}$, $1\frac{1}{2}$, and $3\frac{3}{4}$ yd. Does she have enough for the job? If so, how much will be left over? Yes, $\frac{5}{12}$ yd

Appendix 2 A Table of Squares and Square Roots

N	N^2	$\sqrt{N}$
1	1	1.000
2	4	1.414
3	9	1.732
4	16	2.000
5	25	2.236
6	36	2.449
7	49	2.646
8	64	2.828
9	81	3.000
10	100	3.162
11	121	3.317
12	144	3.464
13	169	3.606
14	196	3.742
15	225	3.873
16	256	4.000
17	289	4.123
18	324	4.243
19	361	4.359
20	400	4.472
21	441	4.583
22	484	4.690
23	529	4.796
24	576	4.899
25	625	5.000
26	676	5.099
27	729	5.196
28	784	5.292
29	841	5.385
30	900	5.477
31	961	5.568
32	1,024	5.657
33	1,089	5.745
34	1,156	5.831
35	1,225	5.916
36	1,296	6.000
37	1,369	6.083
38	1,444	6.164
39	1,521	6.245
40	1,600	6.325
41	1,681	6.403
42	1,764	6.481
43	1,849	6.557
44	1,936	6.633
45	2,025	6.708
46	2,116	6.782
47	2,209	6.856
48	2,304	6.928
49	2,401	7.000
50	2,500	7.071
51	2,601	7.141
52	2,704	7.211
53	2,809	7.280
54	2,916	7.348
55	3,025	7.416
56	3,136	7.483
57	3,249	7.550
58	3,364	7.616
59	3,481	7.681
60	3,600	7.746
61	3,721	7.810
62	3,844	7.874
63	3,969	7.937
64	4,096	8.000
65	4,225	8.062
66	4,356	8.124
67	4,489	8.185
68	4,624	8.246
69	4,761	8.307
70	4,900	8.367
71	5,041	8.426
72	5,184	8.485
73	5,329	8.544
74	5,476	8.602
75	5,625	8.660
76	5,776	8.718
77	5,929	8.775
78	6,084	8.832
79	6,241	8.888
80	6,400	8.944
81	6,561	9.000
82	6,724	9.055
83	6,889	9.110
84	7,056	9.165
85	7,225	9.220
86	7,396	9.274
87	7,569	9.327
88	7,744	9.381
89	7,921	9.434
90	8,100	9.487
91	8,281	9.539
92	8,464	9.592
93	8,649	9.644
94	8,836	9.695
95	9,025	9.747

N	N^2	$\sqrt{N}$	N	N^2	$\sqrt{N}$
96	9,216	9.798	149	22,201	12.207
97	9,409	9.849	150	22,500	12.247
98	9,604	9.899	151	22,801	12.288
99	9,801	9.950	152	23,104	12.329
100	10,000	10.000	153	23,409	12.369
101	10,201	10.050	154	23,716	12.410
102	10,404	10.100	155	24,025	12.450
103	10,609	10.149	156	24,336	12.490
104	10,816	10.198	157	24,649	12.530
105	11,025	10.247	158	24,964	12.570
106	11,236	10.296	159	25,281	12.610
107	11,449	10.344	160	25,600	12.649
108	11,664	10.392	161	25,921	12.689
109	11,881	10.440	162	26,244	12.728
110	12,100	10.488	163	26,569	12.767
111	12,321	10.536	164	26,896	12.806
112	12,544	10.583	165	27,225	12.845
113	12,769	10.630	166	27,556	12.884
114	12,996	10.677	167	27,889	12.923
115	13,225	10.724	168	28,224	12.961
116	13,456	10.770	169	28,561	13.000
117	13,689	10.817	170	28,900	13.038
118	13,924	10.863	171	29,241	13.077
119	14,161	10.909	172	29,584	13.115
120	14,400	10.954	173	29,929	13.153
121	14,641	11.000	174	30,276	13.191
122	14,884	11.045	175	30,625	13.229
123	15,129	11.091	176	30,976	13.266
124	15,376	11.136	177	31,329	13.304
125	15,625	11.180	178	31,684	13.342
126	15,876	11.225	179	32,041	13.379
127	16,129	11.269	180	32,400	13.416
128	16,384	11.314	181	32,761	13.454
129	16,641	11.358	182	33,124	13.491
130	16,900	11.402	183	33,489	13.528
131	17,161	11.446	184	33,856	13.565
132	17,424	11.489	185	34,225	13.601
133	17,689	11.533	186	34,596	13.638
134	17,956	11.576	187	34,969	13.675
135	18,225	11.619	188	35,344	13.711
136	18,496	11.662	189	35,721	13.748
137	18,769	11.705	190	36,100	13.784
138	19,044	11.747	191	36,481	13.820
139	19,321	11.790	192	36,864	13.856
140	19,600	11.832	193	37,249	13.892
141	19,881	11.874	194	37,636	13.928
142	20,164	11.916	195	38,025	13.964
143	20,449	11.958	196	38,416	14.000
144	20,736	12.000	197	38,809	14.036
145	21,025	12.042	198	39,204	14.071
146	21,316	12.083	199	39,601	14.107
147	21,609	12.124	200	40,000	14.142
148	21,904	12.166			

Answers to Self-Tests and Cumulative Review Tests

Chapter 1 Self-Test

1. $a - 5$ **2.** $6m$ **3.** $4(m + n)$ **4.** $\frac{a + b}{3}$ **5.** Commutative property of multiplication **6.** Distributive property **7.** Associative property of addition **8.** 21 **9.** $20x + 12$ **10.** 4^4 **11.** $7b^3$ **12.** $15a$ **13.** $3x^2y$ **14.** $19x + 5y$ **15.** $8a^2$ **16.** a^{14} **17.** $15x^3y^7$ **18.** $2x^3$ **19.** $4ab^3$ **20.** 3 **21.** 65 **22.** 144 **23.** 6 **24.** 10 **25.** 3

Chapter 2 Self-Test

1. 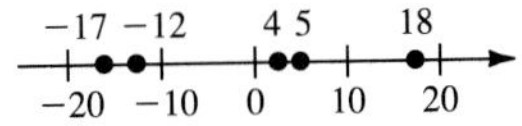

2. 7 **3.** 7 **4.** 11 **5.** 11 **6.** −13 **7.** −3 **8.** −21 **9.** 1 **10.** −6 **11.** −24 **12.** 9 **13.** 0 **14.** −40 **15.** 63 **16.** −27 **17.** −24 **18.** −25 **19.** 3 **20.** −5 **21.** Undefined **22.** −4 **23.** 80 **24.** 144 **25.** 5

Chapter 3 Self-Test

1. No **2.** Yes **3.** 11 **4.** 12 **5.** 7 **6.** 7 **7.** −12 **8.** 25 **9.** 3 **10.** 4 **11.** $-\frac{2}{3}$ **12.** −5 **13.** $\frac{C}{2\pi}$ **14.** $\frac{3y}{B}$ **15.** $\frac{6 - 3x}{2}$

16. 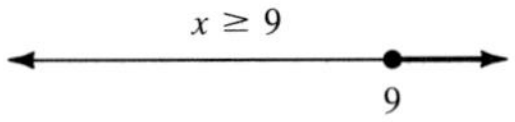

17. 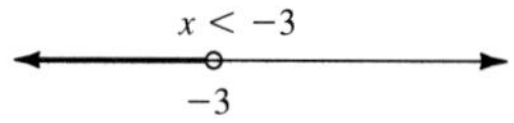

18. 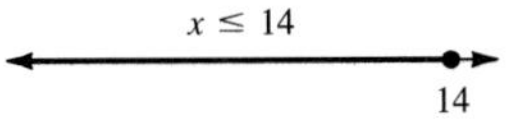

19. 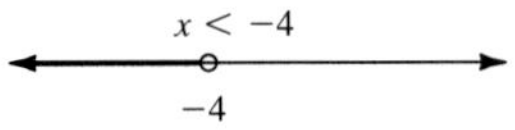

20. 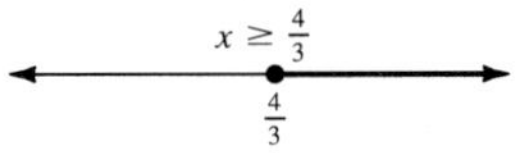

21. 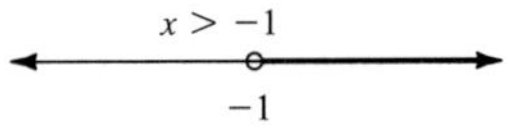

22. 7 **23.** 21, 22, 23 **24.** Steve, 6; Jan, 12; Rick, 17 **25.** 10 in, 21 in

Chapters 1 to 3 Cumulative Test

1. $3(r + s)$ **2.** $\frac{x - 5}{3}$ **3.** Associative property of addition **4.** Distributive Property **5.** $8x^3y^2$ **6.** $5a^2b$ **7.** $12a^2 + 3a$ **8.** $15m^5n^3$ **9.** $5xy$ **10.** 2 **11.** 80 **12.** 7 **13.** 7 **14.** −16 **15.** 4 **16.** 63 **17.** −16 **18.** −18 **19.** 0 **20.** −9 **21.** 13 **22.** 3 **23.** 5 **24.** −24 **25.** $\frac{5}{4}$ **26.** $-\frac{2}{5}$ **27.** 5 **28.** $\frac{I}{Pt}$ **29.** $\frac{2A}{b}$ **30.** $\frac{c - ax}{b}$

31. 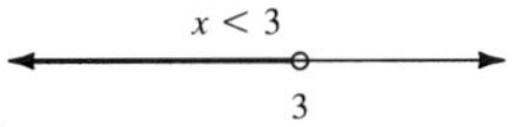

32. 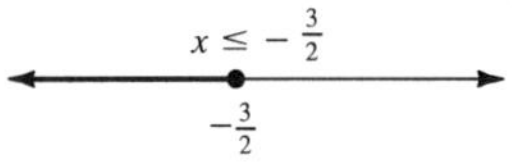

33. 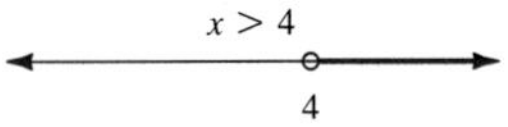

34. 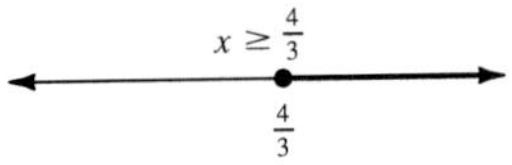

35. 13 **36.** 42, 43 **37.** 7 **38.** $420 **39.** 5 cm, 17 cm **40.** 8 in, 13 in, 16 in

Chapter 4 Self-Test

1. Binomial **2.** Trinomial **3.** $8x^4 - 3x^2 - 7$; 8, −3, −7; 4 **4.** $10x^2 - 12x - 7$ **5.** $7a^3 + 11a^2 - 3a$ **6.** $3x^2 + 11x - 12$ **7.** $b^2 - 7b - 5$ **8.** $7a^2 - 10a$ **9.** $4x^2 + 5x - 6$ **10.** $2x^2 - 7x + 5$ **11.** $15a^3b^2 - 10a^2b^2 + 20a^2b^3$ **12.** $3x^2 + x - 14$ **13.** $a^2 - 49b^2$ **14.** $9m^2 + 12mn + 4n^2$ **15.** $2x^3 + 7x^2y - xy^2 - 2y^3$ **16.** $2x^2 - 3y$ **17.** $4c^2 - 6 + 9cd$ **18.** $x - 6$ **19.** $x + 2 + \frac{10}{2x - 3}$ **20.** $2x^2 - 3x + 2 + \frac{7}{3x + 1}$ **21.** 4 **22.** −2 **23.** 6, 11 **24.** 20 dimes, 25 quarters **25.** 175 mi/h, 225 mi/h

Chapter 5 Self-Test

1. $6(2b + 3)$ **2.** $3p^2(3p - 4)$ **3.** $5(x^2 - 2x + 4)$ **4.** $6ab(a - 3 + 2b)$ **5.** $(a + 5)(a - 5)$ **6.** $(8m + n)(8m - n)$ **7.** $(7x + 4y)(7x - 4y)$ **8.** $2b(4a + 5b)(4a - 5b)$ **9.** $(a - 7)(a + 2)$ **10.** $(b + 3)(b + 5)$ **11.** $(x - 4)(x - 7)$ **12.** $(y + 10z)(y + 2z)$ **13.** $(2x - 1)(x + 8)$ **14.** $(3w + 7)(w + 1)$ **15.** $(4x - 3y)(2x + y)$ **16.** $3x(2x + 5)(x - 2)$ **17.** 3, 5 **18.** −1, 4 **19.** $-1, \frac{2}{3}$ **20.** 0, 3 **21.** $-7, \frac{1}{2}$ **22.** 5, 7 **23.** 3 cm by 11 cm **24.** $\frac{P - 2L}{2}$ **25.** $\frac{b}{a - 1}$

Chapter Six Self-Test

1. 4 **2.** −3, 3 **3.** $\frac{-3x^4}{4y^2}$ **4.** $\frac{4}{a}$ **5.** $\frac{x + 1}{x - 2}$ **6.** $\frac{4p^2}{7q}$ **7.** $\frac{2}{x - 1}$ **8.** $\frac{3}{4y}$ **9.** $\frac{3}{m}$ **10.** a **11.** 2 **12.** 5 **13.** $\frac{17x}{15}$ **14.** $\frac{3s - 2}{s^2}$ **15.** $\frac{4x + 17}{(x - 2)(x + 3)}$ **16.** $\frac{15}{w - 5}$ **17.** $\frac{2}{3x}$ **18.** $\frac{n}{2n + m}$ **19.** 36 **20.** 9 **21.** 2, 6 **22.** 4, 12 **23.** 50 mi/h, 45 mi/h **24.** 6 **25.** 20 ft, 35 ft

Chapters 4 to 6 Cumulative Test

1. $8x^2 - 7x - 4$ **2.** $-a^2 - 3a - 7$ **3.** $w^2 - 8w - 4$ **4.** $28x^3y - 14x^2y^2 + 21x^2y^3$ **5.** $15s^2 - 23s - 28$
6. $6a^3 + a^2b - 4ab^2 + b^3$ **7.** $-x^2 + 2xy - 3y$ **8.** $2x + 4$ **9.** $x^2 - 2x + 2 + \frac{5}{3x + 6}$ **10.** 9
11. 27 \$5 bills, 56 \$10 bills **12.** 120 mi/h, 140 mi/h **13.** $8a^2(3a - 2)$ **14.** $7mn(m - 3 - 7n)$ **15.** $(a + 8b)(a - 8b)$
16. $5p(p + 4q)(p - 4q)$ **17.** $(a - 6)(a - 8)$ **18.** $2w(w - 7)(w + 3)$ **19.** $(3r - 7s)(r + 4s)$ **20.** 4, 5 **21.** -4, 4
22. $-\frac{2}{3}$, 2 **23.** 7 **24.** 5 in by 17 in **25.** $\frac{2S_n - na_n}{n}$ **26.** $-1, \frac{2}{3}$ **27.** $\frac{m}{3}$ **28.** $\frac{a - 7}{3a + 1}$ **29.** $\frac{3}{x}$ **30.** $\frac{1}{3w}$
31. $\frac{8r + 3}{6r^2}$ **32.** $\frac{x + 33}{3(x - 3)(x + 3)}$ **33.** $\frac{12}{y - 4}$ **34.** $\frac{x - 1}{2x + 1}$ **35.** $\frac{n}{3n + m}$ **36.** $\frac{6}{5}$ **37.** $-\frac{9}{2}$, 7 **38.** 2
39. 52 mi/h, 48 mi/h **40.** 125 min

Chapter Seven Self-Test

1. (3, 3), (6, 0) **2.** (4, 0), (3, -2) **3.** (4, 1), (0, 3), (6, 0), (2, 2) **4.** (4, 0), (2, 3), (0, 6), $\left(\frac{2}{3}, 5\right)$
5. (0, -8), (2, -6), (4, -4), (6, -2) **6.** (0, -3), (2, 0), (4, 3), (6, 6) **7.** (4, 5) **8.** (-7, 1) **9.** (0, -4)
10. -12

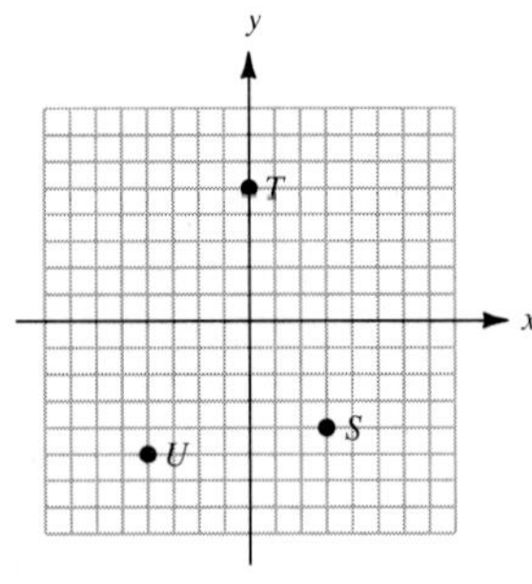

Chapter Eight Self-Test

1. $x + y = 5$
$x - y = 3$

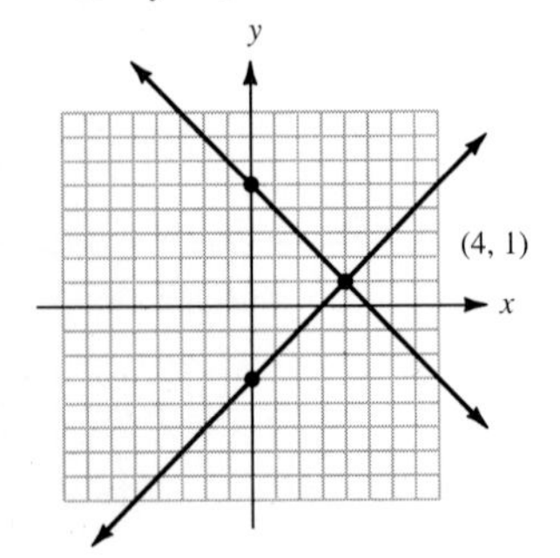

2. $x + 2y = 8$
$x - y = 2$

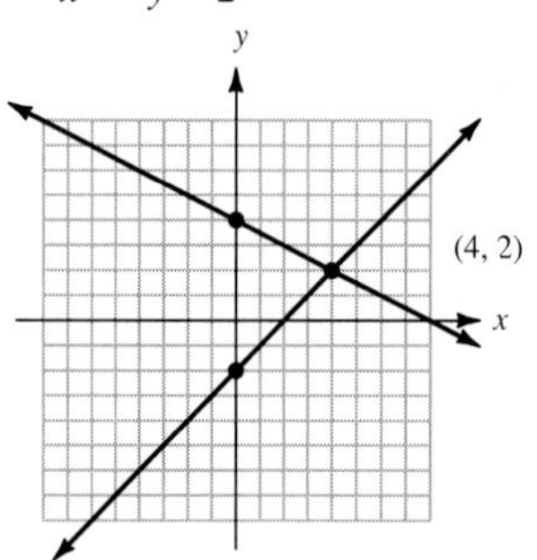

3. $x - 3y = 3$
$x - 3y = 6$

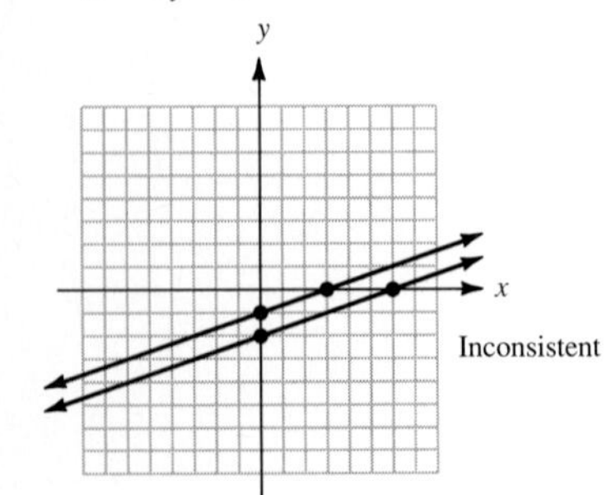

4. $4x - y = 4$
$x - 2y = -6$

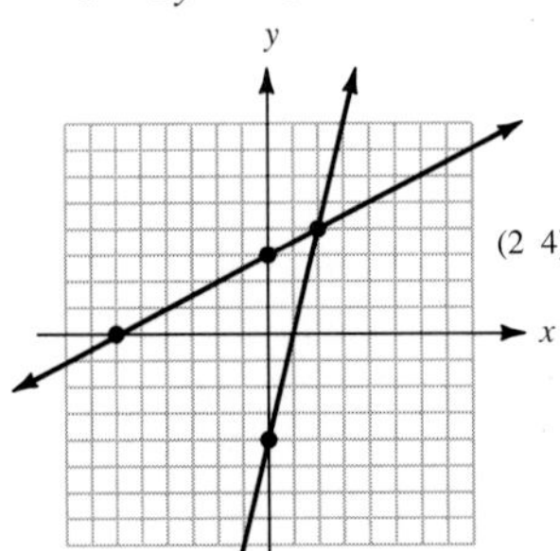

5. (4, 1) **6.** (4, 2) **7.** (1, 3) **8.** $\left(2, \frac{5}{2}\right)$ **9.** Dependent **10.** $\left(\frac{3}{4}, -1\right)$ **11.** (6, 2) **12.** Inconsistent **13.** (2, 6)
14. (6, -3) **15.** (6, 2) **16.** (5, 4) **17.** (-3, 3) **18.** Inconsistent **19.** (3, -5) **20.** (3, 2) **21.** 12, 18
22. 21 m, 29 m **23.** 12 in by 20 in **24.** 12 dimes, 18 quarters **25.** Boat 15 mi/h, current 3 mi/h

Chapters 7 and 8 Cumulative Test

1. (4, 4), (8, 0), (3, 5) **2.** (0, −6), (2, 0), (1, −3) **3.** (4, 0), (0, −8), $\left(\frac{5}{2}, -3\right)$, (1, −6) **4.** (4, 0), (8, −3), (0, 3), (−4, 6)
5. (2, 0), (1, −3), (0, −6), (−1, −9) **6.** (0, 2), (5, 1), (10, 0), (15, −1) **7.** −9.

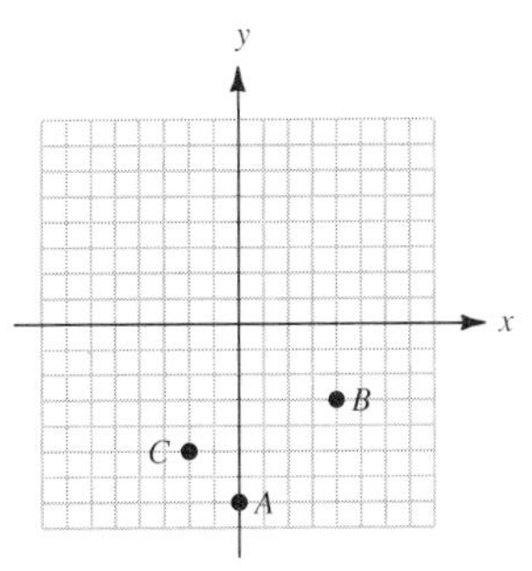

10. $x - y = 5$

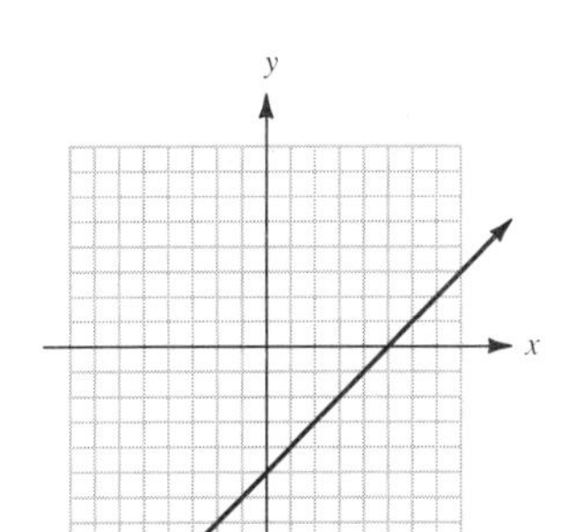

11. $y = \frac{2}{3}x + 3$

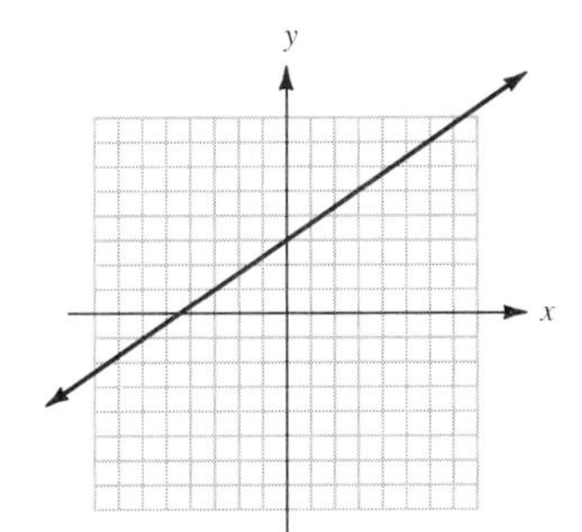

12. $x + 2y = 6$

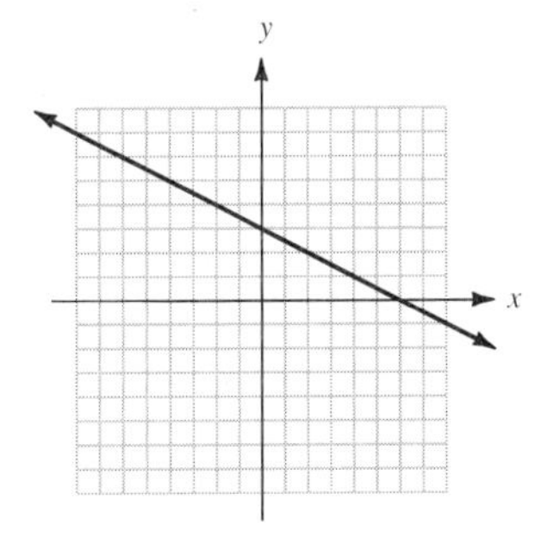

13. $2x - 5y = 10$

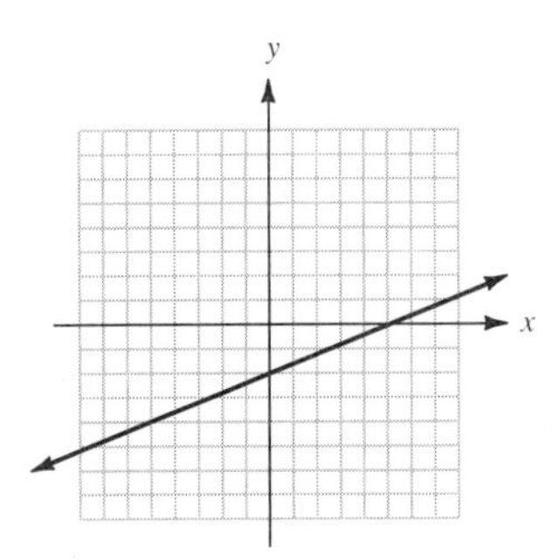

14. $y = -5$

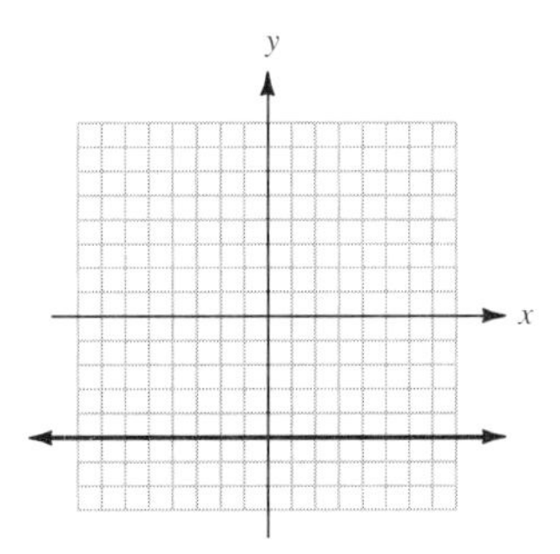

15. $\frac{10}{7}$ **16.** $-\frac{7}{5}$ **17.** $m = -3$, $b = 7$
18. $m = \frac{5}{3}$, $b = -5$

19. $y = 2x - 5$

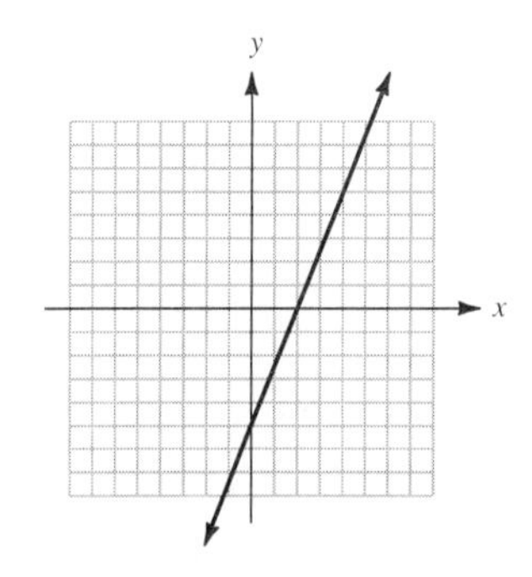

20. $y = -\frac{3}{2}x + 5$

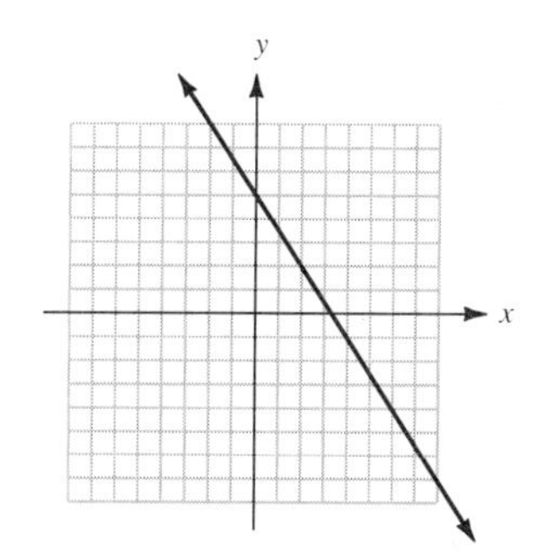

21. $x + 2y < 6$

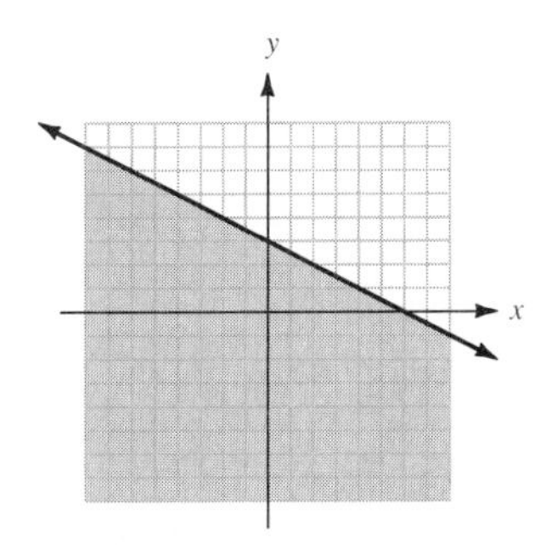

22. $3x - 4y \geq 12$

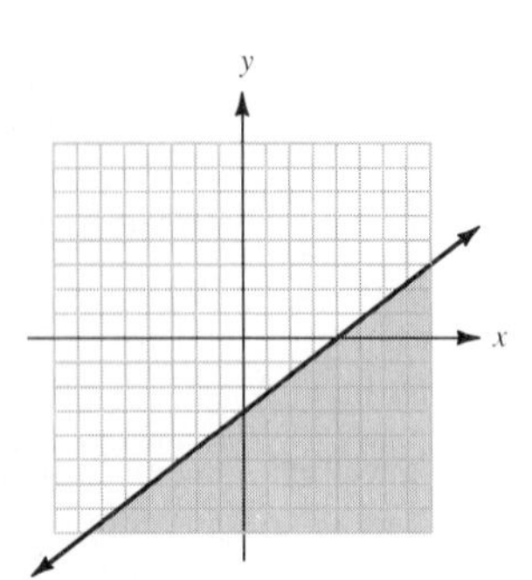

23. $x - y = 2$
$x + 3y = 6$ (3, 1)

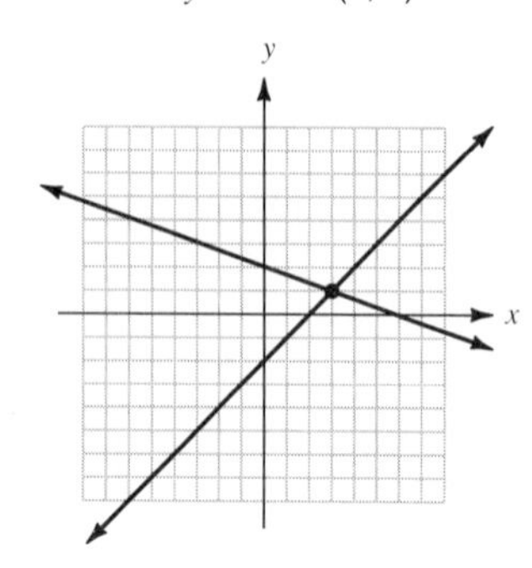

24. $3x + 2y = 6$
$x + 2y = -2$ (4, −3)

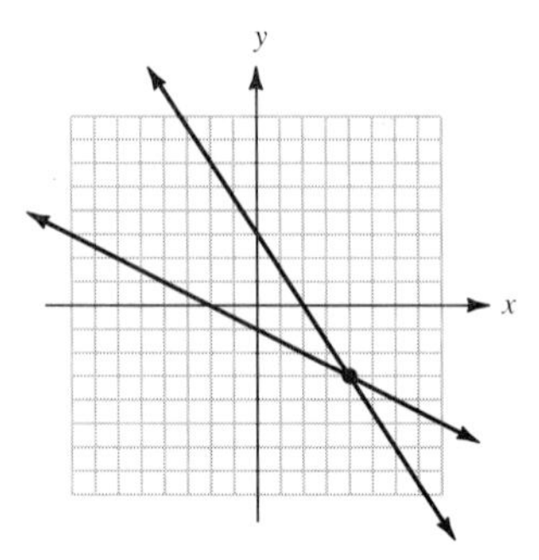

25. $\left(4, \frac{2}{3}\right)$ **26.** $(3, -2)$ **27.** $\left(7, -\frac{5}{2}\right)$ **28.** Dependent **29.** $(5, 0)$ **30.** $(-3, -2)$ **31.** Inconsistent

32. $\left(\frac{3}{2}, -\frac{1}{3}\right)$ **33.** $\left(\frac{2}{5}, -3\right)$ **34.** $\left(\frac{3}{2}, 4\right)$ **35.** 5, 21 **36.** \$4.50, \$1.50 **37.** 325 at \$7, 125 at \$4

38. \$5000 at 6%, \$7000 at 9% **39.** 100 mL of 30%, 200 mL of 60% **40.** Boat: 20 mi/h; current: 5 mi/h

Chapter 9 Self-Test

1. x^{15} **2.** $16a^4b^2$ **3.** y^3 **4.** $\frac{m^{15}}{n^6}$ **5.** 1 **6.** 7 **7.** $\frac{1}{x^7}$ **8.** $\frac{5}{a^4}$ **9.** $\frac{1}{x^2}$ **10.** $\frac{1}{m^8}$ **11.** 11 **12.** 3

13. Not a real number **14.** $5\sqrt{3}$ **15.** $2a\sqrt{6a}$ **16.** $\frac{4}{5}$ **17.** $\frac{\sqrt{5}}{3}$ **18.** $4\sqrt{10}$ **19.** $3\sqrt{3}$ **20.** $3\sqrt{2}$ **21.** $3x\sqrt{2}$

22. $\sqrt{21} - 6$ **23.** $11 + 5\sqrt{5}$ **24.** $\frac{\sqrt{14}}{2}$ **25.** $2\sqrt{11} - 6$

Chapter 10 Self-Test

1. $\pm\sqrt{15}$ **2.** $\pm 2\sqrt{2}$ **3.** $1 \pm \sqrt{7}$ **4.** $\frac{\pm\sqrt{10}}{3}$ **5.** 4, −2 **6.** $\frac{-3 \pm \sqrt{13}}{2}$ **7.** $-1 \pm \sqrt{6}$ **8.** $\frac{5 \pm \sqrt{17}}{4}$ **9.** −1, 3

10. 3 **11.** $\frac{5 \pm \sqrt{33}}{2}$ **12.** $\frac{1 \pm \sqrt{11}}{2}$ **13.** $\frac{1 + \sqrt{33}}{4}$ **14.** $-1 \pm \sqrt{6}$ **15.** 20 **16.** 12 **17.** $3\sqrt{5}$ **18.** $\sqrt{15}$

19. Approximately 9.747 cm

20. $y = x^2 + 4$

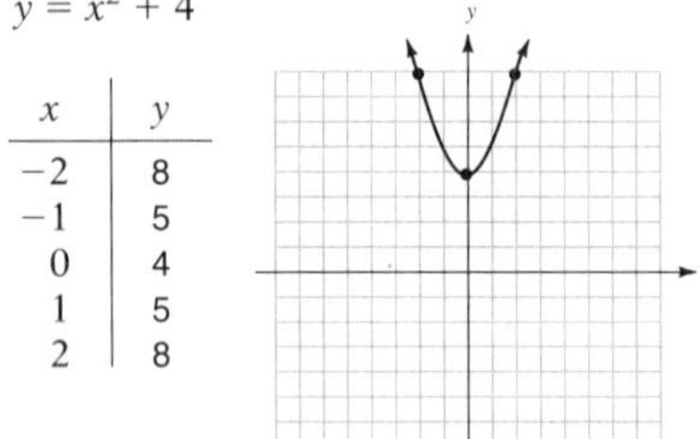

x	y
−2	8
−1	5
0	4
1	5
2	8

21. $y = x^2 - 2x$

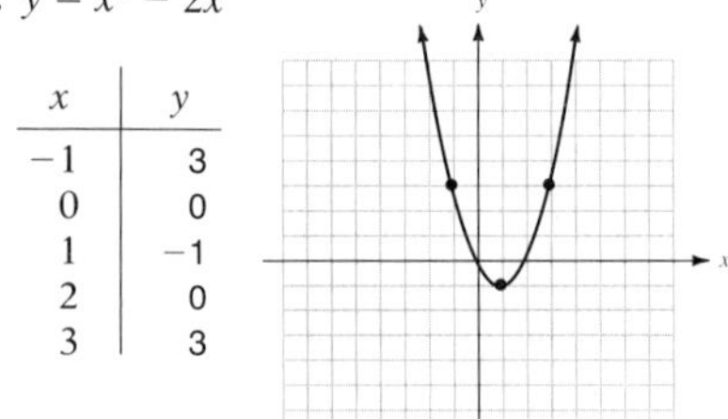

x	y
−1	3
0	0
1	−1
2	0
3	3

22. $y = x^2 - 3$

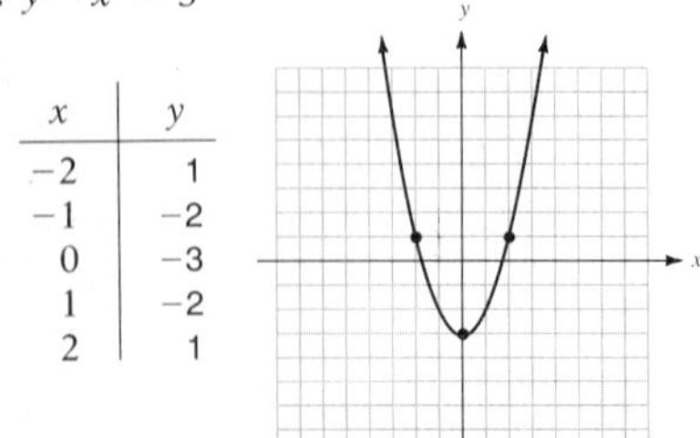

x	y
−2	1
−1	−2
0	−3
1	−2
2	1

23. $y = x^2 + x - 2$

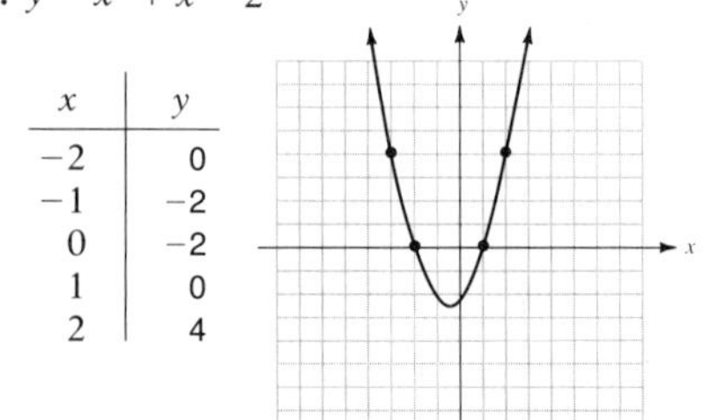

x	y
−2	0
−1	−2
0	−2
1	0
2	4

24. $y = -x^2 + 4$

x	y
−2	0
−1	3
0	4
1	3
2	0

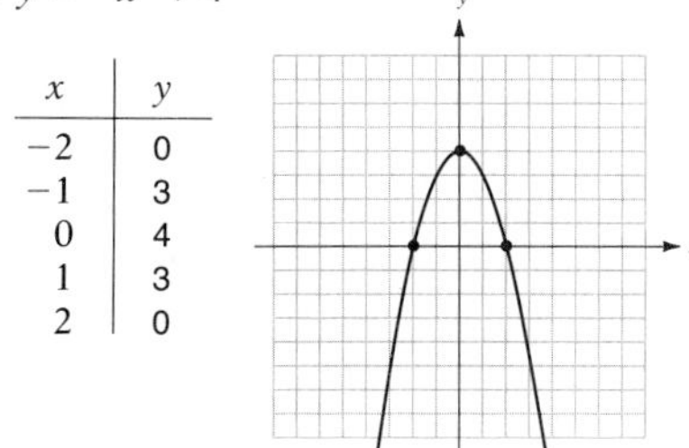

25. $y = -x^2 + 2x$

x	y
−1	−3
0	0
1	1
2	0
3	−3

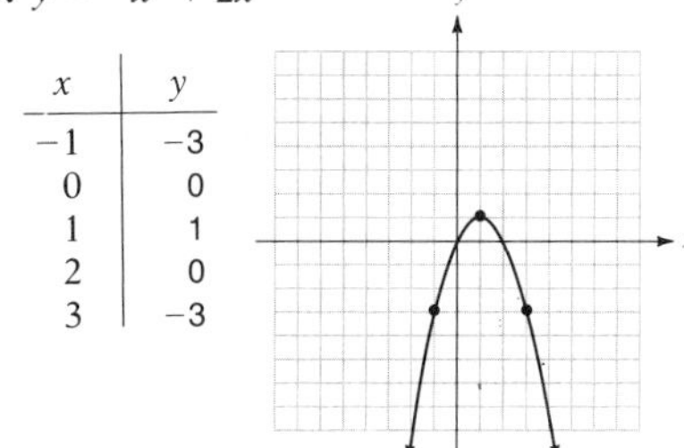

Chapters 9 and 10 Cumulative Test

1. m^{20} **2.** $9x^2y^8$ **3.** $4a^{14}$ **4.** $\frac{r^8}{s^6}$ **5.** 5 **6.** 1 **7.** $\frac{1}{x^8}$ **8.** $\frac{4}{a^5}$ **9.** $\frac{1}{r^3}$ **10.** $\frac{1}{y^{10}}$ **11.** 12 **12.** −12 **13.** Not a real number **14.** −3 **15.** $7x^2\sqrt{2}$ **16.** $5mn\sqrt{6m}$ **17.** $\frac{2a\sqrt{3}}{5}$ **18.** $\frac{y\sqrt{15y}}{3}$ **19.** $6\sqrt{3}$ **20.** $-4a\sqrt{5}$ **21.** $30a\sqrt{3}$ **22.** $2\sqrt{15} - 15$ **23.** $-13 - 2\sqrt{2}$ **24.** $\frac{2x\sqrt{6x}}{3}$ **25.** $2\sqrt{10} + 4$ **26.** $\pm 6\sqrt{2}$ **27.** $-3 \pm 2\sqrt{3}$ **28.** $2 \pm \sqrt{7}$ **29.** 5, −3 **30.** $\frac{5 \pm \sqrt{41}}{4}$ **31.** −4, 6 **32.** $\frac{5 \pm \sqrt{57}}{4}$ **33.** $\frac{1 \pm \sqrt{7}}{2}$ **34.** 2.2 in, 4.6 in **35.** $\sqrt{39}$ **36.** 3.12 ft, 5.12 ft

37. $y = x^2 - 2$

x	y
−2	2
−1	−1
0	−2
1	−1
2	2

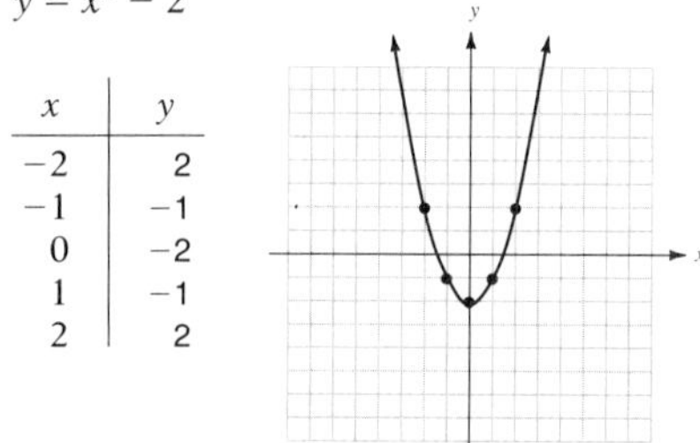

38. $y = x^2 - 4x$

x	y
0	0
1	−3
2	−4
3	−3
4	0

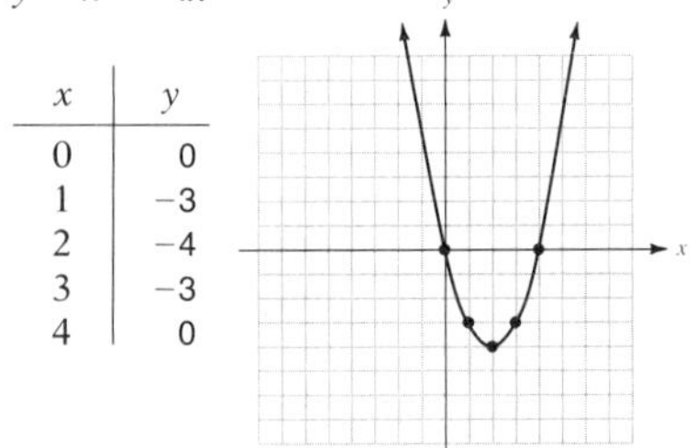

39. $y = x^2 - 3x + 2$

x	y
−1	6
0	2
1	0
2	0
3	2

40. $y = -x^2 + 4$

x	y
−2	0
−1	3
0	4
1	3
2	0

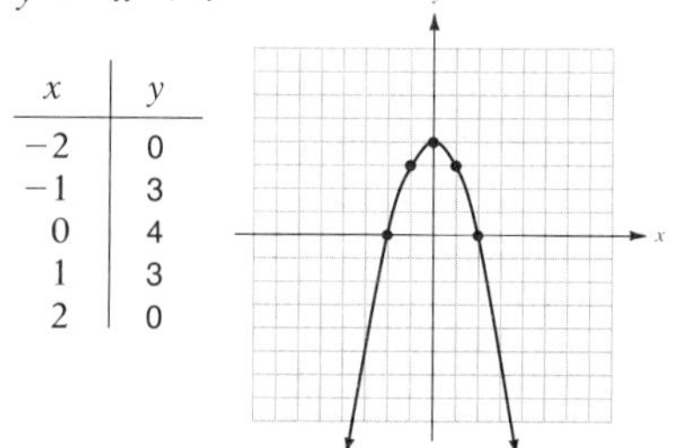

Appendix 1

1. *(a)* Fraction $\frac{3}{8}$, numerator 3, denominator 8; *(b)* fraction $\frac{5}{6}$, numerator 5, denominator 6 **2.** *(a)* $\frac{13}{23}$; *(b)* $\frac{10}{23}$ **3.** *(a)* $\frac{2}{3}, \frac{7}{10}$; *(b)* $\frac{5}{4}, \frac{45}{8}, \frac{7}{7}, \frac{9}{1}, \frac{12}{5}$; *(c)* $2\frac{3}{7}, 3\frac{4}{5}, 5\frac{2}{9}$ **4.** *(a)* $6\frac{5}{6}$; *(b)* 4 **5.** *(a)* $\frac{19}{6}$; *(b)* $\frac{35}{8}$ **6.** *(a)* Not equivalent; *(b)* equivalent **7.** *(a)* $\frac{5}{6}$; *(b)* $\frac{2}{3}$; *(c)* $\frac{7}{9}$; *(d)* $\frac{5}{7}$ **8.** *(a)* 15; *(b)* 32 **9.** *(a)* $\frac{5}{9}, \frac{4}{7}$; *(b)* $\frac{7}{10}, \frac{4}{5}, \frac{5}{6}$ **10.** *(a)* $\frac{4}{24}, \frac{21}{24}$; *(b)* $\frac{36}{120}, \frac{75}{120}, \frac{70}{120}$ **11.** *(a)* $\frac{15}{32}$; *(b)* $\frac{4}{15}$; *(c)* $\frac{1}{6}$; *(d)* $1\frac{1}{2}$; *(e)* $1\frac{5}{14}$; *(f)* $9\frac{3}{5}$; *(g)* $4\frac{1}{2}$; *(h)* $11\frac{1}{3}$ **12.** *(a)* $2\frac{2}{5}$; *(b)* $\frac{2}{3}$; *(c)* 4; *(d)* $1\frac{1}{2}$; *(e)* $\frac{3}{10}$; *(f)* $1\frac{1}{2}$ **13.** \$272 **14.** 52 mi/h **15.** *(a)* $\frac{1}{2}$; *(b)* $1\frac{4}{7}$ **16.** *(a)* $\frac{19}{20}$; *(b)* $1\frac{11}{24}$; *(c)* $\frac{31}{36}$; *(d)* $1\frac{41}{60}$ **17.** *(a)* $\frac{1}{4}$; *(b)* $\frac{13}{30}$; *(c)* $\frac{7}{60}$; *(d)* $\frac{27}{100}$ **18.** *(a)* $9\frac{3}{5}$; *(b)* $10\frac{2}{7}$; *(c)* $7\frac{13}{24}$; *(d)* $9\frac{37}{60}$; *(e)* $10\frac{17}{24}$; *(f)* $4\frac{1}{3}$; *(g)* $1\frac{5}{7}$; *(h)* $4\frac{1}{24}$; *(i)* $2\frac{19}{24}$; *(j)* $1\frac{1}{3}$ **19.** $9\frac{11}{12}$ mi **20.** $1\frac{3}{8}$ in **21.** Yes, $\frac{5}{12}$ yd

INDEX